Foundation Vibration Analysis Using Simple Physical Models

Foundation Vibration Analysis Using Simple Physical Models

John P. Wolf
Swiss Federal Institute of Technology

PTR Prentice Hall
Englewood Cliffs, NJ 07632

Library of Congress Cataloging-In-Publication Data

Wolf, John P.
Foundation vibration analysis using simple physical models / John P. Wolf
p. cm.
Includes bibliographical references and index.
ISBN 0-13-010711-5
1. Foundations—Vibration—Mathematical models. I. Title.
TA775.W57 1994
624.1'5—dc20 93-41384
CIP

Editorial/production supervision: BooksCraft, Inc., Indianapolis, IN
Cover design: Bruce Kenselaar
Acquisitions editor: Michael Hays
Manufacturing manager: Alexis R. Heydt

Prentice-Hall, Inc.
A Paramount Communications Company
Englewood Cliffs, NJ 07632

The publisher offers discounts on this book when ordered in bulk quantities. For more information contact:

Corporate Sales Department
PTR Prentice Hall
113 Sylvan Avenue
Englewood Cliffs, NJ 07632
Phone: 201-592-2863
FAX: 201-592-2249

Printed in the United States of America
10 9 8 7 6 5 4 3 2 1

ISBN 0-13-010711-5

Prentice-Hall International (UK) Limited, *London*
Prentice-Hall of Australia Pty. Limited, *Sydney*
Prentice-Hall Canada Inc., *Toronto*
Prentice-Hall Hispanoamericana, S.A., *Mexico*
Prentice-Hall of India Private Limited, *New Delhi*
Prentice-Hall of Japan, Inc., *Tokyo*
Simon & Schuster Asia Pte. Ltd., *Singapore*
Editora Prentice-Hall do Brasil, Ltda., *Rio de Janeiro*

Contents

Foreword

Building on a solid foundation has traditionally been considered an essential requirement to guarantee that a structure, or for that purpose any human creation, will withstand the attacks of time, the environment, and other human actions, performing safely and reliably its intended function. The fact that not only the strength but also the deformability of the foundation will affect the behavior and performance of a building has also been long recognized. Yet structures and their foundations are often analyzed and designed independently, with little or no communication between structural and geotechnical (once foundation) engineers. Digital computers—which allowed engineers to obtain at a relatively low cost the solution of complex models with many degrees of freedom, and, perhaps more importantly, allowed the storage in a common database of the information corresponding to the different subsystems that constitute a building—were expected at one time to provide the needed integration between these two areas of civil engineering. This was indeed the purpose of the Integrated Civil Engineering System (ICES) conceived by Charles L. Miller in the early sixties, but the goal remains unfulfilled.

The one instance in which the interaction between a structure, its foundation, and the surrounding or underlying soil is properly taken into account is when dealing with dynamic loads. The design of structures to support heavy machinery that may induce vibrations was already recognized as an important practical problem in the twenties. In the thirties, Reissner derived the first analytical solution for the vertical displacements on the surface of a linear elastic, homogeneous, and isotropic halfspace subjected to a harmonic normal stress uniformly distributed over a circular area, following it with the solution for torsional vibrations. The application of these results to the study of a vibrating rigid body represents an approximation; the stress distribution under a rigid foundation would actually be unknown whereas the displacement (or rotation) would be constant. Reissner selected the value of the vertical displacement at the center of the loaded area as representative of the motion of a rigid, massless foundation. Work along these lines was continued and improved upon in the forties by Reissner and Sagoci and by Shekhter, who used the average of the displacements at the center and at the edge of the loaded area to obtain curves of dynamic amplification as a function of a dimensionless frequency and a mass ratio which became widely known and used.

The fifties saw a significant increase both in the number of researchers engaged in this area and in the number of related publications, with important contributions by Arnold, Bycroft, Lorenz, Novak, Pauw, Quinlan, Robson, Savinov, Sung, and Warburton among others. Quinlan and Sung, for instance, considered other stress distributions to study the effect of this simplifying assumption on the results. Bycroft accounted for internal soil damping and considered other types of motion. Warburton studied the response of a foundation on an elastic soil layer of finite depth by opposition to a halfspace, a case of practical importance which exhibits some marked differences in the solution. Not only will the static stiffness be larger, but the soil layer will have natural frequencies of its own that will be reflected in a larger variation of the foundation stiffnesses with frequency (for a soil without internal damping the stiffness would become zero at a resonant frequency). In addition, there will not be any radiation damping below a threshold frequency. Additional studies were conducted in the sixties by Borodatchev, Collins, Elorduy, Gladwell, Hsieh, Kobori, Lysmer, Minai, Novak, Richart, Robertson, Sigalov, Stallybrass, Whitman, and others. Of particular importance is the publication in 1962 of the English edition of Barkan's book, *Dynamics of Bases and Foundations*, as until then the main books on machine foundations had been either in German (Rausch's various editions of *Maschinen Fundamente* and Lorenz's *Grundbau Dynamik*), or in Russian (Barkan's own book, first published in 1948).

The solution of the true mixed boundary value problem, where stresses are specified along the surface of the soil outside of the foundation (a stress-free surface) while displacements are imposed at the base of a rigid and massless body, was addressed by Borodatchev in 1964 for the vertical case. A comprehensive treatment of the problem and an alternative graphical solution for this case were presented by Awojobi and Grootenhuis in 1965, and a numerical solution was obtained the same year by Lysmer. Sigalov extended Borodatchev's work to rocking; Robertson used a series solution, which was extended to other cases by Gladwell in 1968. By 1971, a rigorous solution for a rigid circular foundation on the surface of an elastic halfspace, covering an extended range of dimensionless frequencies, was presented in graphical and tabular form by Veletsos and Wei for coupled horizontal and rocking motions. An independent solution by Luco and Westman was published almost at the same time. Similar results for vertical and torsional excitations and for viscoelastic or hysteretic media were obtained by Veletsos and Nair, and Veletsos and Verbic. All these solutions represent important benchmarks and have greatly contributed to the understanding of the nature and importance of interaction effects. Yet there are very few soil deposits that can be considered as homogeneous and isotropic halfspaces. Elastic moduli of soils will generally vary with depth. With the availability of digital computers and the development of new discrete formulations (finite elements and boundary elements in particular), solutions for rigid foundations on the surface or embedded in a horizontally layered medium follow immediately through the work of Waas, Kausel, Luco, and Dominguez. The dynamic stiffnesses of single piles and pile groups were investigated by Novak, Nogami, Blaney, Kausel, Kaynia, Gomez, Waas, and Wolf, among others. By the late seventies, the capability existed to compute the dynamic stiffnesses of foundations of arbitrary shape in horizontally stratified soil deposits with any desired degree of accuracy, as long as linear elastic behavior could be assumed.

The effect of soil–structure interaction (or foundation flexibility) on the seismic response of buildings was addressed by Martel as early as 1940. In the fifties, Housner and Merritt looked again at this problem, using data recorded at and near a building. In the sixties, a number of contributions by Sandi, Lycan and Newmark, Monge and Rosenberg, and Hashiba and Whitman appeared in the literature. The main effects of the foundation flexibility, changing the effective natural period and the effective damping, were described by Parmelee using a simple model which has been widely used since. Parametric studies along these lines were conducted in the early seventies by Sarrazin and by Bielak. By this time it had become clear that the dynamic soil–structure interaction effects might not be important for regular, flexible buildings on rock or very stiff soil, but that they could be very significant for very stiff and massive structures such as nuclear power plants. Kausel pointed out the need to consider in the seismic case not only the deformation of the foundation and the soil due to the inertia forces in the structure (base shear and overturning moment), which corresponds exactly to the problem of interest for the design of machine foundations, but also the effect of a rigid foundation on a train of seismic waves, filtering out high frequency components of the translations and introducing rotational motions. To account for these effects in a linear analysis, he suggested a three-step or substructure approach. Whitman introduced the terms *inertial* and *kinematic interaction* to refer to these two effects. Studies by Elsabee and Morray confirmed the potential importance of kinematic interaction effects, particularly for embedded foundations. Although much remained to be done to be able to accurately predict all aspects of seismic soil–structure interaction, by the late seventies, after some controversy related to the merits of different analysis procedures, the basic phenomena were also well known and understood for this case. In 1985, John Wolf's book *Dynamic Soil-Structure Interaction* was published. It represented the first rigorous and comprehensive treatment of the topic with applications both to machine foundations and particularly to the seismic problem. The book dealt with solutions for harmonic loads—all that is needed when interested in the steady state response of a machine foundation. To obtain the transient response during start up or shut down of the machine and for seismic applications, it is necessary to solve the problem for many different frequencies and superimpose the results with a Fourier transformation from the frequency to the time domain. This technique has not been widely used by civil engineers in traditional structural dynamics, but it is a powerful one that can provide a considerable insight into the behavior of the solution. It is, however, limited to linear problems, at least in its classical form. The book also emphasized the use of the indirect boundary element method (or source method) in its most general form.

A theoretical formulation needs experimental verification before it will be accepted and used in engineering practice. This is so, in some cases, because of a reluctance to accept any fact that cannot be seen and verified with one's own eyes or because of an innate mistrust and aversion towards mathematical derivations. In more enlightened cases, reluctance stems from the realization that even if a mathematical formulation is correct it may not include all the variables that influence the physical problem, either because of difficulties in determining the values of these variables or because of limitations in the instrumentation used for the experimental measurements. In the case of dynamic soil–structure interaction, one would have to know, for instance, the soil properties at the site and their variation both in space and with the level of strains. Experimental data on dynamics of machine foundations had been gathered in the thirties by Hertwig, Früh and Lorenz, in the fifties by Novak, in the sixties by Fry and Maskimov, and in the seventies by Stokoe and Erden, among others. Through the years many of the researchers before mentioned tried to correlate the experimental results available at the time with predictions of the existing theoretical formulations, and a number of legitimate reasons were offered to explain any differences. Among these, one often used was nonlinear soil behavior. Nonlinear soil effects were discussed for instance by Lorenz, Novak and Robson in the fifties, and by Lorenz, Funston and Hall, and Novak in the sixties. One would expect that for a properly designed machine foundation the dynamic strains induced in the soil by the machine vibrations would be very small, and that therefore linear elasticity would generally apply except at points where there are concentrations of stresses (edges of a mat foundation, near the head of piles). This will not, however, be the case for field tests in which different frequencies and amplitudes of excitation are used, or for unbalanced foundations which have been improperly designed. Nonlinear soil effects will be even more important for the seismic case when dealing with moderate or large earthquakes. In this case, it is necessary to account for the nonlinearities induced in the soil without any structure or foundation (the free field) by the seismic waves, and for the additional nonlinear effects created by the structural vibrations. Nonlinear behavior can also be expected to occur in the structure itself for conventional buildings; according to the design philosophy implicit in most seismic design codes, failure and collapse are to be prevented but not inelastic action. Other types of nonlinearities can also be encountered, such as separation between a mat foundation and the soil (sliding or uplifting), as studied by Meek, Scaletti and Wolf, or separation and gapping near the pile heads, as discussed by Angelides, Nogami and Novak.

Thus, although linear soil–structure interaction analyses are very valuable, they cannot provide by themselves a complete solution to the problem. Seed and Idriss proposed an iterative linearization scheme to account for nonlinear soil behavior in the study of seismic waves propagating vertically through a horizontally stratified soil deposit. This is only an approximation, as pointed out by Constatopoulos, but it provides reasonable and useful results for many practical cases. The extension of the approach to two or three dimensional states of strain, as encountered in soil–structure interaction, is much more questionable. True nonlinear solutions are normally carried out through direct integration of the equations of motion in the time domain. A main difficulty with this approach was the simulation of a semi-infinite domain through the use of appropriate boundary conditions at the edges of a finite domain. The solution of dynamic structure interaction problems in the time domain was the subject of John Wolf's second book, *Soil-Structure-Interaction Analysis in Time Domain*, published in 1988.

Analytical formulations, whether in closed form, as a series expansion, or in integral form, provide rigorous solutions which are often of direct practical value and which can always be used as benchmarks to judge the validity of numerical approaches and computer codes. Properly validated numerical formulations can provide accurate solutions to actual practical problems. However, to gain a good understanding of the physical behavior of the problem, the relative importance of the various parameters involved, and the effect of changes in the values of the different variables, conducting a large number of parametric or sensitivity studies would be necessary. These formulations are therefore more appropriate for final analyses than for preliminary studies or design purposes. Even then, they may only be used for very special structures such as nuclear power plants, much as theory of elasticity or finite element solutions are only used for very special cases in the structural analysis of prismatic members. Simplified models, on the other hand, will provide only approximate solutions; but, by reducing the number of parameters to the few most significant ones, they allow a much clearer visualization of the physical behavior, at least in its fundamental aspects. Moreover, even if these models are approximate, the accuracy provided by good simple models may be sufficient for practical purposes; for example, engineering beam theory, while an approximation, is the mainstay of structural analysis and design.

Unsurprisingly, therefore, since the beginning the same researchers engaged in the derivation of continuous formulations based on the theory of elasticity (or *elastodynamics*, the more fashionable present term) have tried at the same time to explain their results with simple models. Reissner explored the possibility of reproducing his results with a lumped parameter model consisting of a mass, a spring, and a dashpot, but he concluded that their values would have to be functions of frequency, and he could not find simple expressions for them. Similar conclusions were reached by Hertwig and Lorenz trying to match their experimental data with the same type of models. Shekhter, on the other hand, found that her theoretical curves for dynamic amplification as function of mass ratio and dimensionless frequency could be reasonably approximated by a mass–spring–dashpot system replacing the elastic halfspace. Merritt and Housner substituted the foundation by a rotational spring in their 1954 study of seismic soil–structure interaction; Lycan and Newmark in 1961 replaced the foundation by a free mass. In 1965, Fleming, Screwvala and Kodner used a horizontal and rotational spring to simulate interaction effects in swaying and rocking. Lysmer and Richart in 1966, Whitman and Richart in 1967, Hall in 1968, and Whitman in 1969 used again lumped-parameter models with springs, masses, and dashpots. In the early seventies, Veletsos and Meek used a truncated cone to explain successfully some of the basic features of the dynamic stiffnesses and showed that a simple mass–spring–dashpot system was not sufficient to simulate the exact solution over an extended range of frequencies. For small values of the dimensionless frequency a constant mass seems to reasonably reproduce the frequency variation of the stiffnesses, and this had given rise to the concept of an added mass of soil vibrating in phase with the foundation. In fact, such a concept was described in many technical reports and even books, yet Veletsos' results showed that it is incorrect and that lumped-parameter models can be used but may have to be slightly more complicated (not just one mass, one spring, and a dashpot). Veletsos and Verbic also suggested simple expressions for the dynamic stiffnesses.

Following this rich tradition, Wolf has extended the concept of the truncated cones to cover a complete range of dynamic excitations and physical situations. He has followed this by developing more accurate lumped-parameter models to simulate not only the behavior of mat foundations on the surface of an elastic or viscoelastic halfspace but also those of embedded foundations and, in general, foundations in layered media underlain by an elastic halfspace or rigid rock. By considering, finally, the displacements produced by dynamic loads along a horizontal plane, he has provided simple means to compute the dynamic stiffnesses of foundations of arbitrary shape. These three important developments provide what would be, in Wolf's own terms, the equivalent of a strength-of-materials approach to dynamic soil–structure interaction, allowing the designer to conduct analyses with the same basic models and tools used for other purposes. His third book, *Foundation Vibration Analysis Using Simple Physical Models*, represents the results of these efforts. Although this text could be considered a complement to his previous two books, it is in fact self-contained, including not only the derivation of the three types of approximations (truncated cones, lumped-parameter models, and displacement expansions), but also the formulation of the complete dynamic soil–structure interaction problem for machine foundations or seismic applications, using both rigorous and approximate methods. The text is primarily intended for practicing engineers engaged in everyday applications involving dynamic soil–structure interaction. The aim is clearly to provide them with simple yet accurate models they can use in practice. It should also be of interest, however, to other researchers.

Jose M. Roesset
University of Texas at Austin

Preface

In the two previous books by the author (*Dynamic Soil-Structure Interaction*, Prentice-Hall, 1985, and *Soil-Structure-Interaction Analysis in Time Domain*, Prentice-Hall, 1988) the emphasis is placed on *rigorous methods* to model the unbounded soil in the frequency and time domains. These procedures to solve the complicated boundary- and initial-value problem in three-dimensional elastodynamics with sophisticated semi-analytical and numerical techniques with their mathematical complexity obscure the physical insight; these belong more to the discipline of applied computational mechanics than to civil engineering. They will be used primarily for large projects of critical facilities. For the vast majority of all other projects, *simple physical models* developed in this book to represent the unbounded soil can be applied. They easily fit the budget and available time of the project, and need no sophisticated computer code. Use of the simple physical models leads to some loss of precision, but that is more than compensated by their many advantages. As the simple physical models cannot cover all cases, they do not supplant the much more generally applicable rigorous methods, but rather they supplement them.

The physical models consist of the following representations:

1. *Cones.* Translational and rotational truncated semi-infinite single and double cones are based on rod (bar) theory (plane sections remain plane) with the corresponding one-dimensional displacements.
2. *Lumped-parameter models.* With a few springs, dashpots, and masses with frequency-independent coefficients, the unbounded soil is represented by the same type of dynamic model as the structure enabling the same structural dynamics program to be applied. The lumped-parameter model can be conceptionally constructed from cones by assembling their exact discrete-element models in parallel and by calibrating with rigorous solutions.
3. *Prescribed wave patterns in the horizontal plane.* These are one-dimensional body and surface waves on the free surface and cylindrical waves.

Just as engineering beam theory in stress analysis is based on assumed displacement patterns instead of rigorous elasticity solutions, the physical models present a major step towards developing a *strength-of-materials approach to foundation dynamics*. The described procedures satisfy the following requirements:

1. Conceptual clarity and physical insight.
2. Simplicity in physical description (e.g., one-dimensional, not three-dimensional, wave propagation is then solved exactly) and in application, permitting an analysis with a hand calculator in many cases.
3. Sufficient scope of application (shape of foundation, soil profile, embedment, piles).

4. Sufficient accuracy, as demonstrated by comparing the results of the physical models with those of rigorous methods reported in the two textbooks mentioned above.
5. Adequacy to explain physical phenomena.
6. Direct use in engineering practice for everyday design of machine foundations and structures founded on soil subjected to dynamic loads such as earthquakes, explosions, waves, traffic excitations, and so on. Possibility to check the results of more sophisticated analyses.
7. Potential generalization of the concepts with clear links to the rigorous methods.

These simple procedures emerged from the experience gained through the extensive development of the field of foundation vibration and dynamic soil–structure interaction during the past 25 years. The addressed methods are not "primitive"—they were developed to achieve simplicity, which, based on rationality, is the ultimate sophistication in analysis! (A.S. Veletsos).

This book covers almost all types of surface and embedded foundations, including piles. Starting with the simplest case, the surface foundation on a homogeneous halfspace, the procedures expand in steps up to the most complicated case, the foundation embedded in a stratified site consisting of many layers resting on an underlying halfspace. The book commences with an overview in Chapter 1. Chapter 2 examines the foundation on the surface of a homogeneous halfspace. Chapter 3 addresses the foundation on the surface of a single soil layer resting on rigid and flexible rocks, and Chapter 4, the foundation embedded in a halfspace and in a layer resting on rock. The analysis procedures based on cones and on lumped-parameter models—with the coefficients of the springs, dashpots, and masses specified in tables ready for practical application—are developed for all these foundations. The calculation can be performed directly in the familiar time domain, permitting nonlinear behavior of the structure; or especially for machine vibrations, the calculation can be made in the frequency domain (harmonic loading). In addition, both the unexpected overestimation of radiation damping in an equivalent two-dimensional model of a three-dimensional problem (Section 2.8) and the vanishing of radiation damping up to the cutoff frequency for a layer on rigid rock (Section 3.8) are discussed. Using a prescribed cylindrical wave pattern to calculate dynamic-interaction factors, Section 4.9 examines a pile foundation. The extension to a foundation of arbitrary shape on the surface of a homogenous halfspace based on a prescribed wave pattern on the free surface is examined in Chapter 5. Appendix D describes the generalization of the concept of cones to a layered halfspace. The loading can either be applied directly to the basemat or to the structure (as for forces generated from machine vibrations) or be introduced via the soil, as with seismic excitation addressed in Chapter 6. The coupled dynamic system structure–soil is addressed in Chapter 7; in particular, equivalent one-degree-of-freedom systems are derived.

The reader can follow all derivations step by step. The key parameters dominating the response are identified and discussed, and simple examples illustrate the procedures in great detail. Practical examples link the results of the analysis to the design procedure. At the end of each chapter is a summary.

The book will primarily appeal to the geotechnical consultant; it enables the analysis of foundation vibration and of dynamic soil–structure interaction based on simple procedures with valuable physical insight and sufficient engineering accuracy. In addition, the book can serve as text in a first course on soil dynamics, which can be offered in the senior year. The course on which this book is based has been taught for the past few years to final-year undergraduate students in civil engineering at the Swiss Federal Institute of Technology in Lausanne and to first-year graduate students at the same university in Zurich. Traditionally, more than two thirds of the participants are practicing engineers! As a prerequisite some knowledge of elementary structural dynamics is necessary.

The book is based to a large extent on research and development performed in an informal, enthusiastic, and collegial atmosphere during the past four years with Dr. J.W. Meek, who is one of the pioneers of cone models. His role as leader and his significant creative contributions are acknowledged. The author has been influenced over the years by the research on practical methods published by many authorities; to name just a few (in alphabetical order): Professors G. Gazetas, J.M. Roesset, and A.S. Veletsos.

The author gratefully acknowledges the important contributions of colleagues. In particular, the author would like to thank Dr. E. Miranda, A. Paronesso, and Dr. Ch. Song. The financial support of the Swiss Federal Institute of Technology in Lausanne is also acknowledged.

John P. Wolf
Swiss Federal Institute of Technology

List of Tables for Lumped-Parameter Models

HOMOGENEOUS SOIL HALFSPACE

Surface Foundation

Embedded Foundation

SOIL LAYER ON RIGID ROCK

Surface Foundation

Embedded Foundation

1

Introduction

1.1 STATEMENT OF PROBLEM

The fundamental objective of *soil–structure-interaction* analysis [W4, W6] is illustrated in Fig. 1-1. A structure with finite dimensions is embedded in soft soil, which extends to infinity. A specified time-varying load—originating, for example, from rotating machinery—acts on the structure or the dynamic excitation is introduced through the soil as for seismic waves. The dynamic response of the structure interacting with the soil is to be determined.

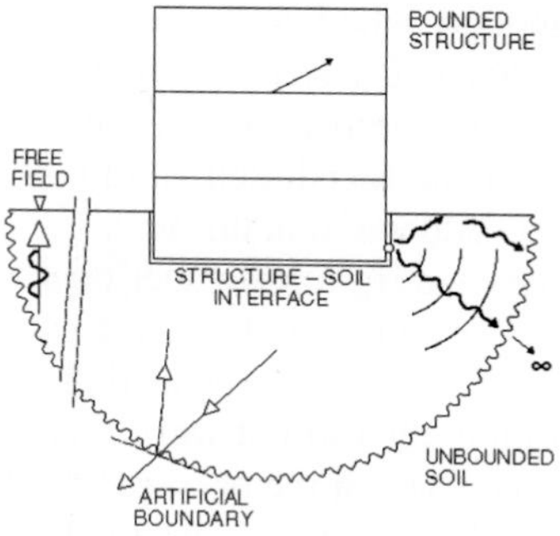

Fig. 1-1 Objective of soil–structure-interaction analysis.

The various aspects are discussed qualitatively in more detail. The *bounded structure* (and, if applicable, an adjacent irregular soil region which can be regarded as part of the structure) can exhibit *nonlinear* behavior. Using the well-established methods of structural dynamics, determining a finite-element model with a finite number of degrees of freedom for the structure is rather straightforward. The corresponding (nonlinear) dynamic equations of motion of the discretized structure can then be formulated, and highly developed methods for solving them (iteratively) are readily available.

In contrast, the *unbounded soil* is assumed to remain *linearly elastic with linear-hysteretic material damping* during the dynamic excitation. This can be justified by noting that the allowable displacements of foundations for satisfactory operation of machines equal fractions of a millimeter. Inelastic deformations are thus ruled out for this type of dynamic loads (in contrast to the static settlements caused by dead load which can be several orders of magnitude larger). It should also be noted that the amplitudes of all (body and surface) waves in three dimensions propagating from the structure–soil interface away through the soil towards infinity decay due to *geometric spreading*. By including a sufficiently large soil region (with nonlinearities) as part of the structure (as discussed above), the remaining soil extending to infinity can be regarded as linear. In addition, worth mentioning is that a nonlinear elasto-plastic analysis of the soil reduces the stress concentrations calculated in the three-dimensional elastic body. The same is also achieved by applying an elastic strength-of-materials approach which is used extensively in this text. The dynamic behavior of the unbounded soil, a semi-infinite domain, is significantly different from that of the bounded structure. The motions on the structure–soil interface emit *waves propagating in the soil in all directions towards infinity* (strictly speaking, no energy is radiated from infinity towards the structure). At the free surface of the soil reflections occur, and in addition refractions arise at the interfaces of a layered soil. This complicated wave pattern radiates energy towards infinity, outside the total dynamic system. The unbounded soil thus acts as an energy sink, leading to *damping (called radiation damping) also in a linear system* (in contrast to a bounded system). To model the unbounded soil for static loading, an *artificial boundary* at a sufficient distance from the structure can be introduced. This leads to a bounded domain for the soil which can be modeled similarly to the structure. The total discretized system, consisting of the structure and the soil can then be analyzed straightforwardly. However, for dynamic loading, this procedure cannot be used. This fictitious boundary would reflect waves originating from the vibrating structure back into the discretized soil region instead of letting them pass through and propagate towards infinity. This need to model the unbounded soil (foundation) medium properly distinguishes soil dynamics from structural dynamics. The total dynamic system of structure and soil thus consists of two parts with radically different dynamic properties. The challenge in performing analyses of dynamic soil–structure-interaction and foundation vibrations consists of modeling the unbounded soil, which is the subject addressed in this text.

The load can act directly on the structure—as from machines operating, impacts, vehicles moving within the structure, ocean waves, wind, and so on. In this *radiation problem*, waves are emitted from the structure–soil interface propagating in the soil towards infinity. Alternatively, the dynamic excitation in the form of waves can arrive through the soil—as from earthquakes, underground explosions, or the passage of vehicles away from the structure. In this *scattering problem* (or diffraction problem), the loading environment must first be specified; that is, the response of the *free field* of the soil (site) must be determined before the actual soil–structure-interaction analysis can be performed. This consists of the spatial and temporal variations of the motion before excavating the soil and superimposing the structure.

In the majority of cases both the structure and the soil will remain *linear*. But linear analysis is also helpful in many more complicated nonlinear areas of application. The results of a nonlinear analysis of certain dynamic systems with strong nonlinearities are similar to those of a linear calculation. This applies for instance to the global results such as the overturning moment when even substantial partial uplift of a structure's basemat (gap along the structure–soil interface) for seismic excitation occurs. Of course, counter examples also exist. A linear analysis can in many cases also serve as a "crutch" when interpreting complicated nonlinear results. The response to the lower-level, more frequent operational design earthquake must in many cases remain linear, although nonlinear procedures might be necessary to address the maximum credible earthquake. In addition, for preliminary design simple approximate linear methods are appropriate. Linear analyses are also needed for the evaluation of field tests performed at low-amplitude response which provide some basic material constants.

The dynamic system described in connection with Fig. 1-1 also includes many practical *nonlinear* cases. Examples are structures that perform in the nonlinear range for high seismic excitation; base isolation systems with friction plates exhibiting strong nonlinear characteristics which have to be considered in design; impact-producing machines activating gaps in pads; local nonlinearities such as the partial uplift of the basemat and the separation occurring between the sidewalls of the base and the neighboring soil in the case of embedded structures; and the highly nonlinear soil behavior arising adjacent to the basemat. A hammer foundation (impact-producing machine) and the partial uplift of a rigid block are investigated as examples of nonlinear soil–structure-interaction analyses in this text.

Some of the methods to be addressed can also be applied to calculate more complicated cases such as the dynamic response of a structure excited through the soil by rotating machinery located in a neighboring structure (structure–soil-structure interaction). The analysis of other dynamic unbounded medium-structure interactions (reservoir–dam interaction for seismic waves, acoustics) lies outside the scope of this text.

1.2 DESIGN OF MACHINE FOUNDATION

For the sake of illustration and to establish the link of foundation vibration to an actual engineering task, the design process of a machine foundation is outlined. A sketch of a typical rigid foundation block (the structure) supporting rotating machinery and resting on the surface of layered soil is presented in Fig. 1-2. The dynamic (centrifugal) load is generated by an unbalanced mass rotating with an eccentricity. This causes the foundation block to vibrate, activating inertial loads. These together with the dynamic machine load will be in equilibrium with the horizontal and vertical forces and the moment applied at the base of the foundation block. These stress resultants act with the other sign on the rigid interface on the surface of the soil, resulting in horizontal and vertical displacements and a rotation. Waves are emitted, propagating towards infinity whereby reflections and refractions in a layered soil will occur. In the case of rotating machinery operating at a constant frequency as shown in Fig. 1-2, the dynamic load, the inertial loads of the foundation block, the stress resultants and displacements at its base, and all generated waves in the soil will be harmonic with the same frequency. For other modes of operation and types of machines, general transients will occur.

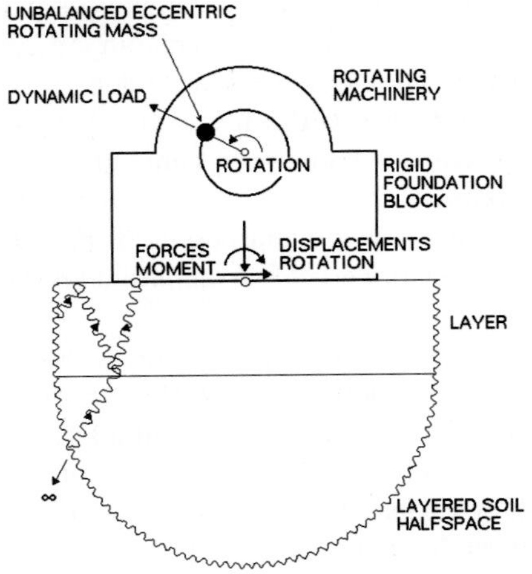

Fig. 1-2 Machine foundation consisting of rigid foundation block with rotating mass of machine resting on layered soil.

The *design procedure of a typical machine foundation* will involve the following steps [G3], which are described qualitatively:

1. Estimate the *magnitude and characteristics of the dynamic loads* in cooperation with the supplier of the machine. The unbalanced forces can either vary *harmonically* (rotating machinery, as discussed above), vary *biharmonically* (superposition of two harmonic motions corresponding to the operational frequency and twice that as in reciprocating machinery), or can be a series of single *pulses* (impact-producing machinery). Foundations for these three types of machines are examined in this text. The reader is referred to these examples for more information concerning the dynamic loads. In addition, general periodic forces and transients have to be addressed in more complicated cases.
2. Establish the *soil profile* (layers resting on halfspace) and determine the low-strain *elastic constants* (shear modulus, Poisson's ratio), the *mass density*, and the *material damping* ratio (viscoelasticity with hysteretic damping).
3. Determine the *design criteria* in cooperation with the client and by considering the relevant codes to guarantee a satisfactory operation of the machine. In general, the allowable peak velocity is specified, which for harmonic motion can be easily formulated as a displacement (or acceleration) amplitude. The induced vibrations should also not be annoying to the workers operating the machine, harmful to precision equipment, or troublesome to persons in adjacent structures.
4. Select, based on experience, the type and the *trial dimensions* of the foundation block.
5. Calculate the *dynamic response* of the trial foundation block of step 4, resting on the soil of step 2, and subjected to the dynamic loads of step 1. Various simplifications and idealizations of the foundation block and the soil are normally introduced. The equations of motion of the dynamic system are established and solved. The key aspect of this foundation vibration analysis is the soil's interaction force-displacement equation defined at the rigid interface on the surface, that is, the relationship between the displacements, rotation, and the interaction forces, moment shown in Fig. 1-2.

The determination of this *dynamic stiffness* of the unbounded soil with specified properties is a difficult task.

6. Check if the calculated response of step 5 meets the design criteria of step 3.
7. If the check of step 6 is not satisfied, repeat steps 4, 5, and 6 until a satisfactory design results.

To construct starting from reality a model which can be analyzed, many idealizations have to be introduced. When analyzing this model in this trial-and-error design process, many uncertainties will still exist, especially in steps 1, 2, and 3. *Parametric studies* varying, for instance, the soil properties of step 2 are thus essential to gain insight in the dynamic behavior leading to an acceptable design. These studies can be performed adequately and economically using simple approximate procedures in the dynamic analysis of step 5. Rigorous analysis methods to model the soil are unnecessary for standard foundations. What is needed are *simple physical models* to represent the soil, in particular to determine its dynamic stiffness. The main goal of this text is to develop such simple physical models for soil.

1.3 EFFECTS OF SOIL–STRUCTURE INTERACTION

To illustrate the salient features of soil–structure interaction, the seismic response of a structure founded on (rigid) rock is compared to that of the same structure with a rigid base embedded in (flexible) soil resting on rock (Fig. 1-3). Again only qualitative statements are, of course, possible at this stage. The incident seismic waves with horizontal motion (called the control motion) propagating vertically in the rock are the same in the two cases. The motions are shown as solid arrows with lengths proportional to the earthquake excitation.

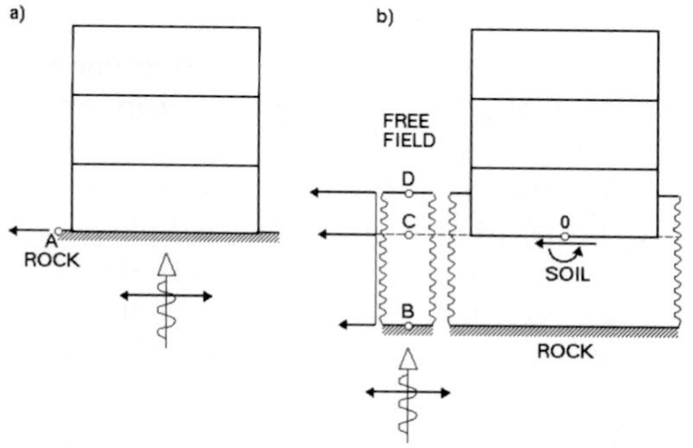

a. Founded on rock.
b. Embedded in soil resting on rock with free field.

Fig. 1-3 Seismic response of structure.

For the structure founded on rock (Fig. 1-3a), the *control motion* in point A can be applied directly to the base of the built-in structure. The input accelerations resulting in the applied horizontal inertial loads will be constant over the height of the structure. During the earthquake, an overturning moment and a transverse shear force acting at the base will develop. As the rock is assumed to be rigid, these two stress resultants will not lead to any (additional) deformations at the base. The resulting horizontal displacement of the base is thus equal to the seismic control motion; no rocking motion arises at the base.

For the structure founded in soft soil (Fig. 1-3b), the motion of the base of the structure in point 0 will be different from the control motion. To gain insight into how the soil affects the dynamic response of the structure, it is convenient to distinguish between the three following effects. First, the *free-field response*, the motion of the site in the absence of the structure and of any excavation, will *differ* from the control motion. If there were no soil on top of the rock in point B, the motion in this fictitious rock outcrop would be equal to the control motion. The presence of the soil layer will reduce the motion in point B. The corresponding waves will propagate vertically through the soil layer, resulting in motions in points C and D which differ from that in B. Points C and D are nodes in the free field which will subsequently lie on the structure–soil interface when the structure has been built. In general, depending on its frequency content, the motion is usually *amplified*, thus resulting in horizontal displacements that increase towards the free surface of the site. Second, excavating and inserting the rigid (massless) base into the site will modify the motion. The base will experience some *average horizontal displacement* and a *rocking* component, called the *effective foundation input motion*. This rigid-body motion will result in accelerations (leading to inertial loads) which will vary over the height of the structure, in contrast to the applied accelerations in the case of a structure founded on rock. Third, the inertial loads applied to the structure will lead to an overturning moment and a transverse shear force acting at point 0. These will cause deformations in the soil and thus, once again, modify the motion at the base.

Figure 1-3 also illustrates the main effects of taking soil–structure interaction into consideration. First, the seismic *input motion* acting on the structure–soil system *will change*. Because of the amplification of the free-field motion, the translational component of this effective foundation input motion will in many cases be larger than the control motion, and, in addition, a significant rocking component will arise for an embedded structure—even for vertically propagating waves with a horizontal motion. Second, the presence of the soil in the final dynamic model will make the system *more flexible*, decreasing the fundamental frequency to a value that will, in general, be significantly below that applicable for the fixed-base structure. The shape of the vibrational mode will also be changed. Its *introduced rocking* of the base will affect the response, especially at the top of a tall structure. Third, the *radiation of energy* of the propagating waves away from the structure will result in an *increase of the effective damping* of the final dynamic system. For a soil site approaching a homogeneous elastic halfspace, this increase will be significant, leading to a strongly reduced response. In contrast, for a soil site consisting of a shallow layer resting on rigid rock, it is possible that no waves propagate away from the structure (if the frequencies of the excited waves lie below the so-called *cutoff frequency* of the layer, which is equal to its fundamental frequency). In this case only the material damping of the soil will act, and no beneficial effect on the seismic response is to be expected. In any soil–structure-interaction analysis it is very important to determine whether the loss of energy by radiation of waves (*radiation damping*) can actually take place.

It is obvious from the many opposing effects that it is, in general, impossible to determine a priori whether the interaction effects will decrease the seismic response or not. If, however, the first effect discussed above (the change of the seismic input motion) is neglected when the interaction analysis is performed, the *response* will in many circumstances (excluding very stiff structures) be *smaller* than the fixed-base response obtained from an analysis that neglects all interaction effects. For this approximate interaction analysis, the control motion is directly used as input to the dynamic analysis.

For loads acting directly on the structure, only the second (reduction of the fundamental frequency) and third (potential increase of the effective damping due to radiation of waves) effects are present.

1.4 CLASSIFICATION OF ANALYSIS METHODS

1.4.1 Direct Method and Substructure Method

How to treat the behavior at infinity of the unbounded soil computationally is the central theme in dynamic soil–structure-interaction analysis. As discussed in connection with Fig. 1-1, waves emitted from the vibrating structure–soil interface will eventually propagate in the soil towards infinity. At a sufficient distance from the structure only outgoing waves are present in the actual interaction analysis (radiation problem). This avoids an infinite energy buildup. Or to formulate it differently, *no incoming waves* propagating from infinity towards the structure exist. Strictly speaking, no energy associated with the waves may radiate from infinity into the structure–soil system. For the simple systems addressed in this text, the actual waves (and not the energy) can be examined. This *radiation condition* will lead to a boundary-value problem (formulated in the frequency domain) for an unbounded domain with a *unique solution*. Two analysis procedures exist: the direct method and the substructure method.

In the *direct method*, illustrated in Fig. 1-4a, the region of the (linear) soil adjacent to the structure–soil interface is also explicitly modeled (e.g., with finite elements, in the same way as the structure) up to the *artificial boundary*. The latter must be introduced, as it is impossible to cover the unbounded soil domain with a finite number of elements with bounded dimensions. Appropriate boundary conditions must be formulated which have to represent the missing soil located on the exterior. Besides modeling the soil's stiffness up to infinity, reflections of the outwardly propagating waves on the artificial boundary must be avoided, which thus acts as a *transmitting boundary*. The "radiation condition" is formulated not at infinity, but on the artificial boundary which is located at a significant distance from the structure. The rigorous boundary condition will be global in both space and time—that is, to advance one time step in a transient analysis requires information over the entire boundary from all previous time stations. As one would intuitively expect, results of sufficient accuracy can be determined in many cases using only approximate boundary conditions. These approximations are *local in space and time*: They use information only from the node or the nearby region of the mesh at the start of the time step or, at most, at a few recent time stations. The boundary conditions must be *independent of frequency* to be able to be used in a transient analysis in the time domain. The viscous dashpots acting independently in the nodes in all directions with constant coefficients (Fig. 1-4a) are such local frequency-independent boundary conditions. A spring with a constant coefficient can be added in parallel to the dashpot. Such a local spring–dashpot system whose frequency-independent coefficients can be determined using truncated semi-infinite cones is developed in this text. Many more sophisticated schemes have been developed to formulate highly absorbing local approximations to the perfect transmitting boundary. Due to the discretization of the soil region adjacent to the structure, a large number of degrees of freedom arise, leading to a significant computational effort. A loss of physical insight also results. It follows that the direct method is not appropriate for standard projects of moderate and small sizes. Not much emphasis is thus placed on this procedure in this text.

In the *substructure method*, illustrated in Fig. 1-4b, the irregular (nonlinear) structure is modeled by an interconnection of masses, dashpots, and (possibly nonlinear) springs or equivalently by finite elements. The dynamic equations of motion of the structure discretized in the nodes located on the structure–soil interface and in its interior follow. The other substructure, the unbounded soil extending to infinity, will be regular (e.g., a layered halfspace) and linear. This allows analytical solutions to be determined that *satisfy exactly the radiation condition* formulated at infinity. With these fundamental solutions (*Green's functions*) a boundary-integral equation can be formulated, called the *boundary-element method* in discretized form, to calculate the interaction force-displacement relationship in the nodes located on the structure–soil interface. This *dynamic stiffness*, which is *global in space and time*, represents the rigorous boundary condition to model the unbounded soil. Combining this force-displacement relationship of the soil with the discretized equations of motion of the structure results in the final system of equations of the total dynamic system. The substructure method thus permits each substructure to be analyzed by the best-suited computational technique. For a rigid basemat of the structure (as for most machine foundations), the degrees of freedom of the structure–soil interface are reduced to those of a rigid body. Especially for this case physical approximations, such as truncated semi-infinite one-dimensional cones, can be introduced to model the unbounded soil. The substructures method of analysis in thus well suited when physical models for the soil are used.

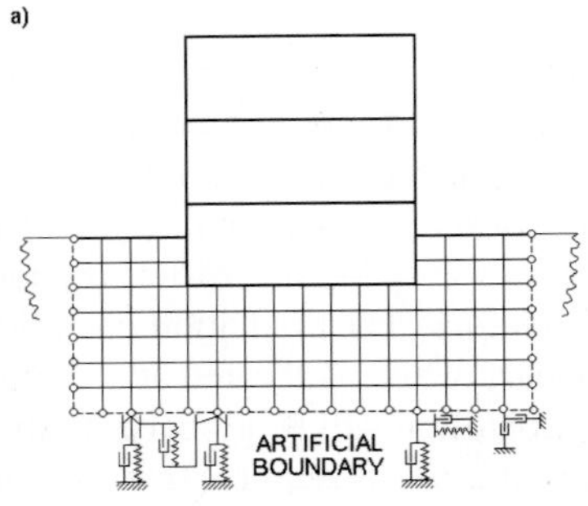

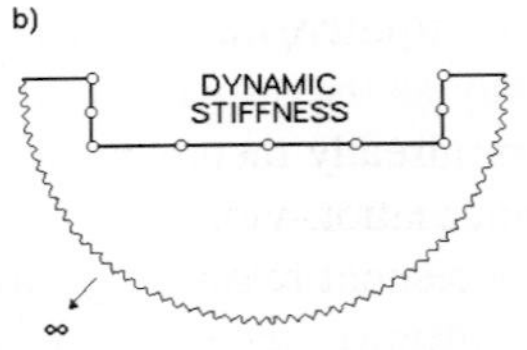

Fig. 1-4 Dynamic system for

a. Direct method of analysis with finite-element mesh of soil and artificial boundary.

b. Substructure unbounded soil with global dynamic stiffness in substructure method of analysis.

1.4.2 Frequency Domain and Time Domain

Another classification addresses the domain in which the analysis is performed: the frequency domain or the time domain. In the *frequency domain* the excitation is decomposed into a Fourier series, and the response is determined independently for each Fourier term corresponding to a specific frequency. In the frequency domain the equations can be manipulated as easily as in the static case. But this formulation has two disadvantages. First, the Fourier transformation from the time to the frequency domain and back again is an abstract concept for the structural or geotechnical analyst who is not accustomed to working in the frequency domain. This mental obstacle has greatly limited the general acceptance of analytical and numerical procedures using frequency-domain formulations. The natural approach is to consider the sequence of developments from one time to the next—that is, to apply the time-domain concept. The other disadvantage is that a frequency-domain analysis can deal only with linear response, thus excluding the important cases described in Section 1.1 with a nonlinear behavior in the structure or the adjacent soil region. (In the so-called hybrid frequency-time-domain analysis, which works mostly in the frequency domain, certain nonlinearities can be considered, affecting only the right-hand side of the equations of motion in this iterative procedure. The stability and an efficient implementation are, however, difficult to achieve.) This limitation means that the analysis does not serve one major purpose of earthquake-engineering analysis—to predict the degree of damage to be expected from the investigated earthquake—because "damage" implies nonlinearity. *Priority is given in this text to time-domain procedures.* The general Fourier transformation to the frequency domain is not addressed; however, the harmonic response (steady state) to a sinusoinal excitation of a specific frequency is examined. This allows valuable physical insight to be gained and covers the important application to foundations of rotating machinery with one operating frequency. The physical models for the unbounded soil are very attractive because the calculation can be performed purely in the time domain. The derivation of the properties of the physical models can also be done in the time domain.

1.4.3 Rigorous Boundary-Element Method and Approximate Simple Physical Model

As already mentioned above, the key aspect of any dynamic soil–structure-interaction or foundation-vibration analysis is the calculation of the fully coupled interaction force-displacement relationship in the nodes located on the structure–soil interface. This dynamic stiffness of the unbounded viscoelastic soil can be determined very accurately based on the *boundary-element method* or on sophisticated finite-element procedures.

The boundary-element method requires a formidable theoretical background. This semi-analytical procedure involves the discretization of the structure–soil interface and uses closed-form analytical solutions incorporating the radiation condition formulated at infinity. Singularities have to be evaluated. A considerable amount of expertise in idealizing the actual dynamic system is necessary [G2], and a significant amount of data preparation has to be performed. The computational expense for just one run is large, making it difficult from an economical point of view to perform the necessary parametric studies and to investigate alternative design schemes. This rigorous boundary-element method can thus provide a false sense of security to the user. The effort to interpret the results is significant. Even some of the most sophisticated codes cannot model certain potentially important aspects such as the partial uplift of the basemat and the separation of the sidewalls during the dynamic excitation.

The boundary-element method with its mathematical complexity obscures the physical insight and belongs more to the discipline of applied computational mechanics than to civil engineering. Engineers tend to be intimidated by these procedures. These methods should only be used for large projects of critical facilities such as nuclear power plants, bunkered military constructions, or dams with the corresponding budget and available time to perform the analyses. For all other projects—the majority—the *simple physical models to represent the unbounded soil* developed in this text can be used. They easily fit the size and economics of the project, and no sophisticated computer code needs to be available. Use of these simple physical models leads to some loss of precision, but this is more than compensated for by their many advantages. It cannot be the aim of the engineer to calculate the complex reality as closely as possible. A well-balanced design, one that is both safe and economical, does not call for rigorous results in a standard project. The accuracy of such results is limited anyway because of the many uncertainties, some of which can never be eliminated.

As the simple physical models cannot cover all cases, they do not supplant the more generally applicable rigorous boundary-element method, but rather they supplement it. It should also be stressed that an improved understanding can be gained from the results of a rigorous analysis, which should thus be developed to enable progress.

Finally, the relationship of these simple physical models to the development of the disciplines of foundation dynamics and soil–structure-interaction analysis is discussed. As with any other new field of science, soil dynamics as part of civil engineering that concentrates on statics was not addressed at all at first; but after significant damage in certain earthquakes had occurred which could not be explained without addressing the vibrations of soil, elastodynamics for soil was developed based on mechanics. General formulations were derived, sophisticated calculations were performed, and as experience increased, the key aspects of the behavior could be identified. This then lead to the development of simplified procedures which, however, still capture the salient features of the phenomena. The physical models for soil dynamics presented in this text are such simplified procedures. They are not the first attempt to capture the physics that has not been properly evaluated yet—on the contrary, they make full use of the experience gained from the rigorous state-of-the-art formulations such as the boundary-element method mentioned above. The physical models are thus not only simple to use and lead to valuable physical insight, but they are also quite dependable, incorporating implicitly much more know-how than meets the eye.

1.5 PHYSICAL MODELING OF UNBOUNDED SOIL

1.5.1 Applications

Simple physical models to represent the unbounded soil can be applied as follows: In certain cases, the effect of the interaction of the soil and the structure on the response of the latter will be negligible and thus need not be considered. This applies, for example, to a flexible high structure with small mass where the influence of the higher modes (which are actually affected significantly by soil–structure interaction) on the seismic response remains small. Exciting the base of the structure with the prescribed earthquake motion is then possible. For loads applied directly to the structure, the soil can in this case be represented by a static spring, or the structure can even be regarded as built-in. In other cases, including many everyday building structures, ignoring the interaction analysis can lead to an overly conservative design. It should be remembered that seismic-design provisions [N1] allow for a significant *reduction* of the equivalent static lateral force (up to 30%) *for soil-structure interaction effects*. For these two categories, to determine if a dynamic interaction analysis is meaningful or not and to calculate the reduction in the response of everyday structures—that is, to perform the actual dynamic analysis—physical models are well suited. They are also appropriate to help the analyst identify the key parameters of the dynamic system for preliminary design, to investigate alternative designs, and to perform parametric studies varying the parameters with large uncertainties such as the soil properties or the contact conditions on the structure–soil interface. Finally, simple models are used to check the results of more rigorous procedures determined with sophisticated computer codes.

1.5.2 Overview

An overview of the physical models to calculate the dynamic stiffness of the unbounded soil used in this text is presented. Other physical models such as to determine the free-field response and the effective foundation input motion for seismic excitation are also examined in this text; they are, however, not addressed in this section. At this stage only a general description without any details, derivations, or references can be provided. The latter are found in the following chapters.

To construct a physical model, physical approximations are introduced which at the same time simplify the mathematical formulation. The latter can then be solved rigorously, in general in closed form.

The simplest case consists of a rigid massless circular basemat, called "disk" in the following, resting on the *surface of a homogeneous soil halfspace*. A translational degree of freedom, for example, the vertical motion, is examined (Fig. 1-5a). To determine the interaction force-displacement relationship of the disk and thus its dynamic stiffness, the disk's displacement (as a function of time) is prescribed; and the corresponding interaction force, the load acting on the disk (as a function of time), is calculated. The halfspace below the disk is modeled as a truncated semi-infinite rod (bar) with its area varying as in a cone with the same material properties. A load applied to the disk on the free surface of a halfspace leads to stresses from geometric spreading acting on an area that increases with depth, which is also the case for the translational cone. By equating the static stiffness of the *translational cone* to that of the disk on a halfspace, the cone's opening angle is calculated. It turns out that the opening angle for a given degree of freedom depends only on Poisson's ratio of the soil. Through the choice of the physical model, the complicated three-dimensional wave pattern of the halfspace with body and surface waves and three different velocities is replaced by the simple one-dimensional wave propagation governed by the one constant dilatational-wave velocity of the conical rod, whereby plane cross sections remain plane. The radiation condition (outwardly propagating waves only) is enforced straightforwardly by admitting waves traveling downwards only. For the horizontal motion a translational cone in shear with the shear-wave velocity is constructed analogously. For the rocking and torsional degrees of freedom, *rotational cones* can be identified using the same concepts (Fig. 1-5b).

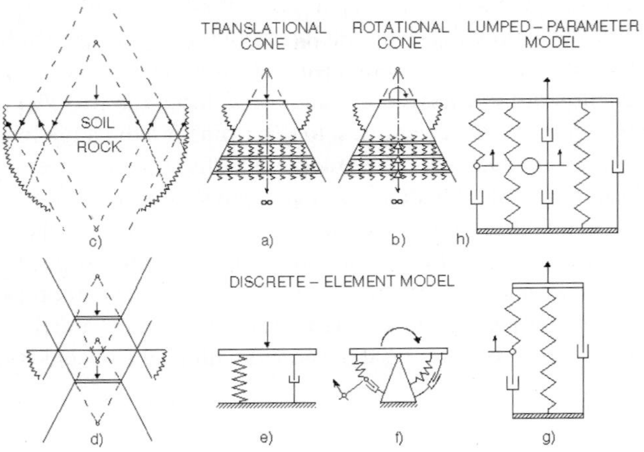

a. Disk on surface of halfspace with truncated semi-infinite translational cone.
b. Disk on surface of halfspace with truncated semi-infinite rotational cone.
c. Disk on surface of soil layer resting on flexible rock halfspace with corresponding cones.
d. Anti-symmetry condition with respect to free surface for disk and mirror-image disk with corresponding double cones to calculate Green's function.
e. Discrete-element model for translational cone.
f. Discrete-element model for rotational cone.
g. Lumped-parameter model consisting of springs and dashpots with one internal degree of freedom corresponding to f.
h. Lumped-parameter model consisting of springs, dashpots and a mass with two internal degrees of freedom.

Fig. 1-5 Physical models to represent dynamic stiffness.

The same cones can also be used to examine a *surface foundation* for a site consisting of a *soil layer resting on flexible rock halfspace*. The vertical degree of freedom is addressed in Fig. 1-5c using the corresponding translational cone. The same approach can be applied for the rotational degrees of freedom. The vertical force applied to the disk produces dilatational waves propagating downwards from the disk. The opening angle of this cone follows again from equating the static stiffness of the truncated semi-infinite cone to that of the homogeneous halfspace with the material properties of the layer. At the interface of the soil layer and the rock, the incident wave will lead to a refracted wave propagating in the rock in the same direction as the incident wave along its own cone (dotted lines). In addition, a reflected wave is created propagating back through the soil layer along the indicated cone (dashed lines) in the opposite upward direction. The latter will reflect back at the free surface and then propagate downwards along the cone shown in Fig. 1-5c. Upon reaching the interface of the soil layer and the rock, a refraction and a reflection again take place, and so on. The waves in the layer thus decrease in amplitude and spread resulting in radiation of energy in the layer in the horizontal direction (in addition to the energy loss through the rock halfspace).

The concepts of cone models can be expanded to the analysis of *embedded cylindrical foundations*. Again, the vertical degree of freedom of a foundation embedded *in a halfspace* is addressed in Fig. 1-5d, but the following argumentation is just as valid for the horizontal, rocking, or torsional ones. To represent a disk within an elastic fullspace, a double-cone model is introduced. Its displacement field defines an approximate Green's function for use in an uncomplicated (one-dimensional) version of the boundary-element method. To enforce the stress-free condition at the free surface of the halfspace (Fig. 1-5d), two identical double cones placed symmetrically (with respect to the free surface) and excited simultaneously by the same forces are considered. Indeed, any halfspace problem amenable to a solution via cone models may also be solved in the fullspace. It is only necessary to augment the actual foundation in the lower halfspace by its mirror image in the upper halfspace. By exploiting principles of anti-symmetry and superposition, the soil's flexibility matrix defined at the disks located within the embedded part of the foundation can be set up. The rest of the analysis follows via conventional matrix methods of structural analysis. The concept can be expanded to a fixed boundary and also to an interface with another halfspace. This permits *cylindrical foundations embedded in a soil layer resting on a rigid or flexible rock halfspace* to be calculated using cones. The methodology points towards a general *strength-of-materials approach to foundation dynamics* using the approximate *Green's functions of the double-cone models.*

A generalization is possible, enabling the dynamic-stiffness coefficients of a *foundation on the surface of or embedded in a layered halfspace* to be calculated. For each layer a dynamic-stiffness matrix based on cones is established. Assembling the dynamic-stiffness coefficient of the underlying halfspace and the dynamic-stiffness matrices of the layers yields that of the layered site.

Returning to the translational and rotational cones of Figs. 1-5a and 1-5b, it should be emphasized that in an actual soil–structure-interaction or foundation vibration analysis the cones are not represented physically by finite elements of a tapered rod. For practical applications it is not necessary to compute explicitly the displacements of the waves propagating along the cone. The attention can be restricted to the interaction force-displacement relationship at the disk. The translational cone's dynamic stiffness can rigorously be represented by the *discrete-element model* shown in Fig. 1-5e. It consists of a spring with the static-stiffness coefficient (of the disk on a halfspace) in parallel with a dashpot with the coefficient determined as the product of the density, the dilatational-wave velocity (for the vertical degree of freedom), and the area of the disk. As the latter is equal to the disk on a halfspace's limit of the dynamic stiffness as the frequency approaches infinity, the cone's dynamic stiffness will be exact for the static case and for the limit of very large frequencies (*doubly asymptotic approximation*). For intermediate frequencies the cone is only an approximation of the disk on a halfspace. One rigorous representation of the rotational cone's dynamic stiffness is shown in Fig. 1-5f. The model again consists of a spring with the static-stiffness coefficient in parallel with a dashpot with as coefficient the exact high-frequency limit of the dynamic stiffness (density times dilatational-wave velocity times disk's moment of inertia for the rocking degree of freedom). An additional internal degree of freedom is also introduced, connected by a spring (with a coefficient equal to minus a third of the static-stiffness coefficient) to the footing and by a dashpot (with a coefficient equal to minus the high-frequency limit) to the rigid support. Again, the rotational cone's dynamic stiffness is doubly asymptotic. This is easily verified by noting that the internal degree of freedom of the discrete-element model is not activated in the two limits.

The model of Fig. 1-5f is shown for a translational degree of freedom in Fig. 1-5g, which forms the starting point to develop systematically a family of consistent *lumped-parameter models*. The direct spring is chosen to represent the static stiffness. The coefficients of the other spring and of the two dashpots are selected so as to achieve an *optimum fit* between the dynamic stiffness of the lumped-parameter model and the corresponding exact value (originally determined by a rigorous procedure such as the boundary-element method). If the direct dashpot is used to represent the high-frequency limit of the dynamic stiffness, the number of coefficients available for the optimum fit is reduced to two. To increase the number of coefficients, and thus the accuracy, several systems of Fig. 1-5g can be placed in parallel. Figure 1-5h shows the lumped-parameter model for three such systems, whereby two of them are combined to form a new system consisting of two springs, one independent dashpot (the two dashpots in series have the same coefficient), and a mass. A total of six coefficients keeping the doubly asymptotic approximation thus results. It will be shown that these six frequency-independent coefficients, which can be determined using curve fitting applied to the dynamic stiffnesses (involving the solution of a linear system of equations only) will be real (but not necessarily positive). The *eight springs, dashpots, and mass* will represent a *stable lumped-parameter model with only two additional internal degrees of freedom*. The fundamental lumped-parameter model of Fig. 1-5g is easy to interpret physically. This physical insight is, however, lost to a large extent in the model shown in Fig. 1-5h. But the latter does allow the analyst to model quite complicated cases, as will be demonstrated, such as a foundation embedded in a soil layer resting on rigid rock. Use is implicitly made of the results obtained with the state-of-the-art formulation which leads to the rigorous dynamic stiffnesses used in the optimum fit. By comparing visibly the dynamic stiffness of the lumped-parameter model with the rigorous solution, the accuracy can be evaluated.

Summarizing, *two types of physical models* are described in connection with Fig.1-5: the translational and rotational cones (truncated semi-infinite single or double cones based on rod [bar] theory with the corresponding one-dimensional displacement and wave propagation) and the lumped-parameter models. The latter can conceptionally be constructed from the former by assembling the exact discrete-element models of the cones in parallel and using calibration with rigorous solutions.

The models shown in Fig. 1-5 prescribe a displacement pattern varying with depth along the axis of the cone. To extend the application, *displacement patterns in the horizontal plane* other than those corresponding to the strength-of-materials assumption that plane cross sections remain plane are introduced in this text. One-dimensional wave propagation is again prescribed.

Two examples follow. For a vertical point load on the surface of an elastic halfspace, the displacement in form of a Green's function may be deduced via nonmathematical physical reasoning, then calibrated with a few constants taken from a rigorous solution. By superposing point loads, an approximate *Green's function* is constructed for a subdisk of radius Δr_0 (Fig. 1-6a). In the near field the displacement amplitude is inversely proportional to the distance r from the center of the loaded source *subdisk* (body-wave), and the wave propagates with a velocity that is slightly less than the Rayleigh-wave velocity c_R. In the far field with a different phase angle, the displacement amplitude decays inversely proportional to the square root of r (surface wave), and the wave propagates with c_R. Arbitrary shaped foundations can be treated as an assemblage of subdisks. The through-soil coupling of neighboring foundations can also be analyzed using subdisks.

To analyze a *pile group*, the *dynamic-interaction factor* describing the effect of the loaded source pile on the receiver pile is applied (Fig. 1-6b). To calculate the interaction factor, as for a horizontal motion of a source pile, it is assumed that dilatational waves propagating with the corresponding velocity c_p are generated in the direction of motion, and that shear waves propagating with the velocity c_s are generated in the perpendicular direction. The amplitudes of both types of these cylindrical waves decay inversely proportional to the square root of the radius.

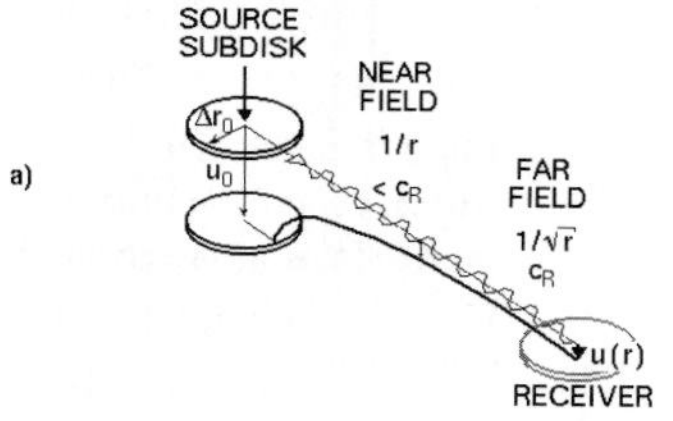

a. Vertical displacement on free surface from loaded source subdisk.

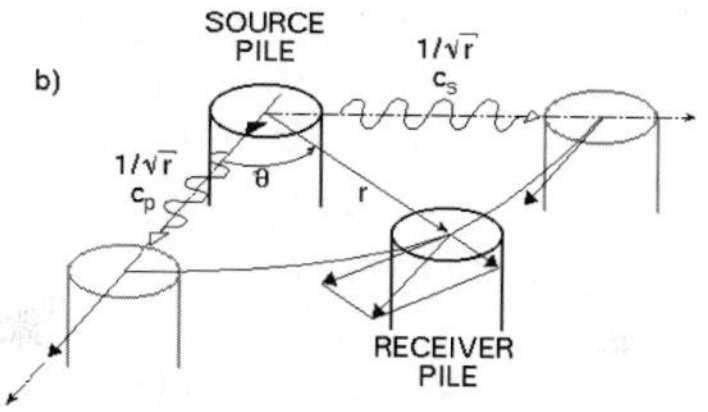

b. Horizontal displacement from loaded source pile.

Fig. 1-6 Displacement patterns in horizontal plane.

1.5.3 Examples

To illustrate the concepts of constructing physical models, some examples with selected results are presented, which allow the accuracy to be evaluated. At this stage not all details of the development of the physical models can be presented. The reader should not get irritated by this and should just glance through the figures. A systematic derivation is of course provided in the corresponding chapters later on in this text.

Harmonic excitation with frequency ω is addressed. Complex-variable notation is used in the following. From the complex response $u(\omega) = Reu(\omega) + iImu(\omega)$, the magnitude is calculated as $\sqrt{Reu^2(\omega) + Imu^2(\omega)}$ and the phase angle as $\arctan[Imu(\omega)/Reu(\omega)]$. Applying a displacement with amplitude $u_0(\omega)$, the corresponding force amplitude $P_0(\omega)$ is formulated as

$$P_o(\omega) = S(\omega)\, u_o(\omega) \tag{1.1}$$

with the (complex) dynamic-stiffness coefficient $S(\omega)$. In foundation dynamics it is appropriate to introduce the dimensionless frequency a_0

$$a_o = \frac{\omega r_o}{c_s} \tag{1.2}$$

with r_0 representing a characteristic length of the foundation, for example, the radius of the disk, and c_s the shear-wave velocity. Using the static-stiffness coefficient K to nondimensionalize the dynamic-stiffness coefficient

$$S(a_o) = K[k(a_o) + ia_o c(a_o)] \tag{1.3}$$

is formulated. The spring coefficient $k(a_0)$ governs the force which is in phase with the displacement, and the damping coefficient $c(a_0)$ describes the force which is 90° out of phase. The dynamic-stiffness coefficient $S(a_0)$ can thus be interpreted as a spring with the frequency-dependent coefficient $Kk(a_0)$ and a dashpot in parallel with the frequency-dependent coefficient $(r_0/c_s)Kc(a_0)$ (Fig. 1-7).

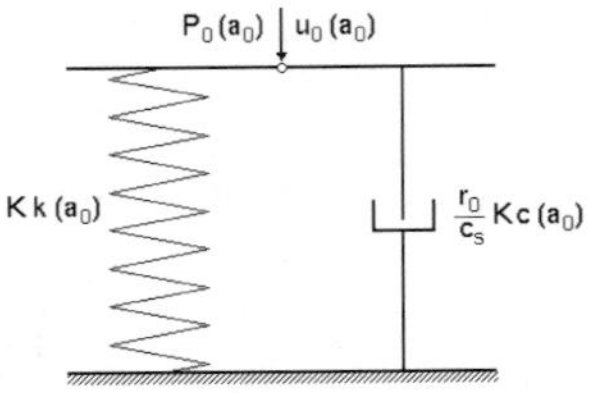

Fig. 1-7 Interpretation of dynamic-stiffness coefficient for harmonic excitation as spring and as dashpot in parallel with frequency-dependent coefficients.

First, the rocking degree of freedom of a rigid disk with radius r_0 (Fig. 1-8a) resting on the surface of an undamped homogeneous soil halfspace with shear modulus G, Poisson's ratio ν and c_s is addressed. The discrete-element model representing exactly the rotational cone (which is a doubly asymptotic approximation of the halfspace) is shown in Fig. 1-8b with $K_2 = 8\,G r_0^3/(3(1 - \nu))$, $C_2 = \rho c_p \pi r_0^4/4$ (ρ = mass density, c_p = dilatational-wave velocity). The accuracy of the corresponding dynamic-stiffness coefficient (Fig. 1-8c) is acceptable. Better agreement is achieved when the coefficients of K_1 and C_1 are determined by an optimum fit based on the exact values, leading to the fundamental lumped-parameter model with the same arrangement of the springs and dashpots. The coefficients K_1, C_1 are presented in the caption.

Second, the vertical and rocking degrees of freedom of a rigid square foundation of length $2b$ resting on the surface of a soil halfspace is investigated. One quadrant is discretized with 7 × 7 subdisks using the Green's function illustrated in Fig. 1-6a. Figure 1-9 shows the dynamic-stiffness coefficients. The agreement is good.

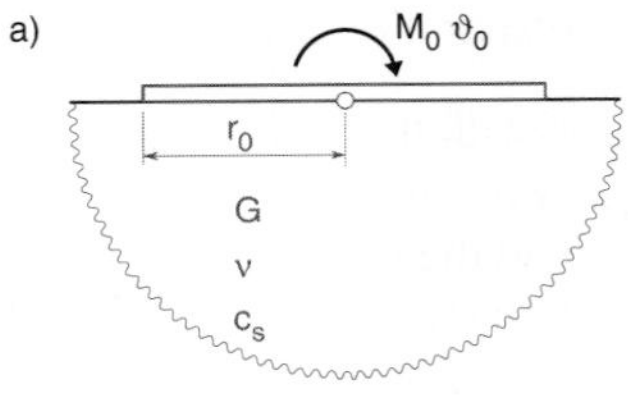

a. Disk with halfspace (ν = 1/3).

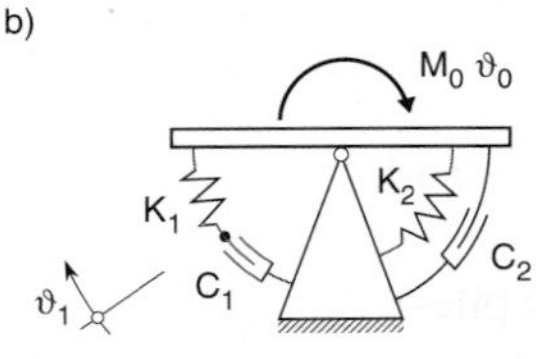

b. Discrete-element model ($K_1/K_2 = -1/3$, $C_1/C_2 = -1$) and fundamental lumped-parameter model ($K_1/K_2 = -0.47$, $C_1/C_2 = -1$).

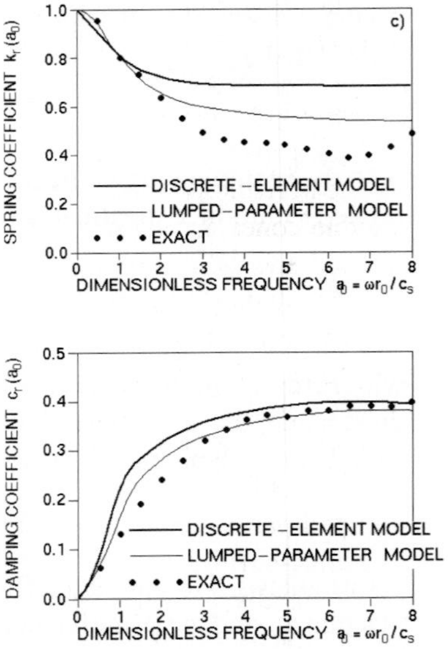

c. Dynamic-stiffness coefficient for harmonic excitation.

Fig. 1-8 Rocking of surface foundation on homogeneous soil halfspace.

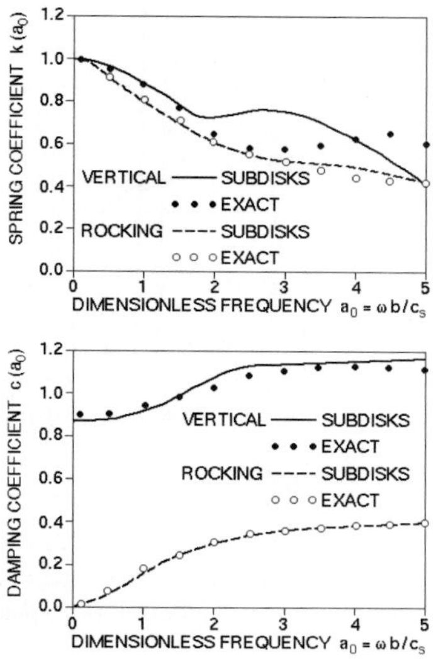

Fig. 1-9 Dynamic-stiffness coefficients of square foundation on surface of homogeneous soil halfspace ($\nu = 1/3$).

Third, the vertical degree of freedom of a rigid disk on the surface of an undamped soil layer of depth d resting on rigid rock (Fig. 1-10a) is examined. The cone model with the wave pattern representing the reflections at the rigid interface and the free surface is shown in Fig. 1-10b. The corresponding dynamic-stiffness coefficient (Fig. 1-10c) yields a smooth approximation in the sense of an average fit to the exact solution, which becomes increasingly irregular. The lumped-parameter model of Fig. 1-10d based on an optimum fit leads to the results shown in Fig. 1-10e.

Fourth, a rigid cylindrical foundation embedded with the depth e in a soil layer resting on rigid rock (Fig. 1-11a) is discussed. The cones are applied to calculate the dynamic-stiffness coefficient of the vertical degree of freedom and the lumped-parameter model that of the horizontal motion. In the embedded part of the foundation (Fig. 1-11b), 8 disks with double cones are selected (two are shown in the figure, one with a solid and one with a dashed line). To enforce approximately the stress-free condition at the free surface and the fixed boundary condition at the base of the layer, mirror images of the disk with the loads acting in the indicated directions with the corresponding double cones (dashed lines) are introduced. The dynamic-stiffness coefficient (Fig. 1-11c) is surprisingly accurate, as can be seen from a comparison with the rigorous solution determined with a very fine mesh of boundary elements. The lumped-parameter model for the coupled horizontal and rocking degrees of freedom is shown in Fig. 1-11d. The coupling term is represented by placing the lumped-parameter model of Fig. 1-5h at the eccentricity e. The agreement for the horizontal dynamic-stiffness coefficient in Fig. 1-11e is good.

Finally, a rigidly capped floating pile group taking pile–soil–pile interaction into consideration is addressed. The 3×3 pile group in a halfspace is shown in Fig. 1-12a (s = distance between axes of two neighboring piles, ℓ = length of pile, $2r_0$ = diameter of pile, E = Young's modulus of elasticity, ρ = density). To model the single pile 25 disks with the corresponding double cones are used. To calculate pile–soil–pile interaction, the dynamic interaction factor based on the sound physical approximation illustrated in Fig. 1-6b—but for

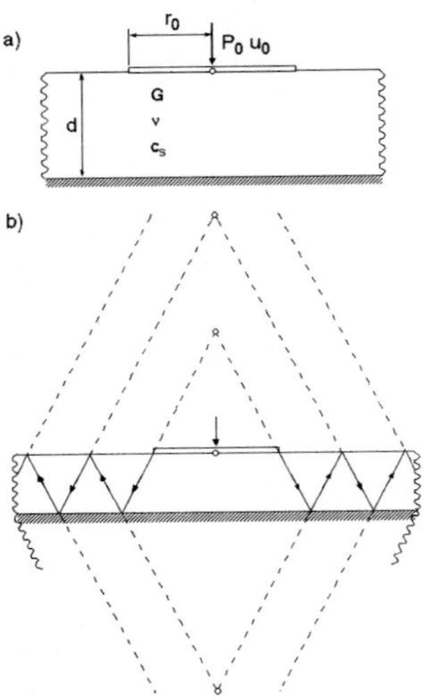

a. Disk with layer built-in at its base ($r_0/d = 1$, $\nu = 1/3$).

b. Cone model.

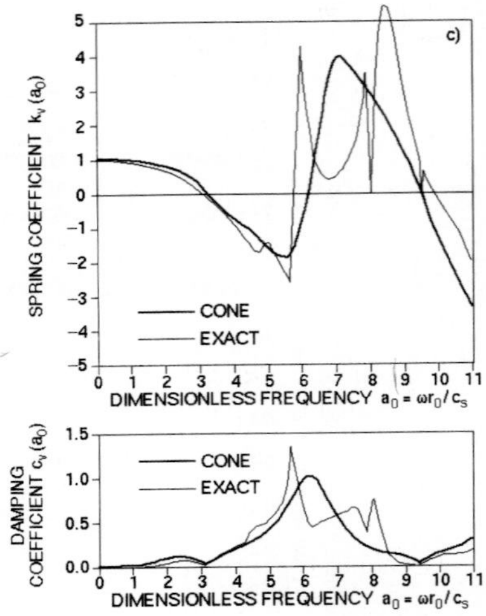

c. Dynamic-stiffness coefficient for harmonic excitation determined with cone model.

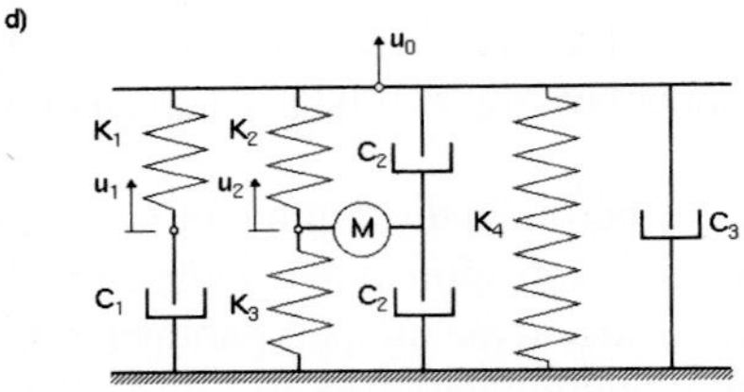

d. Lumped-parameter model ($K_1 = -31.26\ Gr_0$, $K_2 = 5.65\ Gr_0$, $K_3 = -2975.70\ Gr_0$, $K_4 = 10.10\ Gr_0$, $C_1 = -6.20\ Gr_0^2/c_s$, $C_2 = -37.29\ Gr_0^2/c_s$, $C_3 = 43.57\ Gr_0^2/c_s$, $M = -89.68\ Gr_0^3/c_s^2$).

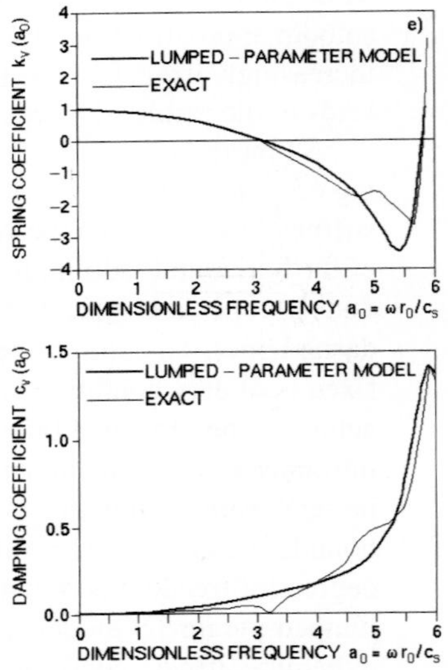

e. Dynamic-stiffness coefficient for harmonic excitation determined with lumped-parameter model.

Fig. 1-10 Vertical motion of surface foundation on soil layer on rigid rock.

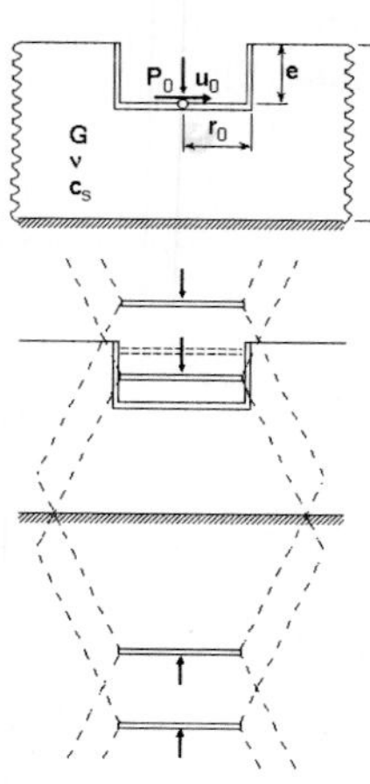

a. Cylindrical foundation with layer fixed at its base ($r_0/e = 1$, $r_0/d = 1/3$, $\nu = 1/3$).

b. Array of disks with double-cone models and mirror-image disks for vertical motion.

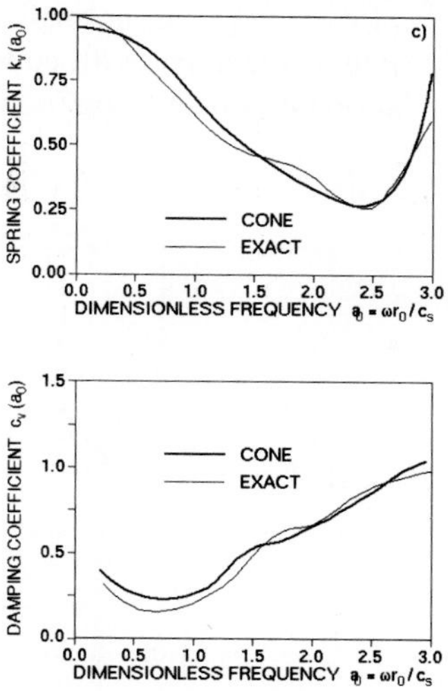

c. Vertical dynamic-stiffness coefficient for harmonic excitation with cone model (5% material damping).

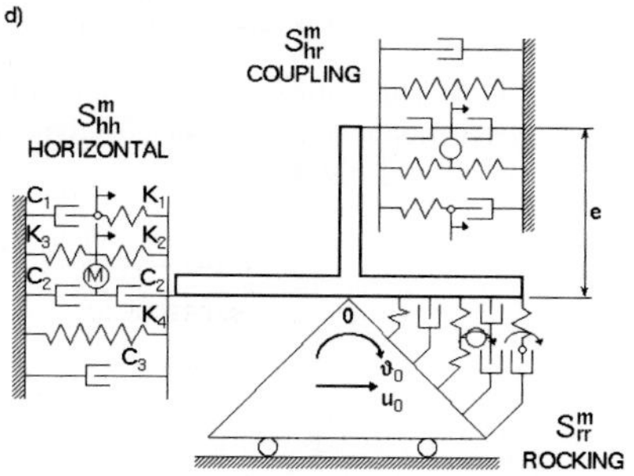

d. Lumped-parameter model with coupling of horizontal and rocking motions (coefficients see Table 4-2).

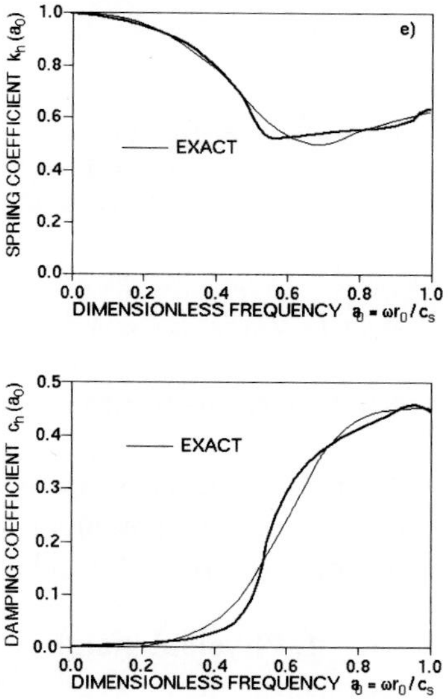

e. Horizontal dynamic-stiffness coefficient for harmonic excitation with lumped-parameter model (undamped).

Fig. 1-11 Foundation embedded in soil layer on rigid rock.

vertical motion—is determined. The dynamic-stiffness coefficient in the vertical direction of the pile group (normalized with the sum of the static-stiffness coefficients of the single piles) calculated with cones is astonishingly accurate (Fig. 1-12b). Even details of the strong dependency on frequency are well represented.

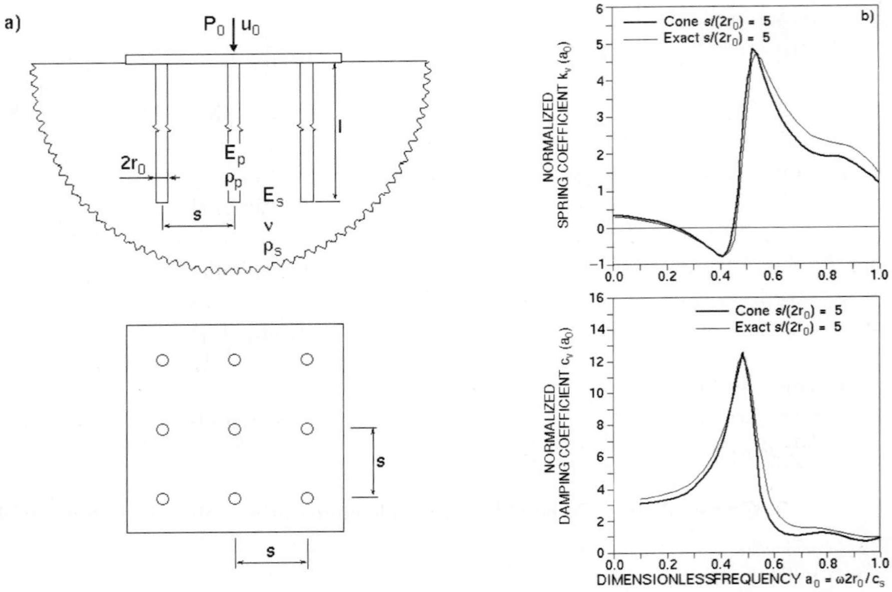

a. Elevation and plan view of 3 × 3 pile group ($s/(2r_0) = 5, \ell/(2r_0) = 15, \nu = 0.4, E_p/E_s = 1000, \rho_p/\rho_s = 1.43$, 5 % material damping).

b. Dynamic-stiffness coefficient for harmonic excitation with cone model and dynamic-interaction coefficient.

Fig. 1-12 Vertical motion of floating pile group in homogeneous soil halfspace.

1.5.4 Requirements

After providing a descriptive overview and examples of results obtained with the physical models used to represent the unbounded soil and analyze foundation vibration, it is appropriate to state some general requirements for successful models.

1. Physical insight. The mathematical complexity of rigorous solutions in elastodynamics often obscures physical insight and intimidates practitioners. By simplifying the physics of the problem, *conceptual clarity* with *physical insight* results. As an example, the calculation of the dynamic stiffness of a disk on the surface of a soil layer resting on rigid rock (Fig. 1-10a)—a complicated three-dimensional mixed boundary-value problem with dispersive waves—is addressed. In the physical model made up of truncated cones (Fig. 1-10b), for which the familiar strength-of-materials theory applies, the wave pattern is clearly postulated. The one-dimensional waves propagate with the dilational-wave velocity (a material constant), reflecting back and forth, spreading and decreasing in amplitude, and thus radiating energy towards infinity in the horizontal direction. Physical insight is more difficult to achieve in a dynamic problem than in a static case.

Based on the physics of the problem, certain features can be introduced better than through some mathematical law which must often be simplified to be able to derive a rigorous solution. Modeling of material damping with causal frictional elements in such a case permits an analysis in the time domain.

2. Simplicity. Due to the simplification of the physical problem, the physical model can be rigorously mathematically solved. The fundamental principles of wave propagation and dynamics are thus satisfied exactly for the simple physical model. Closed-form solutions (even in the time domain) exist for the one-dimensional cones. For instance, to calculate the dynamic stiffness of the disk on the soil layer resting on rigid rock with cones (Fig. 1-10b), the analysis can be performed with a hand calculator as no system of equations is solved; for the embedded cylindrical foundation with cones (Fig. 1-11b), a special-purpose computer code can easily be written; and when lumped-parameter models are applied (Figs. 1-10d, 1-11d), a standard general-purpose structural dynamics program which permits springs, dashpots, and masses as input can be used directly. The *practical application* of the physical models is thus also *simple*, together with the *physics* and the rigorous *mathematical solution*.

3. Generality. To be able to provide engineering solutions to reasonably complicated practical cases and not just to address academic examples, the physical models must reflect the following key aspects of the foundation-soil system for all translational and rotational degrees of freedom [G2].

- *The shape of the foundation–soil (structure–soil) interface*: Besides the circle, the rectangle and the arbitrary shape, which can be modeled as an equivalent disk or directly (Fig. 1-9) without "smearing," can be represented as a three-dimensional case or, if applicable, as a two-dimensional slice of a strip foundation.
- *The nature of the soil profile*: The homogeneous halfspace, the layer resting on a flexible halfspace, and the layer resting on a rigid halfspace as well as the layered halfspace with many layers can be modeled.
- *The amount of embedment*: Surface, embedded (with soil contact along the total height of the wall or only on part of it), and pile foundations can be represented.

The physical models must also allow the calculation of the effective foundation input motion for seismic excitation. They must work well for the static case, for the low- and intermediate-frequency ranges important for machine vibrations and earthquakes, and for the limit of very high frequencies as occurring in impact loads.

4. Accuracy. Due to the many uncertainties addressed earlier, the accuracy of any analysis will always be limited. A deviation of $\pm$ 20% of the results of the physical models from those of the rigorous solution for one set of input parameters is, in general, sufficient. This *engineering accuracy* criterion is, in general, satisfied, as can be verified by examining the dynamic-stiffness coefficients shown in Figs. 1-8 to 1-12. It should also be remembered that for a transient loading such as an earthquake the deviations (with both signs) are "smeared" over the frequency range of the excitation and thus further reduced compared to the largest error for one frequency.

The use of the physical models does indeed lead to some loss of precision compared to applying the rigorous boundary-element procedure or the sophisticated finite-element-based method; however, this is more than compensated by the many advantages discussed in this section.

5. Demonstration of physical features. Besides leading (by construction of the physical model) to physical insight of the mechanisms involved in foundation vibration (item 1), the physical models are also well suited to demonstrate certain unexpected features and to derive further results. Four examples follow.

1. Placing a row of an infinite number of identical vertical point loads (Fig. 1-6a) on the surface of a halfspace as a simple physical model, the vertical dynamic stiffness of a two-dimensional slice of a rigid surface strip foundation can be determined. An analytical solution can be derived. The static stiffness is zero, but the spring coefficient increases abruptly to a more or less constant value for larger frequencies. By contrast, the damping coefficient begins from infinity at zero frequency and then diminishes asymptotically for increasing frequency. The same features exist in the rigorous solution.
2. As derived later on in the text, the radiation damping ratio in a two-dimensional model is significantly larger than in the corresponding three-dimensional case (just the contrary of what is expected intuitively). To model a two-dimensional slice of a strip foundation on a halfplane, wedges can be used which are again based on rod theory, just as cones represent a disk on a halfspace (three-dimensional situation). For both the translational and rotational motions, the damping ratios $\zeta = b_0 c(b_0)/[2k(b_0)]$ of the wedges are significantly larger than those of the cones (Fig. 1-13). The dimensionless frequency parameter is defined as $b_0 = \omega z_0/c$ with z_0 denoting the apex height of the cone or wedge and c the appropriate wave velocity. The multipliers are also specified in the figure.
3. For the frequency range below the cutoff frequency of the soil layer resting on rigid rock, the radiation damping and thus the damping coefficient of the dynamic stiffness vanishes, which is well simulated using cones and lumped-parameter models. See Figs. 1-10c and e, where below the cutoff frequency for dilatational waves, $a_0 = \pi$, $c(a_0)$ is very small.

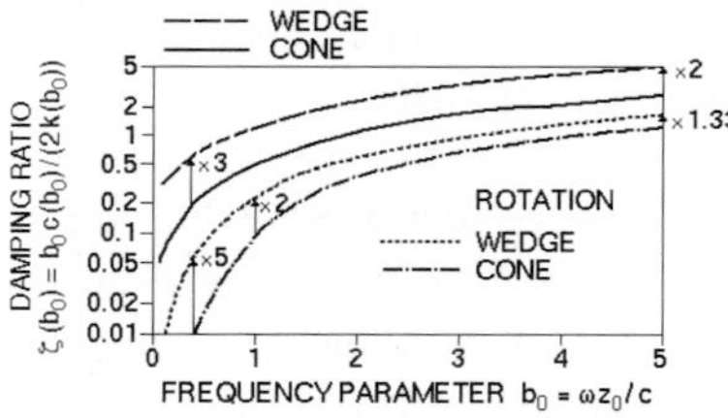

Fig. 1-13 Overestimation of radiation damping ratio of two-dimensional modeling with wedges when compared to that of three-dimensional modeling with cones for translational and rotational motions.

4. The combined structure–soil system can be modeled approximately as an equivalent one-degree-of-freedom system. The corresponding effective natural frequency and damping ratio can be determined by modeling the soil as cones.

6. Suitability for everyday practical foundation-vibration analysis. Especially the ease in use, the sufficient generality and the good accuracy allow the physical models to be applied for foundation vibration and dynamic soil–structure-interaction analyses in a design office.

7. Potential for generalization. The *concepts and certain features of the physical models can be generalized* and the results applied in much more sophisticated calculations. Three examples of such extensions are listed.

1. As already mentioned, the cone models lead to simple Green's functions, and the analysis of an embedded foundation can be interpreted as a straightforward application of the one-dimensional boundary-element method. The rigorous calculation can be performed also with a boundary-element method based on the same concept, the major difference being that the Green's function of the three-dimensional fullspace is used and not the one-dimensional solution derived from the rod theory of cones.
2. A consistent lumped-parameter model for the dynamic-stiffness matrix of any general flexible foundation can be systematically constructed starting from the same fundamental lumped-parameter model (Fig. 1-5g). In general, a large number of these building blocks are assembled in parallel for each coefficient of the matrix to be represented.
3. As will be derived further on in this text, the interaction force-displacement relationship of some physical models will involve convolution integrals which can be evaluated exactly very efficiently using a recursive formulation. The same procedure can also be applied to the corresponding relationship of any general flexible foundation.

Summarizing, the cone models with the prescribed deformation of rod (bar) theory, the lumped-parameter models based on them, and the displacement patterns in the horizontal plane present a major step towards developing a *strength-of-materials approach to foundation dynamics*. The aim is the same as in stress analysis of structural engineering, where, for very complicated skew-curved prestressed concrete bridges, beam theory is applied successively and the general three-dimensional theory of elasticity is not needed. As in stress analysis, each specific case has to be calculated based on the strength-of-materials approach. It is not sufficient just to use tables of dynamic-stiffness coefficients calculated for certain cases based on the rigorous formulation of elastodynamics. As the soil is a three-dimensional body without a dominant axial direction, the strength-of-materials approach, with prescribed displacement behavior taking all essential features into account, will be more difficult to formulate in foundation engineering than in structural engineering. Concluding, the dynamic analyst should always "make things as simple as possible but no simpler" (H. Einstein). Or to state it differently: "*Simplicity that is based on rationality is the ultimate sophistication*" (A.S. Veletsos).

1.6 ORGANIZATION OF TEXT

In the next three chapters physical models to represent various foundation-unbounded soil systems are developed. Methods to determine the interaction force-displacement relationship (dynamic stiffness) and its inverse, the displacement-interaction force equation (dynamic flexibility) for all degrees of freedom are addressed. The cone models are derived, followed by the related lumped-parameter representations. Emphasis is placed on time-domain formulations. Chapter 2 examines the foundation on the surface of a homogeneous soil halfspace, and addresses the consideration of material damping. The unexpected differences in the dynamic behavior of a three-dimensional and a two-dimensional model are critically discussed. Chapter 3 treats the foundation on the surface of a soil layer on rigid rock, and deals with the extension to a flexible rock. Insights on the cutoff frequency are provided. Chapter 4 addresses embedded foundations in the time domain and for harmonic excitation. Methods to deal with a single pile and a pile group for harmonic excitation are developed.

Chapter 5 introduces a simplified vertical dynamic Green's function. The dynamic stiffnesses for harmonic excitation for the vertical and rocking degrees of freedom of rigid and flexible basemats on the surface of a homogeneous soil halfspace are examined.

Chapter 6 addresses seismic excitation. It presents physical models to calculate the free-field response and the effective foundation input motion for surface and embedded foundations (the latter being an extension of the cone models used to determine the dynamic stiffness). The basic equations of motion for the substructure and the direct methods in the time domain and for harmonic excitation are also derived. For cones and lumped-parameter models, the procedure to convert the seismic effective foundation input motion to an equivalent driving load acting on the total dynamic system is discussed. A local frequency-independent transmitting boundary is constructed using cone models.

Chapter 7 investigates the influence of the unbounded soil modeled with cones on the response of a simple one-degree-of-freedom structure. The effective natural frequency, damping ratio and seismic input motion of an equivalent one-degree of freedom system are calculated for a structure founded on a halfspace and on a layer.

In Appendix A, the interaction force-displacement relationship and the Green's function for the cone models are derived. Appendix B systematically develops the consistent lumped-parameter model; and Appendix C develops the corresponding recursive evaluation. In Appendix D, the dynamic-stiffness coefficients of a foundation on the surface of or embedded in a layered halfspace based on cones are calculated.

SUMMARY

1. A (nonlinear) structure excited by a dynamic load interacts with the surrounding (linear) soil. In contrast to the bounded structure, the semi-infinite soil is an unbounded domain with waves propagating towards infinity leading to radiation damping.
2. The design of a typical machine foundation proceeds as follows. The dimensions of the foundation block are modified in an iterative process until the dynamic response of the foundation block subjected to the estimated dynamic loads (harmonic, bi-harmonic, periodic, pulse, general transient) of the machine and resting on the soil with

the established soil profile and viscoelastic material properties meets the selected design criteria. To determine the dynamic stiffness of the unbounded soil used in the dynamic analysis, simple physical models are appropriate. Parametric studies covering the uncertainties are essential.

3. When comparing the response of a structure embedded in flexible soil with that of the same structure founded on rigid rock, first, the presence of the soil makes the dynamic system more flexible; second, the radiation of energy of the propagating waves away from the structure (if occurring) increases the damping. Third, the seismic motion at the base of the structure is affected as follows:

    - The free-field motion differs from the control motion, usually being amplified towards the free surface.
    - The excavation of the soil and insertion of the rigid (massless) base results in averaging of the translation and in additional rocking for a horizontal earthquake.
    - The inertial loads determined from this seismic effective foundation input motion will further change the seismic motion along the base of the structure.

4. In the direct method of analysis, part of the linear regular soil is explicitly modeled up to an artificial boundary. Approximate boundary conditions on this transmitting boundary, which are local in space and time and which are frequency independent, must represent the missing soil on the exterior. In the substructure method of analysis, the interaction force-displacement relationship of the unbounded soil in the nodes on the structure-soil interface is combined with the discretized equations of motion of the structure. This soil's dynamic stiffness, which is global in space and time, is determined with the boundary-element method based on fundamental solutions which satisfy rigorously the radiation condition at infinity. Simple physical models are well suited to represent the unbounded soil in the substructure method.
5. With the exception of the analysis of a machine foundation being excited by harmonic loads with a single operating frequency, dynamic soil–structure-interaction and foundation-vibration analyses should be performed in the time domain. The time-domain analysis familiar to the analyst avoids the abstract transformation to the frequency domain and also permits nonlinear behavior of the structure to be taken into account. Physical models for the unbounded soil can be easily formulated in the time domain.
6. As seismic-design provisions allow for a significant reduction of the equivalent static lateral force for soil–structure interaction effects, analyzing the structure fixed at its base can lead to an overly conservative design.
7. Two types of physical models are introduced: the translational and rotational cones (based on rod theory with one-dimensional wave propagation and leading to the exact stiffness for the static case and the high-frequency limit) and the lumped-parameter models. The latter can conceptionally be constructed from the former by assembling in parallel the exact discrete-element models of the cones and using calibration with rigorous solutions. In addition, physical models can be constructed assuming one-dimensional wave patterns in the horizontal plane. With these physical models for the unbounded soil, all degrees of freedom of a foundation on the surface of or embedded

in a homogeneous halfspace—in a layer resting on flexible or rigid rock and in a layered halfspace as well as pile foundations—can be analyzed.

8. The physical models

- provide physical insight and conceptual clarity
- lead to ease in use (simplicity)
- exhibit sufficient generality
- lead to sufficient engineering accuracy
- can also be used to demonstrate certain unexpected results
- are suitable for everyday practical foundation-vibration analysis in a design office
- have concepts that can be generalized and extended

All these advantages more than compensate for some loss of precision when compared to the rigorous boundary-element procedure and sophisticated finite-element based methods.

9. The physical models present a major step towards developing a strength-of-materials approach to foundation dynamics which is so successful in stress analysis of structural engineering. This allows us to avoid the mathematical complexity (far beyond that typical for civil engineering practice) of rigorous solutions of applied mechanics.

2

Foundation on Surface of Homogeneous Soil Halfspace

The truncated semi-infinite cone model for translation was introduced more than half a century ago [E1], the cone model for rotation considerably later [M2] with an application of the latter contained in Veletsos and Nair [V4]. To elucidate the important phenomenon of radiation damping in two and three dimensions, wedges and cones can be used [G1]. This chapter closely follows several studies [M5, M7, M8, and W15].

In Section 2.1 the concept and the construction of cones to model a foundation on a homogeneous soil halfspace are discussed. The properties of the cones corresponding to the various degrees of freedom of a rigid surface disk are examined, in particular the appropriate propagation velocity and the opening angle of the cones. The translational and the rotational cones are addressed in Sections 2.2 and 2.3. The incorporation of material damping is described in Section 2.4. The reasons why a one-dimensional cone model can represent adequately a disk on a three-dimensional halfspace are discussed in Section 2.5. Generalizing the discrete-element models of the cones leads in Section 2.6 to simple lumped-parameter models of increased accuracy for surface foundations on a homogeneous halfspace. Wedges are introduced in Section 2.7 to model two-dimensional strip foundations. Finally, insight on two-dimensional versus three-dimensional modeling of surface foundations is provided in Section 2.8; in particular, it is demonstrated that in a two-dimensional model radiation damping is grossly overestimated.

2.1 CONSTRUCTION OF CONE MODEL

2.1.1 Elastic Properties and Wave-Propagation Velocities

To a first approximation, unlayered soil may be idealized as a homogeneous linearly elastic, semi-infinite medium with mass density ρ. The soil possesses no material damping (see Section 2.4 for a procedure to incorporate material damping). The stress–strain relationship is specified by two independent elastic constants. In elementary soil mechanics it is traditional to choose as these properties the constrained elastic modulus E_c and Poisson's ratio ν. In soil dynamics it is more customary to specify the shear modulus G and Poisson's ratio ν. When working with cone models, it is convenient to select yet another pair of fundamental properties, the *propagation velocities* c_s *of shear waves* (with the particle motion perpendicular to the direction of propagation) *and* c_p *of dilatational waves* (with the particle motion parallel to the direction of propagation). The elastic properties of each group may be computed from the others using the relationships shown in Table 2-1.

The following discussion addresses the model of the soil, which begins at the bottom surface of the basemat and extends downwards. Conceptually it is convenient to regard the soil–structure interface as a massless rigid basemat which imposes the displacement of the underside of the basemat. The area of the fictitious massless basemat is A_0; its area moment of inertia about the axis of rotation is I_0 (for torsional rotation (twisting), I_0 = polar moment of inertia). An equivalent radius r_0 of a circular disk with the same area as the basemat for translation or the same moment of inertia for rotation can be calculated as an approximation.

The soil is idealized *for each degree of freedom* as a *truncated semi-infinite elastic cone with its own apex height* z_0 (Fig. 2-1). Translational and rotational cones exist. The aspect ratio z_0/r_0 determines the opening angle. The procedure to determine z_0/r_0 is discussed in the next Section 2.1.2. Applying a load on the disk supported on the free surface leads to stresses in the halfspace acting on an area that increases with depth, which is represented approximately in the cone. Such concepts of the "spreading" of the load influence are applied in many areas of structural engineering. The cone is regarded as a rod (bar) with the displacement pattern over the cross section determined by the corresponding value on the axis of the cone. The theory of strength of materials is thus used (plane sections remain plane). The domain of the soil halfspace located outside the cone is disregarded. Depending on the nature of the deformation, it is necessary to distinguish between the translational cone for vertical and horizontal motions and the rotational cone for rocking and torsion. As indicated in Fig. 2-1, the appropriate wave-propagation velocities are $c = c_s$ for the horizontal and torsional cones which deform in shear and $c = c_p$ for the vertical and rocking cones which deform axially. The density ρ of the cone is the same as for the halfspace.

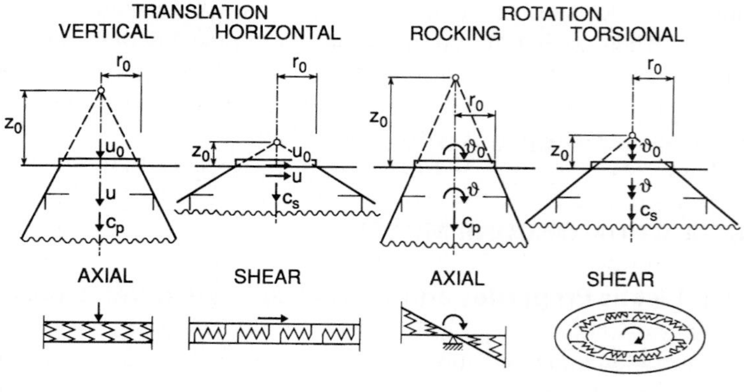

Fig. 2-1 Cones for various degrees of freedom with corresponding apex ratio (opening angle), wave-propagation velocity and distortion.

As for $\nu \rightarrow 0.5$, c_p tends to infinity (Table 2-1), c_p has to be limited for large values of ν. As is discussed in Section 2.1.4, 2 c_s is chosen in this nearly incompressible case.

TABLE 2-1 RELATIONSHIP BETWEEN ELASTIC CONSTANTS FOR SOIL MECHANICS, SOIL DYNAMICS, AND CONE MODEL

		E_c	ν	G	ν	c_s	c_p
E_c	=	E_c		$2G\frac{1-\nu}{1-2\nu}$			ρc_p^2
ν	=		ν		ν		$\frac{\left(\frac{c_p}{c_s}\right)^2 - 2}{2\left(\frac{c_p}{c_s}\right)^2 - 2}$
G	=	$\frac{E_c}{2}\frac{1-2\nu}{1-\nu}$		G		ρc_s^2	
ν	=		ν		ν		$\frac{\left(\frac{c_p}{c_s}\right)^2 - 2}{2\left(\frac{c_p}{c_s}\right)^2 - 2}$
c_s	=	$\sqrt{\frac{E_c}{2\rho}\frac{1-2\nu}{1-\nu}}$		$\sqrt{\frac{G}{\rho}}$		c_s	
c_p	=	$\sqrt{\frac{E_c}{\rho}}$		$\sqrt{2\frac{G}{\rho}\frac{1-\nu}{1-2\nu}}$			c_p

2.1.2 Aspect Ratio

Well-known closed-form solutions exist for the static-stiffness coefficients of a rigid disk resting on the surface of a halfspace. By equating for each degree of freedom the static-stiffness coefficient of the cone as derived below to the corresponding value of the halfspace, the aspect ratio z_0/r_0 is determined. The cone's opening angle is thus not chosen arbitrarily at a fixed value, such as 45°, but is selected in such a way that the behaviors of the disk on the halfspace and the cone in the low-frequency limit, the static case, coincide.

To determine the static-stiffness coefficient K of the translational cone, the vertical motion is examined (Fig. 2-2a), relating the displacement u_0 to the load P_0. With $A_0 = \pi r_0^2$ the area A at depth z equals $A=(z^2/z_0^2)A_0$, where z is measured from the apex. Also, u represents the axial displacement and N the axial force. Formulating the static equilibrium equation of the (upper) infinitesimal element

$$-N + N + N,_z dz = 0 \tag{2–1}$$

and substituting the force-displacement relationship

$$N = E_c A u,_z \tag{2–2}$$

leads to the differential equation

$$u,_{zz} + \frac{2}{z} u,_z = 0 \tag{2–3a}$$

or

$$(zu),_{zz} = 0 \tag{2–3b}$$

Its solution equals

$$zu = c_1 + c_2 z \tag{2–4}$$

with the integration constants c_1, c_2. Enforcing the boundary conditions

$$u(z = z_0) = u_0 \tag{2–5a}$$

$$u(z = \infty) = 0 \tag{2–5b}$$

leads to

$$u = \frac{z_0}{z} u_0 \tag{2–6}$$

With

$$P_0 = -N(z = z_0) = -E_c A_0 u_0,_z \tag{2–7}$$

and substituting the derivative calculated from Eq. 2-6 yields

$$P_0 = \frac{E_c A_0}{z_0} u_0 \tag{2–8}$$

which defines the static-stiffness coefficient of the translational cone in vertical motion as

$$K = \frac{E_c A_0}{z_0} \tag{2–9}$$

For the translational cone in horizontal motion, the analogous derivation involving the displacement u perpendicular to the cone's axis and the shear force leads to

$$K = \frac{G A_0}{z_0} \tag{2–10}$$

The two expressions can be unified by introducing the corresponding wave-propagation velocity c (Table 2-1)

$$K = \frac{\rho c^2 A_0}{z_0} \tag{2–11}$$

where c equals c_p and c_s for the vertical and horizontal motions, respectively.

The rotational static-stiffness coefficient K_ϑ relating the rotation ϑ_0 to the moment M_0 is derived addressing the rocking motion (Fig. 2-2b). With $I_0 = (\pi/4) r_0^4$ the moment of inertia I at depth z equals $I = (z^4/z_0^4)\, I_0$. Here, ϑ denotes the rotation, and M the bending moment. Formulating the static equilibrium equation of the (upper) infinitesimal element

$$-M + M + M,_z dz = 0 \tag{2–12}$$

and substituting the moment-rotation relationship

$$M = E_c I \vartheta,_z \tag{2–13}$$

leads to the differential equation

$$\vartheta,_{zz} + \frac{4}{z}\vartheta,_z = 0 \tag{2–14}$$

To solve this equation

$$\vartheta = z^\alpha \tag{2–15}$$

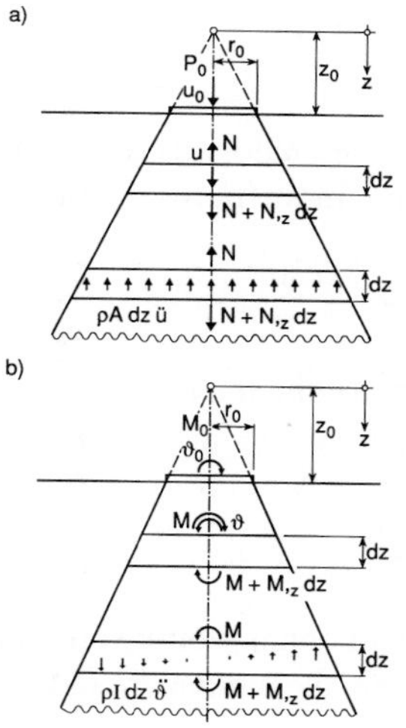

a. Translational cone with nomenclature for vertical motion and static and dynamic equilibrium of infinitesimal element.

b. Rotational cone with nomenclature for rocking motion and static and dynamic equilibrium of infinitesimal element.

Fig. 2-2 Truncated semi-infinite cones.

is assumed with α representing the unknown degree of the polynomial. Substituting Eq. 2-15 in Eq. 2-14 yields

$$\alpha^2 + 3\alpha = 0 \tag{2–16}$$

leading to $\alpha_1 = 0$, $\alpha_2 = -3$. The solution thus equals

$$\vartheta = c_1 + c_2 \frac{1}{z^3} \tag{2–17}$$

Enforcing the boundary conditions

$$\vartheta(z = z_0) = \vartheta_0 \tag{2–18a}$$

$$\vartheta(z = \infty) = 0 \tag{2–18b}$$

determines the integration constants c_1, c_2, resulting in

$$\vartheta = \frac{z_0^3}{z^3}\vartheta_0 \tag{2–19}$$

With

$$M_0 = -M(z = z_0) = -E_c I_0 \vartheta_{0,z} \tag{2–20}$$

and substituting the derivative calculated from Eq. 2-19 yields

$$M_0 = \frac{3E_c I_0}{z_0}\vartheta_0 \tag{2–21}$$

which defines the rocking static-stiffness coefficient of the rotational cone as

$$K_\vartheta = \frac{3E_c I_0}{z_0} \tag{2–22}$$

For the rotational cone in torsional motion, the analogous derivation with the rotation ϑ twisting around the cone's axis and the torsion moment results in

$$K_\vartheta = \frac{3GI_0}{z_0} \tag{2–23}$$

with the polar moment of inertia $I_0 = (\pi/2)r_0^4$. Again, the two expressions can be unified by introducing the corresponding wave-propagation velocity c (Table 2-1)

$$K_\vartheta = \frac{3\rho c^2 I_0}{z_0} \tag{2–24}$$

where c equals c_p and c_s for the rocking and torsional motions, respectively.

To determine the dimensions of the cone, the equivalent radius r_0 is determined first. As already mentioned, the most straightforward approach consists in equating the area A_0 or the moment of inertia I_0 of the basemat of arbitrary shape to the corresponding value of the disk for the translational or rotational cone, resulting in the expressions specified in Table 2-3A. Alternatively, the static-stiffness coefficient of the disk specified in the first column of Table 2-2 is set equal to the formula given for the rectangular foundation or for the foundation of arbitrary shape specified in Table 2-2. For the rectangular foundation with width $2b$ and length $2a$ ($2b \leq 2a$), the expressions are taken from Pais and Kausel [P1]. For the nonrectangular foundation of arbitrary shape, a circumscribed rectangle of width $2b$ and length $2a$ ($2b \leq 2a$) is introduced. Besides depending on a and b, the static stiffnesses of Gazetas [G5] are a function of the area A_0, of the moments of inertia I_{0a} and I_{0b} with respect to the principal axes parallel to the length $2a$ and the width $2b$ and of the polar moment of inertia I_0.

TABLE 2-2 STATIC STIFFNESS OF SURFACE FOUNDATION ON HOMOGENEOUS HALFSPACE

	Disk (r_0)	Rectangle ($2b < 2a$)		Arbitrary Shape ($2b \le 2a$)	
		along (around) lateral axis ($2b$)	along (around) longitudinal axis ($2a$)	along (around) lateral axis ($2b$)	along (around) longitudinal axis ($2a$)
Horizontal	$\frac{8Gr_0}{2-\nu}$	$\frac{Gb}{2-\nu}\left[6.8\left(\frac{a}{b}\right)^{0.65}+0.8\frac{a}{b}+1.6\right]$	$\frac{Gb}{2-\nu}\left[6.8\left(\frac{a}{b}\right)^{0.65}+2.4\right]$	$\frac{2Ga}{2-\nu}\left[2+2.5\left(\frac{A_0}{4a^2}\right)^{0.85}\right](=K_h)$	$K_h-\frac{0.2Ga}{0.75-\nu}\left[1-\frac{b}{a}\right]$
Vertical	$\frac{4Gr_0}{1-\nu}$	$\frac{Gb}{1-\nu}\left[3.1\left(\frac{a}{b}\right)^{0.75}+1.6\right]$		$\frac{2Ga}{1-\nu}\left[0.73+1.54\left(\frac{A_0}{4a^2}\right)^{0.75}\right]$	
Rocking	$\frac{8Gr_0^3}{3(1-\nu)}$	$\frac{Gb^3}{1-\nu}\left[3.73\left(\frac{a}{b}\right)^{2.4}+0.27\right]$	$\frac{Gb^3}{1-\nu}\left[3.2\frac{a}{b}+0.8\right]$	$\frac{2.9G}{1-\nu}I_{0b}^{0.75}\left(\frac{a}{b}\right)^{0.15}$	$\frac{G}{1-\nu}I_{0a}^{0.75}\left(\frac{a}{b}\right)^{0.25}\left[2.4+0.5\frac{b}{a}\right]$
Torsional	$\frac{16Gr_0^3}{3}$	$Gb^3\left[4.25\left(\frac{a}{b}\right)^{2.45}+4.06\right]$		$3.5GI_0^{0.75}\left(\frac{b}{a}\right)^{0.4}\left(\frac{I_0}{b^4}\right)^{0.2}$	

TABLE 2-3A PROPERTIES OF CONE MODELS

Cone Type	Motion	Equivalent Radius r_0	Aspect Ratio $\frac{z_0}{r_0}$	Poisson's Ratio ν	Wave Velocity c	Trapped Mass ΔM
Translations	Horizontal	$\sqrt{\frac{A_0}{\pi}}$	$\frac{\pi}{8}(2-\nu)$	all ν	c_s	0
$K = \rho c^2 A_0/z_0$	Vertical	$\sqrt{\frac{A_0}{\pi}}$	$\frac{\pi}{4}(1-\nu)\left(\frac{c}{c_s}\right)^2$	$\le\frac{1}{3}$	c_p	0
$C = \rho c A_0$				$\frac{1}{3}<\nu\le\frac{1}{2}$	$2c_s$	$2.4\left(\nu-\frac{1}{3}\right)\rho A_0 r_0$

TABLE 2-3A PROPERTIES OF CONE MODELS

Cone Type	Motion	Equivalent Radius r_0	Aspect Ratio $\frac{z_0}{r_0}$	Poisson's Ratio ν	Wave Velocity c	Trapped Mass ΔM
Rotational	Rocking	$\sqrt[4]{\frac{4I_0}{\pi}}$	$\frac{9\pi}{32}(1-\nu)\left(\frac{c}{c_s}\right)^2$	$\leq\frac{1}{3}$	c_p	0
$K_\vartheta = 3\rho c^2 I_0/z_0$				$\frac{1}{3} < \nu \leq \frac{1}{2}$	$2c_s$	$1.2\left(\nu - \frac{1}{3}\right)\rho I_0 r_0$
$C_\vartheta = \rho c I_0$ $M_\vartheta = \rho I_0 z_0$	Torsional	$\sqrt[4]{\frac{2I_0}{\pi}}$	$\frac{9\pi}{32}$	all ν	c_s	0

TABLE 2-3B CONE'S ASPECT RATIO (OPENING ANGLE)

	Poisson's Ratio ν					
	0	$\frac{1}{4}$	$\frac{1}{3}$	0.4	0.45	$\frac{1}{2}$
Horizontal	$\frac{\pi}{4} = 0.785$	$\frac{7\pi}{32} = 0.687$	$\frac{5\pi}{24} = 0.654$	$\frac{\pi}{5} = 0.628$	$\frac{31\pi}{160} = 0.609$	$\frac{3\pi}{16} = 0.589$
Vertical	$\frac{\pi}{2} = 1.571$	$\frac{9\pi}{16} = 1.767$	$\frac{2\pi}{3} = 2.094$	$\frac{3\pi}{5} = 1.885$	$\frac{11\pi}{20} = 1.728$	$\frac{\pi}{2} = 1.571$
Rocking	$\frac{9\pi}{16} = 1.767$	$\frac{81\pi}{128} = 1.988$	$\frac{3\pi}{4} = 2.356$	$\frac{27\pi}{40} = 2.121$	$\frac{99\pi}{160} = 1.944$	$\frac{9\pi}{16} = 1.767$
Torsional	$\frac{9\pi}{32} = 0.884$	$= 0.884$	$= 0.884$	$= 0.884$	$= 0.884$	$= 0.884$

The (apex) *aspect ratio* z_0/r_0 follows for each degree of freedom *from matching the static-stiffness coefficient of the disk* with radius r_0(Table 2-2) *to that of the corresponding cone* (Eqs. 2-11, 2-24). For instance, for the vertical motion, substituting G from Table 2-1,

$$\frac{4\rho c_s^2 r_0}{1-\nu} = \frac{\rho c^2 \pi r_0^2}{z_0} \tag{2–25}$$

results, which yields

$$\frac{z_0}{r_0} = \frac{\pi}{4}(1-\nu)\left(\frac{c}{c_s}\right)^2 \tag{2–26}$$

The other aspect ratios for the horizontal, rocking, and torsional motions are given in Table 2-3A. For the horizontal and torsional motions, c equals c_s for all Poisson's ratios. As explained in depth in Section 2.1.4, for the vertical and rocking motions, c is, in principle, equal to c_p but has to be limited in the nearly incompressible range of $1/3 < \nu \le 1/2$ to the value $2c_s$.The apex aspect ratio z_0/r_0 depends with the exception of the torsional motion *on Poisson's ratio* ν. For $\nu = 0$, 1/4, 1/3, 0.4, 0.45 and 1/2, z_0/r_0 is specified in Table 2-3B. It is interesting to observe that the vertical and rocking cones, in which dilatational waves occur, are slender ($z_0/r_0 > 1$), whereas the horizontal and torsional cones, in which shear waves arise, are squatty ($z_0/r_0 < 1$). The aspects ratios of the cones shown in Fig. 2-1 correspond to $\nu = 1/3$.

Summarizing, as far as the soil's material properties are concerned, to determine the cone the two wave-propagation velocities c_s and c_p (the latter limited in the nearly incompressible case) are needed besides the mass density ρ. In particular, the aspect ratio z_0/r_0 depends on Poisson's ratio ν (which follows from c_p/c_s).

The last column in Table 2-3A is discussed in Section 2.1.4.

2.1.3 High-Frequency Behavior

The high-frequency limit of the dynamic-stiffness coefficient is examined next. For the translational cone in vertical motion, formulating dynamic equilibrium of the (lower) infinitesimal element in Fig. 2-2a taking the inertial loads into account

$$-N + N + N_{,z}dz - \rho A dz\, \ddot{u} = 0 \tag{2–27}$$

and substituting the force-displacement relationship (Eq. 2-2) yields the one-dimensional wave equation in zu

$$(zu)_{,zz} - \frac{(zu)^{\cdot\cdot}}{c_p^2} = 0 \tag{2–28}$$

For loading applied to the disk, only waves propagating in the positive z-direction will exist. The solution of Eq. 2-28 equals

$$zu = z_0 f\left(t - \frac{z - z_0}{c_p}\right) \tag{2–29}$$

It is easy to verify by substitution that this result does indeed satisfy the differential equation of motion.

To calculate the dynamic-stiffness coefficient, the interaction force P_0 is determined for a prescribed u_0. Enforcing the boundary condition

$$u(z = z_0) = u_0 \tag{2–30}$$

yields

$$f(t) = u_0 \tag{2–31}$$

and thus

$$u = \frac{z_0}{z} u_0 \left(t - \frac{z - z_0}{c_p} \right) \tag{2–32}$$

With

$$u,_z = -\frac{z_0}{z^2} u_0 \left(t - \frac{z - z_0}{c_p} \right) - \frac{z_0}{z\, c_p} u_0' \left(t - \frac{z - z_0}{c_p} \right) \tag{2–33}$$

where $u_0'(t - (z - z_0)/c_p)$ denotes differentiation of u_0 with respect to the argument $t - (z - z_0)/c_p$ (for $z = z_0$, $u_0') = \dot{u}_0$).

$$P_0 = -N(z = z_0) = -E_c A_0 u_{0,z} \tag{2–34}$$

is reformulated as

$$P_0 = \frac{E_c A_0}{z_0} u_0 + \rho c_p A_0 \dot{u}_0 \tag{2–35}$$

For the horizontal and vertical motions the interaction force-displacement relationship (Eq. 2-35) can be rewritten in a unified form as

$$P_0 = \frac{\rho c^2 A_0}{z_0} u_0 + \rho c A_0 \dot{u}_0 \tag{2–36}$$

or

$$P_0 = K u_0 + C \dot{u}_0 \tag{2–37}$$

with the constant coefficients of the spring and dashpot specified as

$$K = \frac{\rho c^2 A_0}{z_0} \quad \text{(2–38a)} \qquad\qquad C = \rho c A_0 \quad \text{(2–38b)}$$

Here, c equals c_s and c_p for the horizontal and vertical motions respectively. Note that this simple form of the interaction force-displacement relationship (Eq. 2-37) is valid for all dynamic cases, not just for the high-frequency limit.

For harmonic loading with amplitudes

$$\dot{u}_0(\omega) = i \omega u_0(\omega) \tag{2–39}$$

the interaction force-displacement relationship formulated in amplitudes equals

$$P_0(\omega) = (K + i\omega C)\, u_0(\omega) \tag{2–40}$$

where the coefficient on the right-hand side equals the dynamic-stiffness coefficient

$$S(\omega) = K + i\omega C \tag{2–41}$$

In the limit $\omega \rightarrow \infty$, the second term on the right-hand side dominates, leading to the high-frequency limit of the dynamic-stiffness coefficient

$$S(\omega \rightarrow \infty) = K + i\omega C \simeq i\omega C \tag{2–42}$$

with C specified in Eq. 2-38b. In the time domain, the interaction force is equal to the dash-pot force in the high-frequency limit

$$P_0 \simeq C\,\dot{u}_0 \tag{2–43}$$

This derivation of the force-displacement relationship (Eq. 2-36) is parallel to that described in Appendix A (Sections A1.1 and A1.2).

Turning to the rotational cone in rocking motion, formulating dynamic equilibrium of the (lower) infinitesimal element in Fig. 2-2b considering the inertial loads

$$-M + M + M,_z dz - \rho I dz\, \ddot{\vartheta} = 0 \tag{2–44}$$

and substituting the moment-rotation relationship (Eq. 2-13) leads to the equation of motion in the time domain

$$\vartheta,_{zz} + \frac{4}{z}\vartheta,_z - \frac{\ddot{\vartheta}}{c_p^2} = 0 \tag{2–45}$$

For harmonic excitation with

$$\ddot{\vartheta}(\omega) = -\omega^2 \vartheta(\omega) \tag{2–46}$$

Equation 2-45 formulated in amplitudes equals

$$\vartheta(\omega),_{zz} + \frac{4}{z}\vartheta(\omega),_z + \frac{\omega^2}{c_p^2}\vartheta(\omega) = 0 \tag{2–47}$$

For a prescribed rotation amplitude $\vartheta_0(\omega)$ only waves propagating in the positive z-direction will occur. For the high-frequency limit, the asymptotic solution of Eq. 2-47 will be of the form

$$\vartheta(\omega) = \vartheta_0(\omega)\left(\frac{z_0}{z}\right)^m e^{-i\frac{\omega}{c_p}(z-z_0)} \tag{2–48}$$

with the exponent m to be determined. As $\vartheta(\omega)$ is multiplied by $e^{+i\omega t}$, resulting in a term $e^{+i\omega(t\,-\,(z\,-\,z_0)/c_p)}$, the wave of this solution does indeed propagate in the positive z-direction. Substituting Eq. 2-48 in Eq. 2-47 yields

$$m^2 - 3m + \frac{i\omega z}{c_p}(2m - 4) = 0 \tag{2–49}$$

For $\omega \to \infty$ (actually the frequency parameter $\omega z/c_p \to \infty$), the second term governs. The latter vanishes for $m = 2$. The asymptotic solution (Eq. 2-48) thus equals

$$\vartheta(\omega) = \vartheta_0(\omega)\left(\frac{z_0}{z}\right)^2 e^{-i\frac{\omega}{c_p}(z-z_0)} \tag{2–50}$$

With

$$M_0(\omega) = -E_c I_0 \vartheta_0(\omega)_{,z} \tag{2–51}$$

and substituting the derivative from Eq. 2-50 yields

$$M_0(\omega) = \left(\frac{2\rho c_p^2 I_0}{z_0} + i\omega\rho c_p I_0\right)\vartheta_0(\omega) \tag{2–52}$$

For the rocking and torsional motions with $c = c_p$ and $= c_s$, respectively, the interaction moment-rotation relationship in the high-frequency limit formulated in amplitudes can be rewritten as

$$M_0(\omega) = \left(\frac{2\rho c^2 I_0}{z_0} + i\omega\rho c I_0\right)\vartheta_0(\omega) \tag{2–53}$$

or

$$M_0(\omega) = \left(K_{\vartheta\infty} + i\omega C_\vartheta\right)\vartheta_0(\omega) \tag{2–54}$$

with the constant coefficients of the spring and dashpot

$$K_{\vartheta\infty} = \frac{2\rho c^2 I_0}{z_0} \tag{2–55a}$$

$$C_\vartheta = \rho c I_0 \tag{2–55b}$$

Note that $K_{\vartheta\infty} = 2/3\ K_\vartheta$ (Eq. 2-24). The high-frequency limit of the dynamic-stiffness coefficient is written as

$$S(\omega \to \infty) = K_{\vartheta\infty} + i\omega C_\vartheta \simeq i\omega C_\vartheta \tag{2–56}$$

In the time domain the interaction moment-rotation relationship in the high-frequency limit equals

$$M_0 = K_{\vartheta\infty}\vartheta_0 + C_\vartheta\dot{\vartheta}_0 \simeq C_\vartheta\dot{\vartheta}_0 \tag{2–57}$$

The expressions for C (Eq. 2-38b) and C_ϑ (Eq. 2-55b) are independent of the apex height of the cones z_0. The same relationships for the high-frequency dashpots may be derived simply by assuming that *every surface in contact with the soil possesses an inherent amount of radiation damping equal to density times appropriate wave velocity ρc per unit area* and multiplying this quantity by the contact area A_0 for the translational motions and by the contact moment of inertia I_0 for the rotational motions (Fig. 2-3a). It is conceptually advantageous to regard the impedance ρc as a sort of intrinsic "paint" that coats the basemat–soil interface and thus specifies the damping. The radiation damping ρc per unit area corresponds to *one-dimensional wave propagation perpendicular to the basemat–soil interface*. This is thus the *wave pattern occurring in the high-frequency limit*. The dashpot with the dynamic-stiffness coefficient $i\omega\rho c$ per unit area represents the exact interaction force-displacement relationship (radiation condition) of a prismatic semi-infinite rod for all frequencies, not only for $\omega \to \infty$. This is easily verified by examining the interaction force-displacement relationship of the translational cone, which is exact for all ω's, Eq. 2-37 (or Eq. 2-40). Taking the limit $z_0 \to \infty$, which transforms the cone to a prismatic rod, results in $K = 0$ (Eq. 2-38a). The dashpot remains with a coefficient per unit area equal to ρc (Eq. 2-38b). As shown in Fig. 2-3, the coefficients of the dashpots per unit area are equal to ρc_p in the direction of wave propagation (vertical) and to ρc_s in the two tangential (horizontal) directions. This very important elastodynamical property can be ascertained quite generally (Fig. 2-3b) and appears to reflect an underlying law for high-frequency excitation. The cone thus leads to the exact dynamic-stiffness coefficient of a basemat on the surface of a halfspace in the high-frequency limit. As the aspect ratio of the cone is calculated by matching the static-stiffness coefficient of the cone to that of the disk on an elastic halfspace, a *doubly-asymptotic approximation* results for the cone, correct both for zero frequency (the static case) and the high-frequency limit dominated by the radiation dashpots C and C_ϑ. This explains the accuracy of cone models, even though they transmit axially propagating body waves only. In reality, nonrepresentable Rayleigh surface waves, which propagate perpendicular to the cones' axis, transmit the major portion of energy away from a vibrating basemat in the low- and intermediate-frequency ranges. This aspect is further evaluated in Section 2.5, as are other features of the cone models.

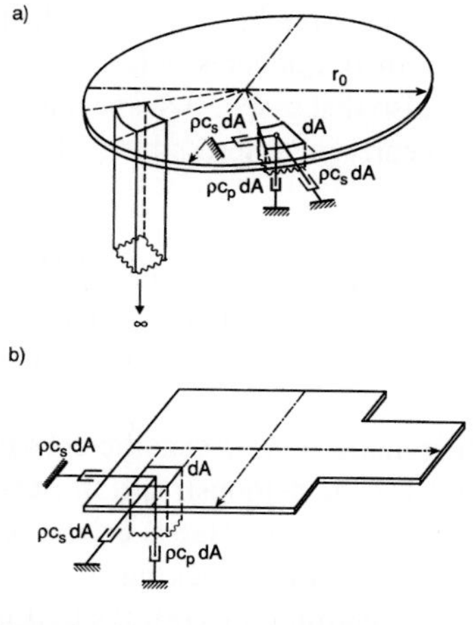

a. Disk foundation.

b. Foundation of arbitrary shape.

Fig. 2-3 High-frequency limit of dynamic stiffness.

The important result from the solution of the one-dimensional wave equation is that an end-excited prismatic semi-infinite rod is dynamically equivalent to a single viscous dashpot. This shows that it is in principle possible to represent the interaction force-displacement relationship of a continuum (having an unlimited number of internal degrees of freedom) by a limited number of discrete elements—in this case only one.

For a loading case with a very high-frequency content only, the soil is modeled with dashpots with the coefficients determined as described above (Fig. 2-3). This represents a simple, but rigorous procedure.

2.1.4 Nearly Incompressible Soil

For the horizontal and torsional degrees of freedom of a surface foundation, *S*-waves occur. The appropriate wave velocity in the cone models is thus c_s, which remains finite for all Poisson ratios ν. The aspect ratio z_0/r_0 follows from the equations specified in Table 2-3A. For the vertical and rocking degrees of freedom producing compression and extension, *P*-waves will dominate for small and intermediate values of ν, resulting in the use of c_p. But c_p tends to infinity (Table 2-1) for ν approaching 0.5. This causes apparently anomalous behavior, not only for cones, but for rigorous solutions as well. Use of c_p for the higher values of ν would overestimate the radiation damping characterized by C (Eq. 2-38b) and C_ϑ (Eq. 2-55b).

It is necessary to develop ways to circumvent these difficulties, because the range of nearly incompressible soil is extremely important in engineering practice. For saturated soil analysed as a single-phase medium, Poisson's ratio is essentially 1/2 due to the near incompressibility of the water filling the pores. Even if entrained air is taken into account (partial saturation), the value of ν will not be smaller than about 0.45. (By comparison, typical values of Poisson's ratio for solid rock and dry sand are around 1/4 and 1/3, respectively).

The reasoning resulting in the modifications for the nearly incompressible case cannot be discussed at this stage; it is deferred to Section 2.5. Based on the rigorous solution for the dynamic stiffness of a rigid disk for all frequencies, whereby the partition of the power among *P*-, *S*- and Rayleigh-waves is also discussed, two special features are necessary for the vertical and rocking motions for nearly incompressible soil with *Poisson's ratio between 1/3 and 1/2*:

1. The appropriate wave velocity dominating the radiation damping is selected as twice the shear-wave velocity and not as the dilatational-wave velocity.
2. A trapped mass which increases linearly with Poisson's ratio is introduced.

Wave velocity. In view of the fact that $c = 2c_s$ yields the correct high-frequency asymptote of damping for both $\nu = 1/3$ and 1/2 (as demonstrated in Section 2.5.2) and in addition provides a "best fit" for small frequencies, this value is used throughout the range of nearly incompressible soil. For both the vertical and rocking motions the appropriate wave velocity c to be used in all calculations regarding the cones is thus chosen as follows

$$c = c_p \qquad \text{for} \qquad \nu \leq \frac{1}{3} \tag{2–58a}$$

$$c = 2\, c_s \qquad \text{for} \qquad \frac{1}{3} < \nu \leq \frac{1}{2} \tag{2–58b}$$

This information is also contained in Table 2-3A. The wave velocity c affects not only the high-frequency radiation dashpots C and C_ϑ but also the static-stiffness coefficients K and K_ϑ and thus the aspect ratios z_0/r_0. The latter are specified for $\nu = 0.4$, 0.45, and 1/2 in Table 2-3B.

Trapped mass. As will become apparent later on (see the dynamic-stiffness coefficients for harmonic excitation, Figs. 2-7 and 2-21), the so-called dynamic-spring coefficient (real part of the dynamic stiffness) of the exact solution exhibits a downward-parabolic tendency. This behavior corresponds to trapped soil beneath the disk, which moves as a rigid body in phase with the disk. A close match is achieved by defining the trapped mass to be

$$\Delta M = \mu\, \rho\, r_0^3 \tag{2–59a}$$

with

$$\mu = 2.4\,\pi\left(\nu - \frac{1}{3}\right) \tag{2–59b}$$

for vertical motion and the trapped mass moment of inertia

$$\Delta M_\vartheta = \mu_\vartheta\, \rho\, r_0^5 \tag{2–60a}$$

with

$$\mu_\vartheta = 0.3\,\pi\left(\nu - \frac{1}{3}\right) \tag{2–60b}$$

for rocking. According to these formulas, the inclusion of trapped mass begins at $\nu = 1/3$ and increases linearly with Poisson's ratio. For perfectly incompressible soil with $\nu = 1/2$, the maximum amount of trapped mass in vertical motion corresponds to an equivalent cylinder of soil with height $0.4\, r_0$; for rocking, the equivalent cylinder is half as tall (concentrated in the middle plane of the disk).

Although these quantities of trapped mass are determined by simple curve-fitting, they have a physical justification. In engineering practice it has been observed that blunt-ended piles may be driven just as easily as piles with pointed tips. The reason, predicted by Prandtl's theory of a blunt punch penetrating into an incompressible plastic medium, is that the pile creates its own pointed tip. As shown in Fig. 2-4a, a cone of trapped soil forms and moves as a rigid body with the pile. Intuitively one would expect the opening angle of the cone, 2α, to be slightly less than 90°. Now for $2\alpha = 80°$ the height of the cone is 1.2 times the radius; and the mass of the trapped soil becomes $0.4\, A_0 r_0$, in agreement with the trapped mass computed from Eq. 2-59 for vertical motion and $\nu = 1/2$.

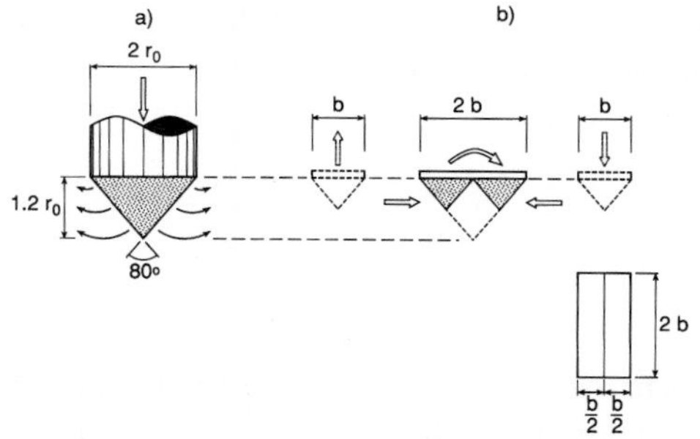

a. Three-dimensional vertical case (pile).

b. Two-dimensional rocking case (elevation and plan view of roof-shaped trapped volume of soil).

Fig. 2-4 Analogy of trapped mass to plastic punching.

The concept of plastic punching also explains why the trapped mass for rocking is half as large. Consider a pair of two-dimensional strips, each having width b, one excited upward, the other downward. If, as pictured in Fig. 2-4b, the strips are brought together, the load couple merges to a rocking moment; and the trapped mass is exactly half that of the vertical case (dashed line). The same result appears logical for the three-dimensional disk, although the geometry is much more complicated. Assuming the basemat in Fig. 2-4b is a square of width $2b$ and depth (length) $2b$, the mass moment of inertia equals

$$\Delta M_\vartheta = b\ 1.2\frac{b}{2}\frac{1}{2}2b\frac{b^2}{4}2\rho = 0.3\ \rho b^5 = 0.155\ \rho\ r_0^5 \tag{2–61}$$

where b is replaced by $b = 0.876\ r_0$, derived from equating the moment of inertias of the square ($I_0 = 4b^4/3$) and of the disk ($I_0 = \pi r_0^4/4$). The formula for the disk's trapped mass moment of inertia (Eq. 2-60) leads for $\nu = 1/2$ to $\Delta M_\vartheta = 0.157 \rho r_0^5$, which is very close to the value derived from the physical interpretation (Eq. 2-61).

As discussed in Section 2.1.3, the cone models represent doubly asymptotic approximations, with the correct dynamic-stiffness coefficients in both the low-frequency and high-frequency limits. The question arises if the limit on the wave velocity and the trapped mass introduced in the nearly incompressible case affect this property. The low-frequency limit (static case) poses no problems as the proportion of the cone is chosen to match the static-stiffness coefficient of the rigorous solution for the disk on the elastic halfspace. For the high-frequency limit the situation is more complicated. The dynamic-stiffness coefficient of a foundation calculated rigorously or with cone models always tends to infinity for high frequencies. Nonetheless, infinity may be approached in two different ways. For perfectly incompressible soil with Poisson's ratio exactly equal to 1/2, the real-valued mass term controls the response, regardless how large or small the dashpot may be. This occurs for both the rigorous solution and for the analysis with cones. On the other hand, if the soil is even to the slightest extent compressible, the trapped-mass term eventually vanishes in the rigorous solution (Section 2.5.2) in contrast to the result with cones. The imaginary-valued damping term dominates in the rigorous solution, which is proportional to frequency and thus also tends to infinity. Practically speaking, it makes no difference whether the real or the imaginary part of the dynamic-stiffness coefficient approaches infinity; in either case, if the applied force or moment is finite, the resulting displacement or rotation always goes to zero for high-frequency excitation.

When performing calculations, the simplest way to account for the trapped mass ΔM and the trapped mass moment of inertia ΔM_ϑ is just to assign them to the basemat. The trapped mass is thus actually shifted over to the structure.

2.2 TRANSLATIONAL CONE

2.2.1 Stiffness Formulation

For the translational cone all relations based on the strength-of-materials approach (rod theory) are mathematically derived in detail in Appendix A.

Interaction force-displacement relationship. For practical applications it is not necessary to compute explicitly the displacement field propagating into the unbounded soil. One may view the soil as a "black box" and restrict attention to the interaction force-displacement relationship of the massless basemat. In the stiffness formulation for the translational cone, the interaction force P_0 is calculated for a specified displacement u_0 (and velocity $\dot{u}_0$) as (Eq. 2-37, Eq. A-14)

$$P_0(t) = K\,u_0(t) + C\dot{u}_0(t) \tag{2–62}$$

with K and C specified in Eq. 2-38, Eq. A-15.

$$K = \frac{\rho c^2 A_0}{z_0} \quad \text{(2–63a)} \qquad\qquad C = \rho c A_0 \quad \text{(2–63b)}$$

Equation 2-62 applies with $c = c_s$ for the horizontal degree of freedom and for $\nu \le 1/3$ with $c = c_p$ for the vertical degree of freedom. For the vertical degree of freedom in the nearly incompressible case ($1/3 < \nu \le 1/2$)

$$P_0(t) = K\,u_0(t) + C\dot{u}_0(t) + \Delta M \ddot{u}_0(t) \tag{2–64}$$

applies with ΔM given in Eq. 2-59a

$$\Delta M = \mu\, \rho r_0^3 \tag{2–65}$$

with μ specified in Eq. 2-59b and $c = 2c_s$. The apex height z_0 follows from Table 2-3A.

From the simple form of Eq. 2-62 it follows that K may be interpreted as the *frequency-independent* coefficient of an ordinary *spring* (the static stiffness), and C as the *frequency-independent* coefficient of an ordinary *dashpot* (Fig. 2-5a). This *discrete-element model represents rigorously the translational cone.* For nearly incompressible soil in the vertical motion (Eq. 2-64), a mass ΔM associated with the disk and vibrating in phase with a frequency-independent coefficient is added (Fig. 2-5b). It is quite remarkable that from outside it is impossible to ascertain whether or not the "black box" model of the soil contains an unbounded cone with an infinite number of internal degrees of freedom or just a discrete-element model with no internal degrees of freedom. To take translational soil–structure interaction into account, it is only necessary to attach to the underside of the structural model the spring with coefficient K, the dashpot with coefficient C, and, if applicable, the mass with coefficient ΔM. Then the complete system may be analyzed directly with a general-purpose computer program.

For harmonic loading, the interaction force-displacement relationship is specified in Eqs. A-16 to A-21. For the nearly incompressible case in vertical motion

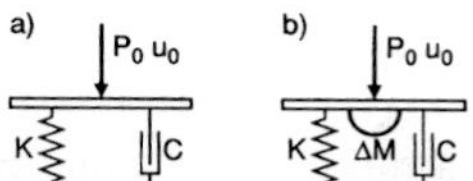

a. Horizontal motion and vertical motion for compressible soil.

b. Vertical motion for nearly incompressible soil.

Fig. 2-5 Discrete-element model for translation.

$$P_0(\omega) = \left(K - \omega^2 \Delta M + i\omega C\right) u_0(\omega) \tag{2–66}$$

results with the *dynamic-stiffness coefficient*

$$S(\omega) = K - \omega^2 \, \Delta M + i\omega \, C \tag{2–67}$$

Using the dimensionless frequency parameter b_0 defined with respect to the properties of the cone

$$b_0 = \frac{\omega \, z_0}{c} \tag{2–68}$$

$S(\omega)$ is formulated as

$$S(b_0) = K\left[k\left(b_0\right) + ib_0 \, c\left(b_0\right)\right] \tag{2–69}$$

with the dimensionless spring and damping coefficients specified as

$$k(b_0) = 1 - \frac{\mu}{\pi} \frac{r_0}{z_0} b_0^2 \quad \text{(2–70a)} \qquad\qquad c\left(b_0\right) = 1 \quad \text{(2–70b)}$$

whereby Eq. 2-65 is used. For $\mu = 0$ the cases of compressible soil in vertical motion and of the horizontal motion are recovered (Eq. A-21).

Dynamic-stiffness coefficient. To examine the horizontal and vertical dynamic-stiffness coefficients of a disk resting on the surface of a halfspace, it is appropriate to use a_0 instead of b_0

$$a_0 = \frac{\omega r_0}{c_s} = \frac{r_0}{z_0} \frac{c}{c_s} b_0 \tag{2–71}$$

Equation 2-69 is then transformed to

$$S(a_0) = K\left[k\left(a_0\right) + ia_0 \, c\left(a_0\right)\right] \tag{2–72}$$

where

$$k(a_0) = 1 - \frac{\mu}{\pi} \frac{z_0}{r_0} \frac{c_s^2}{c^2} a_0^2 \quad \text{(2–73a)} \qquad\qquad c(a_0) = \frac{z_0}{r_0} \frac{c_s}{c} \quad \text{(2–73b)}$$

For the horizontal motion, $c = c_s$ and $\mu = 0$ for all ν. And z_0/r_0 follows as a function of ν from Table 2-3A. For the vertical motion, $c = c_p$ and $\mu = 0$ for $\nu \le 1/3$, and $c = 2\,c_s$ and μ is specified in Eq. 2-59b for $1/3 < \nu \le 1/2$. Again, z_0/r_0 follows from Table 2-3A.

For the horizontal motion, the spring coefficient $k_h(a_0)$ and the damping coefficient $c_h(a_0)$ are plotted versus a_0 for Poisson's ratios $\nu = 0$, 1/3, and 1/2 in Fig. 2-6. The results for the cones are shown as lines. It follows from Eq. 2-73, that $k_h = 1$ and $c_h = z_0/r_0$. The rigorous values of Veletsos and Wei [V1] are also plotted as distinct points. For all three Poisson's ratios the agreement is satisfactory. For the vertical motion, the comparison of the dynamic-stiffness coefficients is shown in Fig. 2-7. The more rigorous results plotted as distinct points are taken from two other studies [V3, L3]. For $\nu = 1/2$ and $\nu = 0.45$, $k_v(a_0)$

is described as a second degree parabola (Eq. 2-73a). The inertial load of the mass ΔM leads to a decrease in $k_v(a_0)$, resulting in negative values for sufficiently large a_0. In the *intermediate- and higher-frequency ranges*, the dynamic-stiffness coefficient is governed by the damping coefficient, as $c(a_0)$ is multiplied by a_0 in contrast to $k(a_0)$ (Eq. 2-72). Both $c_v(a_0)$ and $c_h(a_0)$ of the cone models are *very accurate* in these frequency ranges, whereas in the *lower-frequency range* ($a_0 < 2$) and for $\nu \leq 1/3$, which is of great practical importance, the *cone's results overestimate (radiation) damping to a certain extent*, especially in the vertical motion (Figs. 2-6b and 2-7b).

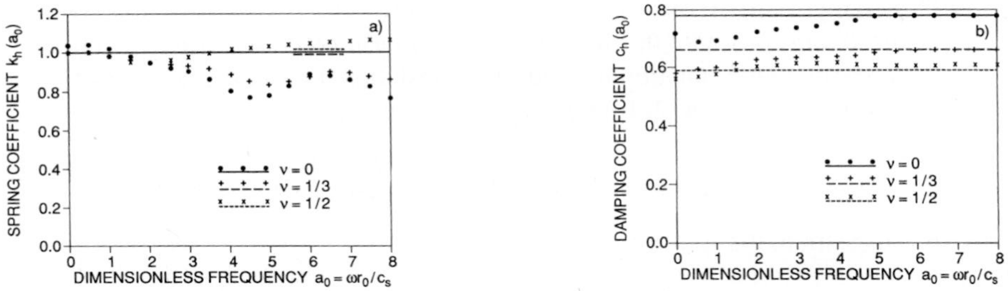

Fig. 2-6 Dynamic-stiffness coefficient for harmonic loading of disk on homogeneous halfspace in horizontal motion for various Poisson's ratios.

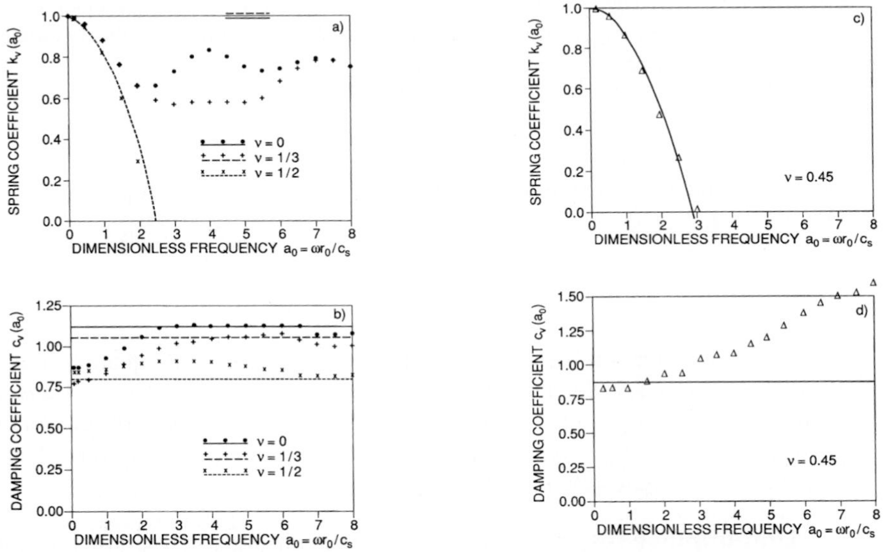

Fig. 2-7 Dynamic-stiffness coefficient for harmonic loading of disk on homogeneous halfspace in vertical motion for various Poisson's ratios.

2.2.2 Flexibility Formulation

Displacement-interaction force relationship. In a flexibility formulation, the interaction force P_0 is regarded as the known input and the corresponding unknown displacement u_0 as the output. Flexibility formulations are particularly well suited for transient machine-vibration problems; these are amenable to simple hand calculation.

First, the vertical motion for the compressible soil ($\nu < 1/3$) and the horizontal motion analyzed in the time domain are discussed. The fundamental solution required for a flexibility formulation is the displacement response as a function of time $u_0(t)$ due to a Dirac-delta impulse $\delta(t)$ of the normalized interaction force P_0/K applied at time zero. It is customary to denote this *unit-impulse response function* by the symbol $h(t)$. To derive the corresponding function $h_1(t)$ for the translational cone, the interaction force-displacement relationship (Eq. 2-36) with $h_1 = u_0$ is written as

$$\frac{z_0}{c}\dot{h}_1 + h_1 = \frac{P_0}{K} = \delta(t) \tag{2–74}$$

By inspection, the solution equals

$$h_1(t) = \frac{c}{z_0} e^{-\frac{c}{z_0}t} \quad t \geq 0 \qquad \text{(2–75a)} \qquad = 0 \quad t < 0 \qquad \text{(2–75b)}$$

which is also derived in Section A1.3 (Eq. A-24).

To derive the displacement-interaction force relationship, $h_1(t)$ is regarded as an influence function. The contribution of the infinitesimal pulse $(1/K)P_0(\tau)d\tau$ acting at time τ to the displacement $du_0(t)$ at time t ($t \geq \tau$) depends on the unit-impulse response function evaluated for the time difference $t - \tau$, $h_1(t - \tau)$

$$du_0(t) = h_1(t-\tau)\frac{P_0(\tau)}{K}d\tau \tag{2–76}$$

The displacement-interaction force relationship is the integral of Eq. 2-76 over all pulses from $0 \leq \tau \leq t$.

$$u_0(t) = \int_0^t h_1(t-\tau)\frac{P_0(\tau)}{K}d\tau \tag{2–77}$$

Another derivation based on the solution of a differential equation in $u_0(t)$ is presented in Section A1.3. Equation 2-77 represents a *convolution integral (Duhamel's integral) of the normalized-interaction force and the unit-impulse response function.* The convolution integral depends on all previous values of the interaction force P_0 and may be regarded as the system's memory of the past.

Unit-impulse response function. The unit-impulse response functions h_1 (Eq. 2-75) multiplied by r_0/c_s of the horizontal and vertical motions are plotted for $\nu = 1/3$ in Fig. 2-8 as a function of the dimensionless time

$$\bar{t} = t\frac{c_s}{r_0} \tag{2–78}$$

defined analogously as a_0 (Eq. 2-71).

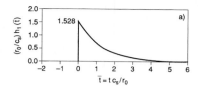

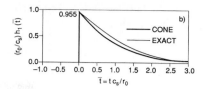

a. Horizontal. **b.** Vertical.

Fig. 2-8 Unit-impulse response function of disk on homogeneous halfspace.

The initial values follow for the horizontal motion from $z_0/r_0 = 0.654$ (Table 2-3B) and with $c = c_s$ as $1/0.654 = 1.528$ and for the vertical motion from $z_0/r_0 = 2.094$ and with $c = c_p$ $(= 2c_s)$ as $2/2.094 = 0.955$. The agreement of the results calculated using the cone models with the exact solution of Veletsos and Verbic [V3] is excellent.

The initial value of the unit-impulse response of a disk on a halfspace can be calculated based on three-dimensional elastodynamics applying the law of conservation of momentum for the first differential time dt $(0 \le t \le dt,\ \lim dt \to 0)$. For instance, for the vertical degree of freedom the excited regions adjacent to the loaded disk are shown in Fig. 2-9. At the beginning of the application of the normalized interaction force P_0/K at $t = 0$, a dilatational wave starts to propagate vertically with the velocity c_p. The corresponding momentum (mass times velocity) vanishes. At the end of the application at time $t = dt$, the wave has traveled the distance $dz = c_p dt$. The corresponding excited mass in the form of a cylinder equals $dm = \rho\, \pi\, r_0^2 c_p\, dt$. The momentum is equal to $dm\, \dot{u}_0(t = dt) = dm\, \dot{u}_0(t = 0^+)$. Formulating the law of conservation of momentum leads to

$$\int_0^{dt} P_0(t)dt = \rho\, c_p\, \pi\, r_0^2\, dt\, \dot{u}_0(t = 0^+) \tag{2–79}$$

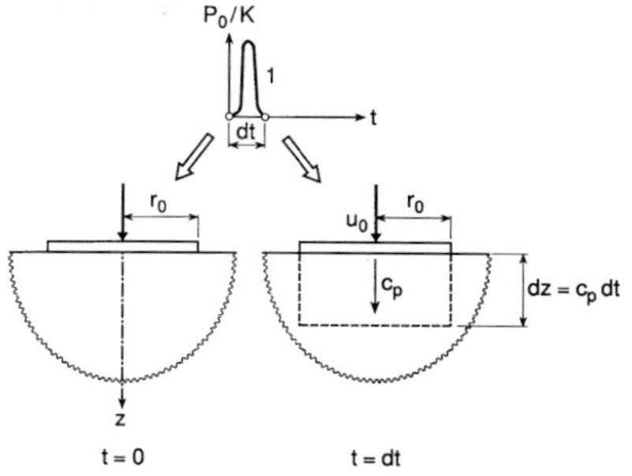

Fig. 2-9 Excited mass for initial values of unit-impulse response function of disk on halfspace in vertical motion.

In passing it is instructive to derive an average interaction force P_0 over the first differential time. Equation 2-79 is divided by dt. The resulting right-hand side represents the product of the radiation damping ρc_p per unit area and the area multiplied by the velocity $\dot{u}_0(t = 0^+)$. This dashpot-force corresponds as expected to one-dimensional wave propagation in the halfspace perpendicular to the disk (see also Fig. 2-3). The corresponding wave pattern is the same as that present in the high-frequency limit (Section 2.1.3). Performing the normalization with the static-stiffness coefficient K of the halfspace (Table 2-2), using the definition of a Dirac-delta function and substituting $u_0(t=0^+) = \dot{u}_0(t=0^+)dt$ transforms Eq. 2-79 to

$$\int_0^{dt} \frac{P_0(t)}{K} dt = 1 = \frac{\rho c_p \pi r_0^2}{K} u_0\,(t=0^+) \tag{2–80}$$

The initial value of the vertical unit-impulse response of a disk on a halfspace equals

$$u_0(t=0^+) = \frac{K}{\rho\, c_p \pi\, r_0^2} \tag{2–81}$$

Note, that the denominator is equal to the radiation dashpot C of the halfspace.

As by construction, K in Eq. 2-81 is equal to the static-stiffness coefficient of the cone (Eq. 2-25),

$$u_0(t=0^+) = \frac{\rho\, c_p^2 \pi\, r_0^2}{z_0\, \rho\, c_p \pi\, r_0^2} = \frac{c_p}{z_0} \tag{2–82}$$

follows. The *initial value of the unit-impulse response* for the halfspace $u_0(t=0^+)$ is thus equal to the corresponding value for the cone $h_1(t=0^+)$ (Eq. 2-75a). The initial peak values at time $t=0^+$ shown in Fig. 2-8 *calculated with the cones are thus exact.* This follows from the fact that the cones enable an error-free representation of the radiation dashpot C, which is proportional to the high-frequency limit of the dynamic-stiffness coefficient.

Recursive evaluation. To determine the complete time history of the response, the convolution integral has to be evaluated starting from $\tau = 0$ for each time station. It is computationally inefficient to calculate these convolution integrals from the definition as specified in Eq. 2-77. (As the unit-impulse response function is an exponential function, it would be possible to transform the convolution integral to a standard integral.) The evaluation may be expedited by the so-called *recursive* procedure. Background on this method is provided in Section C1 of Appendix C. In this elegant approach applied to the convolution integral with $h_1(t)$, a so-called first-order term, derived in Section C2, the *displacement at time station n, u_{0n}, can be expressed as a linear function of the displacement at the previous time station n - 1, u_{0n-1} and of the normalized interaction forces at the same two time stations, P_{0n}/K, P_{0n-1}/K* (Eq. C-17).

$$u_{0n} = a\, u_{0n-1} + b_0 \frac{P_{0n}}{K} + b_1 \frac{P_{0n-1}}{K} \tag{2–83}$$

The recursive coefficients, which depend on the time step Δt (but not on the time station), are specified in Eq. C-24 as

$$a = e^{-\frac{c\Delta t}{z_0}} \tag{2–84a}$$

$$b_0 = 1 + \frac{e^{-\frac{c\Delta t}{z_0}} - 1}{\frac{c\Delta t}{z_0}} \tag{2–84b}$$

$$b_1 = \frac{-e^{-\frac{c\Delta t}{z_0}} + 1}{\frac{c\Delta t}{z_0}} - e^{-\frac{c\Delta t}{z_0}} \tag{2–84c}$$

To facilitate hand calculations, numerical values of the recursion coefficients are presented in the first part of Table 2-4 for a typical range of the dimensionless time increment $c\Delta t/z_0$. In the static case ($P_{0n}/K = u_{0n} = u_0$ for all n) Eq. 2-83 states that the sum of all the a's and b's ($= a + b_0 + b_1$) must add to unity. This fact is convenient for checking the recursion constants prior to computation.

The recursive algorithm is self-starting from $t = 0$ for quiescent initial conditions (i.e., no loading and response for $t < 0$). For $n = 1$ u_{00} and P_{00} are zero, resulting in $u_{01} = b_0 P_{01}/K$. For a periodic excitation convergence from the initial start-up phase to steady-state response occurs quite rapidly, because of the large damping inherent in the cone models. Possible peaks in the response encountered in the starting-up phase are identified automatically in this time-domain analysis, which would not be the case in a calculation for harmonic excitation (frequency domain). It should be emphasized that the recursion relationship Eq. 2-83 is not an approximate numerical integration formula. The results are *exact*, subject only to the assumption that the interaction-force function $P_0(t)$ is a polygon formed by joining the samples P_{0n} by straight lines. The recursive algorithm is unconditionally stable for all values of the time increment Δt. The latter can thus, in many cases, be chosen based only on considering an adequate representation of the interaction forces by piecewise straight lines. The computational savings are obviously enormous for long excitations such as earthquakes. The recursive evaluation makes computations using a hand calculator feasible for machine-foundation vibrations.

Nearly incompressible soil. Next, the vertical motion for nearly incompressible soil ($1/3 < \nu \leq 1/2$) is addressed with $c = 2\, c_s$. The interaction force-displacement relationship is specified in Eq. 2-64.

$$u_0 + \frac{C}{K}\dot{u}_0 + \frac{\Delta M}{K}\ddot{u}_0 = \frac{P_0}{K} \tag{2–85}$$

or after substituting Eqs. 2-63 and 2-65 in

$$\frac{\mu\, z_0^2\, r_0}{\pi\, c^2\, z_0}\ddot{u}_0 + \frac{z_0}{c}\dot{u}_0 + u_0 = \frac{P_0}{K} \tag{2–86}$$

To derive the unit-impulse response function $h_4(t)$, the loading P_0/K is set equal to the Dirac-delta function $\delta(t)$ (the subscripts 2 and 3 are used in connection with the rotational cone),

$$\frac{z_0^2}{\gamma\, c^2}\ddot{h}_4 + \frac{z_0}{c}\dot{h}_4 + h_4 = \delta(t) \tag{2–87}$$

with the abbreviation

$$\gamma = \frac{\pi\, z_0}{\mu\, r_0} \tag{2–88}$$

TABLE 2-4 RECURSION COEFFICIENTS FOR CONVOLUTIONS WITH UNIT-IMPULSE RESPONSE FUNCTIONS

	Convolution with							
		h_1				h_2		
$\frac{c\Delta t}{z_0}$	a	b_0	b_1	a_1	a_2	b_0	b_1	b_2
0.000	1.000000	0.000000	0.000000	2.000000	-1.000000	0.000000	0.000000	-0.000000
0.010	0.990050	0.004983	0.004967	1.970150	-0.970446	0.014900	0.000049	-0.014654
0.020	0.980199	0.009934	0.009868	1.940600	-0.941765	0.029603	0.000194	-0.028632
0.030	0.970446	0.014851	0.014703	1.911350	-0.913931	0.044110	0.000430	-0.041959
0.040	0.960789	0.019736	0.019475	1.882399	-0.886920	0.058424	0.000753	-0.054656
0.050	0.951229	0.024588	0.024182	1.853748	-0.860708	0.072546	0.001159	-0.066746
0.060	0.941765	0.029409	0.028827	1.825395	-0.835270	0.086480	0.001645	-0.078250
0.070	0.932394	0.034197	0.033409	1.797341	-0.810584	0.100277	0.002205	-0.089189
0.080	0.923116	0.038954	0.037929	1.769585	-0.786628	0.113789	0.002837	-0.099583
0.090	0.913931	0.043630	0.042389	1.742127	-0.763379	0.127168	0.003536	-0.109452
0.100	0.904837	0.048374	0.046788	1.714965	-0.740818	0.140368	0.004300	-0.118814
0.120	0.886920	0.057670	0.055409	1.661528	-0.697676	0.166232	0.006007	-0.136091
0.140	0.869358	0.066845	0.063797	1.609267	-0.657047	0.191400	0.007932	-0.151553
0.160	0.852144	0.075899	0.071958	1.558177	-0.618783	0.215887	0.010050	-0.165330
0.180	0.835270	0.084835	0.079895	1.508246	-0.582748	0.239708	0.012337	-0.177543
0.200	0.818731	0.093654	0.087615	1.459467	-0.548812	0.262880	0.014772	-0.188308
0.250	0.778801	0.115203	0.105996	1.342487	-0.472367	0.318068	0.021377	-0.209566
0.300	0.740818	0.136061	0.123121	1.232458	-0.406570	0.369521	0.028500	-0.223909
0.350	0.704688	0.156252	0.139060	1.129176	-0.349938	0.417460	0.035901	-0.232601
0.400	0.670320	0.175800	0.153880	1.032422	-0.301194	0.462099	0.043380	-0.236707
0.450	0.637628	0.194729	0.167643	0.941958	-0.259240	0.503634	0.050773	-0.237126
0.500	0.606531	0.213061	0.180403	0.857540	-0.223130	0.542257	0.057946	-0.234613
0.600	0.548812	0.248019	0.203169	0.705813	-0.165299	0.611481	0.071226	-0.223221
0.700	0.496585	0.280836	0.222579	0.575164	-0.122456	0.671106	0.082625	-0.206438
0.800	0.449329	0.311661	0.239010	0.463506	-0.090718	0.722330	0.091834	-0.186952
0.900	0.406570	0.340633	0.252797	0.368805	-0.067206	0.766222	0.098751	-0.166573
1.000	0.367879	0.367879	0.264241	0.289114	-0.049787	0.803734	0.103419	-0.146479

The solution of this second-order differential equation can be derived by noting that the latter has the same form as the dynamic equation of motion of a one-degree-of-freedom system

$$m\ddot{h}(t) + c\,\dot{h}(t) + k\,h(t) = \delta(t) \tag{2–89}$$

For $c^2 - 4m\,k > 0$, the system is overdamped. This is, in general, satisfied for the cone, as $1 - 4/\gamma > 0$; for instance, for $\nu = 0.45$, $\mu = 0.880$ and $z_0/r_0 = 1.728$ leading to $\gamma = 6.169$. (For $\nu = 0.5$, $\mu = 1.257$, and $z_0/r_0 = 1.571$, resulting in $\gamma = 3.926$; this corresponds to an underdamped system, discussed below.) The unit-impulse response function $h(t)$ is equal to the damped vibration for an initial velocity $\delta(t)\,dt/m = 1/m$

$$h(t) = \frac{1}{\omega_d\,m}\,e^{-\omega_0\zeta t}\sinh\omega_d t \tag{2–90}$$

with the natural frequency ω_0, the damping ratio ζ and the damped frequency ω_d

$$\omega_0 = \sqrt{\frac{k}{m}} \tag{2–91a}$$

$$\zeta = \frac{c}{2\sqrt{km}} \tag{2–91b}$$

$$\omega_d = \omega_0\sqrt{\zeta^2 - 1} \tag{2–91c}$$

$h_4(t)$ follows from Eq. 2-90, comparing Eq. 2-87 to Eq. 2-89

$$h_4(t) = \frac{2c}{z_0}\frac{1}{\sqrt{1-\frac{4}{\gamma}}}\,e^{-\frac{c}{z_0}\frac{\gamma}{2}t}\sinh\frac{c}{z_0}\frac{\gamma}{2}\sqrt{1-\frac{4}{\gamma}}\;t \tag{2–92}$$

In Section C3, $h_4(t)$ is derived via the frequency domain (Eq. C-46).

The displacement-interaction force relationship is formulated in analogy to Eq. 2-77 as

$$u_0(t) = \int_0^t h_4(t-\tau)\frac{P_0(\tau)}{K}\,d\tau \tag{2–93}$$

Again, the recursive algorithm can be used to evaluate efficiently the convolution integral. For a convolution integral with $h_4(t)$, a second-order term, the recursion equation involving three equally spaced time stations is equal to (Eq. C-27)

$$u_{0n} = a_1\,u_{0n-1} + a_2\,u_{0n-2} + b_0\frac{P_{0n}}{K} + b_1\frac{P_{0n-1}}{K} + b_2\frac{P_{0n-2}}{K} \tag{2–94}$$

The recursive coefficients are derived in Section C3. As specified in Eq. C-47

$$a_1 = 2e^{-\frac{c}{z_0}\frac{\gamma}{2}\Delta t}\cosh\frac{c}{z_0}\frac{\gamma}{2}\sqrt{1-\frac{4}{\gamma}}\;\Delta t \tag{2–95a}$$

$$a_2 = -e^{-\frac{c}{z_0}\gamma\,\Delta t} \tag{2–95b}$$

And b_0, b_1, b_2 follow from Eq. C-34 with $r(t) = r_4(t)$ given in Eq. C-48.

It is worth mentioning that the flexibility formulation for the nearly incompressible case can also be used when the foundation block with mass is directly included in the analysis. In this case, ΔM in Eq. 2-85 represents the mass of the foundation block. The horizontal and vertical degrees of freedom can be processed (in the case of the vertical motion for $1/3 < \nu \leq 1/2$, ΔM is equal to the sum of the trapped mass of the soil and the mass of the foundation block). In general, as the mass of the foundation block is larger than the trapped soil part in the nearly incompressible case, the one-degree-of-freedom system will be underdamped; that is, $1 - 4/\gamma < 0$. The equations for $h_4(t)$ (Eq. 2-92), a_1 and a_2 (Eqs. 2-95) and r_4 still apply, whereby $\sqrt{4/\gamma - 1}$ replaces $\sqrt{1 - 4/\gamma}$ and the trigonometric functions sin and cos replace the hyperbolic functions sinh and cosh. Again, this case is also derived systematically in the frequency domain in Section C3 (Eqs. C-43 to C-45). This formulation also applies to the incompressible soil ($\nu = 0.5$) without a foundation block.

Dynamic-flexibility coefficient. Finally, for harmonic loading the *dynamic-flexibility coefficient* F(ω) follows from Eq. 2-67 through inversion

$$u_0(\omega) = F(\omega)\ P_0(\omega) \tag{2–96}$$

with

$$F(\omega) = S(\omega)^{-1} = \frac{K - \omega^2\,\Delta M - i\,\omega\,C}{(K - \omega^2\,\Delta M)^2 + \omega^2\,C^2} \tag{2–97}$$

or after introducing Eqs. 2-63, 2-65, 2-68, and 2-88

$$F(b_0) = \frac{1}{K}\frac{\gamma}{(ib_0)^2 + \gamma\ ib_0 + \gamma} \tag{2–98}$$

results. For $\Delta M = 0$ ($\mu = 0$, $\gamma = \infty$), Eqs. A-27, A-28 follow.

$$F(b_0) = \frac{1}{K}\frac{1}{1 + ib_0} \tag{2–99}$$

2.2.3 Example for Hand Calculation

As a very simple example suited for hand calculation, the vertical response of a disk on the surface of a homogeneous halfspace with $\nu = 1/3$ using the cone model to a rounded triangular pulse of prescribed displacement is analyzed

$$u_0(t) = \frac{u_{0\max}}{2}\left(1 - \cos\frac{2\pi t}{T_0}\right) \qquad 0 \leq t \leq T_0 \tag{2–100}$$

The corresponding velocity equals

$$\dot{u}_0(t) = \frac{\pi}{T_0}\,u_{0\,\max}\,\sin\frac{2\pi\,t}{T_0} \tag{2–101}$$

These excitation quantities are plotted in Figs. 2-10a and 2-10b. The choice of the time increment $\Delta t = T_0/8$ allows the velocity to be adequately represented by a polygon joining the sample points (implicit assumption for the recursive evaluation of convolution integrals). Numerical values of $u_0/u_{0\max}$ and $\dot{u}_0\,\Delta t/u_{0\max}$ are listed in Table 2-5.

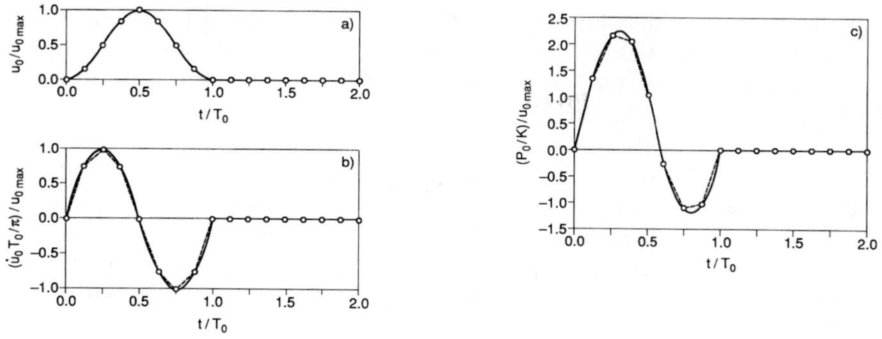

a. Prescribed vertical displacement pulse.

b. Associated vertical velocity.

c. Vertical interaction force.

Fig. 2-10 Example.

TABLE 2-5 EXAMPLE FOR HAND CALCULATION USING TRANSLATIONAL CONE

n	$\dfrac{t}{T_0}$	$\dfrac{u_0}{u_{0\max}}$	$\dfrac{\dot{u}_0\Delta t}{u_{0\max}}$	$\dfrac{P_0}{Ku_{0\max}}$	$\dfrac{u_0}{u_{0\max}}$
0	0.000	0.0000	0.0000	0.0000	0.0000
1	0.125	0.1465	0.2777	1.3097	0.1446
2	0.250	0.5000	0.3927	2.1449	0.4843
3	0.375	0.8535	0.2777	2.0167	0.8228
4	0.500	1.0000	0	1.0000	0.9641
5	0.625	0.8535	-0.2777	-0.3097	0.8271
6	0.750	0.5000	-0.3927	-1.1449	0.4935
7	0.875	0.1465	-0.2777	-1.0167	0.1597
8	1.000	0.0000	0.0000	0.0000	0.0221
9	1.125	0.0000	0.0000	0.0000	0.0174
10	1.250	0.0000	0.0000	0.0000	0.0137
11	1.375	0.0000	0.0000	0.0000	0.0108
12	1.500	0.0000	0.0000	0.0000	0.0085
13	1.625	0.0000	0.0000	0.0000	0.0067
14	1.750	0.0000	0.0000	0.0000	0.0053
15	1.875	0.0000	0.0000	0.0000	0.0042
16	2.000	0.0000	0.0000	0.0000	0.0033

Suppose that the duration of the pulse T_0 expressed in terms of r_0 and c_p happens to be $4r_0/c_p$. Accordingly, the time increment Δt is 0.5 r_0/c_p; and the associated dimensionless parameter $c_p\Delta t/z_0$, which specifies the recursion constants (Eq. 2-84, Table 2-4), is equal to $0.5/(z_0/r_0)$. From Table 2-3B the appropriate aspect ratio z_0/r_0 is 2.094 for vertical motion, leading to $c_p\Delta t/z_0 = 0.2387$.

In the stiffness formulation the interaction force is determined from Eq. 2-37 rearranged as

$$\frac{P_0}{Ku_{0\,\max}} = \frac{u_0}{u_{0\,\max}} + \frac{\dfrac{\dot{u}_0\Delta t}{u_{0\,\max}}}{\dfrac{K}{C}\Delta t} \tag{2–102}$$

As $K = \rho c_p^2 A_0/z_0$ and $C=rc_pA_0$ (Eq. 2-38), the denominator of the second term is identical to the above mentioned parameter $c_p\Delta t/z_0 = 0.2387$; thus

$$\frac{P_0}{Ku_{0\,\max}} = \frac{u_0}{u_{0\,\max}} + 4.1888\,\frac{\dot{u}_0\,\Delta t}{u_{0\max}} \tag{2–103}$$

as listed in Table 2-5 and plotted in Fig. 2-10c.

In the flexibility formulation for vertical motion the interaction force samples $P_0/(Ku_{0\max})$ are regarded to be the input. Ideally the convolution leading to u_0 (Eq. 2-77) should retrieve the original time history of displacement. To effect this convolution, the appropriate recursion coefficients are evaluated as described in Eq. 2-84: $a = 0.787626$, $b_0 = 0.110408$, and $b_1 = 0.101966$. Note that the sum of all the recursion coefficients equals 1. Using these, the last column of Table 2-5 is calculated via Eq. 2-83. A typical value is computed as follows for $n = 4$:

$$\frac{u_0}{u_{0\,\max}} = 0.787626 \cdot 0.8228 + 0.110408 \cdot 1.0000 + 0.101966 \cdot 2.0167 = 0.9641 \tag{2–104}$$

These results are in good agreement with the original values. The small amount of error (3.5%) is due to the fact that the recursive algorithm implicitly assumes the samples $P_0/(Ku_{0\max})$ to be joined to a polygon instead of interpolated by a smooth curve.

2.2.4 Vibration of Machine Foundation

For the sake of illustration, the vertical vibration of a foundation supporting a single-cylinder compressor (and a drive motor) (Fig. 2-11) is addressed. Besides investigating the response from the dynamic load of this reciprocating machine, the same foundation consisting of a concrete prism on the surface of a soil halfspace is also examined for the unbalanced load of a weaver's loom.

Crank mechanism of reciprocating machine. The basic crank mechanism that converts a rotary motion to a reciprocating (translational) motion, or vice versa, is shown in Fig. 2-12. It consists of a piston, which moves vertically within a guiding cylinder, a connecting rod of length ℓ, fixed to the piston and to the crank, and a crank rod of radius r which rotates about the crank shaft with the frequency ω. The mass of the connecting rod can be replaced by two equivalent masses, one moving in translation together with

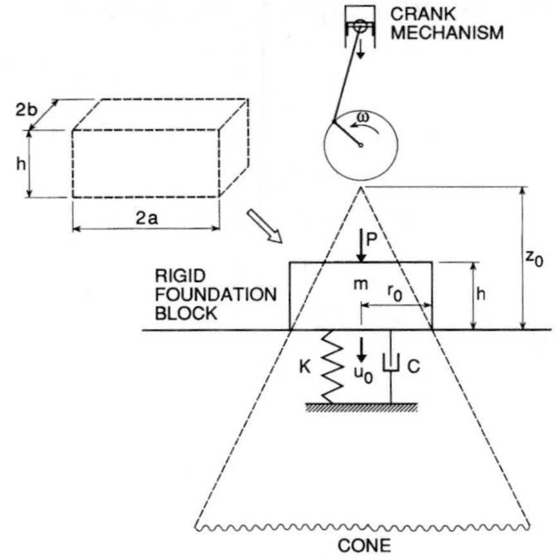

Fig. 2-11 Foundation of single-cylinder compressor (drive motor not shown, not to scale) with dynamic model.

the piston and the other rotating together with the crank rod. The total reciprocating mass m_{rec} and the total rotating mass m_{rot} (Fig. 2-12) follow by adding the contributions of the connecting rod to that of the piston and to that of the crank rod, respectively. The rotating mass m_{rot} will lead to a centrifugal load in the radial direction pointing outwards of magnitude $m_{rot} r\omega^2$, which can be decomposed into a vertical component (in the direction of the piston motion) $m_{rot} r\omega^2 \cos \omega t$ and a horizontal component $m_{rot} r\omega^2 \sin \omega t$. These inertial loads can be eliminated by installing a counter mass with the same mass m_{rot} at an angle of 180°. In the following, only cranks with counter masses are investigated. Thus only the inertial load P of the reciprocating mass m_{rec} remains.

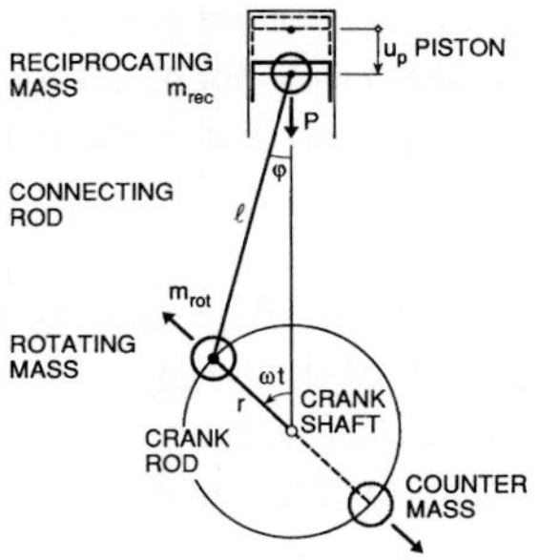

Fig. 2-12 Motion of crank mechanism.

The rotation of the crank with frequency ω causes a displacement u_p of the piston measured from its extreme outward position. The following geometric relationships (Fig. 2-12) exist

$$u_p + \ell \cos \varphi + r \cos \omega t = \ell + r \tag{2–105a}$$

$$\ell \sin \varphi = r \sin \omega t \tag{2–105b}$$

And cos φ follows substituting Eq. 2-105b as

$$\cos \varphi = \sqrt{1 - \sin^2 \varphi} = \sqrt{1 - \frac{r^2}{\ell^2} \sin^2 \omega t} \tag{2–106}$$

As r^2/ℓ^2 is, in general, small compared to 1, a series expansion is applied. Keeping two terms yields

$$\cos\varphi = 1 - \frac{r^2}{2\ell^2}\sin^2\omega t \tag{2–107}$$

Solving Eq. 2-105a for u_p and substituting Eq. 2-107 results in

$$u_p = r(1 - \cos\omega t) + \frac{r^2}{2\ell}\sin^2\omega t \tag{2–108a}$$

or

$$u_p = r(1 - \cos\omega t) + \frac{r^2}{4\ell}(1 - \cos 2\omega t) \tag{2–108b}$$

Differentiating Eq. 2-108b twice with respect to time leads to

$$\ddot{u}_p = r\omega^2(\cos\omega t + \frac{r}{\ell}\cos 2\omega t) \tag{2–109}$$

The inertial load of the reciprocating motion acting in the direction of the piston's motion

$$P = m_{\text{rec}}\ddot{u}_p \tag{2–110}$$

thus equals

$$P = m_{\text{rec}}\, r\omega^2\cos\omega t + m_{\text{rec}}\frac{r^2}{\ell}\,\omega^2\cos 2\omega t \tag{2–111}$$

This unbalanced dynamic load consists of a *primary* component acting at the frequency of rotation ω (first term in Eq. 2-111) and of a *secondary* component acting at twice this frequency 2ω. The amplitude of the secondary load is equal to that of the primary load multiplied by the geometric constant r/ℓ.

Dynamic load of single-cylinder compressor. The mass of the piston equals 8 kg, the mass of the connecting rod = 4 kg, the length of the connecting rod $\ell = 0.3$ m and the crank radius $r = 0.1$ m. The compressor and the driving motor with a total mass = 2750 kg operate at a frequency 9 Hz ($\omega = 2\pi \cdot 9 = 56.55\ \text{s}^{-1}$). The total reciprocating mass m_{rec} thus equals $8 + 0.5 \cdot 4 = 10$ kg. This yields

$$P(t) = 3.198\cos 56.55t + 1.066\cos 113.1t \quad \text{kN} \tag{2–112}$$

The primary and secondary components and the total load are plotted in Fig. 2-13. Obviously, the total load is periodic with a period $1/9 = 0.111$ s. It acts vertically on the foundation block.

Dynamic load of weaver's loom. As a second dynamic load the resultant of the measured vertical reactions of the four supports of a weaver's loom operating at 9 Hz and with the same mass of 2750 kg is processed. The load is expressed as a Fourier series with 20 terms, starting with a term corresponding to 9 Hz (first harmonic) and continuing with terms corresponding to 18 Hz, 27 Hz up to 180 Hz.

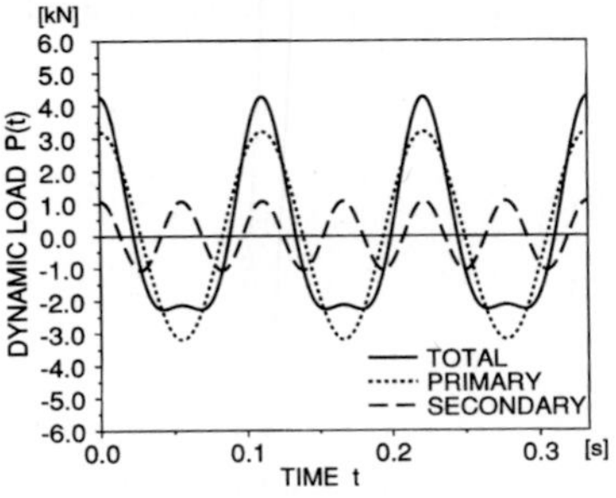

Fig. 2-13 Dynamic vertical load from vertical single-cylinder compressor.

$$P(t) = \sum_{j=1}^{20} |P(\omega_j)| \cos[\omega_j t + \varphi(\omega_j)] \tag{2–113}$$

where $\omega_j = j2\pi9$. The magnitudes $|P(\omega_j)|$ and the phase angles $\varphi(\omega_j)$ are listed in Table 2-6. Figure 2-14 shows the vertical dynamic load of the weaver's loom with a period 1/9 = 0.111 s.

TABLE 2-6 HARMONIC AMPLITUDES OF VERTICAL DYNAMIC LOAD OF WEAVER'S LOOM

Harmonic Number j	Amplitude	
	Magnitude $\|P(\omega_j)\|$ (kN)	Phase Angle $\varphi(\omega_j)$ (degree)
1	1.518	-138
2	1.223	-59
3	0.555	16
4	0.695	126
5	0.580	-74
6	0.232	142
7	0.144	129
8	0.052	43
9	0.122	-131
10	0.107	-38
11	0.063	81
12	0.061	103
13	0.080	144
14	0.048	165
15	0.132	86
16	0.177	137
17	0.052	-73
18	0.029	154
19	0.012	0
20	0.016	0

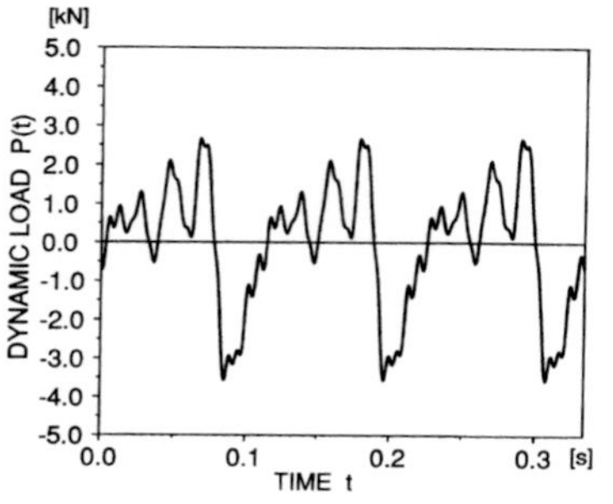

Fig. 2-14 Dynamic vertical load from weaver's loom.

Design criterion. For a machine operating at 9 Hz but also with higher-frequency components present, the allowable value of the peak-to-peak effective displacement amplitude can be selected as 0.05 mm. A less stringent criterion is 0.1 mm. For harmonic loading, the allowable (single) amplitudes are half these values.

Rigid foundation block. The rigid foundation block (Fig. 2-11) made of concrete (density 2500 kg/m^3) has the shape of a prism with length $2a = 2$ m, width $2b = 1.5$ m and height $h = 1$ m. Its mass equals 7500 kg. Adding that of the machine to this value results in the mass $m = 1.025 \cdot 10^4$ kg shown in Fig. 2-11.

Soil. The homogeneous halfspace disregarding material damping with Poisson's ratio $\nu = 1/3$ and mass density $\rho = 1700$ kg/m^3 has a shear modulus $G = 38.25 \cdot 10^6$ N/m^2, resulting in a dilatational-wave velocity $c_p = 300$ m/s (Table 2-1), which is the applicable wave-propagation velocity for the cone model.

The equivalent radius follows as $r_0 = \sqrt{4ab/\pi} = 0.977$ m. From Table 2-3B the aspect ratio of the cone for the vertical motion equals $z_0/r_0 = 2.094$, leading to an apex height $z_0 = 2.046$ m. The coefficients of the vertical spring and dashpot follow from Eq. 2-63 as $K = \rho c_p^2 \pi r_0^2 / z_0 = 2.243 \cdot 10^8$ N/m and $C = \rho c_p \pi r_0^2 = 1.53 \cdot 10^6$ Ns/m.

Dynamic system. The dynamic system shown in Fig. 2-11 with the vertical degree of freedom u_0 consists of a mass m supported by a spring with the coefficient K and in parallel by a dashpot with the coefficient C. The dynamic load P acts on the mass.

The vertical natural frequency ω_v of the (undamped) dynamic system follows as

$$\omega_v = \sqrt{\frac{K}{m}} = 147.92 \text{ s}^{-1} \tag{2–114}$$

leading to 23.54 Hz. The foundation is said to be *high-tuned*, as its natural frequency is higher than the operating frequency. The damping ratio ζ_v at ω_v equals

$$\zeta_v = \frac{\omega_v C}{2K} = 0.504 \tag{2–115}$$

while that at the operating frequency, $\omega = 56.55$ s^{-1}, is calculated as

$$\zeta_v(9 \text{ Hz}) = \frac{\omega C}{2K} = 0.193 \tag{2–116}$$

The free vibration in the vertical direction is thus very strongly damped (over 50%), which is caused by radiation damping.

For both loading cases three dynamic analyses are presented: a stiffness formulation in the time domain, a flexibility formulation in the time domain, and a stiffness formulation for harmonic loading.

Stiffness formulation in time domain. The equation of motion for the system shown in Fig. 2-11 equals

$$m\ddot{u}_0 + C\dot{u}_0 + Ku_0 = P \tag{2–117}$$

The following well-known explicit algorithm based on the Newmark family of methods and with a predictor–corrector scheme is used for the time integration. Starting from the known motion at time $(n - 1)\Delta t$ (that is, u_{0n-1}, $\dot{u}_{0n-1}$, $\ddot{u}_{0n-1}$) the final displacement and the predicted velocity at time $n\Delta t$ are calculated as

$$u_{0n} = u_{0n-1} + \Delta t\dot{u}_{0n-1} + \frac{\Delta t^2}{2}\ddot{u}_{0n-1} \tag{2–118a}$$

$$\tilde{\dot{u}}_{0n} = \dot{u}_{0n-1} + \frac{\Delta t}{2}\ddot{u}_{0n-1} \tag{2–118b}$$

A tilde (~) denotes a predicted value. Based on u_{0n} and $\tilde{\dot{u}}_{0n}$ the forces in the spring Ku_{0n} and in the dashpot $C\tilde{\dot{u}}_{0n}$ can be determined at $n\Delta t$. Formulating the equilibrium equation (Eq. 2-117) leads to the acceleration

$$\ddot{u}_{0n} = \frac{1}{m}\left(P_n - Ku_{0n} - C\tilde{\dot{u}}_{0n}\right) \tag{2–119}$$

The velocity is corrected as

$$\dot{u}_{0n} = \tilde{\dot{u}}_{0n} + \frac{\Delta t}{2}\ddot{u}_{0n} \tag{2–120}$$

which concludes the calculations for the time step. The algorithm is self starting.

For stability of the explicit algorithm the time step Δt must be smaller than the (smallest) natural period of the dynamic system divided by π. This can be expressed as

$$\Delta t < \frac{2}{\omega_v} = 0.0135 \text{ s} \tag{2–121}$$

This Δt is too large to allow an accurate representation of the variation of the load. For the weaver's loom the highest frequency in the load description equals $20 \cdot 9 = 180$ Hz. Selecting 8 time stations in the corresponding period results in $\Delta t = (1/180)/8 \simeq 7 \cdot 10^{-4}$ s. For the single-cylinder compressor the adequate representation of the secondary component with 18 Hz leads to $\Delta t = (1/18)/8 \simeq 7 \cdot 10^{-3}$ s. As the higher harmonics of the weaver's loom load hardly contribute to the response, an accurate description is not necessary. All analyses are thus performed with $\Delta t = 7 \cdot 10^{-3}$ s. In this case the criterion for an accurate description of the load and not that for stability of the explicit algorithm governs.

Starting with vanishing initial conditions, the total vertical displacement u_0 caused by the single-cylinder compressor is plotted in Fig. 2-15a. Due to the high damping present in the system, the transient phase is short (less than half a cycle). Also shown are the contributions of the primary and secondary components, which are used later on for a comparison with the results for harmonic loading (Fig. 2-16a). For the weaver's loom u_0 is presented in Fig. 2-15b.

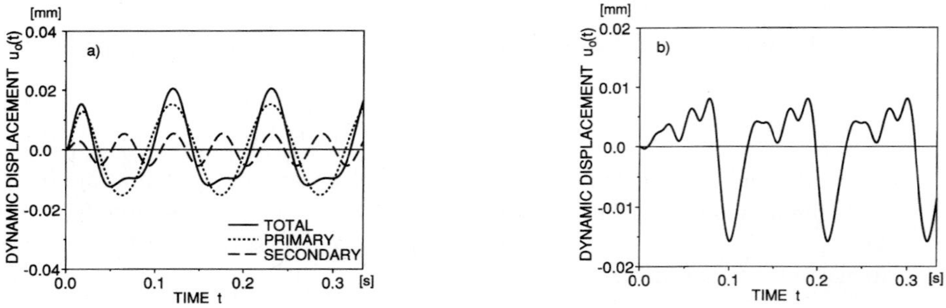

a. Single-cylinder compressor.

b. Weaver's loom.

Fig. 2-15 Vertical displacement from time-domain analyses (stiffness and flexibility formulations).

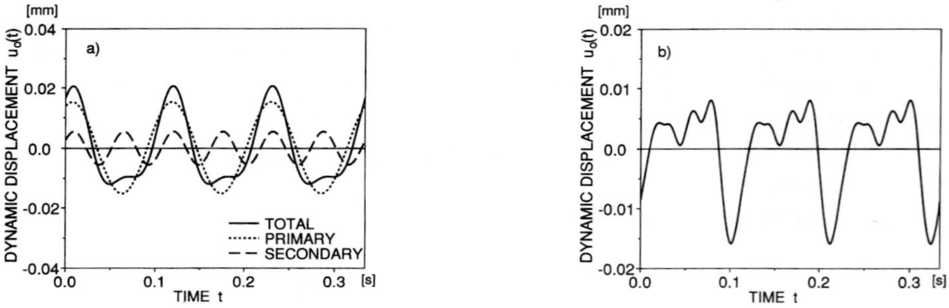

a. Single-cylinder compressor.

b. Weaver's loom.

Fig. 2-16 Vertical displacement from harmonic loading (frequency-domain analysis).

Flexibility formulation in time domain. The convolution integral of the displacement-force relationship (Eq. 2-93) is evaluated recursively as specified in Eq. 2-94. The procedure described at the end of the subsection "Nearly incompressible soil" of Section 2.2.2 applies. The recursive coefficients a_1, a_2 are specified in Eq. C-44, and the ramp function r_4 which is needed to determine b_0, b_1, and b_2 (Eq. C-34) is specified in Eq. C-45. The factor μ defined in Eq. 2-65 (with m replacing ΔM) is determined as $\mu = m/(\rho r_0^3) = 6.465$, leading to $\gamma = (\pi/\mu)z_0/r_0 = 1.017$ (Eq. 2-88). As $1 - 4/\gamma < 0$, the one-degree-of-freedom system is indeed underdamped. Selecting $\Delta t = 7 \cdot 10^{-3}$ s results in $a_1 = 0.742857$, $a_2 = -0.351756$, $b_0 = 0.133594$, $b_1 = 0.396626$, and $b_2 = 0.786787$.

The vertical displacements u_0 calculated using the flexibility formulation coincide with those based on the stiffness formulation (Fig. 2-15).

Stiffness formulation for harmonic loading. Alternative descriptions for harmonic motion are discussed at the end of this Section 2.2.4.

For harmonic loading of frequency ω_j the equation of motion (Eq. 2-117) equals

$$\left(-\omega_j^2 m + i\omega_j C + K\right) u_0(\omega_j) = P(\omega_j) \qquad (2\text{–}122)$$

Solving for the complex amplitude $u_0(\omega_j) = Re u_0(\omega_j) + i Im u_0(\omega_j)$ yields

$$u_0(\omega_j) = \frac{P(\omega_j)}{-\omega_j^2 m + i\omega_j C + K} \qquad (2\text{–}123)$$

The analysis for the single-cylinder compressor is discussed first. For the primary component ($\omega_1 = 56.55$ s^{-1}) with $P(\omega_1) = 3.198$ kN, $u_0(\omega_1) = 1.387 \cdot 10^{-5} - i0.626 \cdot 10^{-5}$ m leading to a magnitude $|u_0(\omega_1)| = \sqrt{1.387^2 + 0.626^2} \cdot 10^{-5} = 1.522 \cdot 10^{-5}$ m and a phase angle $\varphi(\omega_1) = \arctan(-0.626/1.387) = -0.424 = -24.3°$. As the real and imaginary parts of $u_0(\omega_1)$ are positive and negative, respectively, the angle $\varphi(\omega_1)$ is the in the fourth quadrant (and not in the second). The displacement corresponding to the j-th component $u_{0j}(t)$ in the time domain equals

$$u_{0j}(t) = |u_0(\omega_j)| \cos\left[\omega_j t + \varphi(\omega_j)\right] \qquad (2\text{–}124)$$

For the primary component $u_{01}(t) = 1.522 \cdot 10^{-5} \cos(56.55t-0.424)$ m results. The corresponding plot is specified in Fig. 2-16a. The response is periodic with a period 1/9 = 0.111 s. For the secondary component, the procedure is the same ($P(\omega_j) = 1.066$ kN, $\omega_2 = 113.1$s^{-1}). The superposition of the results for the two components leads to the total vertical displacement $u_0(t)$ (Fig. 2-16a).

To automatically incorporate the correct quadrant for the phase angle $\varphi(\omega_j)$, Eq. 2-124 can also be written as

$$u_{0j}(t) = Re u_0(\omega_j)\cos\omega_j t - Im u_0(\omega_j)\sin\omega_j t \qquad (2\text{–}125)$$

as $\cos\varphi(\omega_j) = Re u_0(\omega_j)/|u_0(\omega_j)|$ and $\sin\varphi(\omega_j) = Im u_0(\omega_j)/|u_0(\omega_j)|$. In this formulation $\varphi(\omega_j)$ is replaced by the real and imaginary components $Re u_0(\omega_j)$ and $Im u_0(\omega_j)$.

A somewhat more general formulation is outlined for the analysis of the weaver's loom. Equation 2-122 still applies for each harmonic, whereby the amplitude $P(\omega_j)$ will now be complex

$$P(\omega_j) = |P(\omega_j)| \cos\varphi(\omega_j) + i\,|P(\omega_j)| \sin\varphi(\omega_j) \qquad (2\text{–}126)$$

The magnitude $|P(\omega_j)|$ and the phase angle $\varphi(\omega_j)$ are listed in Table 2-6. Thus, ω_j equals $j56.55$ ($j = 1, \ldots, 20$). The amplitude $u_0(\omega_j)$, a complex number, follows from Eq. 2-123. The magnitude $|u_0(\omega_j)|$ and the corresponding phase angle $\varphi(\omega_j)$ are calculated straightforwardly as described above. The contribution of the j-th harmonic to the response in the time domain $u_{0j}(t)$ is determined from Eq. 2-124 (or 2-125). The total response $u_0(t)$ is equal to the sum over all j's ($j = 1, \ldots, 20$) of $u_{0j}(t)$ (Fig. 2-16b).

The response for the two loading cases satisfies the more stringent design criterion, as the peak-to-peak effective amplitude (Figs. 2-15 and 2-16) is less than 0.05 mm.

Alternative descriptions for harmonic motion. Several alternative ways exist to describe a harmonic motion for a specific frequency ω_j. They are discussed in the following for a load, but apply also for a displacement.

In the *first form*

$$P_j(t) = |P(\omega_j)| \cos\left[\omega_j t + \varphi(\omega_j)\right] \tag{2–127}$$

with $|P(\omega_j)|$ denoting the magnitude—a positive real value, and $\varphi(\omega_j)$, the phase angle, covering the range from -180° to 180° (or from $-\pi$ to π). This form is encountered in Eq. 2-113 or with vanishing $\varphi(\omega_j)$ in Eq. 2-111. As a slight modification, the cos-function in Eq. 2-127 is replaced by a sin-function.

Expanding the cos-function leads to the *second form*

$$P_j(t) = A(\omega_j)\cos\omega_j t + B(\omega_j)\sin\omega_j t \tag{2–128}$$

with the positive or negative real values

$$A(\omega_j) = |P(\omega_j)| \cos\varphi(\omega_j) \tag{2–129a}$$

$$B(\omega_j) = -|P(\omega_j)| \sin\varphi(\omega_j) \tag{2–129b}$$

This form arises when calculating the moment for a three-cylinder machine (see Eq. 2-190b), for example.

As a *third (intermediate) form*

$$P_j(t) = C(\omega_j)e^{i\omega_j t} + C^*(\omega_j)\, e^{-i\omega_j t} \tag{2–130}$$

with $C(\omega_j) = ReC(\omega_j) + iImC(\omega_j)$ being complex and an asterisk denoting the complex conjugate value. Substituting $e^{i\omega_j t} = \cos\omega_j t + i\sin\omega_j t$ results in

$$P_j(t) = 2\,Re\,C(\omega_j)\cos\omega_j t - 2Im\,C(\omega_j)\sin\omega_j t \tag{2–131}$$

which is in the same form as Eq. 2-128 with

$$A(\omega_j) = 2ReC(\omega_j) \tag{2–132a}$$

$$B(\omega_j) = -2\,Im\,C(\omega_j) \tag{2–132b}$$

By defining a complex amplitude $P(\omega_j) = ReP(\omega_j) + iImP(\omega_j) = 2C(\omega_j)$

$$P_j(t) = Re\left[P(\omega_j)e^{i\omega_j t}\right] \tag{2–133}$$

follows as the *fourth form*. Expanding leads to

$$P_j(t) = Re\,P(\omega_j)\cos\omega_j t - Im\,P(\omega_j)\sin\omega_j t \tag{2–134}$$

which is identical to Eq. 2-128 with

$$A(\omega_j) = Re\,P(\omega_j) \tag{2–135a}$$

$$B(\omega_j) = -Im\,P(\omega_j) \tag{2–135b}$$

The complex amplitude is thus given by

$$P(\omega_j) = |P(\omega_j)| \cos\varphi(\omega_j) + i|P(\omega_j)| \sin\varphi(\omega_j) \qquad (2\text{–}136)$$

This form is encountered in Eq. 2-126. It is important to note that the result is equal to the real part of the product of the complex amplitude $P(\omega_j)$ and $e^{i\omega_j t}$ and not just to the real part of $P(\omega_j)$ (Eq. 2-133). To simplify the nomenclature $P_j(t)$ is set equal to the complex function $P(\omega_j)e^{i\omega_j t}$, and it is implicitly implied that at the end of the complex operation the real part only is regarded.

$$P_j(t) = P(\omega_j)\, e^{i\omega_j t} \qquad (2\text{–}137)$$

This complex form provides a powerful and compact way of processing harmonic motion. For instance, to calculate the displacement $u_j(t)$, it is sufficient to multiply $P(\omega_j)$ by the dynamic-flexibility coefficient $F(\omega_j)$ leading to $u(\omega_j)$

$$u(\omega_j) = F(\omega_j)\, P(\omega_j) \qquad (2\text{-}138)$$

The real part of $u(\omega_j)e^{i\omega_j t}$ is equal to $u_j(t)$. This operation is encountered in connection with Eq. 2-123. Differentiation with respect to time is also easy to handle. With

$$u_j(t) = u(\omega_j)\, e^{i\omega_j t} \qquad (2\text{-}139)$$

$\dot{u}_j(t)$ follows as

$$\dot{u}_j(t) = i\,\omega_j u(\omega_j) e^{i\omega_j t} = \dot{u}(\omega_j)\, e^{i\omega_j t} \qquad (2\text{-}140)$$

from which

$$\dot{u}(\omega_j) = i\omega_j\, u(\omega_j) \qquad (2\text{-}141)$$

follows. Differentiation thus amounts to multiplication of the amplitude by $i\omega_j$.

The constants of the four forms [first: $|P(\omega_j)|$, $\varphi(\omega_j)$; second: $A(\omega_j)$, $B(\omega_j)$; third: $ReC(\omega_j)$, $ImC(\omega_j)$; fourth: $ReP(\omega_j)$, $ImP(\omega_j)$] can be calculated from one another using Eqs. 2-129, 2-132, and 2-135. In particular, the complex-amplitude form follows from the magnitude-phase angle form using Eq. 2-136 and going the other way

$$|P(\omega_j)| = \sqrt{ReP(\omega_j)^2 + ImP(\omega_j)^2} \qquad (2\text{-}142a)$$

$$\varphi(\omega_j) = \text{arc}\ \tan\left[ImP(\omega_j)/ReP(\omega_j)\right] \qquad (2\text{-}142b)$$

or more directly (Eq. 2-125)

$$P_j(t) = ReP(\omega_j)\cos\omega_j t - ImP(\omega_j)\sin\omega_j t \qquad (2\text{-}143)$$

applies.

In general, the contributions of several harmonic loads are processed, leading to the sum of a Fourier series

$$P(t) = \sum_j P_j(t) \qquad (2\text{-}144)$$

Each term $P_j(t)$ is processed independently of the others.

2.3 ROTATIONAL CONE

2.3.1 Stiffness Formulation

For the rotational cone, all mathematical relations based on the strength-of-materials approach (rod theory) are systematically derived in detail in Appendix A2.

Interaction moment-rotation relationship. It seems reasonable to expect that the interaction moment-rotation relationship of the rotational cone should involve a rotational spring with the static-stiffness coefficient K_ϑ augmented by a rotational dashpot with the coefficient C_ϑ. This is indeed the case; however, the results are somewhat more complicated. As specified in Eq. A-67,

$$M_0(t) = \underbrace{K_\vartheta \vartheta_0(t) + C_\vartheta \dot{\vartheta}_0(t)}_{M_0^s} - \underbrace{\int_0^t h_1(t-\tau) C_\vartheta \dot{\vartheta}_0(\tau) d\tau}_{M_0^r} \tag{2-145}$$

holds. K_ϑ and C_ϑ are given in Eqs. 2-24 and 2-55b (Eq. A-68).

$$K_\vartheta = \frac{3\rho c^2 I_0}{z_0} \quad \text{(2-146a)} \qquad\qquad C_\vartheta = \rho c I_0 \quad \text{(2-146b)}$$

The first two terms in Eq. 2-145 account for that portion of the moment due to the current values of rotation ϑ_0 and rotational velocity $\dot{\vartheta}_0$ This *instantaneous* contribution to the response is denoted as the singular part M_0^s. The last term with a negative sign is a convolution integral with the rotational velocity $\dot{\vartheta}_0$ (and not the angular rotation ϑ_0) and involves the unit-impulse response function

$$h_1(t) = \frac{c}{z_0} e^{-\frac{c}{z_0}t} \qquad t \geq 0 \tag{2-147a}$$

$$= 0 \qquad t < 0 \tag{2-147b}$$

This is the same simple exponential function encountered in the flexibility formulation of the translational cone (Eq. 2-75). This fortuitous result is a mathematical coincidence. The convolution depends on all previous values of the rotational velocity $\dot{\vartheta}_0$ and may be regarded as the system's memory of the past. This *lingering* portion of the response is called the regular part M_0^r. As the cone model leads to the exact results for the limits of the low and high frequencies (doubly asymptotic behavior, see Section 2.1.3), represented by M_0^s, the convolution integral M_0^r in the equation of the rotational cone is important only in the intermediate-frequency range.

Equation 2-145 applies with $c = c_s$ (and $I_0 = \pi r_0^4/2$) for the torsional degree of freedom and for $\nu \leq 1/3$ with $c = c_p$ (and $I_0 = \pi r_0^4/4$)) for the rocking degree of freedom. For the rocking degree of freedom in the nearly incompressible case ($1/3 < \nu \leq 1/2$),

$$M_0(t) = \underbrace{K_\vartheta \vartheta_0(t) + C_\vartheta \dot{\vartheta}_0(t) + \Delta M_\vartheta \ddot{\vartheta}_0(t)}_{M_0^s} - \underbrace{\int_0^t h_1(t-\tau) C_\vartheta \dot{\vartheta}_0(\tau) d\tau}_{M_0^r} \tag{2-148}$$

applies with the mass moment of inertia ΔM_ϑ specified in Eq. 2-60a

$$\Delta M_\vartheta = \mu_\vartheta \rho r_0^5 \tag{2-149}$$

with μ_ϑ specified in Eq. 2-60b and $c_p = 2 c_s$ (and $I_0 = \pi r_0^4/4$). The apex height z_0 follows from Table 2-3A.

Recursive evaluation. As noted previously in conjunction with the flexibility formulation for the translational cone (Eq. 2-77), it is computationally inefficient to perform convolution directly by numerical quadrature of Duhamel's integral. A more elegant approach to evaluate the convolution integral M_0^r with the first-order term h_1 in Eq. 2-145 (or Eq. 2-148) is to use the *recursion* equation (see Section C1, Eq. C-17)

$$M_{0n}^r = a\, M_{0n-1}^r + b_0 \left(-\, C_\vartheta\, \dot{\vartheta}_{0n}\right) + b_1 \left(-\, C_\vartheta\, \dot{\vartheta}_{0n-1}\right) \tag{2-150}$$

This means that to calculate the *regular part of the moment at time station n*, $M_{0\,n}^r$, only *that at time station n – 1*, $M_{0\,n-1}^r$ and the *rotational velocities at the same two time stations*, $\dot{\vartheta}_{0n}$ and $\dot{\vartheta}_{0n-1}$, *are needed.* The recursive coefficients are specified in Eq. C-24 as

$$a = e^{-\frac{c\Delta t}{z_0}} \tag{2-151a}$$

$$b_0 = 1 + \frac{e^{-\frac{c\Delta t}{z_0}} - 1}{\frac{c\Delta t}{z_0}} \tag{2-151b}$$

$$b_1 = \frac{-e^{-\frac{c\Delta t}{z_0}} + 1}{\frac{c\Delta t}{z_0}} - e^{-\frac{c\Delta t}{z_0}} \tag{2-151c}$$

For a typical range of the time-step parameter $c\Delta t/z_0$ a, b_0, b_1 are listed in the first part of Table 2-4.

The recursion procedure, which is self-starting from $t = 0$ for a quiescent past, is *exact* when the rotational velocity varies piecewise linearly between consecutive time stations.

Discrete-element model. As an alternative to the recursive evaluation of the convolution integral, one may employ physical discrete-element models which incorporate rigorously the convolution implicitly. Two such models are shown in Figs. 2-17a and 2-17b. In the *spring–dashpot model* of Fig. 2-17a the basemat (foundation) node with the degree of freedom ϑ_0 is connected by a rotational spring with the static-stiffness coefficient K_ϑ and in parallel by a rotational dashpot with the high-frequency limit of the radiation damping C_ϑ to a rigid support. An additional internal rotational degree of freedom ϑ_1 is introduced, located within the black box and connected by a rotational spring with a coefficient $-K_\vartheta/3$ to the basemat and by a rotational dashpot with a coefficient $-C_\vartheta$ to the rigid support. There is a conceptual difficulty associated with this model: Springs and dashpots with negative coefficients do not exist in reality. It is thus not actually possible to build this discrete-element model; but to handle such a model mathematically is, of course, no problem. Negative coefficients are avoided in the discrete-element model of Fig. 2-17b [M2] which consists of a mass–dashpot interconnection and resembles a *monkey tail*. The basemat node is attached by a rotational spring with the coefficient K_ϑ to a rigid support. Again, an additional internal rotational degree of freedom ϑ_1, with its own mass moment of inertia M_ϑ,

$$M_\vartheta = \rho\, z_0\, I_0 \tag{2-152}$$

is introduced, which is connected to the basemat node by a rotational dashpot with the coefficient C_ϑ. Unfortunately, this model is not easy to comprehend physically; intuitively one might fear that the mass would eventually fall off! The models of Fig. 2-17 can be used to represent the halfspace in torsional motion for all ν and in rocking motion for compressible soil ($\nu \leq 1/3$). For nearly incompressible soil ($1/3 < \nu \leq 1/2$) a mass moment of inertia with coefficient ΔM_ϑ is introduced in the basemat node with the degree of freedom ϑ_0 (not shown in Fig. 2-17).

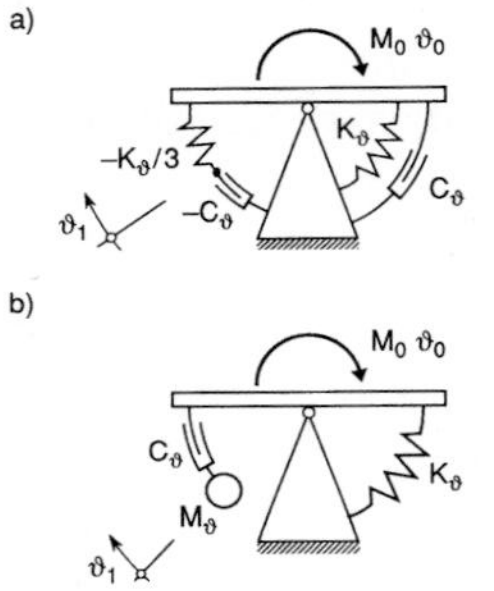

a. Spring–dashpot model.

b. Monkey-tail model.

Fig. 2-17 Discrete-element model with additional internal degree of freedom for rotation (without additional mass moment of inertia for nearly incompressible soil in rocking motion).

The equivalence of the discrete-element models to Eq. 2-145 is verified in the time domain as follows. First, the spring–dashpot model (Fig. 2-17a) is addressed. Formulating the equations of motion in the basemat node and in the node with the additional degree of freedom yields

$$M_0 = K_\vartheta \vartheta_0 + C_\vartheta \dot{\vartheta}_0 - \frac{K_\vartheta}{3}(\vartheta_0 - \vartheta_1) \tag{2-153a}$$

$$-\frac{K_\vartheta}{3}(\vartheta_1 - \vartheta_0) - C_\vartheta \dot{\vartheta}_1 = 0 \tag{2-153b}$$

As $3C_\vartheta/K_\vartheta = z_0/c$ (Eq. 2-146), Eq. 2-153b leads to

$$\frac{z_0}{c}\dot{\vartheta}_1 + \vartheta_1 = \vartheta_0 \tag{2-154}$$

By analogy to Eqs. 2-74 and 2-77,

$$\vartheta_1(t) = \int_0^t h_1(t-\tau)\,\vartheta_0(\tau)\,d\tau \tag{2-155}$$

follows with $h_1(t)$ defined in Eq. 2-75. It is interesting to note that $\vartheta_1(t)$ [which is introduced in the Appendix A as a general function that can be expressed as a convolution integral involving $\vartheta_0(t)$, Eq. A-56] has a clear physical meaning representing the internal degree of freedom of the discrete-element model. Differentiating Eq. 2-155 with respect to t and substituting in Eq. 2-154 results in

$$\vartheta_0 - \vartheta_1 = \frac{z_0}{c}\int_0^t h_1(t-\tau)\,\dot{\vartheta}_0(\tau)\,d\tau \tag{2-156}$$

Equation 2-156 transforms Eq. 2-153a to

$$M_0 = K_\vartheta\,\vartheta_0 + C_\vartheta\dot{\vartheta}_0 - \int_0^t h_1(t-\tau)\,C_\vartheta\,\dot{\vartheta}_0(\tau)\,d\tau \tag{2-157}$$

which is identical to Eq. 2-145.

Next, the monkey-tail model (Fig. 2-17b) is discussed. The equations of motion are equal to

$$M_0 = K_\vartheta\,\vartheta_0 + C_\vartheta\left(\dot{\vartheta}_0 - \dot{\vartheta}_1\right) \tag{2-158a}$$

$$C_\vartheta\left(\dot{\vartheta}_1 - \dot{\vartheta}_0\right) + M_\vartheta\,\ddot{\vartheta}_1 = 0 \tag{2-158b}$$

Substituting C_ϑ (Eq. 2-146b) and M_ϑ (Eq. 2-152) in Eq. 2-158b yields the first-order differential equation in the variable $\dot{\vartheta}_1$.

$$\frac{z_0}{c}\ddot{\vartheta}_1 + \dot{\vartheta}_1 = \dot{\vartheta}_0 \tag{2-159}$$

By analogy to Eqs. 2-74 and 2-77,

$$\dot{\vartheta}_1(t) = \int_0^t h_1(t-\tau)\,\dot{\vartheta}_0(\tau)\,d\tau \tag{2-160}$$

holds with $h_1(t)$ specified in Eq. 2-75. Substitution into Eq. 2-158a yields the desired result

$$M_0 = K_\vartheta\,\vartheta_0 + C_\vartheta\,\dot{\vartheta}_0 - \int_0^t h_1(t-\tau)\,C_\vartheta\,\dot{\vartheta}_0(\tau)\,d\tau \tag{2-161}$$

which coincides with Eq. 2-145. In passing, it is interesting to note that the convolution integral in Eq. 2-161 corresponds to $-C_\vartheta\,\dot{\vartheta}_1$ (Eq. 2-158a).

In practical applications either of the *discrete-element models with one internal degree of freedom* shown in Fig. 2-17 may be attached to the underside of the structural system as an *exact equivalent representation of the rotational cone*. Then the complete system may be analyzed directly with a general-purpose computer program. Both discrete-element models shown in Fig. 2-17 will be used in this text.

Alternative general-purpose computer program implementation. It is possible to select an alternative implementation of the rotational cone with a general-purpose computer program which avoids introducing an additional internal degree of freedom. This model, shown in Fig. 2-18, includes instead of C_ϑ a modified rotational dashpot C_ϑ^* as well as an external moment M_0^* which is calculated from known values of the previous time station. As described in Eqs. 2-145 and 2-150, the stiffness formulation of the rotational cone at time station $n\Delta t$ involves the evaluation of

$$M_{0n} = K_\vartheta \vartheta_{0n} + C_\vartheta \dot{\vartheta}_{0n} + M_{0n}^r \tag{2-162}$$

in which the regular part is determined recursively

$$M_{0n}^r = a\, M_{0n-1}^r - b_0 C_\vartheta \dot{\vartheta}_{0n} - b_1 C_\vartheta \dot{\vartheta}_{0n-1} \tag{2-163}$$

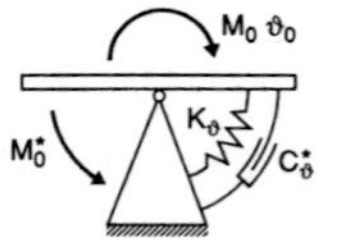

Fig. 2-18 Alternative discrete-element model with modified dashpot and external interaction moment but without additional internal degree of freedom.

The term $b_0 C_\vartheta \dot{\vartheta}_{0n}$ on the right-hand side of the recursion formula poses no problem if the rotational velocity $\dot{\vartheta}_{0n}$ is prescribed a priori. In a soil-structure interaction problem, however, the quantity $\dot{\vartheta}_{0n}$ is unknown at time $n\Delta t$ just as is $M_{0\,n}^r$. Indeed, the interdependence of $\dot{\vartheta}_{0n}$ and M_{0n} is the essence of the interaction phenomenon. Fortunately, this difficulty may be surmounted by simple algebraic manipulation of Eqs. 2-162 and 2-163 into

$$M_{0n} = K_\vartheta\, \vartheta_{0n} + C_\vartheta^*\, \dot{\vartheta}_{0n} + M_{0n}^* \tag{2-164}$$

$$M_{0n}^* = a\, M_{0n-1}^* - b_1^* C_\vartheta\, \dot{\vartheta}_{0n-1} \tag{2-165}$$

with

$$C_\vartheta^* = (1 - b_0)\, C_\vartheta \tag{2-166}$$

$$b_1^* = a\, b_0 + b_1 \tag{2-167}$$

The spring K_ϑ and the modified dashpot C_ϑ^* are incorporated as physical components into the discrete-element model shown in Fig. 2-18. The exterior moment $M_{0\,n}^*$ involves only the known past values $M_{0\,n\text{-}1}^*$ and $\dot{\vartheta}_{0n\text{-}1}$. Again, to represent the nearly incompressible halfspace in rocking, the mass moment of inertia ΔM_ϑ is introduced in the basemat node. This formulation is particularly well suited for use with a general-purpose computer program. At each increment of time it is only necessary to update the external moment $M_{0\,n}^*$, which is treated as part of the load applied to the basemat.

Dynamic-stiffness coefficient. For harmonic loading the interaction moment-rotation relationship is specified in Eqs. A-72 to A-74. For the nearly incompressible case in rocking, the term $-\omega^2 \Delta M_\vartheta$ is added to the *dynamic-stiffness coefficient*. Equation A-73 with b_0 defined in Eq. 2-68 still holds

$$S_\vartheta(b_0) = K_\vartheta \left[k_\vartheta(b_0) + i\, b_0\, c_\vartheta(b_0)\right] \tag{2-168}$$

but with the dimensionless spring and damping coefficients

$$k_\vartheta(b_0) = 1 - \frac{4\mu_\vartheta}{3\pi}\frac{r_0}{z_0} b_0^2 - \frac{1}{3}\frac{b_0^2}{1+b_0^2} \tag{2-169a}$$

$$c_\vartheta(b_0) = \frac{1}{3}\frac{b_0^2}{1+b_0^2} \tag{2-169b}$$

In Eq. 2-60b, μ_ϑ is defined. For $\mu_\vartheta = 0$, the cases of the compressible soil in rocking and of the torsional motion are recovered (Eq. A-74). The dynamic-stiffness coefficient $S_\vartheta(b_0)$ can be interpreted as a rotational spring with the frequency-dependent coefficient $K_\vartheta k_\vartheta(b_0)$ and a rotational dashpot with the frequency-dependent coefficient $(z_0/c)\, K_\vartheta c_\vartheta(b_0)$ in parallel (Fig. 2-19).

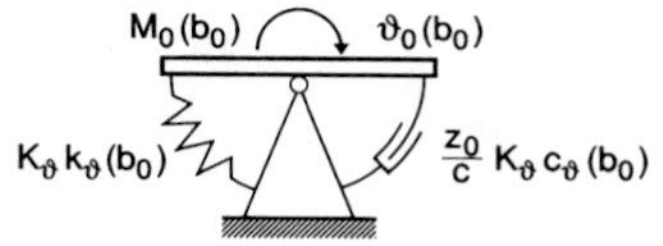

Fig. 2-19 Interpretation of dynamic-stiffness coefficient for harmonic excitation as spring and as dashpot in parallel with frequency-dependent coefficients.

To examine the torsional and rocking dynamic-stiffness coefficients of a disk resting on the surface of a halfspace, b_0 is replaced by a_0 defined in Eq. 2-71. Equation 2-168 is then transformed to

$$S_\vartheta(a_0) = K_\vartheta \left[k_\vartheta(a_0) + i\, a_0\, c_\vartheta(a_0)\right] \tag{2-170}$$

where

$$k_\vartheta(a_0) = 1 - \frac{4}{3}\frac{\mu_\vartheta}{\pi}\frac{z_0}{r_0}\frac{c_s^2}{c^2} a_0^2 - \frac{1}{3}\frac{a_0^2}{\left(\frac{r_0 c}{z_0 c_s}\right)^2 + a_0^2} \tag{2-171a}$$

$$c_\vartheta(a_0) = \frac{z_0\, c_s}{3 r_0\, c}\frac{a_0^2}{\left(\frac{r_0 c}{z_0 c_s}\right)^2 + a_0^2} \tag{2-171b}$$

For the torsional motion, $c = c_s$ and $\mu_\vartheta = 0$. For all ν, $z_0/r_0 = 0.884$ follows from Table 2-3A. For the rocking motion, $c = c_p$ and $\mu_\vartheta = 0$ for $\nu \le 1/3$ and $c = 2\, c_s$ and μ_ϑ is specified in Eq. 2-60b for $1/3 < \nu \le 1/2$. As a function of ν, z_0/r_0 follows from Table 2-3A.

For the torsional motion the spring coefficient $k_t(a_0)$ and the damping coefficient $c_t(a_0)$ are plotted versus a_0 in Fig. 2-20. The cone's results are shown as lines. The rigorous values of Veletsos and Nair [V4] are plotted as distinct points. For the rocking motion the comparison of the dynamic-stiffness coefficients for Poisson's ratios $\nu = 0$, 1/3, 0.45, and 0.5 is shown in Fig. 2-21. The rigorous results plotted as distinct points are taken from Veletsos and Wei [V1] and Luco and Mita [L3]. In general, the agreement is satisfactory for all Poisson's ratios, but there is room for improvement. The inertial load of the mass moment of inertia ΔM_ϑ present for $\nu = 0.5$ leads to a negative k_r for $a_0 > 4.8$. In the *higher-frequency range* ($a_0 > 4$) $c_t(a_0)$ and $c_r(a_0)$ of the cone models are *very accurate*, while in the *lower-frequency range*, which is important in practical applications, the *cone's results overestimate (radiation) damping* for $\nu \leq 1/3$ to a certain extent. This is the same tendency as encountered for the translational cones (Figs. 2-6, 2-7).

2.3.2 Flexibility Formulation

Rotation-interaction moment relationship. The interaction moment M_0 is regarded as the known input and the corresponding unknown rotation ϑ_0 as the output. The torsional motion and the rocking motion for the compressible soil ($\nu \leq 1/3$) in the time domain are addressed in the following.

The fundamental solution required to derive a flexibility formulation is the rotation response as a function of time $\vartheta_0(t)$ due to a Dirac-delta impulse $\delta(t)$ of the normalized interaction moment M_0/K_ϑ applied at time zero. The derivation presented in Section A2.3 leads to the following *unit-impulse response function* for the rotational cone (Eq. A-81).

$$h_2(t) = \frac{c}{z_0} e^{-\frac{3}{2}\frac{c}{z_0}t} \left(3 \cos \frac{\sqrt{3}}{2}\frac{c}{z_0}t - \sqrt{3} \sin \frac{\sqrt{3}}{2}\frac{c}{z_0}t \right) \qquad t \geq 0 \qquad \text{(2-172a)}$$

$$= 0 \qquad t < 0 \qquad \text{(2-172b)}$$

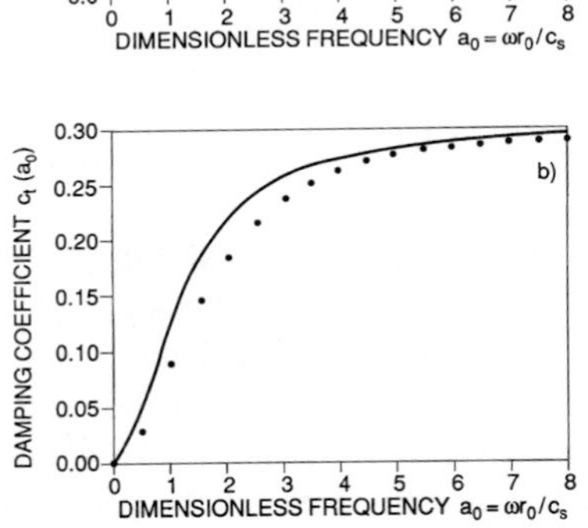

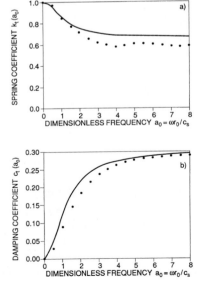

Fig. 2-20 Dynamic-stiffness coefficient for harmonic loading of disk on homogeneous halfspace in torsional motion.

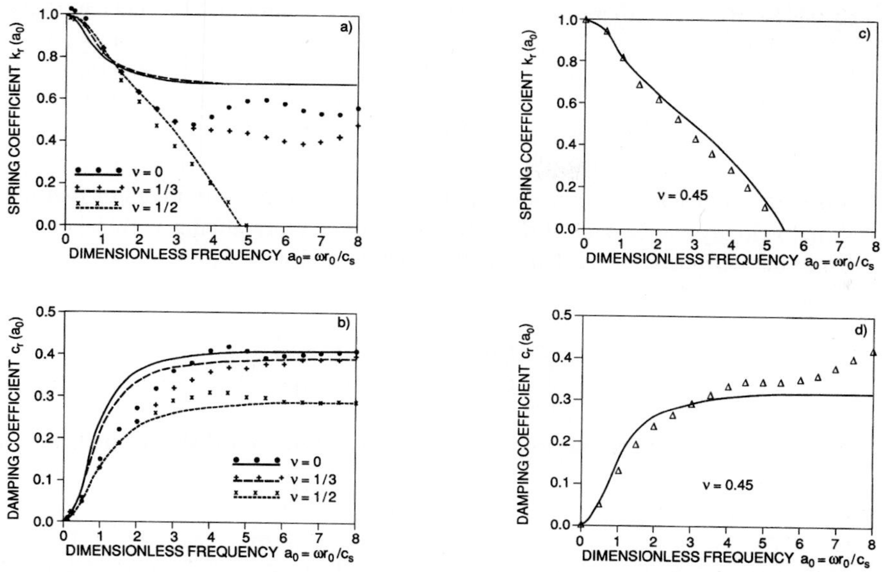

Fig. 2-21 Dynamic-stiffness coefficient for harmonic loading of disk on homogeneous halfspace in rocking motion for various Poisson's ratios.

The rotation-interaction moment relationship is specified as the *convolution integral (Duhamel's integral) of the interaction moment and the unit-impulse response function* (Eq. A-82).

$$\vartheta_0(t) = \int_0^t h_2(t-\tau)\,\frac{M_0(\tau)}{K_\vartheta}\,d\tau \qquad (2\text{-}173)$$

The response, which depends on all previous values of M_0, is purely of the lingering (regular) type without an instantaneous (singular) part.

Unit-impulse response function. The unit-impulse response function h_2 (Eq. 2-172) of the rocking motion multiplied by r_0/c_s is plotted for $\nu = 1/3$ in Fig. 2-22 as a function of the dimensionless time $\bar{t}$ defined in Eq. 2-78. The initial value follows from z_0/r_0=2.356 (Table 2-3B) and with $c = c_p$ (= $2c_s$) as $2 \cdot 3/2.356 = 2.546$. The agreement of the result calculated using the cone model with the exact solution of Veletsos and Verbic [V3] is good. In contrast to the unit-impulse functions h_1 of the translational degrees of freedom (Fig. 2-8), that of the rotational degree of freedom h_2 changes sign, as in the free vibration of an underdamped system. The (radiation) damping ratio corresponding to a rotation will thus be smaller than that of a translation.

The initial value of $h_2(t)$ for the disk on the surface of an elastic halfspace can be determined rigorously based on three-dimensional elastodynamics applying the law of conservation of momentum for the first differential time dt. Proceeding as for the translational cone (Fig. 2-9), a dilatational wave has traveled vertically the distance $dz = c_p\,dt$ at the end of dt for the rocking degree of freedom. The corresponding excited mass moment of inertia in the form of a cylinder equals $\rho(\pi r_0^4/4)c_p\, dt$. Formulating the law of conservation of momentum yields

$$\int_0^{dt} M_0(t)dt = \rho c_p \frac{\pi r_0^4}{4} dt\, \dot{\vartheta}_0(t = 0^+) \tag{2-174}$$

where the right-hand side is equal to the momentum (mass moment of inertia times rotational velocity) at the end of the differential time. Normalizing with K_ϑ of the halfspace, using the definition of the Dirac-delta function and with $\vartheta_0(t = 0^+) = \dot{\vartheta}_0(t = 0^+)\, dt$ transforms Eq. 2-174 to

$$\int_0^{dt} \frac{M_0(t)}{K_\vartheta} dt = 1 = \frac{\rho c_p}{K_\vartheta} \frac{\pi r_0^4}{4} \vartheta_0(t = 0^+) \tag{2-175}$$

The initial value of the unit-impulse rocking response for a disk on a halfspace thus equals

$$\vartheta_0(t = 0^+) = \frac{K_\vartheta}{\rho c_p \frac{\pi r_0^4}{4}} \tag{2-176}$$

The denominator is equal to C_ϑ.

As by construction K_ϑ is equal to the static-stiffness coefficient of the cone (Eq. 2-146a), it follows from Eq. 2-176

$$\vartheta_0(t = 0^+) = \frac{\frac{3\rho c_p^2 \pi r_0^4}{z_0\, 4}}{\rho c_p \frac{\pi r_0^4}{4}} = 3\frac{c_p}{z_0} \tag{2-177}$$

The initial value of the unit-impulse response for a disk on a halfspace $\vartheta_0(t = 0^+)$ is thus equal to the corresponding value for the cone $h_2(t = 0^+)$ (Eq. 2-172a). The initial peak value at time $t = 0^+$ shown in Fig. 2-22 calculated with the rotational cone is thus exact. This follows from the fact that the cone enables an error-free representation of the high-frequency limit of the dynamic-stiffness coefficient—of the radiation dashpot C_ϑ.

Recursive evaluation. To evaluate the convolution integral (Eq. 2-173) efficiently, the *recursive evaluation* involving only a few discrete samples of the immediate

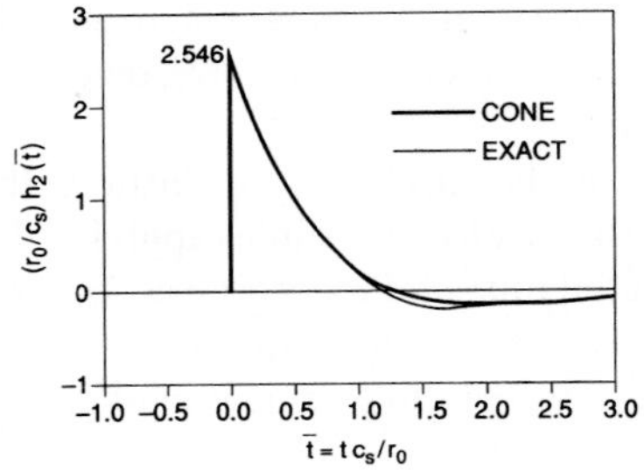

Fig. 2-22 Unit-impulse response function of disk on homogeneous halfspace for rocking motion.

past is used (see Section C1 for background). To calculate a convolution with $h_2(t)$, a second-order term addressed in Section C3, the recursion equation with three equally spaced time stations is equal to (Eq. C-27)

$$\vartheta_{0n} = a_1 \vartheta_{0n-1} + a_2 \vartheta_{0n-2} + b_0 \frac{M_{0n}}{K_\vartheta} + b_1 \frac{M_{0n-1}}{K_\vartheta} + b_2 \frac{M_{0n-2}}{K_\vartheta} \tag{2-178}$$

The *rotation at time station n, ϑ_{0n}, is expressed as a linear function of the rotations at the two previous time station n – 1 and n – 2,* ϑ_{0n-1} and ϑ_{0n-2}, *and of the normalized interaction moments at the same three time stations* M_{0n}/K_ϑ, M_{0n-1}/K_ϑ and M_{0n-2}/K_ϑ. The recursive coefficients, which are a function of the time step Δt (but not of the time station), are derived in Section C3. As specified in Eq. C-39

$$a_1 = 2\, e^{-\frac{3}{2}\frac{c\Delta t}{z_0}} \cos \frac{\sqrt{3}}{2}\frac{c\Delta t}{z_0} \tag{2-179a}$$

$$a_2 = -\, e^{-3\frac{c\Delta t}{z_0}} \tag{2-179b}$$

From Eq. C-34 b_0, b_1, b_2 follow, with $r(t) = r_2(t)$ given in Eq. C-40. Numerical values of the recursion coefficients are specified in the second part of Table 2-4 for a typical range of the dimensionless time increment $c\Delta t/z_0$. The sum of all recursion coefficients $a_1+a_2+b_0+b_1+b_2$ add up to unity, which may be verified by examining any row of Table 2-4.

As discussed for the translational cone in Section 2.2.2 the recursive algorithm, which is self-starting for quiescent initial conditions, is *exact*, subject to the assumption that the interaction moment varies linearly piecewise between consecutive time stations.

Dynamic-flexibility coefficient. For the rocking motion in the case of nearly incompressible soil ($1/3 < \nu \leq 1/2$) the stiffness formulation for harmonic loading described by Eqs. 2-168 and 2-169 can be inverted (Eq. A-83)

$$\vartheta_0(b_0) = F_\vartheta(b_0)\, M_0(b_0) \tag{2-180}$$

where the dynamic-flexibility coefficient

$$F_\vartheta(b_0) = S_\vartheta^{-1}(b_0) = \frac{1}{K_\vartheta} \frac{1}{1 - \frac{1}{3}\frac{b_0^2}{1+b_0^2} - \frac{1}{\gamma_\mu} b_0^2 + \frac{1}{3}\, i b_0 \frac{b_0^2}{1+b_0^2}} \tag{2-181}$$

with

$$\gamma_\mu = \frac{3\pi}{4\mu_\vartheta} \frac{z_0}{r_0} \tag{2-182}$$

Canceling with $1 - ib_0$ results in

$$F_\vartheta(b_0) = \frac{1}{K_\vartheta} \frac{1 + ib_0}{1 + ib_0 + \left(\frac{1}{3} + \frac{1}{\gamma_\mu}\right)(ib_0)^2 + \frac{1}{\gamma_\mu}(i\, b_0)^3} \tag{2-183}$$

For $\Delta M_\vartheta = 0$ ($\mu_\vartheta = 0$, $\gamma_\mu = \infty$) the cases of the compressible soil for rocking and of the torsional motion are recovered (Eq. A-84).

Equation 2-183 can be used as the starting point to derive a flexibility formulation in the time domain for rocking when the soil is nearly incompressible. The unit-impulse response function is the inverse Fourier transform of Eq. 2-183, as described in Eq. C-4. The convolution integral in the flexibility formulation can be evaluated recursively, whereby the recursion coefficients can be derived as explained in Appendix C. The formulation becomes, however, quite complicated. For instance, the roots of a cubic equation (denominator in Eq. 2-183) have to be calculated. As the flexibility formulation is anyway less important than the fully developed stiffness formulation with the convenient discrete-element models, the case of rocking for the nearly incompressible soil is not further pursued in this section.

2.3.3 Example for Hand Calculation

The various computational procedures, in particular the recursive techniques, may be demonstrated by calculating the rocking response of a disk on the surface of a homogeneous halfspace with $\nu = 1/3$ to a rounded triangular pulse of prescribed rotation

$$\vartheta_0(t) = \frac{\vartheta_{0\max}}{2}\left(1 - \cos\frac{2\pi t}{T_0}\right) \qquad 0 \le t \le T_0 \tag{2-184}$$

The corresponding angular velocity equals

$$\dot{\vartheta}_0(t) = \frac{\pi}{T_0}\,\vartheta_{0\max}\,\sin\frac{2\pi t}{T_0} \tag{2-185}$$

These excitation quantities are plotted in Figs. 2-23a and 2-23b. The choice of the time increment $\Delta t = T_0/8$ allows the rocking velocity to be adequately represented by a polygon joining the sample points (implicit assumption for the recursive evaluation of convolution integrals). Numerical values of $\vartheta_0/\vartheta_{0\max}$ and $\dot{\vartheta}_0\Delta t/\vartheta_{0\max}$ are listed in Table 2-7.

Suppose that the duration of the pulse T_0, expressed in terms of the r_0 and c_p, happens to be $4r_0/c_p$. Accordingly, the time increment Δt is $0.5\, r_0/c_p$; and the associated dimensionless parameter $c_p\Delta t/z_0$, which specifies the recursion constants (Eqs. 2-151 and 2-179, Table 2-4) is equal to $0.5/(z_0/r_0)$. From Table 2-3B the appropriate aspect ratio z_0/r_0 is 2.356 for rocking, leading to $c_p\Delta t/z_0 = 0.2122$.

In the stiffness formulation for rocking the singular part of the moment is determined from Eq. 2-145, rearranged as

$$\frac{M_0^s}{K_\vartheta\,\vartheta_{0\max}} = \frac{\vartheta_0}{\vartheta_{0\max}} + \frac{\dfrac{\dot{\vartheta}_0\Delta t}{\vartheta_{0\max}}}{\dfrac{K_\vartheta}{C_\vartheta}\Delta t} \tag{2-186}$$

Since $K_\vartheta = 3\rho c_p^2 I_0/z_0$ and $C_\vartheta = \rho c_p I_0$ (Eq. 2-146), the denominator is equal to $3c_p\Delta t/z_0 = 3 \cdot 0.2122 = 0.6366$, and thus

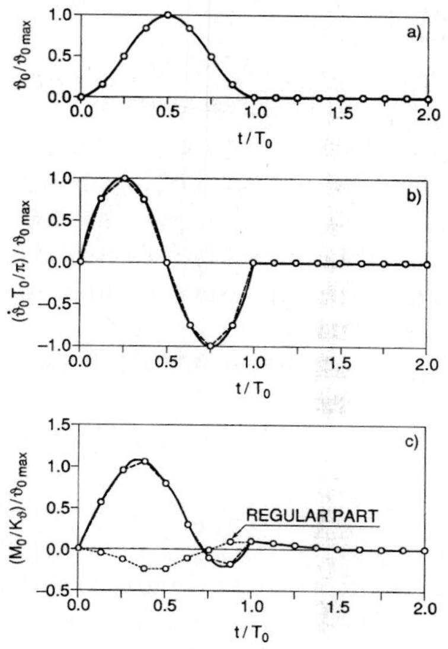

a. Prescribed rotational rocking pulse.

b. Associated rotational rocking velocity.

c. Rocking interaction moment.

Fig. 2-23 Example.

TABLE 2-7 EXAMPLE FOR HAND CALCULATION USING ROTATIONAL CONE

n	$\frac{t}{T_0}$	$\frac{\vartheta_0}{\vartheta_{0\max}}$	$\frac{\dot{\vartheta}_0 \Delta t}{\vartheta_{0\max}}$	$\frac{M_0^s}{K_\vartheta \vartheta_{0\max}}$	$\frac{M_0^r}{K_\vartheta \vartheta_{0\max}}$	$\frac{M_0}{K_\vartheta \vartheta_{0\max}}$	$\frac{\vartheta_0}{\vartheta_{0\max}}$
0	0.000	0.0000	0.0000	0.0000	0.0000	0.0000	0.0000
1	0.125	0.1465	0.2777	0.5827	-0.0432	0.5395	0.1493
2	0.250	0.5000	0.3927	1.1169	-0.1362	0.9807	0.4937
3	0.375	0.8535	0.2777	1.2897	-0.2102	1.0795	0.8371
4	0.500	1.0000	0.0000	1.0000	-0.2103	0.7897	0.9817
5	0.625	0.8535	-0.2777	0.4173	-0.1269	0.2904	0.8448
6	0.750	0.5000	-0.3927	-0.1169	-0.0013	-0.1182	0.5075
7	0.875	0.1465	-0.2777	-0.2897	0.0990	-0.1907	0.1679
8	1.000	0.0000	0.0000	0.0000	0.1203	0.1203	0.0247
9	1.125	0.0000	0.0000	0.0000	0.0973	0.0973	0.0124
10	1.250	0.0000	0.0000	0.0000	0.0787	0.0787	0.0047
11	1.375	0.0000	0.0000	0.0000	0.0636	0.0636	0.0001
12	1.500	0.0000	0.0000	0.0000	0.0515	0.0515	-0.0023
13	1.625	0.0000	0.0000	0.0000	0.0416	0.0416	-0.0033
14	1.750	0.0000	0.0000	0.0000	0.0337	0.0337	-0.0036
15	1.875	0.0000	0.0000	0.0000	0.0272	0.0272	-0.0033
16	2.000	0.0000	0.0000	0.0000	0.0220	0.0220	-0.0029

$$\frac{M_0^s}{K_\vartheta \vartheta_{0\max}} = \frac{\vartheta_0}{\vartheta_{0\max}} + 1.5708 \frac{\dot{\vartheta}_0 \Delta t}{\vartheta_{0\max}} \tag{2-187}$$

as listed in Table 2-7.

Next, the regular part of the moment is addressed. Just as for the singular part, the expression for $M_0^r/(K_\vartheta \vartheta_{0\max})$ also includes the factor 1.5708. For this convolution with h_1 the appropriate recursion constants are $a = 0.808798$, $b_0 = -1.5708 \cdot 0.098980 = -0.155478$, $b_1 = -1.5078 \cdot 0.092222 = -0.144862$. Note that the constant factor -1.5708 is multiplied a priori with the b's. This enables the column $\dot{\vartheta}_0 \Delta t/\vartheta_{0\max}$ of Table 2-7 to be used directly as input to the recursive algorithm.

The sum of $M_0^s/(K_\vartheta \vartheta_{0\max})$ and $M_0^r/(K_\vartheta \vartheta_{0\max})$ of Table 2-7 yields the total moment $M_0/(K_\vartheta \vartheta_{0\max})$ shown as a solid line in Fig. 2-23c. It may be observed that the contribution of the regular part (dotted curve and the entire lingering tail for $t > T_0$) is small but not negligible.

To conclude the example, the moment samples $M_0/(K_\vartheta \vartheta_{0\max})$ are regarded as input for the flexibility formulation. The convolution leading to ϑ_0 (Eq. 2-173) should retrieve the original time history of rotation. The reader is encouraged to consult Table 2-4, evaluate the recursion constants for this convolution with h_2, $a_1 = 1.430258$, $a_2 = -0.529078$, $b_0 = 0.276710$, $b_1 = 0.016322$, $b_2 = -0.194212$, and recompute the numerical values in the last column of Table 2-7. The error is practically negligible (< 2%) because the solid curve in Fig. 2-23c is very well approximated by the dashed polygon.

2.3.4 Vibration of Machine Foundation

The same machine foundation addressed in Section 2.2.4 for vertical motion (Fig. 2-11) is examined for rocking motion (Fig. 2-24). The dynamic loading consists of a moment arising from a three-cylinder compressor with cranks at 120°. The coupling of the rocking degree of freedom with the horizontal one through the foundation block is also discussed.

Dynamic load of three-cylinder compressor with cranks at 120°. Figure 2-25 shows the horizontal crank axis with three cylinders placed vertically. The three cranks with counter masses are arranged at 120°. The resulting unbalanced dynamic load is determined by superposition of the contributions of the three cranks. The unbalanced vertical inertial force of the first cylinder P_1 with a crank angle ωt specified in Eq. 2-111 (m_{rec} = total reciprocating mass, r = crank radius, ℓ = length of connecting rod) equals

$$P_1 = m_{rec} r \omega^2 \cos \omega t + m_{rec} \frac{r^2}{\ell} \omega^2 \cos 2\omega t \tag{2-188a}$$

that of the second cylinder P_2 with a crank angle $\omega t + 2\pi/3$

$$P_2 = m_{rec} r \omega^2 \cos\left(\omega t + \frac{2\pi}{3}\right) + m_{rec} \frac{r^2}{\ell} \omega^2 \cos 2\left(\omega t + \frac{2\pi}{3}\right) \tag{2-188b}$$

and that of the third cylinder P_3 with a crank angle $\omega t + 4\pi/3$

$$P_3 = m_{rec} r \omega^2 \cos\left(\omega t + \frac{4\pi}{3}\right) + m_{rec} \frac{r^2}{\ell} \omega^2 \cos 2\left(\omega t + \frac{4\pi}{3}\right) \tag{2-188c}$$

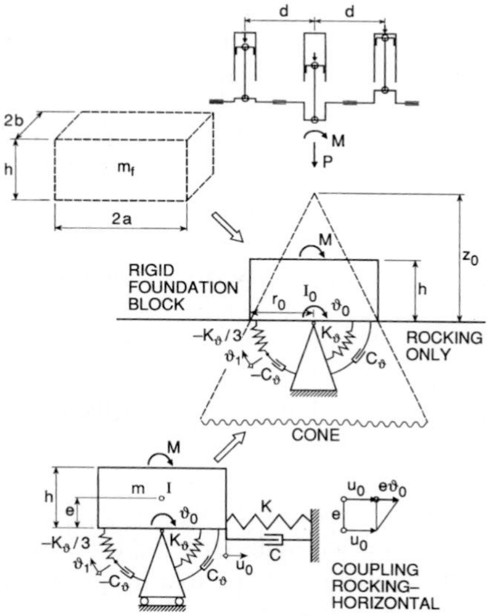

Fig. 2-24 Foundation of three-cylinder compressor (not to scale) with dynamic models.

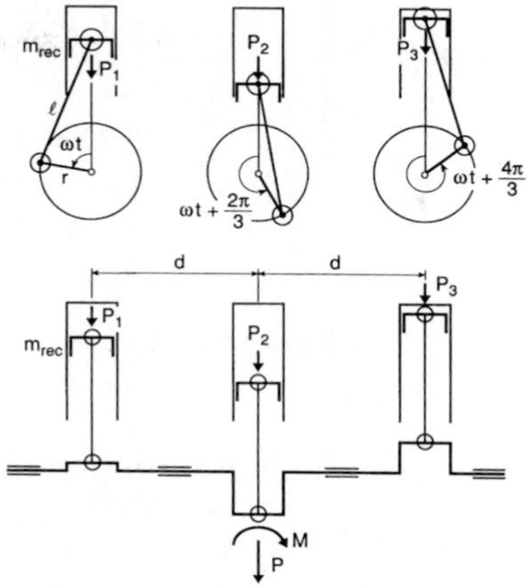

Fig. 2-25 Three-cylinder compressor with cranks at 120°.

Using the well-known expressions of a cos-function of an argument consisting of a sum, P_2 and P_3 can be expressed as a function of cos ωt and sin ωt (primary component) and of cos $2\omega t$ and sin $2\omega t$ (secondary component). P_1, P_2 and P_3 lead to the following resultants

$$P = P_1 + P_2 + P_3 \tag{2-189a}$$

$$M = dP_1 - dP_3 \tag{2-189b}$$

with d = distance between the center line of the cylinders. Substituting leads to

$$P \quad = 0 \tag{2-190a}$$

$$\begin{aligned} M \quad &= m_{\text{rec}} r\omega^2 d\left(\frac{3}{2}\cos\omega t - \frac{\sqrt{3}}{2}\sin\omega t\right) + m_{\text{rec}}\frac{r^2}{\ell}\omega^2 d\left(\frac{3}{2}\cos 2\omega t + \frac{\sqrt{3}}{2}\sin 2\omega t\right) \\ &= \sqrt{3}\, m_{\text{rec}} r\omega^2 d\cos\left(\omega t + \frac{\pi}{6}\right) + \sqrt{3}\, m_{\text{rec}}\frac{r^2}{\ell}\,\omega^2 d\cos 2\left(\omega t - \frac{\pi}{12}\right) \end{aligned} \tag{2-190b}$$

The three-cylinder machine with cranks at 120° is completely balanced in the vertical direction ($P = 0$). A moment M consisting of primary and secondary components with different phase angles arises (Eq. 2-190b).

As discussed in Section 2.2.4 the compressor and the driving motor with a total mass = 2750 kg operate at a frequency 9 Hz ($\omega = 2\pi \cdot 9 = 56.55\text{s}^{-1}$). With $r = 0.1$ m, $\ell = 0.3$ m, $m_{\text{rec}} = 10$ kg and $d = 0.5$ m, Eq. 2-190b yields

$$M(t) = 2.769\cos(56.55t + 0.524) + 0.923\cos(113.1t - 0.524)\ \ kN \tag{2-191}$$

The primary and secondary components and the total moment with a period 1/9 = 0.111 s are plotted in Fig. 2-26.

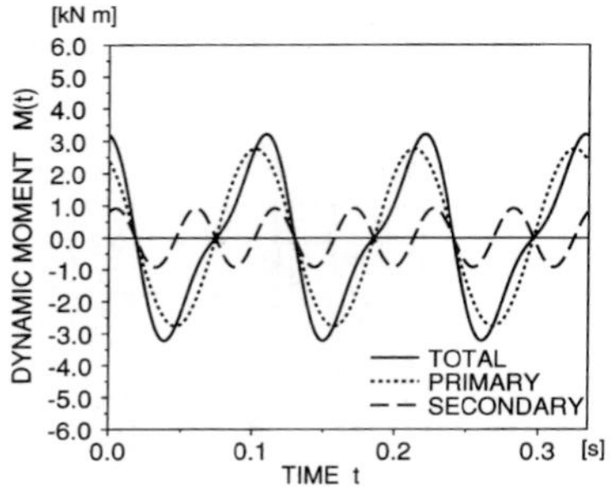

Fig. 2-26 Dynamic moment from three-cylinder compressor.

Design criterion. As before the peak-to-peak effective displacement amplitude is selected as 0.05 mm, with a less stringent criterion being 0.1 mm.

Rigid foundation block. The rigid foundation block (Fig. 2-24) made of concrete (density 2500 kg/m^3) has the shape of a prism with length $2a$ = 2 m, width $2b$ = 1.5 m, and height h = 1 m. Its mass m_f equals 7500 kg. Adding that of the machine to this value results in the mass $m = 1.025 \cdot 10^4$ kg. The mass moment of inertia of the foundation block with respect to the center of gravity $I_f = m_f/12[(2a)^2 + h^2] = 3.125 \cdot 10^3$ kg m^2. Assuming an eccentricity of the machine equals $h/2$, its contribution will be $0.5^2 \cdot 2.75 \cdot 10^3$, leading to a total value $I = 3.813 \cdot 10^3$ kg m^2. The mass moment of inertia with respect to the center of the foundation block–soil interface equals $I_0 = I + (h/2)^2\, m = 6.375 \cdot 10^3$ kg m^2 (Fig. 2-24).

Soil. The homogeneous halfspace disregarding material damping with Poisson's ratio $\nu = 1/3$ and mass density $\rho = 1700$ kg/m^3 has a shear modulus $G = 38.25 \cdot 10^6$ N/m^2, resulting in a shear-wave velocity $c_s = 150$ m/s and in a dilatational-wave velocity $c_p = 300$ m/s (Table 2-1).

The equivalent radius for the rocking motion follows from equating the moments of inertia $r_0 = \sqrt[4]{2b\,(2a)^3/(3\pi)} = 1.062$ m. From Table 2-3B the aspect ratio of the rocking cone equals $z_0/r_0 = 2.356$, leading to an apex height $z_0 = 2.503$ m. The coefficients of the rocking spring and dashpot follow from Eq. 2-146 as $K_\vartheta = 3\rho c_p^2(\pi r_0^4/4)/z_0 = 1.834 \cdot 10^8$ Nm and $C_\vartheta = \rho c_p \pi r_0^4/4 = 5.101 \cdot 10^5$ Nms.

For the coupled system the horizontal motion must also be addressed (Fig. 2-24). The equivalent radius follows as $r_0 = \sqrt{4ab/\pi} = 0.977$ m. From Table 2-3B $z_0/r_0 = 0.654$, leading to $z_0 = 0.639$ m. The coefficients of the horizontal spring and dashpot follow from Eq. 2-63 as $K = \rho c_s^2 \pi r_0^2/z_0 = 1.796 \cdot 10^8$ N/m and $C = \rho c_s \pi r_0^2 = 7.650 \cdot 10^5$ Ns/m.

Uncoupled dynamic system with rocking motion only. Neglecting the coupling between the rocking and horizontal motions, the dynamic system shown in the center of Fig. 2-24 applies. The only motion of the foundation block taken into consideration consists of rocking ϑ_0 around the center of the foundation block–soil interface with a mass moment of inertia I_0. To the underside of this rigid foundation block, the discrete-element model of the rocking cone shown in Fig. 2-17a is added. Besides the rotation ϑ_0 an additional internal degree of freedom with the rotation ϑ_1 is introduced. As there is only one mass moment of inertia, the system has one dynamic degree of freedom and thus one natural frequency. The coefficients of the springs and dashpots are specified in Fig. 2-24. The loading consists of a moment M acting on the block.

The dynamic system can also be interpreted for harmonic excitation as a one-degree-of-freedom system with a mass moment of inertia I_0 and a spring with the frequency-dependent coefficient $K_\vartheta k_\vartheta(b_0)$ and in parallel a dashpot with the frequency-dependent coefficient $(z_0/c_p)K_\vartheta c_\vartheta(b_0)$ (Fig. 2-27). The rocking natural frequency ω_r of the rigid block on flexible soil (undamped system) follows as

$$\omega_r = \sqrt{\frac{K_\vartheta k_\vartheta(b_0)}{I_0}} \left(= 151.26 \text{ s}^{-1}\right) \tag{2-192}$$

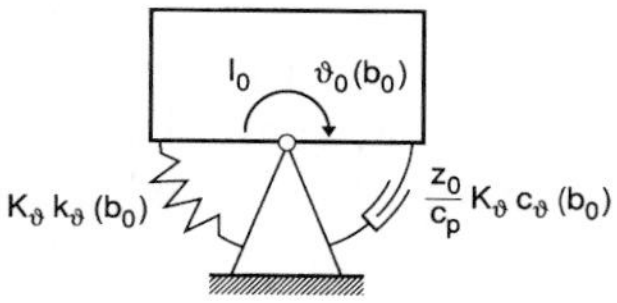

Fig. 2-27 Interpretation of uncoupled dynamic system for harmonic motion as a one-degree-of-freedom system with spring and dashpot in parallel with frequency-dependent coefficients.

The spring coefficient $K_\vartheta k_\vartheta(b_0)$ (Eq. 2-168) with $k_\vartheta(b_0) = 1 - (1/3)b_0^2/(1 + b_0^2)$ (Eq. 2-169a) depends on $b_0 = \omega z_0/c_p$ evaluated at $\omega = \omega_r$. Equation 2-192 is solved iteratively. Starting with $\omega_r = 0$ and thus $k_\vartheta = 1$, this first iteration leads to $\omega_r = 169.61 \text{ s}^{-1}$, which is used to calculate b_0 for the next iteration. Convergence is reached after five iterations, leading to $\omega_r = 151.26 \text{ s}^{-1}$ which corresponds to 24.07 Hz. For rocking motion the foundation is said to be *high-tuned*, as its natural frequency is larger than the operating frequency of the machine. The damping ratio ζ_r at ω_r equals the ratio of the imaginary part of the dynamic-stiffness coefficient (Eq. 2-168) to twice its real part both evaluated at ω_r

$$\zeta_r = \frac{b_0\, c_\vartheta(b_0)}{2\, k_\vartheta(b_0)} = 0.163 \tag{2-193}$$

with $c_\vartheta(b_0) = (1/3)b_0^2/(1 + b_0^2)$ (Eq. 2-169b) and $b_0 = \omega_0 z_0/c_p$. The corresponding value at the operating frequency of 9 Hz is determined as ζ_r (9 Hz) = 0.015. The free vibration in the rocking motion is damped significantly less (0.163) than that in the vertical direction (0.504, Eq. 2-115), whereby the natural frequencies are similar (24.07 Hz versus 23.54 Hz). This is a direct consequence of the different amounts of radiation damping, which result from geometric spreading of waves generated at the foundation block–soil interface [G3]. For *vertical motion* all points of this interface *move in phase* and the waves will *reach long distances* away from the foundation, resulting in a large loss of wave energy and thus *large radiation damping*. In contrast, for *rocking motion*, two points symmetrically located on opposite sides of the interface will emit waves that are *180° out of phase* that tend to *cancel each other at large distances*, leading to *small radiation damping*.

Uncoupled stiffness formulation in time domain. With the nomenclature specified in the center of Fig. 2-24 the two rotational equations of motion are written as

$$I_0 \ddot{\vartheta}_0 + C_\vartheta \dot{\vartheta}_0 + K_\vartheta \vartheta_0 - \frac{K_\vartheta}{3}(\vartheta_0 - \vartheta_1) = M \tag{2-194a}$$

$$-C_\vartheta \dot{\vartheta}_1 - \frac{K_\vartheta}{3}(\vartheta_1 - \vartheta_0) = 0 \tag{2-194b}$$

For use in an explicit algorithm they are reformulated as

$$\ddot{\vartheta}_0 = \frac{M - \frac{2}{3} K_\vartheta \vartheta_0 \; - \frac{K_\vartheta}{3} \vartheta_1 - C_\vartheta \dot{\vartheta}_0}{I_0} \tag{2-195a}$$

$$\dot{\vartheta}_1 = \frac{K_\vartheta (\vartheta_0 - \vartheta_1)}{3C_\vartheta} \tag{2-195b}$$

For ϑ_0 the predictor–corrector relations of Eqs. 2-118 and 2-120 apply, while for ϑ_1, no prediction and correction are formulated. Starting from the known motion at time $(n - 1)\,\Delta t$—that is, ϑ_{0n-1}, $\dot{\vartheta}_{0n-1}$, $\ddot{\vartheta}_{0n-1}$, ϑ_{1n-1}, $\dot{\vartheta}_{1n-1}$—the final rotations ϑ_{0n}, ϑ_{1n} and the predicted rotational velocity $\tilde{\dot{\vartheta}}_{0n}$ at time $n\Delta t$ are calculated as

$$\vartheta_{0n} = \vartheta_{0n-1} + \Delta t\, \dot{\vartheta}_{0n-1} + \frac{\Delta t^2}{2} \ddot{\vartheta}_{0n-1} \tag{2-196a}$$

$$\vartheta_{1n} = \vartheta_{1n-1} + \Delta t\, \dot{\vartheta}_{1n-1} \tag{2-196b}$$

$$\tilde{\dot{\vartheta}}_{0n} = \dot{\vartheta}_{0n-1} + \frac{\Delta t}{2} \ddot{\vartheta}_{0n-1} \tag{2-196c}$$

A tilde (~) denotes a predicted value. Based on these values, the rotational acceleration $\ddot{\vartheta}_{0n}$ and the rotational velocity $\dot{\vartheta}_{1n}$ follow from the equilibrium equations Eq. 2-195 formulated at time $n\Delta t$. The rotational velocity is corrected as

$$\dot{\vartheta}_{0n} = \tilde{\dot{\vartheta}}_{0n} + \frac{\Delta t}{2} \ddot{\vartheta}_{0n} \tag{2-197}$$

which concludes the calculations for the time step. The algorithm is self-starting.

For stability of the explicit algorithm the time step Δt must be smaller than the (smallest) natural period divided by π. This is expressed as

$$\Delta t < \frac{2}{\omega_r} = 0.0132\ s \tag{2-198}$$

To achieve an adequate representation of the dynamic load, eight time stations in the period corresponding to the secondary component of 18 Hz are selected. This results in $\Delta t = (1/18)/8 \simeq 7 \cdot 10^{-3}$ s which is more stringent than the Δt calculated with the stability criterion (Eq. 2-198). The analysis is performed with $\Delta t = 7 \cdot 10^{-3}$ s, starting with vanishing initial conditions.

The total vertical displacement at the edge of the foundation block caused by rocking $a\vartheta_0$ is plotted in Fig. 2-28a. The transient phase is short, but not to the same extent as for the vertical motion (Fig. 2-15a), as the damping ratio for rocking motion is not as large. Also shown are the contributions of the primary and secondary components, which are used in the next subsection for a comparison with the results for harmonic loading (Fig. 2-28b).

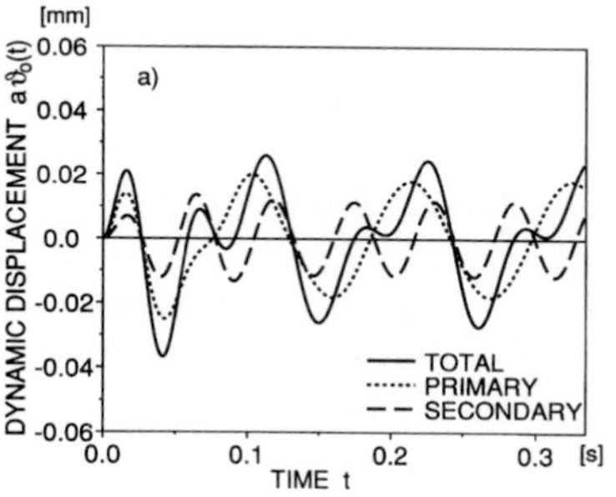

a. Time domain.

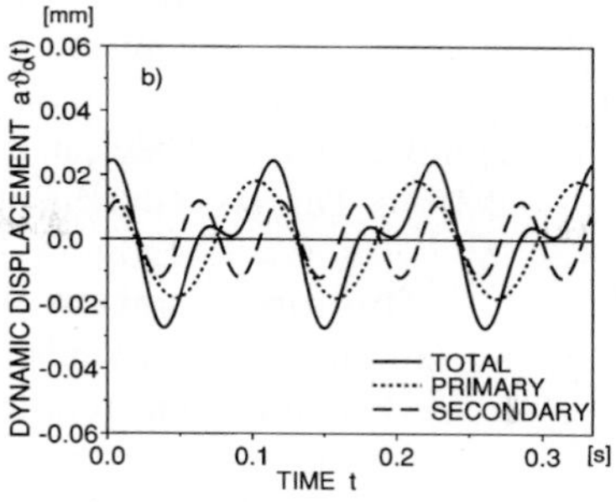

b. Frequency domain.

Fig. 2-28 Vertical displacement at edge of foundation block neglecting coupling of rocking motion with horizontal motion.

Uncoupled stiffness formulation for harmonic loading. For harmonic loading of frequency ω_j the rocking equation of motion equals

$$\left[-\omega_j^2 I_0 + S_\vartheta(\omega_j)\right] \vartheta_0(\omega_j) = M(\omega_j) \tag{2-199}$$

with the dynamic-stiffness coefficient $S_\vartheta(\vartheta_j)$ specified with $\mu_\vartheta = 0$ in Eqs. 2-168 and 2-169. Solving for the complex amplitude $\vartheta_0(\omega_j)$ yields

$$\vartheta_0(\omega_j) = \frac{1 + ib_0}{1 + ib_0 + \left(\frac{1}{3} + \frac{1}{\gamma_\mu}\right)(ib_0)^2 + \frac{1}{\gamma_\mu}(ib_0)^3} \frac{M(\omega_j)}{K_\vartheta} \tag{2-200}$$

with

$$\gamma_\mu = \frac{z_0^2}{c_p^2} \frac{K_\vartheta}{I_0} \tag{2-201}$$

By considering I_0 to represent the mass moment of inertia ΔM_ϑ encountered in the nearly incompressible rocking cone, Eq. 2-200 is the flexibility formulation for harmonic motion specified by Eqs. 2-180 and 2-183.

When analyzing the three-cylinder compressor, the primary and secondary components of the moment loading are processed separately. Various methods differing in details exist. See the description of alternative formulations for harmonic motion at the end of Section 2.2.4. When the procedure used for the vertical motion where the phase angle of the vertical load vanishes (Eq. 2-112) is followed (Section 2.2.4), the phase of the moment (0.524 for the primary component and –0.524 for the secondary component, Eq. 2-191) is added to the phase angle of the response (Eq. 2-124, but for ϑ_{0j}). Alternatively, the cos and sin terms of each component of the load with vanishing phase angles can be processed separately. The response to a sin-term will be a sin-term (analogous to Eq. 2-124). As a third method, after determining the magnitude and the phase angle of the moment load from Eq. 2-191, a complex amplitude can be constructed as in Eq. 2-126.

The vertical displacements at the edge of the foundation block $a\vartheta_0$ for the primary and secondary components and the superposition of the two are plotted in Fig. 2-28b.

As an alternative, the two equations of motion (Eq. 2-194) can be formulated for harmonic excitation. They can then be solved for the amplitudes $\vartheta_0(\omega)$ and $\vartheta_1(\omega)$ as a function of $M(\omega)$. This procedure is illustrated for the coupled motion further on.

Dynamic system with coupling of rocking and horizontal motions. As the center of the foundation block–soil interface is at a distance $e = 0.5h$ measured in the vertical direction from the center of gravity of the foundation block, coupling of the rocking motion with the horizontal motion occurs. The dynamic system is shown at the bottom of Fig. 2-24. It consists of the discrete-element model for rocking as before and in addition of that for the horizontal motion with the displacement u_0, defined at the foundation block–soil interface, as shown in Fig. 2-5a.

Addressing the motions ϑ_0 and u_0, the two equations of motion formulated at the center of gravity of the foundation block for harmonic loading are

$$m\ddot{u}_0(\omega) + em\ddot{\vartheta}_0(\omega) + S(\omega)\,u_0(\omega) = 0 \tag{2-202a}$$

$$I\ddot{\vartheta}_0(\omega) - eS(\omega)\,u_0(\omega) + S_\vartheta(\omega)\,\vartheta_0(\omega) = M(\omega) \tag{2-202b}$$

The horizontal displacement amplitude at the center of gravity of the foundation block equals $u_0(\omega) + e\vartheta_0(\omega)$. To derive a symmetric system, Eq. 2-202b is replaced by the sum of Eq. 2-202b and Eq. 2-202a multiplied by e, which corresponds to formulating the rotational equation of motion with respect to the center of the foundation block–soil interface, where the degrees of freedom are defined.

$$m\ddot{u}_0(\omega) + em\ddot{\vartheta}_0(\omega) + S(\omega)\,u_0(\omega) = 0 \tag{2-203a}$$

$$em\ddot{u}_0(\omega) + (I + e^2m)\,\ddot{\vartheta}_0(\omega) + S_\vartheta(\omega)\,\vartheta_0(\omega) = M(\omega) \tag{2-203b}$$

Substituting the horizontal (Eq. 2-67) and rocking dynamic-stiffness coefficients (Eq. 2-168) yields

$$S(\omega)u_0(\omega) = Ku_0(\omega) + C\dot{u}_0(\omega) \tag{2-204a}$$

$$S_\vartheta(\omega)\,\vartheta_0(\omega) = K_\vartheta k_\vartheta(b_0)\vartheta_0(\omega) + \frac{z_0}{c_p}K_\vartheta c_\vartheta(b_0)\,\dot{\vartheta}_0(\omega) \tag{2-204b}$$

where $k_\vartheta(b_0)$ and $c_\vartheta(b_0)$ are specified in Eq. 2-169. The right-hand side of Eq. 2-204b is interpreted physically in Fig. 2-27, consisting of a spring and a dashpot with frequency-dependent coefficients. As can be seen from Eq. 2-203 using Eq. 2-204, the coefficients of the acceleration amplitudes $\ddot{u}_0(\omega)$, $\ddot{\vartheta}_0(\omega)$ determine the frequency-independent mass matrix $[M]$, the coefficients of the displacement amplitudes $u_0(\omega)$, $\vartheta_0(\omega)$ the frequency-dependent stiffness matrix $[K]$ and the coefficients of the velocity amplitudes $\dot{u}_0(\omega)$, $\dot{\vartheta}_0(\omega)$ the frequency-dependent damping matrix $[C]$.

$$[M] = \begin{bmatrix} m & em \\ em & I+e^2m \end{bmatrix} \tag{2-205a}$$

$$[K(\omega)] = \begin{bmatrix} K & \\ & K_\vartheta k_\vartheta(b_0) \end{bmatrix} \tag{2-205b}$$

$$[C(\omega)] = \begin{bmatrix} C & \\ & \frac{z_0}{c_p} K_\vartheta c_\vartheta(b_0) \end{bmatrix} \tag{2-205c}$$

Equation 2-203 is then written as

$$[M]\begin{Bmatrix} \ddot{u}_0(\omega) \\ \ddot{\vartheta}_0(\omega) \end{Bmatrix} + [C(\omega)]\begin{Bmatrix} \dot{u}_0(\omega) \\ \dot{\vartheta}_0(\omega) \end{Bmatrix} + [K(\omega)]\begin{Bmatrix} u_0(\omega) \\ \vartheta_0(\omega) \end{Bmatrix} = \begin{Bmatrix} 0 \\ M(\omega) \end{Bmatrix} \tag{2-206}$$

This foundation block–soil system does not have classical modes; that is, the transformation to modal coordinates does not decouple the damping terms, as is demonstrated below. As an approximation, natural frequencies ω_i and the corresponding mode shapes $\{\varphi_i\}$ (i = 1, 2) can be determined by solving the eigenvalue problem

$$\left([K(\omega_i)] - \omega_i^2[M]\right)\{\phi_i\} = 0 \tag{2-207}$$

Setting the determinant $|[K(\omega_i)] - \omega_i^2\ [M]| = 0$ results in

$$\omega_{1,2}^2 = \frac{1}{2}\left(1 + \frac{e^2 m}{I}\right)\left(\omega_h^2 + \omega_r^2\right)\left[1 \mp \sqrt{1 - \frac{4}{1+\frac{e^2 m}{I}}\frac{\omega_h^2\,\omega_r^2}{(\omega_h^2+\omega_r^2)^2}}\right] \tag{2-208}$$

$$\{\phi_1\} = u_{01}\begin{Bmatrix} 1 \\ \frac{1}{e}\left(\frac{\omega_h^2}{\omega_1^2} - 1\right) \end{Bmatrix} \quad \text{(2-209a)} \qquad \{\phi_2\} = u_{02}\begin{Bmatrix} 1 \\ \frac{1}{e}\left(\frac{\omega_h^2}{\omega_2^2} - 1\right) \end{Bmatrix} \quad \text{(2-209b)}$$

with the uncoupled natural frequencies for the horizontal

$$\omega_h = \sqrt{\frac{K}{m}} \tag{2-210a}$$

and rocking motions

$$\omega_r = \sqrt{\frac{K_\vartheta k_\vartheta(\omega_{1,2})}{I + e^2 m}} = \sqrt{\frac{K_\vartheta k_\vartheta(\omega_{1,2})}{I_0}} \tag{2-210b}$$

Equation 2-210b is analogous to Eq. 2-192, but dependent on a different frequency (either ω_1 or ω_2). The factors u_{01} and u_{02} are the undetermined amplitudes of the first and second mode shapes.

From Eq. 2-210a, $\omega_h = 132.37\ s^{-1}$, which corresponds to 21.07 Hz. Iterating on ω_r as specified in Eq. 2-210b using Eq. 2-208, convergence is reached after four iterations for the fundamental frequency $\omega_1 = 111.24$ s^{-1} leading to 17.70 Hz. Analogously, $\omega_2 = 230.36$ s^{-1} (36.68 Hz) follows.

It is of interest to calculate the damping ratio of the uncoupled horizontal motion ζ_h at ω_h analogous to Eq. 2-115

$$\zeta_h = \frac{\omega_h C}{2K} = 0.282 \tag{2-211}$$

As an approximation, the fundamental frequency of the coupled system can be determined using the uncoupled natural frequencies only as

$$\frac{1}{\omega_1^2} = \frac{1}{\omega_h^2} + \frac{1}{\omega_r^2} \tag{2-212}$$

with ω_r specified in Eq. 2-210b. After iterating, $\omega_1 = 101.31$ s^{-1}, which corresponds to 16.12 Hz. This value is quite close to that of the coupled system (17.70 Hz). To establish the condition under which this approximation is valid, ω_1^2 determined from Eq. 2-212 is set equal to the rigorous expression, Eq. 2-208. This leads to the condition

$$\frac{\omega_1^2}{\left(1 + \frac{e^2 m}{I}\right)\left(\omega_h^2 + \omega_r^2\right)} = 0 \tag{2-213}$$

which is approximately satisfied for

$$\omega_1^2 \ll \left(1 + \frac{e^2 m}{I}\right)\left(\omega_h^2 + \omega_r^2\right) \tag{2-214}$$

For the machine foundation, this inequality is formulated as $1.24 \cdot 10^5 < 7.00 \cdot 10^5$.

The mode shapes of the coupled system (Eq. 2-209) are

$$\{\phi_1\} = u_{01}\begin{Bmatrix} 1 \\ 0.832 \end{Bmatrix} \quad \text{(2-215a)} \qquad \{\phi_2\} = u_{02}\begin{Bmatrix} 1 \\ -1.340 \end{Bmatrix} \quad \text{(2-215b)}$$

$\{\varphi_2\}$ involves more rocking motion than does $\{\varphi_1\}$.

The off-diagonal term of the transformed damping matrix in modal coordinates equals substituting Eqs. 2-205c and 2-209

$$\{\phi_1\}^T [C(\omega)] \{\phi_2\} = C + \frac{z_0}{e^2 c_p}\left(\frac{\omega_h^2}{\omega_1^2} - 1\right)\left(\frac{\omega_h^2}{\omega_2^2} - 1\right) K_\vartheta c_\vartheta(b_0) \tag{2-216}$$

which does indeed not vanish.

Finally, a damping ratio for the fundamental mode is determined on the basis of an energy consideration. For hysteretic damping, the dissipated energy is proportional to the product of the damping ratio and the strain energy. Adding the dissipated energies of the uncoupled horizontal and rocking motions, and setting this value equal to that of the equivalent one-degree-of-freedom system using the same total strain energy lead to

$$\zeta_1 = \frac{\zeta_h \frac{Ku_0^2}{2} + \zeta_r \frac{K_\vartheta k_\vartheta(b_0)\vartheta_0^2}{2}}{\frac{Ku_0^2}{2} + \frac{K_\vartheta k_\vartheta(b_0)\vartheta_0^2}{2}} \tag{2-217}$$

Applying a horizontal load acting at the center of gravity of the foundation block, formulating equilibrium and neglecting the damping terms leads to

$$\vartheta_0 = \frac{e\,K}{K_\vartheta\,k_\vartheta(b_0)}\,u_0 \tag{2-218}$$

Substituting Eq. 2-218 in Eq. 2-217 and introducing the approximation discussed in connection with Eq. 2-212 yields

$$\zeta_1 = \frac{\omega_1^2}{\omega_h^2}\zeta_h + \frac{\omega_1^2}{\omega_r^2}\zeta_r \tag{2-219}$$

To determine the damping ratio for the fundamental mode of the coupled system, the contributions of the damping ratios of the uncoupled horizontal and rocking motions with corresponding weights are added. Evaluating ζ_1 for ω_1 (with ζ_h also calculated for ω_1) leads to $\zeta_1 = 0.210$, whereby the horizontal motion contributes 0.167 and the rocking motion 0.043.

Coupled stiffness formulation in time domain. With the nomenclature specified at the bottom of Fig. 2-24, the two rotational and one translational equations of motion are written as

$$I\ddot{\vartheta}_0 + C_\vartheta\,\dot{\vartheta}_0 + K_\vartheta\vartheta_0 - \frac{K_\vartheta}{3}\,(\vartheta_0-\vartheta_1) - C\,e\,\dot{u}_0 - K\,e\,u_0 = M \tag{2-220a}$$

$$-\,C_\vartheta\,\dot{\vartheta}_1 - \frac{K_\vartheta}{3}\,(\vartheta_1-\vartheta_0) = 0 \tag{2-220b}$$

$$m\left(\ddot{u}_0 + e\,\ddot{\vartheta}_0\right) + C\,\dot{u}_0 + Ku_0 = 0 \tag{2-220c}$$

For use in an explicit algorithm they are reformulated as

$$\dot{\vartheta}_1 = \frac{K_\vartheta}{3C_\vartheta}\left(\vartheta_0-\vartheta_1\right) \tag{2-221a}$$

$$\ddot{\vartheta}_0 = \frac{M - \frac{2}{3}K_\vartheta\vartheta_0 - C_\vartheta\,\dot{\vartheta}_0 - \frac{K_\vartheta}{3}\vartheta_1 + Ke\,u_0 + C\,e\,\dot{u}_0}{I} \tag{2-221b}$$

$$\ddot{u}_0 = \frac{-Ku_0 - C\dot{u}_0}{m} - e\,\ddot{\vartheta}_0 \tag{2-221c}$$

For ϑ_1 no prediction and correction are formulated, while for ϑ_0, u_0 the predictor–corrector relations of Eqs. 2-118 and 2-120 apply. Starting from the known motion at time $(n-1)\Delta t$—that is, ϑ_{0n-1}, $\dot{\vartheta}_{0n-1}$, $\ddot{\vartheta}_{0n-1}$, ϑ_{1n-1}, $\dot{\vartheta}_{1n-1}$, u_{0n-1}, $\dot{u}_{0n-1}$, $\ddot{u}_{0n-1}$—the final rotations and displacement ϑ_{0n}, ϑ_{1n}, u_{0n} and the predicted velocities $\tilde{\dot{\vartheta}}_{0n}$, $\tilde{\dot{u}}_{0n}$ at time $n\Delta t$ are calculated as

$$\vartheta_{0n} = \vartheta_{0n-1} + \Delta t\, \dot{\vartheta}_{0n-1} + \frac{\Delta t^2}{2}\, \ddot{\vartheta}_{0n-1} \tag{2-222a}$$

$$\vartheta_{1n} = \vartheta_{1n-1} + \Delta t\, \dot{\vartheta}_{1n-1} \tag{2-222b}$$

$$u_{0n} = u_{0n-1} + \Delta t\, \dot{u}_{0n-1} + \frac{\Delta t^2}{2}\, \ddot{u}_{0n-1} \tag{2-222c}$$

$$\tilde{\dot{\vartheta}}_{0n} = \dot{\vartheta}_{0n-1} + \frac{\Delta t}{2}\, \ddot{\vartheta}_{0n-1} \tag{2-223a}$$

$$\tilde{\dot{u}}_{0n} = \dot{u}_{0n-1} + \frac{\Delta t}{2}\, \ddot{u}_{0n-1} \tag{2-223b}$$

A tilde (~) denotes a predicted value. Based on these values, the rotational velocity $\dot{\vartheta}_{1n}$ and the accelerations $\ddot{\vartheta}_{0n}$, $\ddot{u}_{0n}$ follow from the equilibrium equations Eq. 2-221 formulated at time $n\Delta t$. The two predicted velocities are corrected as

$$\dot{\vartheta}_{0n} = \tilde{\dot{\vartheta}}_{0n} + \frac{\Delta t}{2}\, \ddot{\vartheta}_{0n} \tag{2-224a}$$

$$\dot{u}_{0n} = \tilde{\dot{u}}_{0n} + \frac{\Delta t}{2}\, \ddot{u}_{0n} \tag{2-224b}$$

This concludes the calculations for the time step. The algorithm is self-starting.

As for the dynamic system neglecting coupling, eight time stations in the period corresponding to the secondary component of the load (18 Hz) are selected. This leads to a time step $\Delta t = (1/18)/8 \simeq 7 \cdot 10^{-3}$ s, which is used for the analysis. The criterion for stability of the explicit algorithm (Eq. 2-198) with the highest natural frequency of the system leads to

$$\Delta t < \frac{2}{\omega_2} = 0.0087 \text{ s}^{-1} \tag{2-225}$$

a value which is slightly larger and thus does not govern.

Starting with vanishing initial conditions, the total vertical displacement at the edge caused by rocking $a\vartheta_0$ is calculated and plotted in Fig. 2-29a. The result of the analysis neglecting coupling of the horizontal and rocking motions of Fig. 2-28a is also shown as a dotted line. Including coupling increases the vertical response, whereby the corresponding horizontal displacement u_0 remains small (Fig. 2-29b).

As another loading case, a dynamic moment with the same function of time as the load of the weaver's loom (Fig. 2-14) is examined. The moment is equal to $P(t)$ defined in Eq. 2-113 and in Table 2-6 multiplied by an eccentricity equal to 1m. As can be seen from the vertical edge displacement plotted in Fig. 2-30, neglecting coupling again underestimates the actual response.

The two loading cases do not satisfy the more stringent design criterion of a peak-to-peak amplitude equal 0.05 mm.

Coupled stiffness formulation for harmonic loading. For harmonic loading of frequency ω_j the three equations of motion, Eq. 2-220, are formulated as

$$\left(-\omega_j^2 I + i\,\omega_j\, C_\vartheta + \frac{2}{3} K_\vartheta\right) \vartheta_0(\omega_j) + \frac{K_\vartheta}{3}\, \vartheta_1(\omega_j) - \left(i\,\omega_j Ce + Ke\right)\, u_0(\omega_j) = M(\omega_j) \tag{2-226a}$$

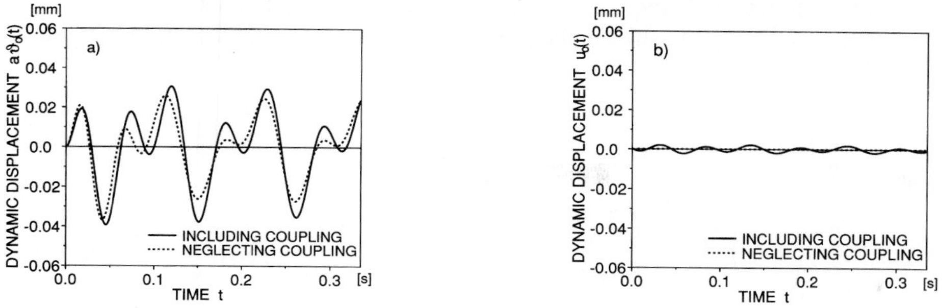

a. Vertical displacement at edge of foundation block.

b. Horizontal displacement.

Fig. 2-29 Response analyzed in time domain including coupling of rocking motion with horizontal motion.

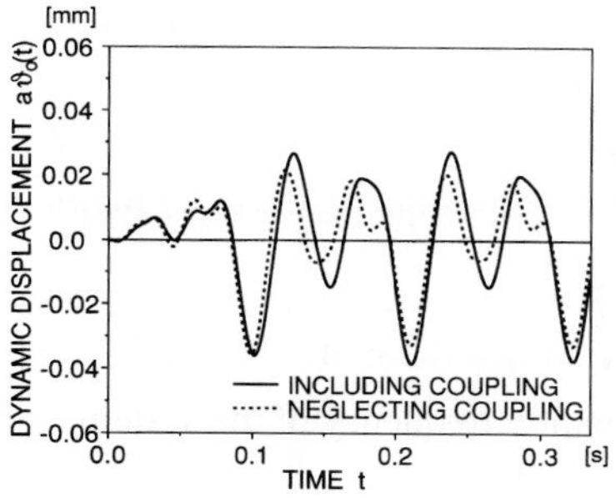

Fig. 2-30 Vertical displacement at edge of foundation block including coupling of rocking motion with horizontal motion for dynamic moment of weaver's loom analyzed in time domain.

$$\frac{K_\vartheta}{3}\,\vartheta_0(\omega_j) + \left(-i\,\omega_j\,C_\vartheta - \frac{K_\vartheta}{3}\right)\vartheta_1(\omega_j) = 0 \tag{2-226b}$$

$$-\omega_j^2\,me\vartheta_0(\omega_j) + \left(-\omega_j^2 m + i\,\omega_j\,C + K\right)\,u_0(\omega_j) = 0 \tag{2-226c}$$

They can be solved for the amplitudes $\vartheta_0(\omega_j)$ and $u_0(\omega_j)$ for a specified amplitude $M(\omega_j)$.

Alternatively, the interaction force-displacement and the interaction moment-rotation relationships of the cones involving the dynamic-stiffness coefficients $S(\omega_j)$ (Eqs. 2-69, 2-70) and $S_\vartheta(\omega_j)$ (Eqs. 2-168, 2-169) can be used when formulating the one rocking equation and the horizontal equation of motion.

$$\left[-\omega_j^2 I \;+\; S_\vartheta(\omega_j)\right]\vartheta_0(\omega_j) - S(\omega_j)\,e\,u_0(\omega_j) = M(\omega_j) \tag{2-227a}$$

$$-\;\omega_j^2\,me\vartheta_0(\omega_j) + \left[-\omega_j^2 m + S(\omega_j)\right]\,u_0(\omega_j) = 0 \tag{2-227b}$$

Again, $\vartheta_0(\omega_j)$ and $u_0(\omega_j)$ follow as a function of $M(\omega_j)$.

The analysis procedure is the same as described for the uncoupled case.

The vertical displacements at the edge of the foundation block $a\vartheta_0$ for the primary and secondary components of the three-cylinder compressor and the superposition of the two are plotted in Fig. 2-31. This result can be compared to the response of the analysis for harmonic loading (frequency-domain) neglecting coupling in Fig. 2-28b and to the result of the analysis performed in the time domain including coupling in Fig. 2-29a.

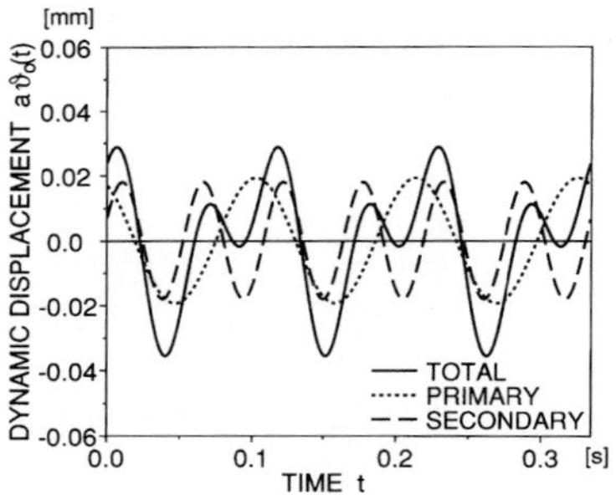

Fig. 2-31 Vertical displacement at edge of foundation block from harmonic loading (frequency-domain analysis) including coupling of rocking motion with horizontal motion.

Dynamic system of torsional motion. To be able to compare the natural frequencies and corresponding damping ratios for all degrees of freedom of the rigid foundation block on the soil halfspace, the torsional motion is also addressed.

The polar mass moment of inertia of the foundation block $I = (m_f/12)[(2a)^2 + (2b)^2] = 3.906 \cdot 10^3$ kg m^2. The equivalent radius follows from equating the polar moments of inertia $r_0 = \sqrt[4]{2a\, 2b\left[(2a)^2+(2b)^2\right]/(6\pi)} = 0.999\ m$. From Table 2-3A the aspect ratio of the torsional cone $z_0/r_0 = 0.884$, leading to an apex height $z_0 = 0.883$ m. The coefficients of the torsional spring and dashpots follow from Eq. 2-146 as $K_\vartheta = 3\rho c_s^2(\pi r_0^4/2)/z_0 = 2.031 \cdot 10^8$ Nm and $C_\vartheta = \rho c_s \pi r_0^4/2 = 3.985 \cdot 10^5$ Nms. The dynamic system with the torsional degree of freedom ϑ_0 consists of a polar mass moment of inertia I supported by the discrete-element model of Fig. 2-17a working in torsion with its own internal torsional degree of freedom ϑ_1.

The torsional natural frequency ω_t of the rigid block on flexible soil (undamped system) follows as

$$\omega_t = \sqrt{\frac{K_\vartheta k_\vartheta(b_0)}{I}} \quad \left(= 204.32\ s^{-1}\right) \tag{2-228}$$

As in the case of rocking motion (Eq. 2-192), the spring coefficient $K_\vartheta\, k_\vartheta\,(b_0)$ (Eq. 2-168) with $k_\vartheta\,(b_0) = 1 - (1/3)b_0^2/(1 + b_0^2)$ (Eq. 2-169a) depends on $b_0 = \omega z_0/c_s$ evaluated at $\omega = \omega_t$. Equation 2-228 is thus solved iteratively. Convergence is reached after five iterations, resulting in $\omega_t = 204.32$ s^{-1}, which corresponds to 32.52 Hz. Analogous to Eq. 2-193, the damping ratio ζ_t at ω_t is calculated as

$$\zeta_t = \frac{b_0 c_\vartheta(b_0)}{2\, k_\vartheta(b_0)} = 0.148 \tag{2-229}$$

with $c_\vartheta(b_0) = (1/3)b_0^2/(1 + b_0^2)$ specified in Eq. 2-169b and $b_0 = \omega_t\, z_0/c_s$. The free vibration in the torsional motion is damped less (0.148) than that in the uncoupled horizontal motion (0.282, Eq. 2-211). This difference is the same as already discussed in connection with the rocking and vertical motions (see after Eq. 2-193). For a vast range of parameters, in particular of the mass (or mass moment of inertia) of the foundation block, the translational motions are damped significantly more than the rotational ones. The damping ratio of the vertical motion, which is larger than for the horizontal one, can reach very high values. Especially the damping ratio of the rocking motion can be quite small for a slender foundation block ($h > 2a$, resulting in a large I_0 and thus a small ω_r).

2.4 MATERIAL DAMPING

2.4.1 Correspondence Principle Applied for Harmonic Motion

The cones addressed in Sections 2.2 and 2.3 and other continuum models of foundations are analyzed under the a priori assumption that the soil is a perfectly elastic medium which dissipates energy only by radiation of waves towards infinity. The second important source of energy dissipation—material damping—will also, in general, be present.

Material damping involves a frictional loss of energy. This so-called linear-hysteretic loss of energy is independent of frequency. The only change when considering material damping consists of modifying the elastic constants used in the elastic analysis. Material damping can be introduced into the solution for harmonic loading by using the so-called *correspondence principle*. The latter states that the damped solution incorporating an energy loss per cycle is obtained from the elastic one by multiplying the elastic constants by the complex factor $1 + 2i\zeta_g$

$$G \rightarrow G(1 + 2i\zeta_g) \tag{2-230a}$$

$$E_c \rightarrow E_c(1 + 2i\zeta_g) \tag{2-230b}$$

where ζ_g is the material damping ratio (subscript g for ground). It follows from Table 2-1, that Poisson's ratio ν is not modified. As an alternative description of the amount of material damping, the coefficient of friction $\tan\delta$ can be used, whereby δ denotes a sort of angle of mobilized internal friction for the soil. Since $\zeta_g = 0.5 \tan\delta$, both measures of material damping are alternative representations of the same physical entity; and it is a matter of choice which of them is preferable. The $\tan\delta$ representation is better from a soil-mechanics point of view, because the maximum possible upper limit of δ is the angle of response of sand slopes. Nevertheless, current practice seems to favor the ζ_g representation. The physical interpretation of ζ_g is rather artificial. If a rigid block were mounted on soil without radiation damping, the ratio of critical damping for translational vibration of the coupled block–soil system resulting solely from friction in the ground would be ζ_g. The more prevalent ζ_g representation of material damping is employed in this text except when the $\tan\delta$ representation facilitates physical insight.

As an example of applying the correspondence principle for harmonic excitation, the dynamic-stiffness coefficients of the cone models are addressed. Both the spring and damping coefficients are modified. The complex elastic constants affect the wave velocities as

$$c_s \rightarrow c_s\sqrt{1 + 2i\zeta_g} \simeq c_s(1 + i\zeta_g) \tag{2-231a}$$

$$c_p \rightarrow c_p\sqrt{1 + 2i\zeta_g} \simeq c_p(1 + i\zeta_g) \tag{2-231b}$$

and the dimensionless frequency as

$$a_0 \rightarrow a_0^* = \frac{a_0}{\sqrt{1 + 2i\zeta_g}} \simeq a_0(1 - i\zeta_g) \tag{2-232}$$

In these expressions, $\zeta_g^2 \ll 1$ is assumed. The dynamic-stiffness coefficients for the undamped case of the translational and rotational cones are specified in Eqs. 2-72, 2-73, and Eqs. 2-170, 2-171. For a damped system, a subscript ζ is added

$$S_\zeta(a_0) = K\left[k_\zeta(a_0) + ia_0 c_\zeta(a_0)\right] \tag{2-233}$$

Applying the correspondence principle to the expression of the undamped case results in

$$S(a_0^*) = S_\zeta(a_0) = K(1 + 2i\zeta_g)\left[k(a_0^*) + ia_0^* c(a_0^*)\right] \tag{2-234}$$

with a_0^* given in Eq. 2-232. The material-damping ratio thus affects the dynamic-stiffness coefficient in three ways. First, the static-stiffness coefficient used to nondimensionalize the expression is multiplied by $1 + 2i\zeta_g$. Second, a_0^* is substituted for a_0 and third, $k(a_0)$ and $c(a_0)$ are replaced using the same functions by their damped (complex) counterparts $k(a_0^*)$ and $c(a_0^*)$.

As for the cone models an analytical expression exists for the dynamic-stiffness coefficient (Eq. 2-72 with Eq. 2-73), incorporation of material damping for harmonic excitation is easy to perform. However, to gain physical insight, the dynamic-stiffness coefficient of the damped system is approximately expressed as a linear function of that of the undamped system at the same frequency with the material-damping ratio appearing in the coefficients as follows. Assuming that the complex damped values $k(a_0^*)$ and $c(a_0^*)$ can be set equal to their corresponding (real) undamped values $k(a_0)$ and $c(a_0)$ and using Eq. 2-232, Eq. 2-234 is written as

$$S_\zeta(a_0) = K(1 + 2i\zeta_g)\left[k(a_0) + ia_0(1 - i\zeta_g)c(a_0)\right] \tag{2-235}$$

Comparing with Eq. 2-233

$$k_\zeta(a_0) = k(a_0) - \zeta_g a_0\, c(a_0) \tag{2-236a}$$

$$c_\zeta(a_0) = c(a_0) + 2\frac{\zeta_g}{a_0} k(a_0) \tag{2-236b}$$

results. As expected, the *damping coefficient increases* with the increase being proportional to ζ_g. At the same time the *spring coefficient decreases*. Of course, Eq. 2-236 can also be used for practical calculations.

As a further approximation, the effect of material damping on the spring coefficient is suppressed. Neglecting the second term on the right-hand side of Eq. 2-236a leads to

$$S_\zeta(a_0) = K\left[k(a_0) + ia_0\left(c(a_0) + 2\frac{\zeta_g}{a_0} k(a_0)\right)\right] \tag{2-237}$$

or

$$S_\zeta(a_0) = Kk(a_0)\left[1 + 2i\zeta(a_0) + 2i\zeta_g\right] \tag{2-238}$$

with the radiation-damping ratio of the case without material damping

$$\zeta(a_0) = \frac{a_0\, c(a_0)}{2k(a_0)} \tag{2-239}$$

It follows from Eq. 2-238 that for this approximation the spring coefficient is not affected by material damping and in the damping coefficient, material damping does not modify radiation damping, the two damping ratios being additive. This form of the dynamic-stiffness coefficient will be used to derive simple expressions for the properties of an equivalent one-degree-of-freedom system modeling the dynamic interaction of a structure with the soil (Section 7.2).

In the remainder of this section, which is based on Meek and Wolf [M10], the correspondence principle is applied directly to the discrete-element models of the cones, enabling Voigt-type material damping (viscoelasticity) to be introduced. Each original spring is augmented by a dashpot; and each original dashpot is augmented by a mass, attached in a special way. Going a step farther, more realistic linear-hysteretic damping is represented by replacing the augmenting dashpots and masses by frictional elements. The analysis, which is effected solely in the time domain, is illustrated by an example from earthquake engineering.

2.4.2 Correspondence Principle Applied to Discrete-Element Model

The amplitudes of the forces in the spring(s) and dashpot(s) of the discrete-element models (Figs. 2-5, 2-17) for unit-distortional amplitudes are affected as follows when introducing material damping (Eqs. 2-66, 2-63, 2-231)

$$\rho\, c^2 \rightarrow \rho\, c^2(1 + 2i\zeta_g) \tag{2-240}$$

$$i\omega\,\rho\, c \rightarrow i\omega\,\rho\, c(1 + i\zeta_g) \tag{2-241}$$

with the appropriate wave velocity c corresponding to the predominant motion.

Although experiments verify that the damping ratio ζ_g does not depend on the excitation frequency, the concept of material damping was developed historically from the theory of linear viscoelasticity, which presupposes a frequency variation. In the classical *Voigt model of viscoelasticity the damping ratio ζ_g is presumed to be linearly proportional to frequency*, with $\zeta_g = \zeta_0\omega/\omega_0$. It is convenient to select ω_0 as the circular frequency at which the dimensionless frequency of soil dynamics, $a_0 = \omega r_0/c_s$ (r_0 = radius of the foundation) is unity. Correspondingly, ζ_0 is the value of ζ_g at $a_0 = 1$. For Voigt viscoelasticity Eqs. 2-240 and 2-241 take the form

$$\rho c^2 \rightarrow \rho c^2 \left(1 + i\omega \frac{2\zeta_0}{\omega_0}\right) \tag{2-242a}$$

$$i\omega\rho\, c \rightarrow \rho c\left(i\omega - \omega^2 \frac{\zeta_0}{\omega_0}\right) \tag{2-243a}$$

On the right-hand sides of these expressions various powers of $i\omega$ appear. For harmonic motion it is well known that terms of the dynamic-stiffness coefficient not multiplied by a power of $i\omega$ correspond to springs. Terms involving $i\omega$ correspond to dashpots, and terms containing $-\omega^2$ correspond to masses. Taking this into account, Eqs. 2-242a and

2-243a have a simple physical interpretation, pictured in Fig. 2-32. According to Eq. 2-242a, *Voigt viscoelasticity augments each original elastic spring K by an additional dashpot* $C = 2\,(\zeta_0/\omega_0)K$, connected in parallel, resulting in the force

$$P = \frac{2\zeta_0}{\omega_0} K \left(\dot{u}_0 - \dot{u}_1 \right) \tag{2-242b}$$

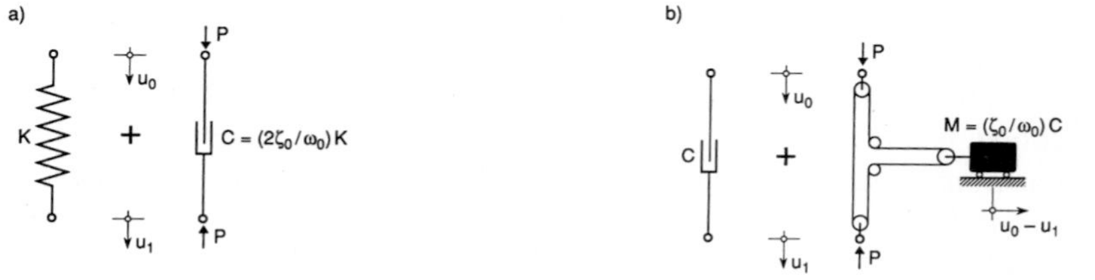

a. Original spring with augmenting dashpot. **b.** Original dashpot with augmenting pulley mass.

Fig. 2-32 Augmenting elements to represent Voigt viscoelasticity.

From Eq. 2-243a *each original elastic dashpot C is augmented by an additional mass* $M = (\zeta_0/\omega_0)C$. This additional mass is also connected in parallel to the original elastic element (the dashpot), leading to the force

$$P = \frac{\zeta_0}{\omega_0} C\left(\ddot{u}_0 - \ddot{u}_1 \right) \tag{2-243b}$$

As shown in Fig. 2-32b, the connection is effected by *pulleys*, which subject the mass to the difference of the motion of the top (subscript 0) and bottom (subscript 1) nodes. The model with the pulleys is an important new concept, which makes it possible to *realize Voigt viscoelasticity mechanically in the time domain*, not analytically in the frequency domain. It is only necessary to attach the augmented discrete-element assemblage to the underside of the (possibly nonlinear) structure and integrate directly in the time domain with a general-purpose computer program.

The necessary modifications are illustrated in Fig. 2-33 for the cone model of a disk foundation on the elastic halfspace (Fig. 2-33a corresponding to Fig. 2-5a for translational motion, Fig. 2-33b corresponding to the monkey-tail model for rotational motion in Fig. 2-17b). Observe that, as shown in Fig. 2-33a, pulleys are unnecessary if the original dashpot is fixed at the far end. The mass can in this case be attached directly to the disk. It is important to note that the inclusion of viscoelasticity augments only the original elastic springs and dashpots. If the discrete-element model includes a mass such as the "monkey tail" M_ϑ in Fig. 2-33b, then that is not modified.

A discrete-element model incorporating Voigt-type material damping can be constructed straightforwardly also for the other spring–dashpot discrete-element model for rotational motion (Fig. 2-17a). The corresponding augmented discrete-element model is shown in Fig. 2-33c. The augmenting masses are assigned to the nodes. Formulating equilibrium in the two nodes yields

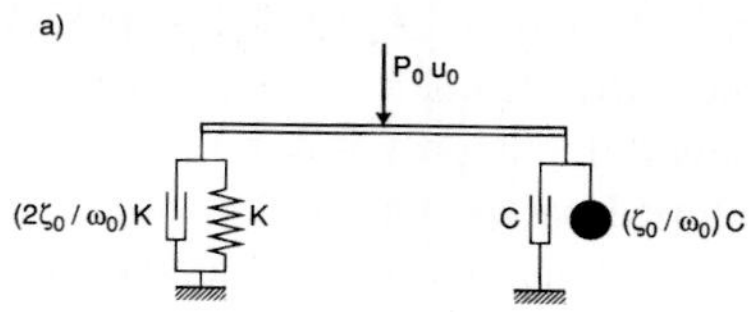

a. Translation.

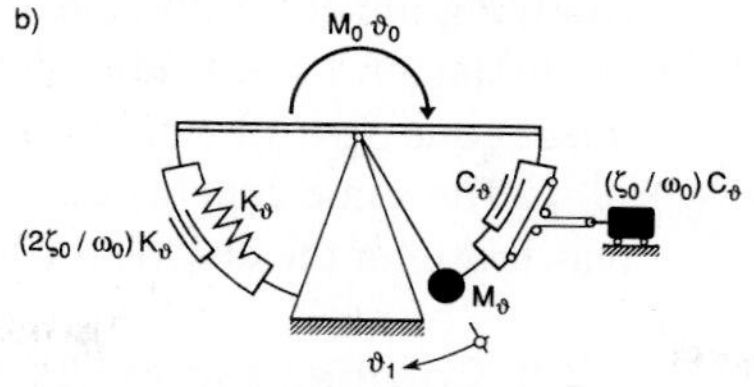

b. Monkey-tail model for rotation.

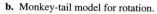

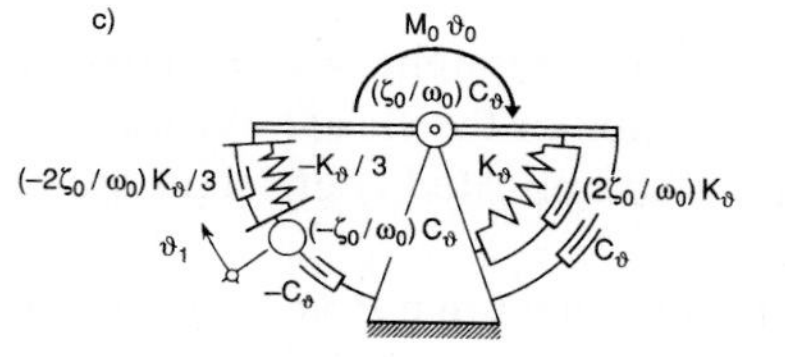

c. Spring-dashpot model for rotation.

Fig. 2-33 Discrete-element representation of Voigt viscoelastic cone models.

$$M_0 = -\frac{K_\vartheta}{3}(\vartheta_0 - \vartheta_1) - \frac{2}{3}\frac{\zeta_0}{\omega_0}K_\vartheta\left(\dot{\vartheta}_0 - \dot{\vartheta}_1\right) + K_\vartheta\,\vartheta_0 + \frac{2\zeta_0}{\omega_0}K_\vartheta\,\dot{\vartheta}_0 + C_\vartheta\,\dot{\vartheta}_0 + \frac{\zeta_0}{\omega_0}C_\vartheta\,\ddot{\vartheta}_0 \qquad (2\text{-}244)$$

$$-\frac{K_\vartheta}{3}(\vartheta_1 - \vartheta_0) - \frac{2}{3}\frac{\zeta_0}{\omega_0}K_\vartheta\left(\dot{\vartheta}_1 - \dot{\vartheta}_0\right) - C_\vartheta\,\dot{\vartheta}_1 - \frac{\zeta_0}{\omega_0}C_\vartheta\,\ddot{\vartheta}_1 = 0 \qquad (2\text{-}245)$$

To check this augmented discrete-element model, the dynamic-stiffness coefficient for harmonic loading is determined. Substituting K_ϑ and C_ϑ (Eq. 2-146) in Eq. 2-245 and using the definition of b_0 (Eq. 2-68) yields with $\zeta_0^2 \ll 1$.

$$\vartheta_1(b_0) = \frac{1 + i\zeta b_0}{1 + i(1+\zeta)b_0}\,\vartheta_0(b_0) \qquad (2\text{-}246)$$

where the abbreviation

$$\zeta = \frac{c}{\omega_0\, z_0}\,\zeta_0 \qquad (2\text{-}247)$$

is introduced. Substitution of Eq. 2-246 in Eq. 2-244 results in

$$\frac{M_0(b_0)}{\vartheta_0(b_0)} = S_{\vartheta\zeta}(b_0) = K_\vartheta\left[1 + 2i\zeta b_0 - \frac{1}{3}\frac{(1+i\zeta b_0)b_0^2}{1 + i(1+\zeta)b_0}\right] \qquad (2\text{-}248)$$

The dynamic-stiffness coefficient for the elastic rotational cone equals (Eqs. 2-168 and 2-169)

$$S_\vartheta(b_0) = K_\vartheta\left[1 - \frac{1}{3}\frac{b_0^2}{1 + b_0^2} + \frac{ib_0}{3}\frac{b_0^2}{1+b_0^2}\right] \qquad (2\text{-}249)$$

Applying the correspondence principle—multiplying K_ϑ by the factor $1 + i\omega 2\zeta_0/\omega_0$ and b_0 by the factor $1 - i\omega\zeta_0/\omega_0$—leads to the dynamic-stiffness coefficient of the damped rotational cone $S_{\vartheta\zeta}(b_0)$, which for $\zeta_0^2 \ll 1$ coincides with the expression in Eq. 2-248.

The same conclusions also hold for all lumped-parameter models more complicated than those for the disk on the halfspace, which are systematically developed in Appendix B.

2.4.3 Comparison of Models with Linear-Hysteretic Damping and Voigt Viscoelasticity

As discussed in Section 2.4.1, material damping affects the spring and damping coefficients $k(a_0)$ and $c(a_0)$ of the dynamic-stiffness coefficient. Straightforward application of the correspondence principle to the analytical expression of the elastic dynamic-stiffness coefficient for harmonic loading yields the more complicated coefficient including material damping (Eq. 2-234).

Abandoning the complex notation, the exciting force $P_0(a_0)$ causing a harmonic time history of the displacement $u_0(a_0)$ sin ωt and the corresponding velocity $u_0(a_0)\omega$ cos ωt can be expressed (with $(r_0/c_s)Kc(a_0)u_0(a_0)\omega \cos \omega t = a_0 Kc(a_0)u_0(a_0) \cos \omega t$) in the following standard form

$$\frac{P_0(a_0)}{Ku_0(a_0)} = k(a_0) \sin \omega t + a_0 c(a_0) \cos \omega t \tag{2-250}$$

where $k(a_0)$ and $c(a_0)$ are the same dimensionless spring and damping coefficients of the undamped case addressed above.

A vertically excited disk on an elastic halfspace with $\nu = 1/3$ is examined. Based on the elastic cone model the dynamic-stiffness coefficient is formulated with (Eqs. 2-72, 2-73)

$$k = 1 \tag{2-251a}$$

$$c = \frac{z_0}{r_0}\frac{c_s}{c_p} \tag{2-251b}$$

The aspect ratio z_0/r_0 follows from Table 2-3B. Equation 2-251 is equivalent to a constant spring and a constant dashpot, as confirmed by the solid straight lines of k and c in Fig. 2-34. When Voigt viscoelasticity is introduced, the dynamic-stiffness coefficient is modified as (Section 2.4.1)

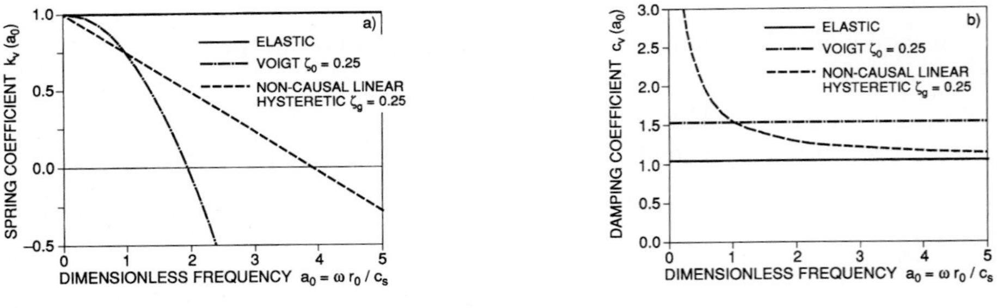

Fig. 2-34 Dynamic-stiffness coefficient of vertically excited elastic, Voigt viscoelastic and noncausal linear-hysteretic cones.

$$K\left(1+i\omega\frac{2\zeta_0}{\omega_0}\right)\left[1+ia_0\left(1-\frac{i\omega\zeta_0}{\omega_0}\right)\frac{z_0}{r_0}\frac{c_s}{c_p}\right]\simeq$$
$$K\left[1-\frac{\zeta_0}{\omega_0}\frac{z_0}{r_0}\frac{c_s}{r_0}\frac{c_s}{c_p}a_0^2+ia_0\left(\frac{z_0}{r_0}\frac{c_s}{c_p}+\frac{2\zeta_0}{\omega_0}\frac{c_s}{r_0}\right)\right] \tag{2-252}$$

This leads to

$$k_\zeta(a_0)=1-\frac{\zeta_0}{\omega_0}\frac{z_0}{r_0}\frac{c_s}{r_0}\frac{c_s}{c_p}a_0^2 \tag{2-253a}$$

$$c_\zeta(a_0)=\frac{z_0}{r_0}\frac{c_s}{c_p}+\frac{2\zeta_0}{\omega_0}\frac{c_s}{r_0} \tag{2-253b}$$

This result can also be derived from the augmented discrete-element model of Fig. 2.33a. For $\zeta_0 = 0.25$ the result is plotted as dashed–dotted curves. The spring is augmented by a mass, leading to the downward-parabolic variation of $k_\zeta(a_0)$. In addition, the damping coefficient $c_\zeta(a_0)$, still constant for all frequencies, is increased by a factor of about 50%. The decrease of the spring coefficient and the increase of the damping coefficient are qualitatively correct tendencies but do not agree quantitatively with experiments. In particular, the damping is underestimated for low frequencies and overestimated for high frequencies. It is easy to circumvent this disadvantage in the frequency domain by postulating that ζ_0 in Eqs. 2-242a and 2-243a is replaced by $\zeta_g\omega_0/\omega$. Physically speaking, this makes the augmenting dashpots and masses in Fig. 2-33 inversely proportional to frequency. Although such frequency-dependent elements seem to pose no conceptual problems, they cannot be built in reality. Nevertheless, if such impossible elements are permitted, the frequency-independent linear-hysteretic response as observed in experiments results. It appears paradoxical that impossible elements are verified by real tests, and the next Section 2.4.4 will be devoted to resolving this aspect. First, however, attention is focused on the harmonic response computed mathematically for the fictitious augmenting dashpots and masses with coefficients inversely proportional to frequency. From Fig. 2-33a, or substituting $\zeta_0 = \zeta_g\,\omega_0/\omega$ in Eq. 2-253, yields

$$k_\zeta(a_0)=1-\zeta_g\frac{z_0}{r_0}\frac{c_s}{c_p}a_0 \tag{2-254a}$$

$$c_\zeta(a_0)=\frac{z_0}{r_0}\frac{c_s}{c_p}+\frac{2\zeta_g}{a_0} \tag{2-254b}$$

These results are shown for $\zeta_g = 0.25$ as dashed curves in Fig. 2-34. By comparison to Voigt viscoelasticity, the spring coefficient $k_\zeta(a_0)$, a linear function in a_0, decays less rapidly with a_0; and the damping coefficient $c_\zeta(a_0)$, a hyperbolic function of a_0, is greater for low frequencies and less for high frequencies. By definition, the Voigt and constant-hysteretic models coincide at the dimensionless frequency $a_0 = 1$.

It is well known that for a linear system the inverse Fourier transform of the harmonic response yields the transient response to a Dirac impulse (see Eq. A-29). If the dynamic-stiffness coefficient corresponding to Eq. 2-254 is Fourier-transformed, something strange happens. This response begins before the excitation. This so-called noncausal behavior is obviously impossible and confirms that the augmenting dashpots and pulley masses inversely proportional to frequency do not exist in reality. In spite of this, the harmonic response will be shown to be correct. No paradox is involved: The spring and damping coefficients of the noncausal linear-hysteretic system turn out to match those of a causal nonlinear-hysteretic system (averaged over one cycle of response). But this goes beyond simply making linear-hysteretic damping realizable in the time domain. Rather, the goal is to enable incorporation of more realistic nonlinear-hysteretic behavior.

2.4.4 Frictional Material Damping

Frictional discrete-element model. The formal procedure of augmenting each elastic spring and dashpot in the discrete-element model (lumped-parameter model) of the foundation by an additional element is very attractive, both conceptually and computationally. To capture realistic frequency-independent hysteretic material damping, it would be desirable to preserve the analogy and simply replace the augmenting dashpot and pulley–mass associated with Voigt viscoelasticity by suitable causal nonlinear elements. On the microscopic level, material damping arises from the mutual rubbing and sliding of soil particles on each other. Obviously, then, the sought-after nonlinear elements should involve friction.

The appropriate *frictional element* with the clamping force F *which augments an elastic spring* is shown in Fig. 2-35. Its force-displacement relationship is

$$P = |F| \tan\delta \; \mathrm{sgn}(\dot{u}_0 - \dot{u}_1) \tag{2-255}$$

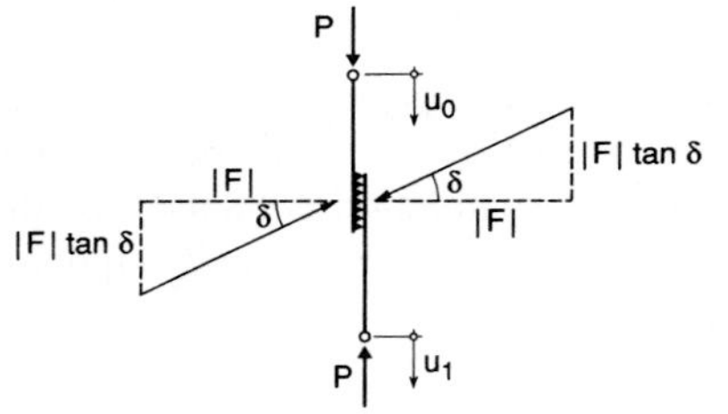

Fig. 2-35 Frictional element which augments an elastic spring.

in which the so-called "signum" function sgn( ) returns the algebraic sign of the argument. For example, sgn($A \sin \omega t$) yields a square wave of amplitude +1 (not +A) for $0 < \omega t < \pi$ and amplitude –1 (not –A) for $\pi < \omega t < 2\pi$. If the argument happens to vanish, sgn(0) is defined to be zero. Note that in Eq. 2-255 the argument of the signum function is the difference of the velocities $\dot{u}_0 - \dot{u}_1$, by contrast to the difference of the displacements $u_0 - u_1$, which determines the force in the original elastic spring. In other words, the argument of this signum function for the augmenting frictional element is the time derivative of the quantity which determines the force in the original elastic element. (This general rule will be used later to define the frictional element which augments an elastic dashpot). Equation 2-255 for the frictional element should be compared to Eq. 2-242b for Voigt viscoelasticity.

In Fig. 2-35 the restraining force P in the augmenting element is seen to result from the absolute value of the clamping force F, multiplied by the coefficient of friction $\tan\delta$. (This is one of those instances in which the $\tan\delta$ representation of material damping leads to better physical insight than the ζ_g representation). Now if the clamping force F were a constant, Eq. 2-255 would correspond to classical Coulomb friction. This primitive assumption is, however, illogical. The transverse clamping force should somehow be related to the magnitude of the direct force $K|u_0 - u_1|$ in the original elastic spring, more generally, to the past time history of the spring force. In some sense the clamping force should capture the "memory" of the direct spring force. The following simple memory law is illustrated in Fig. 2-36. The clamping force F is the larger of the magnitude of either the current spring force or its last peak value (not the absolute peak value). Equation 2-255 becomes

$$P = K\,\overline{|\,u_0 - u_1\,|}\tan\delta \operatorname{sgn}(\dot{u}_0 - \dot{u}_1) \tag{2-256}$$

in which the bar above $u_0 - u_1$ denotes the short-term memory just described. In the special case of harmonic motion, $\overline{|\,u_0 - u_1\,|}$ converges to the repetitive peak amplitude in the steady state, as shown by the middle portion of Fig. 2-36. This definition of the short-term memory law is justified by the agreement of the results for harmonic motion of noncausal linear-hysteretic and causal frictional systems.

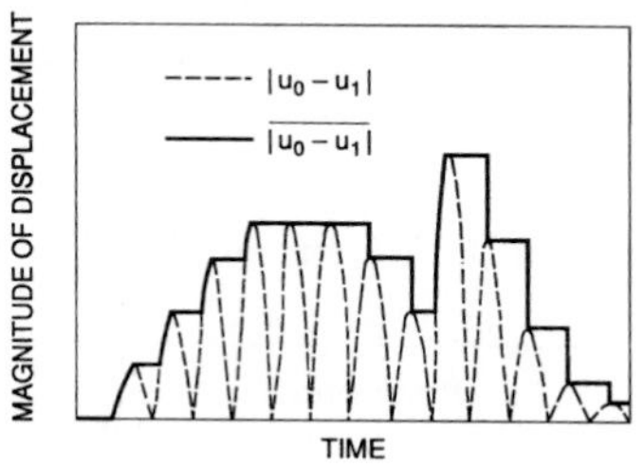

Fig. 2-36 Short-term memory law.

Equation 2-256 has special properties. It is at the same time both nonlinear and linear. Because of the memory law, the absolute value and the signum function, the time history of P resembles neither the time history of $u_0 - u_1$ nor that of $\dot{u}_0 - \dot{u}_1$. In this sense, Eq. 2-256 is nonlinear. On the other hand, if $u_0 - u_1$ is doubled, P is also doubled (provided that tan δ remains unchanged). In this sense, Eq. 2-256 is linear.

Attention is next directed to the appropriate *frictional element which augments an elastic dashpot.* In this case there is no physical picture corresponding to Fig. 2-35, but the force-displacement relationship may be written by analogy to Eq. 2-256 as

$$P = C\,\overline{|\,\dot{u}_0 - \dot{u}_1\,|}\,0.5\tan\delta \operatorname{sgn}(\ddot{u}_0 - \ddot{u}_1) \tag{2-257}$$

which involves an additional time derivative of u_0 and u_1 and replaces tan δ by 0.5 tan δ (see Eq. 2-241). The bar notation above $\dot{u}_0$ - $\dot{u}_1$ implies that the short-term memory law is now applied to the velocity difference. Equation 2-257 for the frictional element should be compared with Eq. 2-243b for Voigt viscoelasticity.

Equations 2-256 and 2-257 are causal. They depend only on current and previous values of u_0 and u_1 and their derivatives and may be easily evaluated in time-domain integration procedures. Needless to say, such frictional elements are not included in the libraries of general-purpose computer programs and must be specially generated.

The *friction elements* defined by Eqs. 2-256 and 2-257 *are just nonlinear versions of the augmenting dashpot* in Fig. 2-32a *and the pulley–mass* in Fig. 2-32b. Since inclusion of the augmenting dashpot and the pulley–mass leads automatically to the proper representation of Voigt viscoelasticity, systematic replacement of these by the frictional elements appears to be justified. Although this approach is logical, the existence of a frictional correspondence principle remains a postulate which cannot be proven rigorously. Nevertheless, a convincing partial proof is provided by the fact that for harmonic motion frictional material damping indeed agrees with linear-hysteretic viscoelasticity, as shown below.

Harmonic motion of translational cone. Suppose the frictional discrete-element model of a translational cone is subjected to vertical harmonic motion with a displacement history $u_0(\omega)$ sin ωt, a velocity history $\omega u_0(\omega)$ cos ωt, and an acceleration history $-\omega^2 u_0$ sin ωt. Using Eqs. 2-256 and 2-257 with $u_1 = \dot{u}_1 = \ddot{u}_1 = 0$ and by analogy to Fig. 2-33a, the force-displacement relationship may be written by inspection as

$$\frac{P_0(\omega)}{Ku_0(\omega)} = \sin\omega t - \frac{\omega C}{K}\zeta_g \mathrm{sgn}(\sin\,\omega t) + \frac{\omega C}{K}\cos\,\omega t + 2\,\zeta_g\,\mathrm{sgn}(\cos\,\omega t) \tag{2-258}$$

For soil with $\nu = 1/3$ the term $\omega C/K$ may be computed from the properties of the cone model (Table 2-3B) to be $(z_0/r_0)(c_s/c_p)a_0 = 1.047\,a_0$. This value will replace $\omega C/K$ in Eq. 2-260. In addition, the square-wave signum functions may be expanded into their respective Fourier series. From elementary calculus

$$\mathrm{sgn}(\sin\,\omega t) = \frac{4}{\pi}\sin\omega t + \underbrace{\frac{4}{\pi}\left(\frac{1}{3}\sin 3\omega t + \frac{1}{5}\sin 5\omega t + \dots\right)}_{harmonics} \tag{2-259a}$$

$$\mathrm{sgn}(\cos\,\omega t) = \frac{4}{\pi}\cos\,\omega t + \underbrace{\frac{4}{\pi}\left(-\frac{1}{3}\cos\,3\omega t + \frac{1}{5}\cos\,5\omega t - \dots\right)}_{harmonics} \tag{2-259b}$$

If Eq. 2-259 is substituted into Eq. 2-258, omitting the harmonics for the time being, the following result is obtained

$$\frac{P_0(\omega)}{Ku_0(\omega)} = \left(1 - \frac{4}{\pi}\zeta_g\;1.047\,a_0\right)\sin\omega t + a_0\left(1.047 + \frac{4}{\pi}\zeta_g\frac{2}{a_0}\right)\cos\,\omega t \tag{2-260}$$

By comparison to Eq. 2-250, the spring and damping coefficients are identified as

$$k_\zeta(a_0) = 1 - \frac{4}{\pi}\zeta_g\;1.047\,a_0 \tag{2-261a}$$

$$c_\zeta(a_0) = 1.047 + \frac{4}{\pi}\zeta_g\frac{2}{a_0} \tag{2-261b}$$

If the factor $4/\pi = 1.273$ preceding ζ_g were unity, Eq. 2-261 would be identical to Eq. 2-254, and Eq. 2-261 would correspond exactly to the dashed curves in Fig. 2-34 for linear-hysteretic viscoelasticity. This implies that *causal nonlinear frictional material damping is essentially equivalent to noncausal linear-hysteretic viscoelastic damping* if the "frictional" value of ζ_g is taken to be the fraction $\pi/4 = 0.785$ of that for linear-hysteretic viscoelasticity. This fact, shown here for translational motion, also holds for rotation, as will be demonstrated subsequently. (Due to the uncertainty inherent in the specification of ζ_g, the difference between the factor 0.785 and unity has minor practical significance.)

The harmonics in Eq. 2-259, neglected up to now, have no counterparts in the purely linear Eq. 2-250. The question arises whether the harmonics should be associated with the spring or the damping coefficient. The answer becomes obvious when one recalls that the essence of the damping coefficient is to dissipate energy. The work done per period (energy dissipated) is by definition the time integral of the force times the velocity $\omega u_0 \cos \omega t$ (see Eq. B-14). Because of the orthogonality of sines and cosines, the harmonics do not dissipate energy; the damping is concentrated in the fundamental frequency and is accounted for exclusively by Eq. 2-261b. The conclusion is that the harmonics modify the spring property only, changing the shape of the time history of the force such that the response becomes causal. In this sense Eq. 2-261a for the spring coefficient is an approximate relationship which omits the effects of the harmonics.

Harmonic motion of rotational cone. During earthquakes, structures resting on flexible soil are subjected to rocking. As a rule, this rocking motion is the predominant interaction effect, particularly for stiff structures with height to width ratios greater than one, such as reactor-containment buildings, bridge piers, silos, and the like. For the rocking motion, the damping coefficient starts at $a_0 = 0$ from zero (Fig. 2-21). Thus for low-frequency rocking (values of $a_0 < 1$), very little energy is dissipated by the radiation of waves towards infinity, so that the relative importance of material damping is much greater than in translation. To investigate this effect, sinusoidal rocking motion $\vartheta_0 \sin \omega t$ is applied to the discrete-element model of the rotational cone (analogous to Fig. 2-33b, but with frictional elements defined by Eqs. 2-256 and 2-257 in place of the augmenting dashpot and pulley–mass). The system is integrated numerically in the time domain until the steady state is reached. For $\nu = 1/3$ the dynamic-stiffness coefficient is presented in Fig. 2-37. As before, the spring and damping coefficients are computed from the fundamental terms in a Fourier-series expansion of this response and do not include the effects of harmonics. Therefore, the damping coefficient is exact, but the spring coefficient involves an approximation. The ζ_g values shown in the figure are those for the associated noncausal linear-hysteretic system. The frictional ζ_g values are $\pi/4 = 0.785$ times as large. As for the translational case, the frictional damping coefficient coincides with that for linear-hysteretic viscoelasticity. By contrast, the frictional spring coefficient is slightly less than that for linear hysteresis. Nevertheless, it is clear that probably because the harmonics are neglected as far as sinusoidal motion is concerned, frictional material damping and noncausal linear-hysteretic viscoelasticity are for all practical purposes identical. The same qualitative tendencies as for translational motion are observed: *Material damping leads to a pseudo-hyperbolic increase in the low-frequency damping coefficient and a pseudo-linear decrease in the high-frequency spring coefficient.*

Figure 2-38 shows the time history of the moment M_0 of one period of response. Although the frictional response exhibits jump discontinuities when the velocity changes sign, a smoothed version of this irregular curve corresponds well with the shifted sine wave for linear-hysteretic viscoelasticity. The figure elucidates the way frictional damping works. As the rotation ϑ_0 (and hence the spring force) increases, the friction is additive and increases the stiffness. By contrast, when the rotation ϑ_0 decreases, the friction changes sign and reduces the stiffness dramatically.

Figure 2-38 is computed for a large amount of material damping ($\zeta_g = 0.25$, i.e., $\tan \delta = 0.5$), in order to emphasize the influence of internal friction. This amount of material

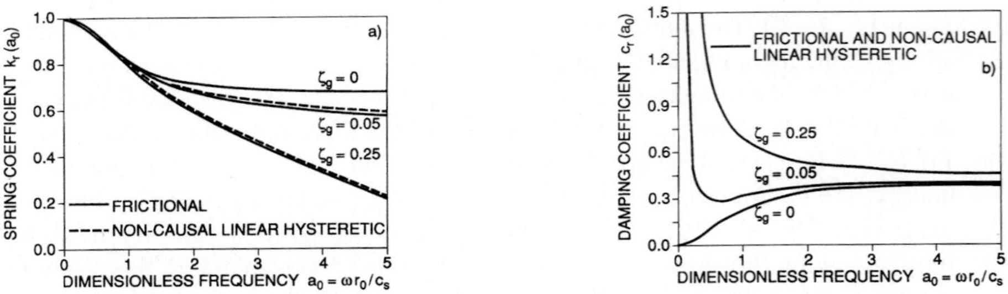

Fig. 2-37 Dynamic-stiffness coefficient of rocking cone with noncausal linear-hysteretic damping and frictional material damping.

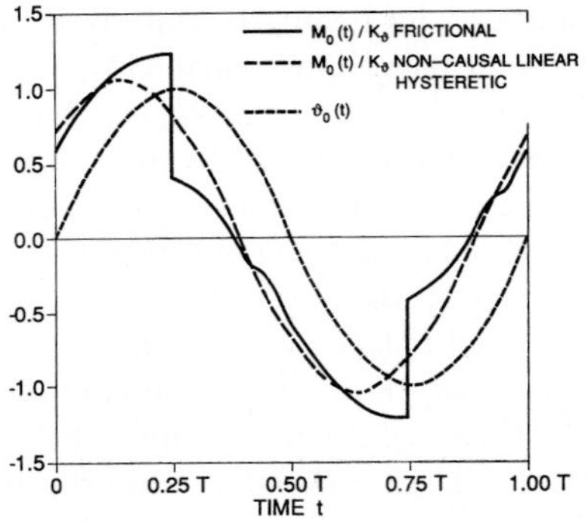

Fig. 2-38 One period of harmonic response of rocking cone with noncausal linear-hysteretic damping and frictional material damping.

damping is appropriate for high strain levels on the order of 0.01. For the rocking cone the maximum strain ε under the edge of the basemat can be calculated for the static case as follows (Appendix A2):

$$\varepsilon = \frac{M_0 \, r_0}{\rho c_p^2 \, I_0} \tag{2-262a}$$

$$\vartheta_0 = \frac{M_0}{K_\vartheta} = \frac{M_0 \, z_0}{3\rho c_p^2 \, I_0} \tag{2-262b}$$

Eliminating M_0 and with $z_0 \simeq 2r_0$ (Table 2-3B)

$$\varepsilon = \frac{3}{2} \vartheta_0 \tag{2-263}$$

results. The maximum strain ε under the edge of the basemat is about 1.5 times the angle of rotation ϑ_0. (Proceeding analogously for the translational cone in vertical motion with $z_0 \simeq 2r_0$, $\varepsilon = 0.5u_0/r_0$ follows and in horizontal motion with $z_0 = (2/3)r_0$, $\varepsilon = \gamma = 1.5u_0/r_0$ with the corresponding displacement of the basemat u_0.) An upper bound for the strain-dependent amount of material-damping ratio is therefore

$$\zeta_g = 0.25 \frac{1.5}{0.01} \bar{\vartheta}_0 = 37.5\, \bar{\vartheta}_0 \qquad (2\text{-}264)$$

in which $\bar{\vartheta}_0$ denotes the short-term memory value of ϑ_0. According to this equation (which for emphasis presupposes that ζ_g depends on the maximum strain at the edge, not on some smaller average strain under the basemat), Fig. 2-38 with $\zeta_g = 0.25$ would correspond to steady-state harmonic rocking of magnitude (1/150) rad = 0.38°. In engineering practice, however, such sinusoidal motion usually results from machinery vibrations which typically induce angles of rotation several orders of magnitude smaller. A glance at the curves for $\zeta_g = 0.05$ in Fig. 2-37 verifies that material damping is usually negligible and need not to be considered for such machinery-vibration problems. By contrast, large-amplitude rocking does occur in strong-motion earthquakes. In an earthquake problem it is definitely appropriate to include material damping in the analysis and update the value of ζ_g according to Eq. 2-264 after each time step. ζ_g is thus proportional to the strain amplitude of the response, but independent of the frequency. This procedure is illustrated by the example in the next Section 2.4.5.

2.4.5 Example of Seismic Analysis

To illustrate the earthquake analysis, a simplified model of a reactor building is considered. The structure is idealized as a rigid cylinder with radius $r_0 = 20$ m and height $h = 60$ m. For a typical average mass density 700 kg/m^3 the total mass is $m = 5.28 \cdot 10^7$ kg. If the mass is uniformly distributed, the centroid is located at the half height $h_c = 30$ m; and the mass moment of inertia at the base level is $I_0 = 6.86 \cdot 10^{10}$ kgm^2. The structure rests on the surface of soft sandy soil with Poisson's ratio $\nu = 1/3$, mass density $\rho = 2 \cdot 10^3$ kg/m^3, shear-wave velocity $c_s = 100$ m/s, and thus dilatational-wave velocity $c_p = 200$ m/s. (The soil in this example is artificially weakened to emphasize the influence of interaction.) From the properties of cone models (Eq. 2-146a) the static-rotational stiffness coefficient of the foundation is computed to be $K_\vartheta = 6.40 \cdot 10^{11}$ Nm. Because the reactor building is taller than it is wide, the response will be dominated by rocking; and attention is limited to this component of motion. The seismic excitation consists of a horizontal acceleration $\ddot{u}^g(t)$ arising from vertically propagating *S*-waves.

Before performing the transient earthquake analysis, it is useful to estimate the key dynamic characteristics of the coupled soil–structure system from the harmonic response of the foundation, Fig. 2-37. The natural frequency (Eq. 2-192), estimated as $\sqrt{0.9K_\vartheta/I_0}$) turns out to be 2.9 s^{-1} (0.46 Hz), which corresponds to the value of $a_0 = 2.9 \cdot 20/100 = 0.58$. Figure 2-37a confirms that at $a_0 = 0.58$ the spring coefficient k_r is indeed 0.9, yielding $0.9K_\vartheta$, so the natural frequency is correct without iteration. The damping ratio of the coupled system, including both radiation and material damping, may be calculated as $a_0 c_r(a_0)/[2k_r(a_0)]$ (Eq. 2-193). If material damping is neglected, Fig. 2-37b yields $c_r(0.58) = 0.125$, leading to the small damping ratio $0.58 \cdot 0.125/(2 \cdot 0.9) = 0.04$. By contrast, if material damping with $\zeta_g = 0.25$ is included, the c_r-coefficient is about 7.5 times greater, which increases the damping ratio to the large value 0.3. By virtue of these parameters, the earthquake response may be predicted to be essentially a sine wave with a period of 1/0.46 = 2.2 s, decaying slowly in the absence of material damping but very rapidly in its presence.

The analysis procedure for seismic excitation is systematically addressed in Chapter 6. The basic equation of motion for the rigid cylindrical structure on soil for a prescribed horizontal acceleration $\ddot{u}^f(t) = \ddot{u}^g(t)$with the rocking motion $\vartheta_0(t)$ as the only degree of freedom can be formulated by inspection (Fig. 2-39). Formulating the horizontal and rotational equilibrium equations in the center of gravity of the rigid cylindrical structure yields

$$m\ddot{u}^g + m\, h_c\, \ddot{\vartheta}_0 + P_0 = 0 \qquad (2\text{-}265)$$

$$I\ddot{\vartheta}_0 + M_0 - h_c\, P_0 = 0 \qquad (2\text{-}266)$$

with the corresponding mass moment of inertia I ($I = I_0 - h_c^2$ m). Eliminating the horizontal interaction force of the soil P_0 leads to

$$(I + h_c^2\, m)\, \ddot{\vartheta}_0 + M_0 + mh_c\ddot{u}^g = 0 \qquad (2\text{-}267a)$$

or equivalently

$$\ddot{\vartheta}_0 + \frac{M_0}{I_0} = -\frac{m\, h_c}{I_0}\, \ddot{u}^g \qquad (2\text{-}267b)$$

in which M_0 is the interaction moment applied to the soil. The last term of Eq. 2-267a is the inertial load acting through the centroidal height h_c.

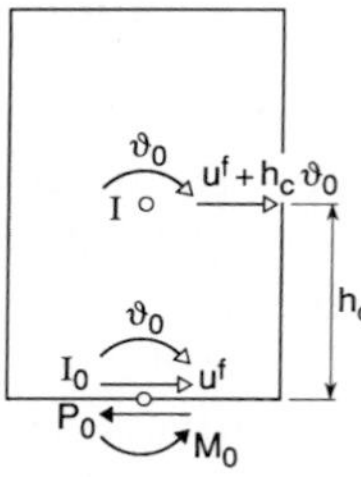

Fig. 2-39 Nomenclature of rigid cylindrical structure.

The coupled system is excited by the strong-motion portion of the Parkfield earthquake (27 June 1966, N65E-Cholame-Shandon record) with a maximum ground acceleration of 0.51 g and a maximum ground displacement of 0.29 m. Because the natural frequency is very low, the response is sensitive to the ground displacement (and not to the ground velocity or the acceleration). The displacement trace is indicated in Fig. 2-40a by a dashed–dotted curve, normalized by the factor in Eq. 2-267b. The displacement time history for the Parkfield earthquake is seen to be a simple half-cycle pulse, the duration of which (2.3 s) corresponds almost exactly to the period of the coupled system. As a result an amplification factor of about 1.3 may be expected in the absence of material damping.

The analysis is effected in the time domain based on an explicit algorithm with a predictor–corrector scheme. The procedure is analogous to that for the rigid foundation block of Section 2.3.4 (Eqs. 2-194 to 2-197). The forces in the frictional elements (Eqs. 2-256 and 2-257), which are straightforwardly determined based on the predicted values, lead to contributions in the equilibrium equations.

Figures 2-40a and 2-40b show, respectively, the computed rotation $\vartheta_0(t)$ and base moment $M_0(t)$. The solid curves correspond to a preselected constant value of material damping, $\zeta_g = 0.25$, whereas the dashed curves are calculated with strain-dependent mate-

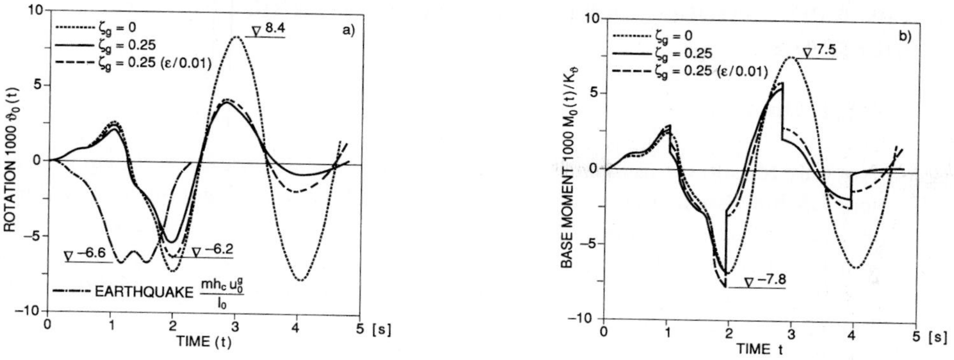

a. Rotation.

b. Base moment.

Fig. 2-40 Effect of frictional material damping on earthquake response of rigid cylindrical structure.

rial damping, as discussed in conjunction with Eq. 2-264. The good agreement implies that it is unnecessary to estimate the damping ratio in advance; strain dependent material damping adjusts itself automatically to the level of response and does not introduce any sort of anomalous behavior. The time histories of rotation of Fig. 2-40a are exactly as predicted. The system oscillates with a period of 2.2 s, regardless whether material damping is included or not. In the absence of material damping, the motion is so lightly damped that the peak rotation occurs after the excitation has ended. The amplification factor is 8.4/6.6 = 1.27. The presence of material damping prevents the build-up of resonance, so that the maximum response occurs during the earthquake. The amplification factor is reduced to 6.2/6.6 = 0.94. The plots of the base moment in Fig. 2-40b are similar in form to those for the rotation but exhibit the same sort of jump discontinuities previously observed in the harmonic motion, Fig. 2-38. It is very important to notice that the inclusion of material damping does not in this case decrease the peak base moment by comparison to the purely elastic analysis; rather, a slight increase is observed (factor 7.8 instead of 7.5). The physical explanation, previously ascertained in the discussion of the harmonic response, is important enough to warrant repeating. During monotonically increasing motion, friction adds to the stiffness of the foundation. The stiffer the foundation, the greater is its reaction force or moment. To put this point into perspective, it should be remembered that seismic damage may occur either due to the short-term peak response or due to the gradual degradation of structural components during long-term oscillations. In the first instance, material damping may be a detrimental effect; however, in the second instance it is certainly beneficial. As shown in Fig. 2-40, internal friction brakes the lingering motion very quickly after the earthquake. The structure literally "grinds to a halt."

This rather simple discrete-element model with causal frictional elements may well yield a better approximation of reality than a complicated "rigorous" analysis by the boundary-element method with noncausal linear-hysteretic damping or with some other sophisticated description of material damping.

2.5 WHY A CONE MODEL CAN REPRESENT THE ELASTIC HALFSPACE

2.5.1 Object

Not only for surface foundations on a homogeneous halfspace, as is shown in Sections 2.2 and 2.3, but also on a layered soil, for embedded foundations and piles (as will be demonstrated in Chapter 3, Appendix D, and Chapter 4) cone models yield good agreement with exact results. Nevertheless, cone models are not generally used in engineering practice. Practitioners and researchers presumably have reservations against cone models for the following three main reasons:

- They are based on strength of materials, not the rigorous theory of the elastic halfspace.
- The portion of the halfspace outside the cone is neglected.
- Cone models cannot represent the influence of Rayleigh surface waves.

In this section, which is based on Meek and Wolf [M7, M8], these potential objections are addressed in a systematic fashion and proven step by step to be unfounded. It turns out that cone models indeed incorporate and provide valuable insight into all the salient features of rigorous solutions; the aspects omitted by cones are revealed to be of minor physical importance. Attention is focused on the fundamental case of a disk foundation on the surface of a homogeneous halfspace consisting of compressible soil with Poisson's ratio ν not exceeding 1/3. For nearly incompressible soil with $1/3 < \nu \leq 1/2$ special considerations are required for the vertical and rocking motions (Section 2.1.4), which are also examined in this section.

2.5.2 Strength-of-Materials Versus Rigorous Theory of Halfspace

Cone models were originally derived by intuitive, physically motivated reasoning. The rigorous theory of the homogeneous halfspace was not considered explicitly. Nevertheless, many aspects of the cones turn out to be implicit in the classical static Boussinesq solutions. The following discussion addresses the cases of vertical translation and rocking because these motions are easiest to visualize; however, all conclusions are just as applicable by analogy to the cases of horizontal translation and torsion.

Static case. Consider first a static vertical load P applied on the surface of a homogeneous halfspace and the vertical displacement along the ray $u(z,\varphi)$ it produces at depth z and angle of inclination φ to the z-axis, as shown in Fig. 2-41a. The directions along which the angle φ remains constant are denoted henceforth as "rays." For given values of Poisson's ratio ν and inclination φ, the displacement u along the ray can depend only on the load P, the depth z, and Young's modulus E. For linearly elastic behavior the displacement is obviously directly proportional to P and inversely proportional to E

$$u = \frac{\alpha P}{E z^n} \tag{2-268a}$$

in which α is a dimensionless constant, initially unknown; and n is an exponent which prescribes the decay of the displacement with depth. For consistency of dimensions the only possible choice for n is unity, so that the *displacement is inversely proportional to the depth*

$$u = \frac{\alpha P}{Ez} \tag{2-268b}$$

a. Inclined ray.

b. Infinitesimal ray element.

Fig. 2-41. Vertical load on surface of halfspace.

By the same token, dimensionless analysis reveals that the vertical stress σ at depth z and angle of inclination φ must be independent of E and equal to

$$\sigma = \frac{\beta P}{z^2} \tag{2-269}$$

in which β is another dimensionless constant, initially unknown. Equation 2-269 is satisfied if the *area A of the ray is proportional to the square of the depth* z^2(Fig. 2-41a). These most salient features of the exact Boussinesq solutions are thus derivable by logical reasoning without recourse to differential equations; the rigorous analysis *only* provides the numerical values of the constants α and β.

Dynamic case (radiation criterion). Although Eq. 2-268b is a result from statics, the decay law $1/z$ for the displacement due to a point load has important implications for the dynamic case as well. The decay law determines whether or not radiation damping exists for low-frequency vertical or rocking motion. In order for waves to penetrate to infinity and transmit energy, the displacement magnitude must die off at large distances in a special way [M4]. The appropriate law is derived by considering the conservation of radiated energy as the waves pass through successive control surfaces of ever-increasing distances R measured along the ray. It is well known that the rate of energy transmission (power) per unit area is proportional to the square of the displacement's magnitude $|u(R)|$. This result is derived in Appendix B (Eq. B-15). The total radiated rate of energy (power) N is equal to the product of the power per unit area and the area $A(R)$ of the surface at infinity $R \to \infty$ (i.e., the far-field boundary surface)

$$N \propto |u(R)|^2 A(R) \tag{2-270}$$

N remains constant if and only of

$$|u(R)| \propto \frac{1}{\sqrt{A(R)}} \tag{2-271}$$

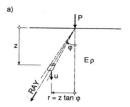

This so-called *radiation criterion* states that *the displacement magnitude (amplitude) must thus decay at infinity* ($R \to \infty$) *in inverse proportion to the square root of the area of the surface at infinity* for radiation of energy to occur. The surface at infinity is shown in Fig. 2-42. For body waves (*P*- and *S*-waves) the surface at infinity is a large hemisphere with area $2\pi z^2$. Hence, if the displacement decays with $1/z$, as is true for a single vertical load, radiation is possible starting at zero frequency ω. As is discussed in connection with Eq. B-15, $N > 0$ implies that the damping coefficient c_v must be positive. The equivalent vertical damping coefficient c_v does indeed have a finite positive value in the limit $\omega \to 0$, as is observed in Figs. 2-7b and 2-7d for all Poisson's ratios both for the rigorous solution and for the cone model. Based on the *decay law* of the *static displacement*, it is possible, to determine if the *zero-frequency damping coefficient is zero or not* (see also the criterion for the existence of a cutoff frequency, Section 3.8.4).

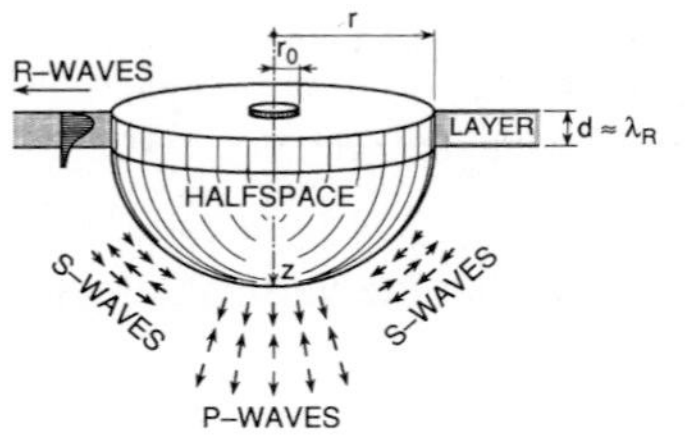

Fig. 2-42 Surface at infinity.

Next consider a pair of equal and opposite vertical loads acting on the free surface separated by the distance a. This couple is the most primitive version of the rocking case. Now the static displacement at large radial distances r is proportional to the difference of two 1/radial-distance terms

$$\frac{1}{r} - \frac{1}{r+a} = \frac{a}{r^2 + ar} \approx \frac{a}{r^2} \tag{2-272}$$

The resultant proportionality to $1/r^2$ implies that the displacement dies off too rapidly for waves to reach infinity (stronger than $1/r$). As a result, there is no energy transmission, and thus radiation damping in the limit $\omega \to 0$; the equivalent damping coefficient for rocking c_r must start from zero, as indeed it does in Figs. 2-21b and 2-21d, both for the rigorous solution and for the cone model for all Poisson's ratios. Again, important insights have been obtained solely by logical reasoning without solving differential equations. (Also for the dynamic case, the $1/\sqrt{A}$ decay law (Eq. 2-271) can be applied to demonstrate that for any non-zero frequency, the damping coefficient c_r of the rocking cone is finite. As shown in Eq. A-70, the rotation amplitude ϑ exhibits a term which is proportional to $1/z^2$ for $\omega \neq 0$, leading to a vertical displacement amplitude $u = r\vartheta$ with $r = z \tan \varphi$ (Fig. 2-41a). The displacement amplitude u is thus proportional to $1/z$, indicating that radiation of energy occurs for any $\omega \neq 0$).

Conical nature of rigorous solution. The rigorous static displacement field produced by a square array of four point loads, one pair acting upward, the other pair acting downward, is pictured in Fig. 2-43. The figure calculated for Poisson's ratio $\nu = 1/4$ demonstrates that comparable response in the vertical middle plane (for example the location of u_{max}) is observed along the directions of the rays. The behavior is inherently conical, spreading from an apex above the surface of the soil. Within the central cone the basic assumption of strength-of-materials, namely that plane sections remain plane, holds at least approximately. The formula $u = r\vartheta$ (with $r = z \tan \varphi$ and ϑ = the angle of rocking) is exact along the axis; however, the accuracy gradually decreases as the radius r increases. The heavy diagonal lines correspond to the boundaries of the rocking cone for $\nu = 1/4$ with the apex height $z_0 = 1.988\ r_0 \simeq 2r_0$ (Table 2-3B). The opening angle of the cone is chosen a priori to match the rigorous static-stiffness coefficient of a rocking disk on the elastic half-space. Notice that the cone model exactly parallels the direction of spreading and encompasses the region of large displacement. The rigorous solution confirms the original ad hoc assumption that the radius of the cone must increase linearly with depth and not with some higher power of z.

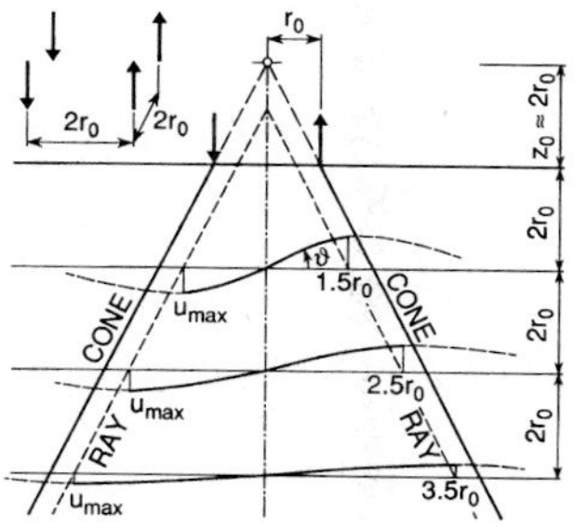

Fig. 2-43 Vertical displacements in middle plane produced by array of four static point loads on halfspace.

In summary, the rigorous Boussinesq results have verified the correctness of the following key aspects of cones:

- *For low-frequency motion, the equivalent damping for a translational motion is finite, but the equivalent rocking damping starts from zero.*
- *The apex of the cone is located above the surface of the soil.*
- *The radius of the cone increases linearly with depth.*
- *The assumption "plane sections remain plane" is (approximately) valid.*

Equation of motion for unified cone model. Because the behavior along a particular ray obeys very simple laws, it seems appropriate to formulate the equation of equilibrium for an infinitesimal element in this "special" direction, as shown in Fig. 2-41b. If shear stresses along the sides of the element are neglected (a typical assumption in strength of materials), vertical equilibrium requires

$$(\sigma A)_{,z} = \rho\, A\, \ddot{u} \tag{2-273a}$$

whereas for rotational equilibrium about the axis

$$(\sigma A\, r)_{,z} = \rho\, A\, r\, \ddot{u} \tag{2-273b}$$

in which ρ = the mass density of the soil, A = the horizontal cross-sectional area of the ray, and σ = the vertical normal stress, related to the vertical normal strain ε by the elastic constitutive law, $\sigma = E_c\varepsilon$ (E_c = constrained modulus). Substitution of the proportionality relationships $r \propto z$, $A \propto z^2$ and introduction of the velocity of dilatational-wave propagation $c_p = \sqrt{E_c/\rho}$ result in

$$\frac{1}{z^2}(z^2\varepsilon)_{,z} = \frac{\ddot{u}}{c_p^2} \qquad \textit{(translation)} \tag{2-274a}$$

$$\frac{1}{z^3}(z^3\varepsilon)_{,z} = \frac{\ddot{u}}{c_p^2} \qquad \textit{(rotation)} \tag{2-274b}$$

Next, the strength-of-materials assumption is used to derive expressions for the strain ε. If plane sections remain plane, as is the case within the cone

$$\varepsilon = u_{,z} \qquad \textit{(translation)} \tag{2-275a}$$

$$\varepsilon = r\vartheta_{,z} \qquad \textit{(rotation)} \tag{2-275b}$$

It should be emphasized that the strain is calculated in the vertical direction, not along the inclined direction of the ray. Since the vertical displacement due to rotation is given by

$$u = r\vartheta = \vartheta z \tan \varphi \tag{2-276a}$$

the expression for the vertical strain due to rotation may be written in terms of u as

$$\varepsilon = u_{,z} - \frac{u}{z} \tag{2-276b}$$

Insertion of this expression for the strains into the equilibrium relationships leads finally to

$$u_{,zz} + \frac{2}{z}u_{,z} = \frac{\ddot{u}}{c_p^2} \qquad \textit{(translation)} \tag{2-277a}$$

$$u_{,zz} + \frac{2}{z}\left(u_{,z} - \frac{u}{z}\right) = \frac{\ddot{u}}{c_p^2} \qquad \textit{(rotation)} \tag{2-277b}$$

A comparison of Eqs. 2-275a, 2-276b, 2-277a, and 2-277b reveals that both the translational and the rotational cones are described by the same governing equation of motion, namely

$$u_{,zz} + \frac{2\varepsilon}{z} = \frac{\ddot{u}}{c_p^2} \tag{2-278}$$

If formulated in terms of the vertical displacement u, both cones are mathematically identical. This interesting new result implies that the translational and rotational cones, previously regarded as separate entities, coalesce to a single unified cone model. Of course, substitution of the relationship $u = \vartheta z \tan \varphi$ yields the more familiar formulation of the rocking cone in terms of the angle of rotation ϑ as (Eq. A-44)

$$\vartheta_{,zz} + \frac{4}{z}\vartheta_{,z} = \frac{\ddot{\vartheta}}{c_p^2} \tag{2-279}$$

The governing equation (Eq. 2-278) for the unified cone model would be a simple wave equation if the second term $2\varepsilon/z$ were not present. For harmonic motion spreading waves have the familiar form $u(z,t) = u_0(z_0/z)^m \sin \omega(t - z/c_p)$ (see Eq. 2-48), which implies that the jth-order derivate of u with respect either to time t or space z contains a term proportional to ω^j. Since the strain involves at most the first derivative of u (see Eqs. 2-275a and 2-276b), it becomes small by comparison to $u_{,zz}$ and $\ddot{u}$ for high frequencies. Therefore, in the limit $\omega \to \infty$ the second term in Eq. 2-278 may be neglected. Thus for high frequencies the unified cone model is described by the same simple wave equation as a prismatic bar without spreading. A well-known result for the prismatic bar is that it is dynamically equivalent to a dashpot with a coefficient determined by multiplying the cross-sectional area (basemat–soil interface) by the specific damping ρc per unit contact area. The same thus holds true in the high-frequency limit for the cone models as already derived in Section 2.1.3 (Eq. 2-43 with Eq. 2-38b, and Eq. 2-57 with Eq. 2-55b). For rigorous solutions it applies as well. The cone models are thus asymptotically correct for high frequencies.

In the other limit, for zero frequency, the angle of opening of a cone executing a particular sort of motion is constructed to exactly match the corresponding rigorous static-stiffness coefficient of the rigid disk on the elastic halfspace. In summary

- *The cones are thus doubly-asymptotic solutions, correct for both very low and very high frequencies.*

Wave velocity and trapped mass for vertical and rocking motions in case of nearly incompressible soil. For a nearly incompressible soil ($1/3 < \nu \leq 1/2$), the two special features which prove necessary as described in Section 2.1.4 are the limit of $2c_s$ on the appropriate wave velocity c and the introduction of a trapped mass. These are substantiated in the following.

The rigorous solutions of the dynamic-stiffness coefficients of a disk for vertical and rocking motions are shown in Figs. 2-44 and 2-45 respectively. The curves in Figs. 2-44a and 2-45a define the customary frequency-dependent spring coefficients $k_v(a_0)$ and $k_r(a_0)$, which when multiplied by the respective static-stiffness coefficients K and K_ϑ yield the frequency-dependent dynamic springs. Similarly, the curves in Figs. 2-44b and 2-45b define the frequency-dependent dampers $C_v(a_0)$ and $C_r(a_0)$, normalised by the impedance expressions $\rho c_s A_0$ for vertical motion and $\rho c_s I_0$ for rocking (area $A_0 = \pi r_0^2$, moment of inertia $I_0 = \pi r_0^4/4$). The normalisation of the dampers with respect to impedance facilitates physical interpretation and is therefore preferable to the more traditional normalisation of the dampers with respect to the static-stiffness coefficients. To be able to investigate the asymptotic value of damping for $\omega \to \infty$ for perfectly incompressible soil, the vertical case is calculated for $\nu = 1/2$ up to $a_0 = 40$.

From the left side of Fig. 2-44b it follows for low-frequency excitation that the appropriate value of c to be inserted in the formula for the vertical damper

$$C_v(a_0) = \rho c A_0 \tag{2-280}$$

is always smaller than c_p. For example, although $c_p = \infty$ for $\nu = 1/2$, the effective value of c in the limit $a_0 \to 0$ is only 2.14 c_s. To a less radical extent, the same is true for the other values of Poisson's ratio: for $\nu = 1/3$, $c = 1.50c_s$ instead of $c_p = 2.0c_s$ and for $\nu = 0.45$, $c = 1.86c_s$ instead of $c_p = \sqrt{11}c_s = 3.32c_s$. From the right side of Fig. 2-44b an apparent contradiction in the high-frequency limit is observed. On the one hand, for nearly incompressible

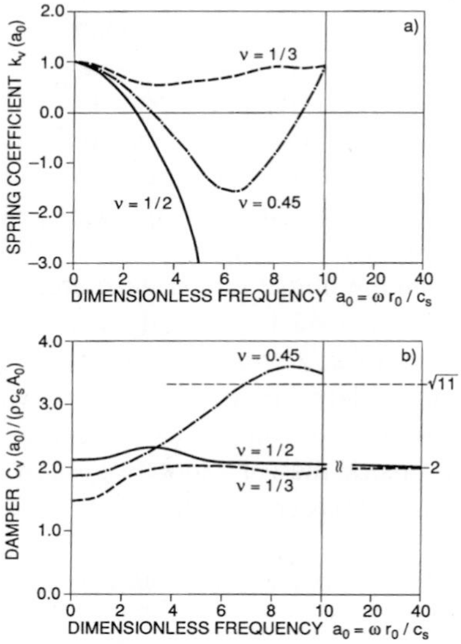

Fig. 2-44 Rigorous dynamic-stiffness coefficient of disk on homogeneous halfspace in vertical motion for various Poisson's ratios.

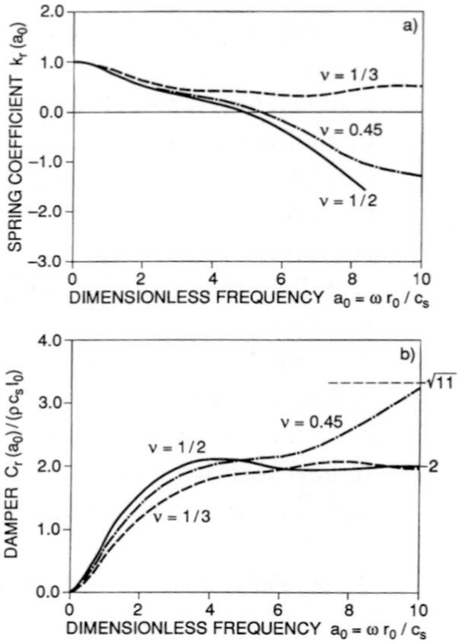

Fig. 2-45 Rigorous dynamic-stiffness coefficient of disk on homogeneous halfspace in rocking motion for various Poisson's ratios.

soil (for example $\nu = 0.45$) the high-frequency limit of c is indeed c_p. By contrast, for perfectly incompressible soil the high-frequency limit of c, verified by the computations out to $a_0 = 40$, tends not to $c_p = \infty$ but only to $2c_s$.

This special behavior of the high-frequency response for $\nu = 1/2$ appears quite puzzling—indeed paradoxical. The explanation turns out to be associated with the nature of the spring coefficients pictured in Figs. 2-44a and 2-45a. Notice that for $\nu = 1/3$ the curves remain positive for all values of a_0. However, for the higher values of Poisson's ratio, the spring coefficients exhibit a downward-parabolic tendency (greater for the vertical case than in rocking). If one recalls the expression for the dynamic-stiffness coefficient of a spring–mass–damper combination (with spring coefficient K, mass M, and damping coefficient C),

$$S(\omega) = K - \omega^2 M + i\omega C \tag{2-281}$$

it is immediately clear that proportionality to the square of frequency (leading to the downward-parabolic tendency) implies that the spring is augmented by a mass "trapped" under the basemat (and moving in phase with it) because the soil is nearly or perfectly incompressible. It suffices to observe that the downward-parabolic tendency eventually ceases for $\nu = 0.45$ and high values of a_0. Indeed, for all values of Poisson's ratio except exactly 1/2 the trapped mass gradually melts away and the curve of k_v comes back up, corresponding to a spring alone. For $\nu = 0.45$ the return of k_r to positive values is not yet observed for $a_0 = 10$, but it is obvious that the curve has reached its turning point. By contrast, for perfectly incompressible material with $\nu = 1/2$ the downward-parabolic tendency remains, in spite of how large a_0 becomes, with $k_v(a_0 = 40) = -270$.

The insights resulting from the preceding study of rigorous solutions for nearly incompressible soil may be summarised as follows:

- *For values of Poisson's ratio between 1/3 and 1/2 the low-frequency specific radiation damping for vertical and rocking motion is around* $2\rho c_s$.
- *For Poisson's ratio < 1/3 no trapped mass is present.*
- *For Poisson's ratio > 1/3 trapped mass is observed, leading to a downward-parabolic component in the spring coefficient.*
- *For Poisson's ratio < 1/2 the high-frequency specific radiation damping for vertical and rocking motions tends to* ρc_p*, and the trapped mass vanishes.*
- *For Poisson's ratio = 1/2 the high-frequency specific radiation damping for vertical and rocking motions tends to* $2\rho c_s$*, and trapped mass is present.*

The construction of the vertical and rocking cone models for the nearly incompressible case described in Section 2.1.4 is compatible with these conclusions.

2.5.3 Neglect of Portion of Halfspace Outside Cone

Static case. For a homogeneous halfspace with Poisson's ratio $\nu = 1/3$, Fig. 2-46 shows how the static vertical displacement under a vertically loaded rigid disk varies with depth according to the classical Boussinesq theory. The numerical factors give the ratio of the vertical displacement at depth z compared to the corresponding vertical displacement of the surface directly above. The heavy diagonal line is the boundary of the vertical cone.

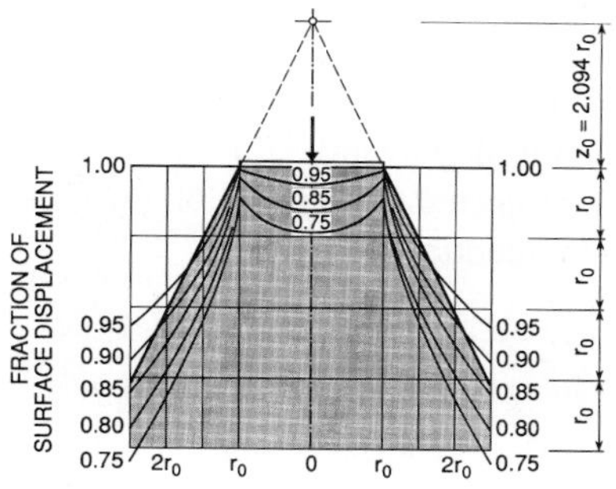

Fig. 2-46 Ratio of underground vertical displacement to surface vertical displacement for statically loaded rigid disk on halfspace.

It is interesting to observe that the unshaded region of soil above the cone is scarcely deformed. Because this portion of the halfspace is practically unstressed, it contributes almost nothing to the vertical stiffness of the system and may be conceptually excavated, leaving only the central cone to take the load. The figure verifies that the region of soil exterior to the cone may indeed be neglected without a serious contradiction of physical reality—yet another principal reservation against cone models proves to be unfounded.

Dynamic case. The conclusion ascertained for the static Boussinesq solution is also valid for dynamic excitation. Figure 2-47 is a polar diagram that shows the relative magnitude of shear waves (*S*-waves) and dilatational waves (*P*-waves) at different angles from a vertical harmonic point load. The results, taken from Miller and Pursey [M12] for a halfspace with Poisson's ratio $\nu = 1/3$, are shaded to illustrate which portions of the half-space transmit more than half the maximum radiated power for the two sorts of body waves. (As the power is proportional to the square of the magnitude of the displacement, the half-power displacement ratio is 0.707.) The power of *P*-waves is maximum directly beneath the source and diminishes to 50% at an angle around 45° from the axis (dilatational window). By contrast, the power of *S*-waves is concentrated in the shear window between about 30° and 45° from the axis. The unshaded region of the halfspace transmits a relatively small fraction of the radiated power towards infinity, except for the Rayleigh wave, which propagates horizontally along the surface; however, as indicated by the dashed lines, the Rayleigh wave forms at an appreciable distance from the source. This phenomenon is discussed in Section 2.5.4. Figure 2-47 is reminiscent of the radiation pattern of a radio or radar antenna [G2]. If the source is not a point load, but rather a disk with finite radius r_0, the shape of the lobes depends upon the ratio of r_0 to the wavelength of the excitation. The figure shows the radiation pattern in the low-frequency limit, in which r_0 is small in comparison to the wavelength. For higher frequencies, in which the wavelength is of the same order of magnitude as the radius, the aperture width of the dilatational window is less. In the very-high-frequency limit, in which the wavelength is much smaller than the radius, the *P*-waves are focused to a "searchlight beam" directly downward from the disk; the half-space indeed behaves as a prismatic bar. Regardless of the frequency, the appreciable displacement magnitudes of body waves are located within the central cone.

Estimation of cone's aspect ratio. The near equality of the displacements of a point on the soil surface and a point of the cone's periphery directly underneath enables estimation of the cone's aspect ratio z_0/r_0, which in turn specifies the static-stiffness coefficient. Suppose waves travel along the surface with an average velocity c_h, in which the subscript h (horizontal) is appended to distinguish this average surface velocity from the velocity c for the waves propagating vertically downward. If as shown in Fig. 2-48, the nodes for horizontal and vertical waves are projected to coincide on the cone's periphery, the aspect ratio and the velocity ratio must be identical:

$$\frac{z_0}{r_0} = \frac{c}{c_h} \tag{2-282}$$

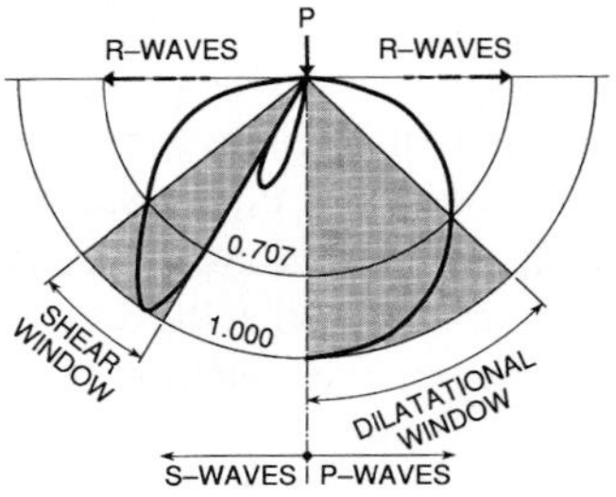

Fig. 2-47 Polar diagram of relative displacement magnitudes of body waves produced by harmonic point load on surface of halfspace.

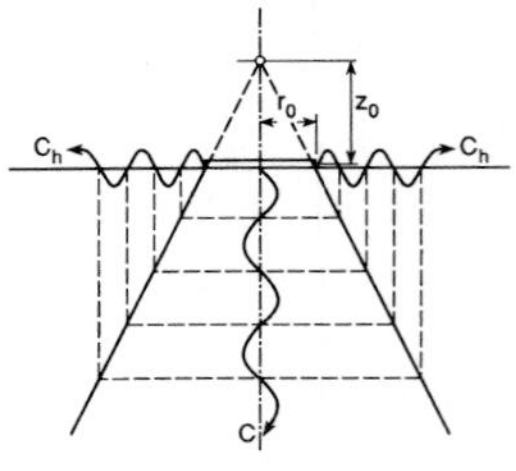

Fig. 2-48 Relationship between ratio of propagation velocities in vertical to horizontal directions and cone's aspect ratio.

For vertical and rocking motions and values of Poisson's ratio up to 1/3, the appropriate vertical velocity c is the dilatational-wave velocity c_p. By contrast, for horizontal motion and torsion, the appropriate vertical velocity c becomes the shear-wave velocity c_s. For all four components of motion it is also easy to estimate the average surface velocity c_h by physical reasoning (Fig. 2-49). The simplest case is torsion, in which only shear waves exist; therefore, there is no choice for c_h except c_s. The next-simplest case is vertical motion; here the value of c_h should be c_R, the velocity of Rayleigh surface waves. The same holds true for the rocking case, which is just a special sort of vertical motion, half upwards, half downwards. Finally, the case of horizontal motion is addressed. In this instance there are two sorts of waves along the surface. In the direction of motion dilatational waves emanate with velocity c_p, whereas in the perpendicular direction shear waves propagate with velocity c_s. Thus it appears reasonable to select as the average velocity of horizontal waves the mean value $c_h = (c_p + c_s)/2$. For these choices of c and c_h the approximate value of the cone's aspect ratio z_0/r_0 given by Eq. 2-282 is computed in Table 2-8 and compared with the exact result determined by setting the static-stiffness coefficient of the cone equal to the rigorous static-stiffness coefficient of a rigid disk on the elastic halfspace (Table 2-3B). The agreement is quite remarkable, particularly when the rigorous results for the vertical and rocking cases are averaged. In the "worst" case, torsion, the deviation is only 11.6%; for the other components of motion the deviation is less than 10%!

The possible objection that it is empirical or artificial to calibrate the aspect ratio of the cones to match the rigorous static-stiffness coefficients of rigid disks is unfounded. Indeed, just the opposite is true. If the rigorous static-stiffness coefficients were not yet known, it would be possible to estimate them to a high degree of accuracy by using the concept of cones. The inverse reasoning also holds: If the cones are calibrated via the rigorous static-stiffness coefficients, their proportions are compatible with the proper velocities in the horizontal direction. The rather amazing fact that the cones, formulated explicitly in the vertical direction, also capture implicitly the horizontal behavior, is doubtless a significant reason for their inherent accuracy. The arguments in this section have provided additional insights into why cones work.

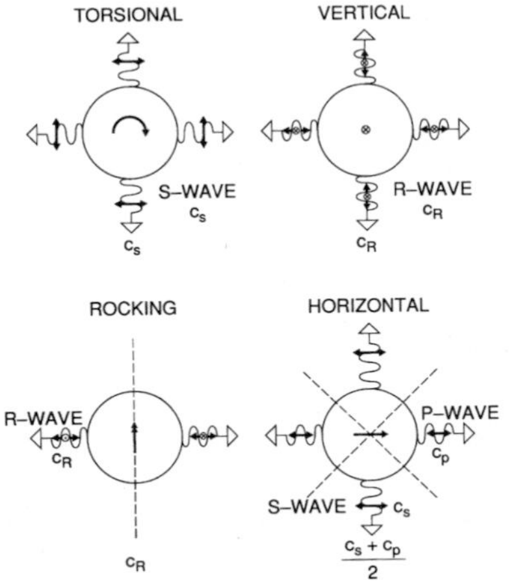

Fig. 2-49. Dominant wave pattern and average horizontal wave velocity at surface for various degrees of freedom.

TABLE 2-8 COMPARISON OF ASPECT RATIOS DETERMINED FROM PROPAGATION VELOCITIES AND STATIC STIFFNESS (EXACT)

		Poisson's Ratio ν		
		0	$\frac{1}{4}$	$\frac{1}{3}$
	$\frac{c_p}{c_s}$	1.414	1.732	2.000
	$\frac{c_R}{c_s}$	0.874	0.919	0.933
Torsional	$\frac{c}{c_s}$	1.000	1.000	1.000
	$\frac{c_h}{c_s}$	1.000	1.000	1.000
	$\frac{z_0}{r_0} = \frac{c}{c_h}$	1.000	1.000	1.000
	$\frac{z_0}{r_0}$exact	0.884	0.884	0.884
Vertical	$\frac{c}{c_s}$	1.414	1.732	2.000
	$\frac{c_h}{c_s}$	0.874	0.919	0.933
	$\frac{z_0}{r_0} = \frac{c}{c_h}$	1.619	1.885	2.144
	$\frac{z_0}{r_0}$exact	1.571	1.767	2.094
Rocking	$\frac{c}{c_s}$	1.414	1.732	2.000
	$\frac{c_h}{c_s}$	0.874	0.919	0.933
	$\frac{z_0}{r_0} = \frac{c}{c_h}$	1.619	1.885	2.144
	$\frac{z_0}{r_0}$exact	1.767	1.988	2.356
Average $\frac{z_0}{r_0}$ exact vertical/rocking		1.669	1.878	2.225
Horizontal	$\frac{c}{c_s}$	1.000	1.000	1.000
	$\frac{c_h}{c_s}$	1.207	1.366	1.500
	$\frac{z_0}{r_0} = \frac{c}{c_h}$	0.829	0.732	0.667
	$\frac{z_0}{r_0}$exact	0.785	0.687	0.654

- *It is physically correct to assume that the region of the soil halfspace outside the cone is essentially inactive and may be neglected,*
- *The proportions of the cones are compatible with the proper wave-propagation velocities in the horizontal direction for all motions.*

2.5.4 Inability of Cone to Represent Rayleigh Wave

By virtue of their geometry, translational cone models can only represent body waves with a decay in displacement magnitude proportional to $1/z$. Indeed, the conceptual excavation of the soil region exterior to the cone removes the surface layer through which Rayleigh waves must travel to infinity. For Rayleigh waves the surface at infinity in Fig. 2-42 is a flat cylinder with height d equal to the depth of penetration, approximately one Rayleigh wave length λ_R, and a very large radius r. The area of the surface at infinity is therefore $2\pi r\lambda_R$, proportional to r. Hence, the displacement magnitude of the Rayleigh wave, which dies off inversely with the square root of the surface at infinity (Eq. 2-271), must be proportional to $1/\sqrt{r}$, as is well known. This typical decay law for Rayleigh waves seems to contradict the main postulate of Section 2.5.3, namely that the displacement of the surface is essentially the same as that of the cone below. The fact that the displacement of the cone dies off with depth in proportion to $1/z$ implies that the displacement of the surface above should decay with $1/r$, not $1/\sqrt{r}$. The apparent dichotomy between cone models and Rayleigh waves, perhaps the major source of skepticism against the use of cones, may be elucidated and then resolved by considering rigorous solutions.

Miller and Pursey [M13] calculate the relative amounts of power (rate of energy) radiated by shear, dilatational, and Rayleigh waves when a vertical point load P_0 acts on the surface of an elastic halfspace with Poisson's ratio $\nu = 1/4$. Rayleigh waves account for the lion's share, 67%, followed by shear waves with 26%. A minimal 7% of the power is transmitted by dilatational waves, which are the only sort of waves that the vertical cone incorporates. At first glance one might well conclude that the vertical cone only represents 7% of reality. This is certainly one of the key reasons why many researchers and practitioners have been reluctant to accept cone models. Nevertheless, the generally accepted notion of the inordinate importance of Rayleigh waves turns out upon closer examination to be a misleading halftruth.

In the analysis of Miller and Pursey [M13] the point load is idealized as the limiting case of a very large constant vertical stress applied on the surface of a very small flexible disk (membrane) with radius r_0. Because the radius vanishes in the limit, the customary dimensionless frequency of elastodynamics, $a_0 = \omega r_0/c_s$, tends to zero, regardless how large the dimensional frequency ω may be. By contrast, if the disk foundation does not shrink to a point, the criterion $a_0 = 0$ is fulfilled only for $\omega = 0$; and the partition of power reported by Miller and Pursey [M13] corresponds unambiguously to the static case.

For a more complete understanding of the relative importance of the three sorts of waves, it is necessary to compute the partition of power for higher values of a_0. To do this, the integrals formulated in Miller and Pursey [M13] are reevaluated in Meek and Wolf [M7] after multiplying the integrand by an additional factor that accounts for the destructive interference of waves on the hemispherical boundary surface at infinity. At locations on this boundary not directly below the load, the waves originating from different portions of the membrane arrive at different instants of time; they are no longer exactly in phase, as would be the case for $a_0 = 0$. The details of the analysis [M7] are not presented here, as only the results are needed to discuss the importance of Rayleigh waves.

The power N (averaged over a period) transmitted through the disk is specified in Eq. B-15, which is repeated here for convenience, adding the subscript 0.

$$N(\omega) = \frac{\omega}{2} \overset{*}{u}_0(\omega)\ Im\big(S(\omega)\big)\, u_0(\omega) \tag{2-283}$$

An asterisk as superscript denotes the complex conjugate value. Introducing a_0 and substituting

$$u_0(a_0) = \frac{P_0(a_0)}{S(a_0)} \tag{2-284}$$

with the dynamic-stiffness coefficient

$$S(a_0) = K\big[k(a_0) + ia_0\, c(a_0)\big] \tag{2-285}$$

results in

$$N(a_0) = \frac{c_s}{2r_0} \frac{a_0^2\, P_0^2(a_0)\, c(a_0)}{K\big[k^2(a_0) + a_0^2 c^2(a_0)\big]} \tag{2-286}$$

The low-frequency limit is proportional to a_0^2 and the high-frequency limit is independent of a_0. For $\nu \neq 0.5$

$$N(a_0 \to \infty) = \frac{c_s}{2r_0} \frac{P_0^2(a_0)}{Kc(a_0 \to \infty)} \tag{2-287}$$

As

$$a_0 Kc(a_0 \to \infty) = \omega\rho\, c_p A_0 \tag{2-288}$$

applies (Section 2.1.3), the high-frequency limit of the power is formulated as ($\nu \neq 0.5$)

$$N(a_0 \to \infty) = \frac{P_0^2}{2\rho\, c_p A_0} \tag{2-289}$$

As for $\nu = 0.5$, $k(a_0)$ is proportional to a_0^2, $N(a_0 \to \infty)$ vanishes (Eq. 2-286).

The results for the average radiated power (rate of energy) and the partition of power among the three types of waves are presented in Figs. 2-50, 2-51, and 2-52 for Poisson's ratio $\nu = 1/3$, 0.45, and 1/2 respectively. A salient feature of all three figures is immediately obvious: As a_0 increases from zero, the relative importance of Rayleigh waves (which indeed account for more than half the power at $a_0 = 0$ for all Poisson's ratios) diminishes rapidly. At a value of a_0 about 3.5 (and at certain higher values), the Rayleigh wave transmits no power whatever! Furthermore, whenever P-waves exist (Figs. 2-50 and 2-51), they transmit the major portion of power as a_0 becomes large. Indeed, it may be proven that in the limit $a_0 \to \infty$ the total power is transmitted by P-waves alone. Very high-frequency vibrations can radiate only vertically downward without spreading; in effect, they are confined within the prismatic "rod" of soil directly under the foundation. Due to the increased directionality of high-frequency waves the entire rest of the halfspace is inactive. Of course, in perfectly incompressible soil ($\nu = 1/2$) no P-waves exist; and the power is therefore split between the remaining S- and R-waves, with at least 2/3 of the power transmitted by S-waves. This behavior is illustrated in Fig. 2-52.

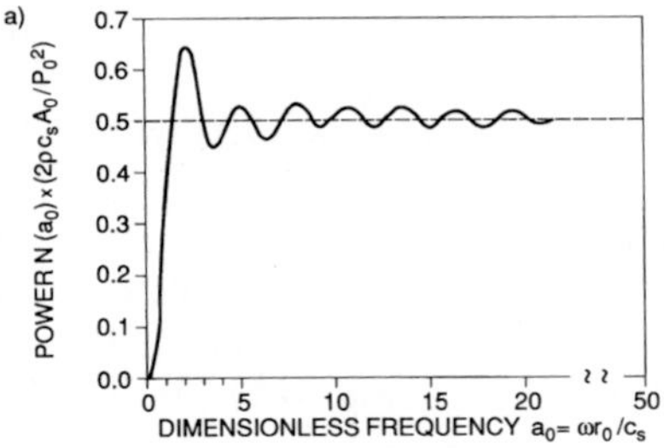

a. Total power.

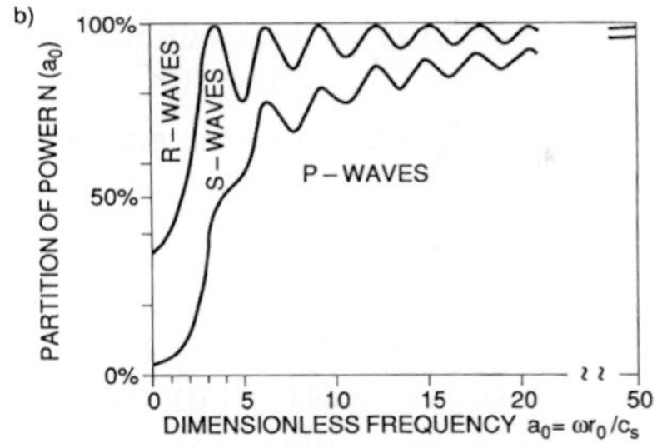

b. Partition of power among the three waves.

Fig. 2-50 Power for vertical motion as function of dimensionless frequency, Poisson's ratio = 1/3.

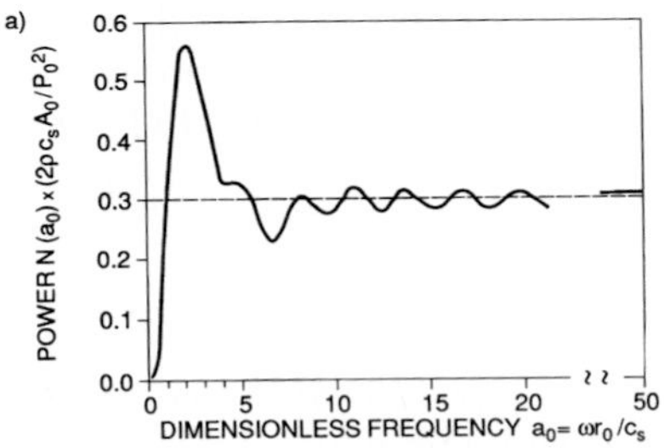

a. Total power.

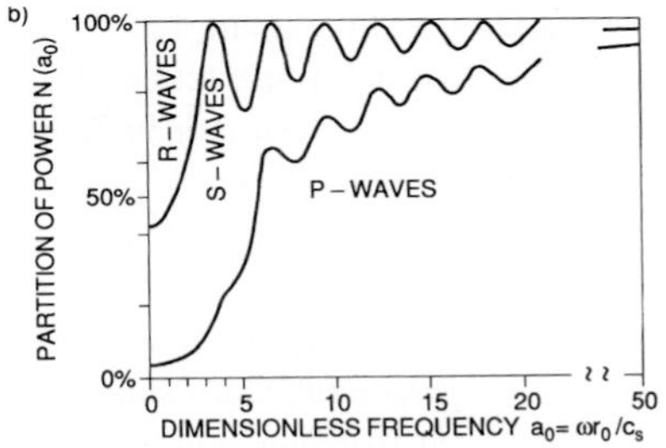

b. Partition of power among the three waves.

Fig. 2-51 Power for vertical motion as function of dimensionless frequency, Poisson's ratio = 0.45.

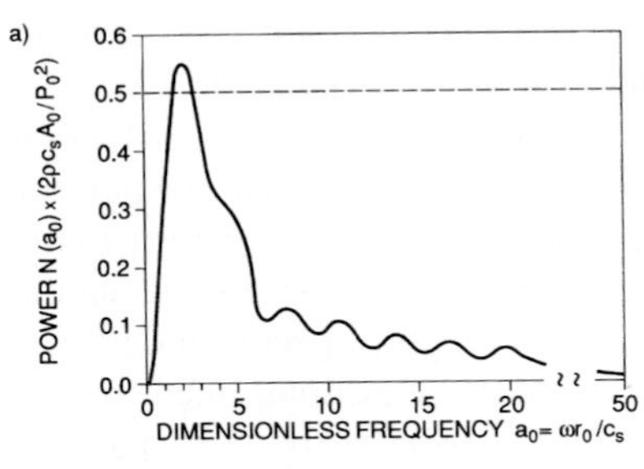

a. Total power.

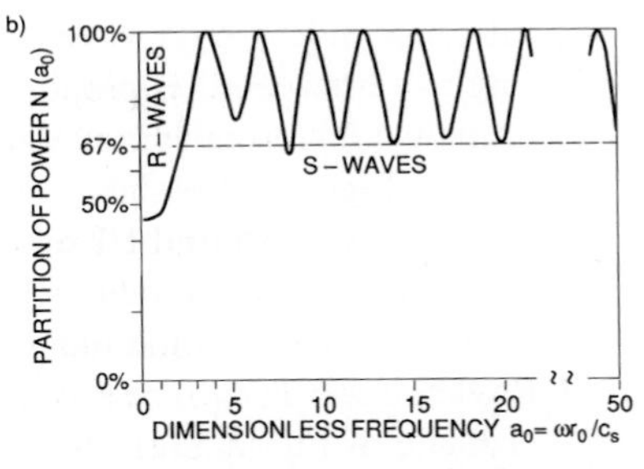

b. Partition of power among the three waves.

Fig. 2-52 Power for vertical motion as function of dimensionless frequency, Poisson's ratio = 1/2.

The physical reason why the importance of Rayleigh waves diminishes with frequency is quite simple. The depth of penetration of the Rayleigh wave (Fig. 2-42) is about one wavelength beneath the surface (skin effect). The power propagates towards infinity through cylinders of this depth and ever-growing radius. Because the wavelength is inversely proportional to the frequency, the cylindrical "surface at infinity" becomes flatter and flatter as the frequency increases; and the window through which the power must pass closes to a circular slit.

Further insight is provided by the curves for the average power N, Figs. 2-50a, 2-51a, and 2-52a. Starting from zero, these curves attain their maxima at about $a_0 = 2$, then decay. For the cases $\nu = 0.45$ (Fig. 2-51a) and $\nu = 1/2$ (Fig. 2-52a) the hump near $a_0 = 2$ is especially pronounced. The added mass associated with nearly incompressible soil, discussed previously at the end of Section 2.5.3 in connection with Figs. 2-44 and 2-45 produces a resonance peak. If, on the other hand, Poisson's ratio is 1/3 (Fig. 2-50a) there is no added mass; and without mass an appreciable resonance is not observed.

The values N presented are nondimensionalized with the factor $2\rho c_s A_0 / P_0^2$. In the high-frequency range of a_0 greater than about 5, the values thus tend to the asymptotic limit c_s/c_p (Eq. 2-289).

Now suppose Poisson's ratio is very close but not exactly equal to 1/2, for example, $\nu = 0.499$ with $c_p = 22.4\ c_s$. As c_p appears in the denominator, the power computed via Eq. 2-289 will be small, tending in the limit $\nu = 1/2$ to zero in accordance with Fig. 2-52a. As far as the resulting power is concerned, there is no contradiction between the case $\nu = 0.499$ (in which the damper is $22.4\ \rho c_s A_0$ but the trapped mass vanishes for very large values of a_0) and the case $\nu = 1/2$ (in which the damper is only $2\rho c_s A_0$ but trapped mass is present). Thus, we can conclude the following:

- *Above* a_0 *about 3.5 the portion of power transmitted by Rayleigh waves in the half-space becomes small; provided* c_p *is defined (i.e.,* $\nu \neq 1/2$*), the power is concentrated in P-waves which propagate vertically downward through the prismatic rod of soil directly under the foundation.*
- *As Poisson's ratio tends to 1/2, the power transmitted in high-frequency vibration approaches zero.*

The arguments above do not yet reconcile the concept of cone models with the significance of Rayleigh waves in the range of a_0 up to 3.5. To this end, it is necessary to distinguish between the so-called near and far fields of wave propagation. Figure 2-53 is a graphical representation of the vertical point-load solution [M12, M13] for Poisson's ratio $\nu = 1/4$. Figure 2-53a shows the variation of the magnitude of the surface displacement with distance away from the load, and Fig. 2-53b presents the corresponding variation of the phase angle. These relationships are discussed in depth in Section 5.1.2. Along the abscissas of the diagrams the radial distance r is normalized with respect to the wavelength of Rayleigh waves λ_R. The dimensionless measure of the surface displacement selected as the ordinate of Fig. 2-53a, $|u| 2\pi G r / ((1 - \nu) P)$ (in which G denotes the shear modulus of the soil), includes r in the numerator. This means that the horizontal, left-hand portion of the double-logarithmic plot corresponds to a decay in magnitude proportional to $1/r$, which is appropriate for cones. By contrast, the sloping, right-hand portion implies the decay law is proportional to $1/\sqrt{r}$, which is typical for Rayleigh waves. The intercept of the two straight-line asymptotes, located at $r = 0.423\ \lambda_R$, defines (somewhat arbitrarily) the boundary between the near and far fields. Roughly speaking, the far field begins about half a wavelength away from the load. Figure 2-53b shows that the effective velocity (phase velocity) of waves along the surface is essentially the Rayleigh-wave velocity c_R in both the near and far fields—more correctly, slightly less than c_R in the near field. To understand the essence of Rayleigh waves, a biological analogy is helpful. The Rayleigh wave is "conceived" at the source, "born" at the far-field boundary, and "lives" in the far field. The near field is the "pregnancy" region, in which the Rayleigh wave gradually develops.

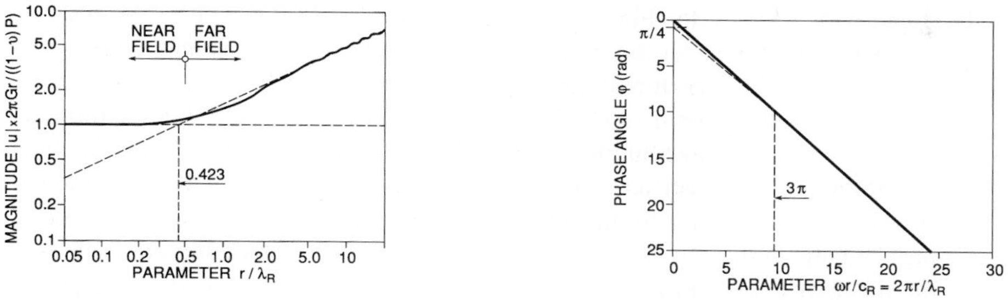

a. Magnitude. **b.** Phase angle.

Fig. 2-53 Vertical displacement of surface produced by harmonic point load on surface of halfspace with Poisson's ratio = 1/4.

Near zero frequency most of the radiated power is concentrated in Rayleigh waves; but the Rayleigh wavelength, which is inversely proportional to the frequency, is very large. The cone is located well within the far-field boundary $0.5\ \lambda_R$. In the interior near-field region the decay laws of cones, $1/r$ along the surface and $1/z$ with depth, are valid. As mentioned before, the effective horizontal propagation velocity c_h is only slightly less than the Rayleigh-wave velocity c_R, which prevails in the far field.

As the frequency increases, the far-field boundary moves inward toward the source; but at the same time the portion of power transmitted by Rayleigh waves diminishes according to Fig. 2-50. For a point load, the distance to the far-field boundary—half a Rayleigh wavelength—is $\pi c_R/\omega$ or about $2.9\ c_s/\omega$ for soil with Poisson's ratio between 1/4 and 1/3. In other words, the far-field boundary is located at radius r_0 from the point source when the dimensionless parameter a_0 is 2.9. By comparison, from Fig. 2-50 the Rayleigh wave associated not with a point load at the center of a membrane but rather with constant stress over the entire circular area of contact transmits no power when $a_0 = 3.5$. The rather small deviation between $a_0 = 2.9$ and 3.5 may be attributed to the difference in the stress distributions, so it may be concluded that the energy of the Rayleigh wave essentially vanishes when the far-field boundary has moved in to the edge of the foundation. These arguments imply that either

- *no Rayleigh waves exist in the portion of the halfspace occupied by the cones (low-frequency excitation), or*
- *the energy transmitted by Rayleigh waves is negligible (high-frequency excitation).*

As a result, the last "disadvantage" of cone models, their inability to represent Rayleigh waves, proves upon closer examination to be physically correct.

2.6 SIMPLE LUMPED-PARAMETER MODELS

2.6.1 Standard Lumped-Parameter Model

The discrete-element models shown in Fig. 2-5 are the exact representations of the translational cone. No additional internal degree of freedom is introduced. K and C are equal to the exact coefficients of the static stiffness and the damping, respectively, and a mass M is also introduced for the nearly incompressible soil in Fig. 2-5b.

As a slight generalization, but keeping the simplicity, the so-called *standard lumped-parameter model* shown in Fig. 2-54 can be used, which also covers the rotational motions. The static-stiffness coefficient of the soil's halfspace is equal to the coefficient of the direct spring K (denoted as K_ϑ for the rotational degrees of freedom in Sections 2.1 to 2.3 addressing the cone model), which leads to the exact result for static loading. The coefficients of the dashpot C and mass (or mass moment of inertia) M, thus *two free parameters*, are selected to reproduce as closely as possible the actual response of the total dynamic system (including the structure regarded as a rigid block with mass connected to the disk) in the low- and medium frequency ranges. Contrary to the concept of the substructure method, the curve-fitting technique is thus applied to the total system's dynamic-stiffness coefficient (disk with mass and soil) and not to that of the soil alone. The mass or mass moment of inertia of the soil is added to that of the structure in the foundation node, leading to a reduction of the spring coefficient of the dynamic-stiffness coefficient (see Eqs. 2-73a). This added mass does not mean that an identifiable mass of the soil really exists that moves with the same amplitude and in phase with the disk over the whole range of frequency. It is a totally fictitious quantity which serves only to obtain a better fit between the dynamic-stiffness coefficient of the lumped-parameter model and that of the actual soil, (in both cases taking the influence of the disk's mass representing that of the structure into consideration). In the higher-frequency range this added mass will lead to a significant error in the dynamic-stiffness coefficient's spring coefficient.

Fig. 2-54 Standard lumped-parameter model for

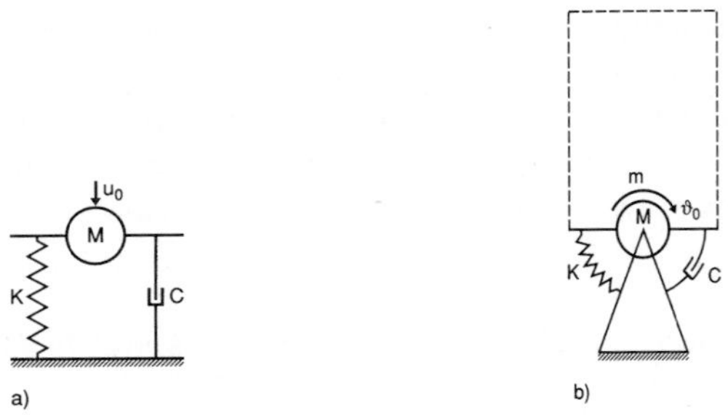

a. Translational motion.

b. Rotational motion.

As expected, slightly different coefficients of this simple ad hoc spring–dashpot–mass model are derived depending on the details of the curve-fitting procedure [W3, R1]. The following nomenclature is used: m denotes the mass of the rigid disk (representing the inertia effects of the structure) for the two translational degrees of freedom, the mass moment of inertia for the rocking motion and the polar mass moment of inertia for the torsional motion with respect to the center of the contact area of the disk of radius r_0 and the soil; G, ν, and ρ are the shear modulus, Poisson's ratio, and mass density of the soil. As already mentioned, the spring coefficient K is equal to the static-stiffness coefficient. The dashpot coefficient C and the mass M (which is a mass moment of inertia for the two rotational degrees of freedom) are specified based on the dimensionless coefficients γ and μ as

$$C = \frac{r_0}{c_s} \gamma K \tag{2-290a}$$

$$M = \frac{r_0^2}{c_s^2} \mu K \tag{2-290b}$$

where $c_s = \sqrt{G/\rho}$ is the shear-wave velocity. K, γ, and μ are listed in Table 2-9. For the two rotational degrees of freedom, the ratio of the mass moments of inertia of the disk and of the soil, $m/(r_0^5\rho)$, arises in the formulas for γ. For the translational degrees of freedom, γ and μ are independent of ν.

TABLE 2-9 STATIC STIFFNESS AND DIMENSIONLESS COEFFICIENTS OF STANDARD LUMPED-PARAMETER MODEL FOR DISK WITH MASS ON HOMOGENEOUS HALFSPACE

		Dimensionless Coefficients of	
	Static Stiffness K	Dashpot γ	Mass μ
Horizontal	$\frac{8Gr_0}{2-\nu}$	0.58	0.095
Vertical	$\frac{4Gr_0}{1-\nu}$	0.85	0.27
Rocking	$\frac{8Gr_0^3}{3(1-\nu)}$	$\frac{0.3}{1+\frac{3(1-\nu)m}{8r_0^5\rho}}$	0.24
Torsional	$\frac{16Gr_0^3}{3}$	$\frac{0.433}{1+\frac{2m}{r_0^5\rho}}\sqrt{\frac{m}{r_0^5\rho}}$	0.045

As this simplest possible lumped-parameter model for the soil consists only of a one-degree-of-freedom system (Fig. 2-54), the agreement of its dynamic-stiffness coefficient $S(a_0)$ as a function of frequency with that of the actual halfspace is limited. With

$$S(\omega) = K - \omega^2 M + i\omega C \tag{2-291}$$

or with the dimensionless frequency $a_0 = \omega r_0/c_s$

$$S(a_0) = K\left[k(a_0) + ia_0\, c(a_0)\right] \tag{2-292}$$

follows with the dimensionless spring coefficient $k(a_0)$ and damping coefficient $c(a_0)$

$$k(a_0) = 1 - \mu\, a_0^2 \tag{2-293a}$$

$$c(a_0) = \gamma \tag{2-293b}$$

$c(a_0)$ is constant, which is not particularly well suited to represent the increase from zero occurring for the two rotational motions (Figs. 2-20b, 2-21b, and 2-21d). Also, $k(a_0)$ is represented by a parabola which restricts the use of this standard lumped-parameter model to the low- and intermediate-frequency ranges ($a_0 < 1.5$). Note also that the high-frequency limit of the dashpot equals $(r_0/c_s)\gamma K$ which leads for the horizontal motion to $8\gamma r_0^2 \rho c_s/(2 - \nu)$. With $\gamma = 0.58$ the dashpot per unit area is equal to $8\gamma/[\pi(2 - \nu)]\ \rho c_s = 1.477/(2 - \nu)\ \rho c_s$ which differs from the exact value $C/(\pi r_0^2) = \rho c_s$ (Eq. 2-38b). The standard lumped-parameter model is thus not a doubly-asymptotic approximation, in contrast to the cone model.

Frictional material damping can be introduced as described in Section 2.4.4.

2.6.2 Fundamental Lumped-Parameter Model

The fundamental *lumped-parameter model* with, as will become apparent, *four free parameters* can be derived starting with the general arrangement of springs, dashpots, and masses of the discrete-element model of the rotational cone with one additional internal degree of freedom (Fig. 2-17). The corresponding spring–dashpot and monkey-tail models are shown for translational motion in Figs. 2-55a and 2-55b (without the additional mass). The coefficient K of the direct spring is set equal to the static-stiffness coefficient of the disk on the surface of the soil's halfspace, resulting in an exact result for static loading. Four free parameters—the dimensionless coefficients of K_1, C_1, C_0, M_0 for the spring–dashpot model, or of C_1, M_1, C_0, M_0 for the monkey-tail model—are selected so as to achieve an optimum fit between the dynamic-stiffness coefficient of the fundamental lumped-parameter model and the corresponding exact value of the disk on the soil's halfspace.

For each degree of freedom (horizontal, vertical, rocking, torsional), a fundamental lumped-parameter model independent of the others and acting in the corresponding direction is introduced. Addressing the monkey-tail model (Fig. 2-55b), the dashpots C_0, C_1 and the masses or mass moments of inertia M_0, M_1 are specified based on the dimensionless coefficients γ_0, γ_1, μ_0, μ_1 as

a. Spring-dashpot model. **b.** Monkey-tail model.

Fig. 2-55 Fundamental lumped–parameter model.

$$C_0 = \frac{r_0}{c_s}\,\gamma_0\,K \tag{2-294a}$$

$$C_1 = \frac{r_0}{c_s}\,\gamma_1\,K \tag{2-294b}$$

$$M_0 = \frac{r_0^2}{c_s^2}\,\mu_0\,K \tag{2-294c}$$

$$M_1 = \frac{r_0^2}{c_s^2}\,\mu_1\,K \tag{2-294d}$$

K and γ_0, γ_1, μ_0, μ_1 for a massless rigid disk of radius r_0 resting on the surface of an elastic homogeneous halfspace of shear modulus G, Poisson's ratio ν and shear-wave velocity c_s are specified in Table 2-10 [W5]. The equations are derived making use of the values for discrete values of ν specified in Veletsos and Verbic [V2]. The coefficients γ_0, γ_1, μ_0, μ_1 depend—with the exception of the torsional motion—on ν. Some elements of the monkey-tail model shown in Fig. 2-55b can be missing. Negative coefficients do not arise. For the torsional motion, additional values corresponding to an improved fundamental lumped-parameter model which are very accurate are specified in parentheses (see Section 2.6.3).

TABLE 2-10 STATIC STIFFNESS AND DIMENSIONLESS COEFFICIENTS OF FUNDAMENTAL LUMPED-PARAMETER MODEL (MONKEY-TAIL ARRANGEMENT) FOR DISK ON HOMOGENEOUS HALFSPACE

	Static Stiffness	Dimensionless Coefficients of				
		Dashpots		Masses		
	K	γ_0	γ_1		μ_0	μ_1
Horizontal	$\frac{8Gr_0}{2-\nu}$	$0.78 - 0.4\,\nu$	—		—	—
Vertical	$\frac{4Gr_0}{1-\nu}$	0.8	$0.34 - 4.3\,\nu^4$	$\nu < \frac{1}{3}$	0	$0.4 - 4\,\nu^4$
				$\nu > \frac{1}{3}$	$0.9\left(\nu - \frac{1}{3}\right)$	
Rocking	$\frac{8Gr_0^3}{3\,(1-\nu)}$	—	$0.42 - 0.3\,\nu^2$	$\nu < \frac{1}{3}$	0	$0.34 - 0.2\,\nu^2$
				$\nu > \frac{1}{3}$	$0.16\left(\nu - \frac{1}{3}\right)$	
Torsional	$\frac{16Gr_0^3}{3}$	—	0.29		—	0.2
		(0.017)	(0.291)		—	(0.171)

The dynamic-stiffness coefficient for harmonic loading at the foundation node relating the displacement amplitude $u_0(\omega)$ to the applied load amplitude $P_0(\omega)$ is calculated for the monkey-tail model as follows. Formulating the equilibrium equations in the internal node and in the foundation node leads to

$$-\omega^2 M_1 u_1(\omega) + i\omega C_1\left[u_1(\omega) - u_0(\omega)\right] = 0 \tag{2-295}$$

$$-\omega^2 M_0 u_0(\omega) + i\omega\left(C_0+C_1\right) u_0(\omega) - i\omega C_1 u_1(\omega) + K u_0(\omega) = P_0(\omega) \tag{2-296}$$

Eliminating $u_1(\omega)$ from Eqs. 2-295 and 2-296 results in

$$P_0(\omega) = K\left[1 - \frac{\dfrac{\omega^2 M_1}{K}}{1 + \dfrac{\omega^2 M_1^2}{C_1^2}} - \frac{\omega^2 M_0}{K} + i\omega\left(\frac{M_1}{C_1}\frac{\dfrac{\omega^2 M_1}{K}}{1 + \dfrac{\omega^2 M_1^2}{C_1^2}} + \frac{C_0}{K}\right)\right] u_0(\omega) \tag{2-297}$$

The coefficient on the right-hand side represents the dynamic-stiffness coefficient $S(\omega)$, which can be written in the familiar form

$$S(a_0) = K\left[k(a_0) + i a_0\, c(a_0)\right] \tag{2-298}$$

After substituting Eq. 2-294 in Eq. 2-297, the dimensionless dynamic spring and damping coefficients follow as

$$k(a_0) = 1 - \frac{\mu_1 a_0^2}{1 + \dfrac{\mu_1^2}{\gamma_1^2} a_0^2} - \mu_0\, a_0^2 \tag{2-299a}$$

$$c(a_0) = \frac{\mu_1}{\gamma_1}\frac{\mu_1 a_0^2}{1 + \dfrac{\mu_1^2}{\gamma_1^2} a_0^2} + \gamma_0 \tag{2-299b}$$

The coefficients γ_1 and μ_1 appear in both coefficients.

For the vertical motion, the dimensionless spring and damping coefficients k_v and c_v, determined from Eq. 2-299 using the values of Table 2-10, are plotted as lines as a function of a_0 in Fig. 2-56. The rigorous values reported in Veletsos and Verbic [V2] are also shown as distinct points. For all three Poisson's ratios the agreement is good. For $\nu > 1/3$, the fundamental lumped-parameter model consists of all elements shown in Fig. 2-55b.

For rocking, k_r and c_r for the fundamental lumped-parameter model, plotted as lines (Fig. 2-57) are compared to the rigorous results of Veletsos and Wei [V1]. For $\nu = 0$ and $\nu = 1/3$, the monkey-tail model consists, in addition to the static rotational spring, only of the mass moment of inertia M_1 with its own degree of freedom and the connecting rotational dashpot with the coefficient C_1. These two elements are well suited to model the decrease of k_r and the increase of c_r with increasing a_0. As c_r is zero for $a_0 \rightarrow 0$, the rotational dashpot with the coefficient C_0 is not present (Eq. 2-299b). (In Fig. 1-8c k_r and c_r for $\nu = 1/3$ are also plotted. The results for the (fundamental) lumped-parameter model are more accurate than

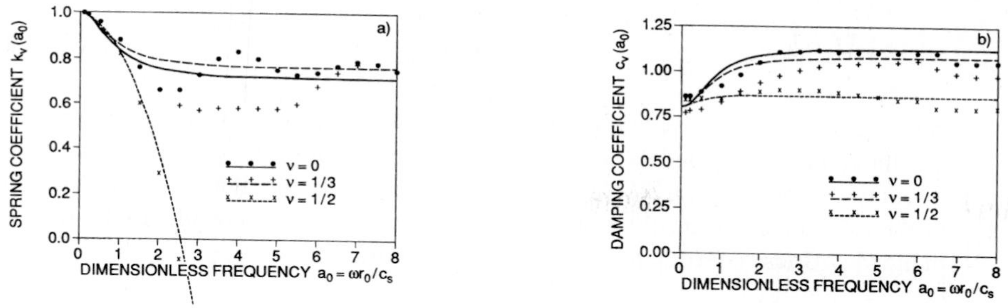

Fig. 2-56 Dynamic-stiffness coefficient of disk on homogeneous halfspace in vertical motion for various Poisson's ratios calculated with fundamental lumped-parameter model.

those for the discrete-element model of the rotational cone. See Fig. 2-21 for the results of the cone model for other ν.) For $\nu > 1/3$, the mass moment of inertia M_0 is also present; it is attached to the disk and thus vibrates in phase. This element is responsible for the strong decrease of k_r, which even becomes negative for sufficiently large a_0 (Eq. 2-299a). The torsional motion is examined in Section 2.6.3.

The high-frequency limit of the damping coefficient $c(a_0 \to \infty)$ equals $\gamma_1 + \gamma_0$ (Eq. 2-299b). The corresponding dashpot per unit area is calculated, for the horizontal motion, as $(r_0/c_s)(\gamma_1 + \gamma_0)K/(\pi r_0^2) = 8(\gamma_1 + \gamma_0)/[\pi(2 - \nu)]\rho c_s = (1.99 - 1.02\nu)/(2 - \nu)\rho c_s$ which differs only slightly from the exact value ρc_s (which equals 0.99 for $\nu = 1/3$ and for $\nu = 0.5$). The fundamental lumped-parameter model is thus *almost a doubly-asymptotic approximation.*

The dimensionless coefficients specified in Table 2-10 apply to the monkey-tail model shown in Fig. 2-55b. For the spring–dashpot model of Fig. 2-55a the elements are expressed by their own dimensionless coefficients as

$$K_1 = \bar{\kappa}\, K \tag{2-300a}$$

$$C_0 = \frac{r_0}{c_s}\, \bar{\gamma}_0\, K \tag{2-300b}$$

$$C_1 = \frac{r_0}{c_s}\, \bar{\gamma}_1\, K \tag{2-300c}$$

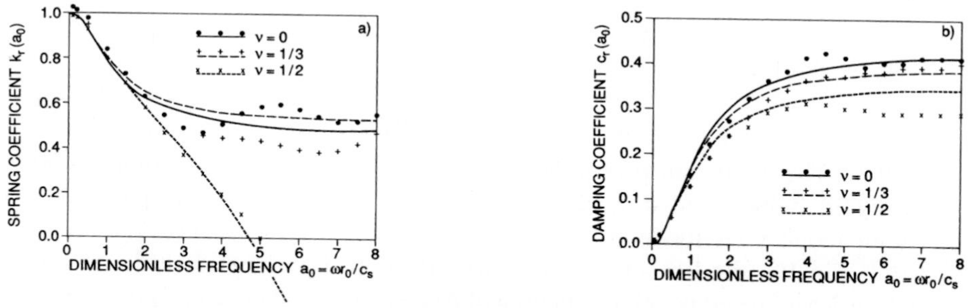

Fig. 2-57 Dynamic-stiffness coefficient of disk on homogeneous halfspace in rocking motion for various Poisson's ratios calculated with fundamental lumped-parameter model.

$$M_0 = \frac{r_0^2}{c_s^2} \bar{\mu}_0 K \tag{2-300d}$$

An overbar ($\bar{\ }$) denotes a coefficient of the spring–dashpot model. As the spring–dashpot and the monkey-tail models are equivalent, the coefficients $\bar{\kappa}$, $\bar{\gamma}_0$, $\bar{\gamma}_1$, $\bar{\mu}_0$ can be expressed as a function of γ_0, γ_1, μ_0, μ_1. This is achieved by equating the dynamic-stiffness coefficients of the two models. Canceling with $1 - i(\mu_1/\gamma_1)a_0$ in Eq. 2-298 results in the following normalized stiffness coefficient for the monkey-tail model

$$\frac{S(a_0)}{K} = \frac{1 - (\gamma_0 + \gamma_1)\dfrac{\mu_1}{\gamma_1} a_0^2 + i\left(\dfrac{\mu_1}{\gamma_1} + \gamma_0\right) a_0}{1 + i\dfrac{\mu_1}{\gamma_1} a_0} - \mu_0 a_0^2 \tag{2-301}$$

Formulating the equilibrium equations for harmonic loading in the internal node and in the foundation node of the spring–dashpot model of Fig. 2-55a leads to

$$i\omega C_1 u_1(\omega) + K_1 u_1(\omega) - K_1 u_0(\omega) = 0 \tag{2-302}$$

$$-\omega^2 M_0 u_0(\omega) + i\omega\, C_0 u_0(\omega) + (K + K_1)\, u_0(\omega) - K_1 u_1(\omega) = P_0(\omega) \tag{2-303}$$

Eliminating $u_1(\omega)$ from Eqs. 2-302 and 2-303 and substituting Eq. 2-300 results in the normalized dynamic-stiffness coefficient for the spring–dashpot model.

$$\frac{S(a_0)}{K} = \frac{1 - \dfrac{\bar{\gamma}_0 \bar{\gamma}_1}{\bar{\kappa}} a_0^2 + i\left(\bar{\gamma}_0 + \bar{\gamma}_1 + \dfrac{\bar{\gamma}_1}{\bar{\kappa}}\right) a_0}{1 + i\dfrac{\bar{\gamma}_1}{\bar{\kappa}} a_0} - \bar{\mu}_0 a_0^2 \tag{2-304}$$

Setting the coefficient of ia_0 in the denominator, $\bar{\gamma}_1/\bar{\kappa}$, equal to the corresponding value of Eq. 2-301, μ_1/γ_1, and proceeding analogously for the coefficients of a_0^2 and ia_0 in the numerator and for the last coefficient leads to a system of four equations with four unknowns. After solution

$$\bar{\kappa} = -\frac{\gamma_1^2}{\mu_1} \tag{2-305a}$$

$$\bar{\gamma}_0 = \gamma_0 + \gamma_1 \tag{2-305b}$$

$$\bar{\gamma}_1 = -\gamma_1 \tag{2-305c}$$

$$\bar{\mu}_0 = \mu_0 \tag{2-305d}$$

follow. These relations permit the construction of the spring–dashpot model of Fig. 2-55a using Eq. 2-300 based on the dimensionless coefficients γ_0, γ_1, μ_0, μ_1 specified in Table 2-10 of the monkey-tail model of Fig. 2-55b.

Finally, the link of the fundamental lumped-parameter model to the consistent lumped-parameter model formulation of Appendix B (and the corresponding recursive procedure of Appendix C) is established. The dynamic-stiffness coefficient of the fundamental lumped-parameter model given in Eq. 2-301 is rewritten as

$$\frac{S(a_0)}{K} = 1 - \frac{\gamma_1^2}{\mu_1} + ia_0\left(\gamma_0+\gamma_1\right) + \frac{\frac{\gamma_1^3}{\mu_1^2}}{ia_0 + \frac{\gamma_1}{\mu_1}} - \mu_0 a_0^2 \tag{2-306}$$

The last term on the right-hand side represents the contribution of the mass M_0 with coefficient μ_0 which is not addressed in Appendix B and can thus be disregarded in the following discussion. The first two terms are the singular part as described in Eq. B-3 (see also Eq. 2-299), leading to

$$k = 1 - \frac{\gamma_1^2}{\mu_1} \tag{2-307a}$$

$$c = \gamma_0 + \gamma_1 \tag{2-307b}$$

where k and c are equal to the dimensionless coefficients of the direct spring κ and dashpot γ (Eq. B-17, Fig. B-1). The third term represents a first-order term $A/(ia_0 - s)$ of the regular part, yielding

$$A = \frac{\gamma_1^3}{\mu_1^2} \tag{2-308a}$$

$$s = -\frac{\gamma_1}{\mu_1} \tag{2-308b}$$

which results in $\gamma = \gamma_1$ and $\mu = \mu_1$ (Eq. B-22) for the monkey-tail model of Fig. B-2b. Equation 2-306 is thus in the form of Eq. B-12 with one first-order term. Assembling the discrete-element models of the singular part and the first-order term leads to a direct spring with the coefficient $(1 - \gamma_1^2/\mu_1 + \gamma_1^2/\mu_1)K = K$, to a direct dashpot with the coefficient $(r_0/c_s)(\gamma_0 + \gamma_1 - \gamma_1)K = (r_0/c_s)\gamma_0 K$ and to a mass $M_1 = (r_0/c_s)^2\mu_1 K$ with its own internal degree of freedom connected to the foundation node with a dashpot with the coefficient $C_1 = (r_0/c_s)\gamma_1 K$. This is exactly the same model as shown in Fig. 2-55b. The fundamental lumped-parameter model thus fits into the formulation with $M = 1$ of the *consistent lumped-parameter models* discussed in Appendix B, which allows a systematic generalization.

Further lumped-parameter models without increasing the number of internal degrees of freedom could be constructed. For instance, a spring could be attached between the internal node and the foundation node in parallel to the dashpot with coefficient C_1 in Fig. 2-55b. Such a model would, however, not be compatible with the consistent formulation of Appendixes B and C.

Frictional material damping can be considered in the lumped-parameter model as discussed in Section 2.4.4.

2.6.3 Torsional Motion of Disk with Polar Mass Moment of Inertia on Halfspace

For the sake of illustration, the rigid disk of radius r_0 with a polar mass moment of inertia m (the structure) resting on an undamped elastic halfspace of shear modulus G and mass density ρ (the soil) loaded by a torsional moment T (Fig. 2-58a) is analyzed using the different simple lumped-parameter models to represent the soil.

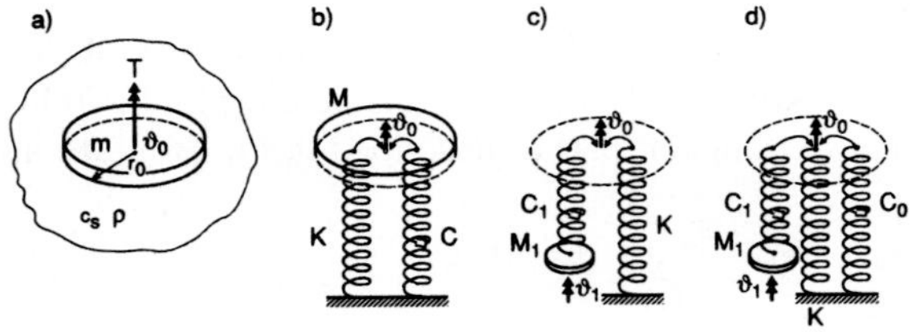

a. Rigid disk with polar mass moment of inertia on halfspace loaded by torsional moment.

b. Standard lumped-parameter model with added polar mass moment of inertia.

c. Discrete-element model and fundamental lumped-parameter model with additional internal degree of freedom.

d. Improved fundamental lumped-parameter model with additional degree of freedom.

Fig. 2-58 Simple lumped-parameter models.

$$T(t) = \frac{T_0}{2}\left[1 - \cos\left(2\pi\frac{t}{t_0}\right)\right] \qquad 0 < t < t_0 \qquad \text{(2-309a)}$$

$$T(t) = 0 \qquad t > t_0 \qquad \text{(2-309b)}$$

This loading is a rounded triangular pulse with a zero value at $t = 0$, a maximum value T_0 at $t = t_0/2$ and which is again zero at $t = t_0$. For the actual calculation, $m = 0.3\rho r_0^5$ and $t_0 = r_0/c_s$ are selected (c_s = shear-wave velocity).

Four calculations are performed. First, the standard lumped-parameter model shown in Fig. 2-58b is selected for the soil where a polar mass moment of inertia M is added to that of the disk, but no internal degree of freedom is introduced. Using the values in Table 2-9 and Eq. 2-290 leads to

$$K = \frac{16}{3}Gr_0^3 \qquad \text{(2-310a)}$$

$$C = 0.148\frac{r_0}{c_s}K = \frac{1}{1 + \frac{2m}{r_0^5\rho}}\sqrt{Km} \qquad \text{(2-310b)}$$

$$M = 0.045\frac{r_0^2}{c_s^2}K = 0.24r_0^5\rho \qquad \text{(2-310c)}$$

The static-stiffness coefficient K also applies to the other models. Second, the discrete-element model (Fig. 2-58c) of the rotational cone with an internal degree of freedom is applied. It follows from Fig. 2-17b and Eqs. 2-146b and 2-152 that $C_1 = C_\vartheta$ and $M_1 = M_\vartheta$ and with $z_0/r_0 = 0.884$ (Table 2-3B), resulting in

$$C_1 = 0.295\frac{r_0}{c_s}K \qquad \text{(2-311a)}$$

$$M_1 = 0.260 \frac{r_0^2}{c_s^2} K \qquad (2\text{-}311b)$$

Third, the fundamental lumped-parameter model of Table 2-10 with $\gamma_1 = 0.29$ and $\mu_1 = 0.2$ yields with Eq. 2-294 (Fig. 2-58c)

$$C_1 = 0.290 \frac{r_0}{c_s} K \qquad (2\text{-}312a)$$

$$M_1 = 0.200 \frac{r_0^2}{c_s^2} K \qquad (2\text{-}312b)$$

Fourth, the improved fundamental lumped-parameter using the coefficients γ_0, γ_1, μ_1 specified in parentheses in Table 2-10 leads to (Fig. 2-58d)

$$C_0 = 0.017 \frac{r_0}{c_s} K \qquad (2\text{-}313a)$$

$$C_1 = 0.291 \frac{r_0}{c_s} K \qquad (2\text{-}313b)$$

$$M_1 = 0.171 \frac{r_0^2}{c_s^2} K \qquad (2\text{-}313c)$$

The torsional natural frequency of the dynamic system with the improved fundamental lumped-parameter model follows as

$$\omega_t = \sqrt{\frac{K\, k(a_0)}{m}} \qquad (2\text{-}314)$$

where $k(a_0)$ is specified in Eq. 2-299a and with $a_0 = \omega_t r_0/c_s$. Equation 2-314 is solved iteratively leading to $\omega_t = 3.29 c_s/r_0$ and $a_0 = 3.29$. This large value is caused by the small mass ratio $m/(\rho r_0^5) = 0.3$. [For instance, for the machine foundation block in Section 2.3.4, $m/(\rho r_0^5) = 2.3$, leading to $a_0 = 1.36$]. The damping ratio at ω_t is calculated as

$$\zeta_t = \frac{a_0 c(a_0)}{2k(a_0)} = 0.666 \qquad (2\text{-}315)$$

with $c(a_0)$ specified in Eq. 2-299b. The very large torsional damping ratio arises because of the large a_0.

The torsional dynamic-stiffness coefficients of the massless disk on the homogeneous halfspace for the standard lumped-parameter model (Eq. 2-293), the discrete-element model, the fundamental and improved fundamental lumped-parameter models (Eq. 2-299) are compared with the rigorous results of Veletsos and Nair [V4] denoted as exact points in Fig. 2-59. The standard lumped-parameter model can only be used for $a_0 < 2$. The fundamental lumped-parameter model is more accurate than the discrete-element model. The improved fundamental lumped-parameter model leads to an excellent agreement of the damping coefficient c_t not only in the range of curve fitting ($0 < a_0 < 4$) performed as described in Section B1.2, but also up to $a_0 = 8$ and to limited deviations of the spring coefficient k_t in the higher-frequency range, where the latter is not important.

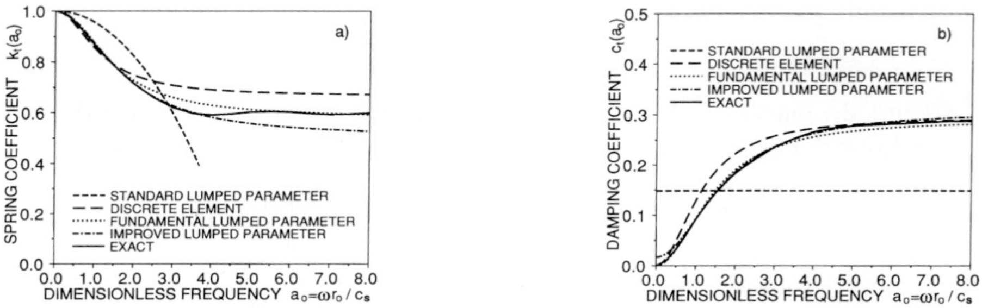

Fig. 2-59 Dynamic-stiffness coefficient for harmonic loading of disk on homogeneous halfspace in torsional motion.

The equations of motion of the rigid disk with the polar mass moment of inertia and the discrete-element model or the fundamental lumped-parameter models (Figs. 2-58c and 2-58d) are formulated as

$$m\ddot{\vartheta}_0 + C_0\,\dot{\vartheta}_0 + C_1\left(\dot{\vartheta}_0 - \dot{\vartheta}_1\right) + K\vartheta_0 = T \tag{2-316a}$$

$$M_1\ddot{\vartheta}_1 + C_1\left(\dot{\vartheta}_1 - \dot{\vartheta}_0\right) = 0 \tag{2-316b}$$

For the model of Fig. 2-58c, C_0 vanishes. For the time integration, an explicit algorithm with a predictor–corrector scheme is applied (see Section 2.2.4). Starting from the known motion at time $(n - 1)\Delta t$—that is ϑ_{0n-1}, $\dot{\vartheta}_{0n-1}$, $\ddot{\vartheta}_{0n-1}$, ϑ_{1n-1}, $\dot{\vartheta}_{1n-1}$, $\ddot{\vartheta}_{1n-1}$—the final rotations ϑ_{0n}, ϑ_{1n}, and the predicted rotational velocities $\tilde{\dot{\vartheta}}_{0n}$, $\tilde{\dot{\vartheta}}_{1n}$ at time $n\Delta t$ are calculated as

$$\vartheta_{0n} = \vartheta_{0n-1} + \Delta t\dot{\vartheta}_{0n-1} + \frac{\Delta t^2}{2}\ddot{\vartheta}_{0n-1} \tag{2-317a}$$

$$\tilde{\dot{\vartheta}}_{0n} = \dot{\vartheta}_{0n-1} + \frac{\Delta t}{2}\ddot{\vartheta}_{0n-1} \tag{2-317b}$$

with analogous relations for ϑ_{1n}, $\tilde{\dot{\vartheta}}_{1n}$. A tilde (~) denotes a predicted value. The accelerations $\ddot{\vartheta}_{0n}$, $\ddot{\vartheta}_{1n}$ follow from Eqs. 2-316 formulated at time $n\Delta t$

$$\ddot{\vartheta}_{0n} = \frac{T_n - K\vartheta_{0n} - C_0\tilde{\dot{\vartheta}}_{0n} - C_1\left(\tilde{\dot{\vartheta}}_{0n} - \tilde{\dot{\vartheta}}_{1n}\right)}{m} \tag{2-318a}$$

$$\ddot{\vartheta}_{1n} = \frac{C_1}{M_1}\left(\tilde{\dot{\vartheta}}_{0n} - \tilde{\dot{\vartheta}}_{1n}\right) \tag{2-318b}$$

The rotational velocities are corrected as

$$\dot{\vartheta}_{0n} = \tilde{\dot{\vartheta}}_{0n} + \frac{\Delta t}{2}\ddot{\vartheta}_{0n} \tag{2-319}$$

and analogously for $\dot{\vartheta}_{1n}$. This concludes the calculations for a time step.

For stability of the explicit algorithm (Eq. 2-121), the time step Δt is subjected to

$$\Delta t < \frac{2}{\omega_t} = 0.607\,\frac{r_0}{c_s} \tag{2-320}$$

To achieve a very accurate description of the load, 16 time steps are selected in t_0, leading to $\Delta t = t_0/16 = 0.0625 c_s/r_0$, which is more stringent than the Δt of the stability criterion.

The rotation ϑ_0 is plotted in nondimensional form as a function as $\bar{t} = tc_s/r_0$ for the various models of the soil in Fig. 2-60. The exact solution of Wolf [W6] which is determined in the frequency domain using the rigorous dynamic-stiffness coefficient is also shown. The standard lumped-parameter model, while predicting the maximum response (Fig 2-60a) well where the static effect dominates, underestimates damping considerably (damping coefficient $\gamma = 0.148$ instead of 0.247 at the natural frequency of the improved fundamental lumped-parameter model), but results in an acceptable (undamped) natural frequency $\omega_t = \sqrt{K/(m+M)} = 3.14\, c_s/r_0$. The corresponding damping ratio ζ_t of this one-degree-of-freedom system equals 0.232, resulting in a damped frequency $\omega_d = \omega_t \sqrt{1-\zeta_t^2} = 3.05\, c_s/r_0$. This value differs from the damped frequency of the improved fundamental lumped-parameter model $\omega_d = 2.454\, c_s/r_0$ (Eqs. 2-314, 2-315). Obviously, the dominant range of a_0 in this problem lies outside the range where the standard lumped-parameter model is applicable. The discrete-element model corresponding to the cone model overestimates damping (Fig. 2-60b) as expected, which is visible in the free-vibration phase of the motion. The fundamental lumped-parameter models are very accurate, the improved model yielding slightly better results (Fig. 2-60a).

Fig. 2-60 Torsional rotation of disk for

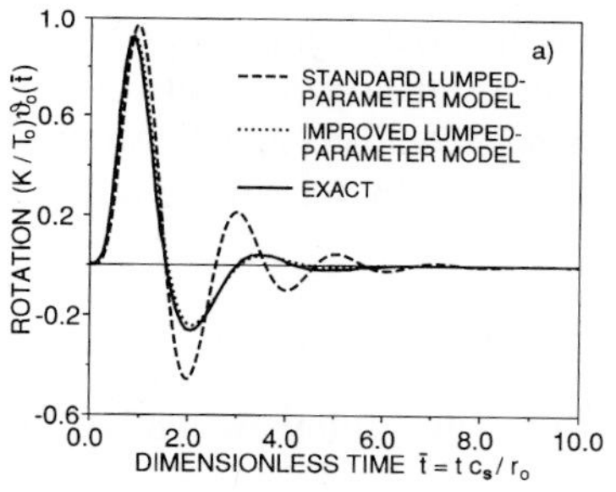

a. Standard lumped-parameter model and improved fundamental lumped-parameter model.

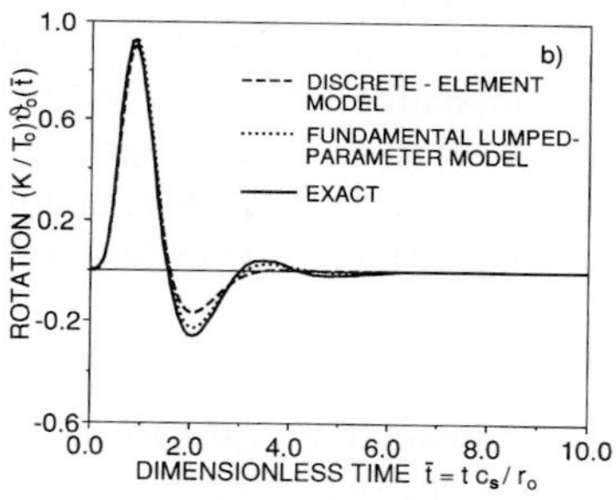

b. Discrete-element model and fundamental lumped-parameter model.

2.7 TRANSLATIONAL AND ROCKING WEDGES

2.7.1 Dynamic-Stiffness Coefficient

Analogous to cones in the three-dimensional case, truncated semi-infinite wedges can be used to model approximately the dynamic-stiffness coefficients of a two-dimensional strip foundation on the surface of a homogeneous soil halfplane. The translational wedge can represent either the horizontal or the vertical motion, whereas the rotational wedge models the rocking motion.

Two-dimensional practical applications are not as common as three-dimensional ones. As described in Section 2.8, it is not possible to construct an equivalent two-dimensional model to adequately represent a three-dimensional situation, not even for quite slender foundations. In addition, the derivations must be performed in the frequency domain and simple discrete-element models to represent the wedges exactly do not exist. For all these reasons, wedges are not addressed in this section as thoroughly as cones are in Sections 2.1 to 2.3. This section is based on Wolf and Meek [W15].

The derivation of the dynamic-stiffness coefficients for harmonic excitation can be unified for cones and wedges as follows. The cone or wedge with apex height z_0, depth z, and surface area A_0 is shown for the vertical degree of freedom in Fig. 2-61. In the horizontal direction the area A is plotted instead of the width of the cone or wedge, which always increases linearly with depth. The cone or wedge experiences a vertical or a horizontal excitation with amplitude $u(\omega)$, the value at the surface being denoted as $u_0(\omega)$. In Fig. 2-61 the argument ω is omitted for the sake of conciseness. For rotational motion the same notation may be preserved if A_0 is interpreted as the moment of inertia of the foundation about its axis and $u_0(\omega)$ is taken to be the angle of rocking or twist. For the cone–wedge family the cross section increases with depth according to the power law $A = A_0(z/z_0)^n$, with z measured from the apex. The values of the exponent n = 1, 2, 3, 4 correspond in turn to the two-dimensional wedge in translation, the three-dimensional cone in translation, the two-dimensional wedge in rocking, and the three-dimensional cone in rotation (rocking and torsion). (The prismatic bar is described by $n = 0$.) If c denotes the appropriate wave velocity (c equals shear-wave velocity c_s for the horizontal and torsional motions, equals dilatational-wave velocity c_p for the vertical and rocking motions, with a limit $2c_s$ in the nearly incompressible case; see Section 2.1.4) and ρ stands for the mass density, the quantity ρc^2 is equal to the corresponding elastic modulus (shear modulus G and constrained modulus E_c, respectively). The equilibrium equation for harmonic motion of the infinitesimal element (Fig. 2-61) including the inertial load is

$$-N(\omega) + N(\omega) + N(\omega)_{,z}\, dz + \rho\, A\, dz\, \omega^2\, u(\omega) = 0 \tag{2-321}$$

Substitution of the force-displacement relationship (with $N(\omega)$ denoting the amplitude of the internal force or moment)

$$N(\omega) = \rho\, c^2\, A\, u\,(\omega)_{,z} \tag{2-322}$$

leads to the governing equation of motion

$$u(\omega)_{,zz} + \frac{n}{z}\, u(\omega)_{,z} + \frac{\omega^2}{c^2}\, u(\omega) = 0 \tag{2-323}$$

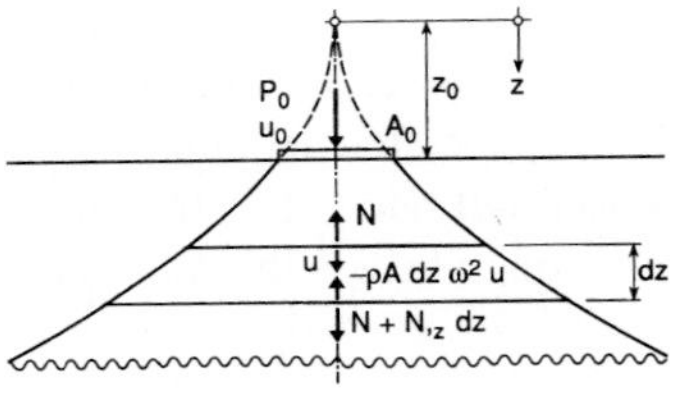

Fig. 2-61 Truncated semi-infinite cone or wedge with equilibrium of infinitesimal element.

Before specifying the general solution of Eq. 2-323, the two limiting cases of $\omega \to 0$ and $\omega \to \infty$ are addressed. For the static case ($\omega = 0$) the solution of

$$u_{,zz} + \frac{n}{z} u_{,z} = 0 \tag{2-324}$$

is assumed as

$$u = z^{\alpha} \tag{2-325}$$

Insertion of Eq. 2-325 into Eq. 2-324 leads to the following quadratic equation for the unknown degree α of the polynomial

$$\alpha^2 + (n-1)\,\alpha = 0 \tag{2-326}$$

The two roots for α are $\alpha_1 = 0$, $\alpha_2 = -(n-1)$, so that

$$u = c_1 + c_2 \frac{1}{z^{n-1}} \tag{2-327}$$

The boundary condition

$$u(z = \infty) = 0 \tag{2-328a}$$

implies $c_1 = 0$, whereas the boundary condition

$$u(z = z_0) = u_0 \tag{2-328b}$$

is satisfied when

$$u = u_0 \left(\frac{z_0}{z}\right)^{n-1} \tag{2-329}$$

The force amplitude is

$$P_0 = -N(z = z_0) = -\rho c^2 A_0 u_{0,z} \tag{2-330}$$

in which the derivative $u_{0,z}$ is calculated from Eq. 2-329, leading to

$$P_0 = (n-1)\frac{\rho c^2 A_0}{z_0} u_0 \tag{2-331}$$

which defines the static-stiffness coefficient

$$K = (n-1)\frac{\rho c^2 A_0}{z_0} \tag{2-332}$$

For the high-frequency limit ($\omega = \infty$), the asymptotic solution of Eq. 2-323 for outwardly propagating waves takes the form

$$u(\omega) = u_0(\omega)\left(\frac{z_0}{z}\right)^m e^{-i\frac{\omega}{c}(z-z_0)} \tag{2-333}$$

with the exponent m to be determined. As $u(\omega)$ is multiplied by $e^{+i\omega t}$, producing a term $e^{i\omega(t-(z-z_0)/c)}$, the wave of this solution does indeed propagate in the positive z-direction. Substituting Eq. 2-333 into Eq. 2-323 yields

$$\frac{c}{i\omega}\frac{m(m+1-n)}{z} + (2m-n) = 0 \tag{2-334}$$

This criterion must be fulfilled. For $\omega \rightarrow \infty$ the second term tends to dominate and vanishes only for $m = n/2$. The first term then equals $-n^2/4 + n/2$ which vanishes identically for $n = 0$ (prismatic rod) and for $n = 2$ (translational cone). For these two values of n the asymptotic solution is valid for all ω. The asymptotic solution valid for $\omega \rightarrow \infty$ for any n thus equals

$$u(\omega) = u_0(\omega)\left(\frac{z_0}{z}\right)^{\frac{n}{2}} e^{-i\frac{\omega}{c}(z-z_0)} \tag{2-335}$$

The dynamic-stiffness coefficient in the high-frequency limit follows from

$$P_0(\omega) = -\rho c^2 A_0 u_0(\omega)_{,z} \tag{2-336}$$

as

$$S(\omega \rightarrow \infty) = \frac{n}{2}\frac{\rho c^2 A_0}{z_0} + i\omega\rho c A_0 \tag{2-337}$$

or

$$S(\omega \rightarrow \infty) = K_\infty + i\omega C \tag{2-338}$$

with

$$K_\infty = \frac{n}{2}\frac{\rho c^2 A_0}{z_0} \tag{2-339a}$$

$$C = \rho c A_0 \tag{2-339b}$$

The factor $n/2$ appearing in the spring term reflects the fact that the amplitude in Eq. 2-335 decays in proportion to $z^{-n/2}$ and not in proportion to $z^{-(n-1)}$, as in the static case. The high-frequency dashpot has the familiar specific damping ρc, which is the same for all values of n.

For intermediate frequencies the general solution of Eq. 2-323 may be written in terms of half-order Hankel functions of the second kind. The general expression for the dynamic-stiffness coefficient is

$$S(b_0) = \frac{b_0 H^{(2)}_{\frac{n+1}{2}}(b_0)}{H^{(2)}_{\frac{n-1}{2}}(b_0)}\frac{\rho c^2 A_0}{z_0} \tag{2-340}$$

in which the dimensionless frequency parameter b_0 is defined with respect to the properties of the cone or wedge

$$b_0 = \frac{\omega z_0}{c} \tag{2-341}$$

To discuss the properties of $S(b_0)$, it is appropriate to decompose the expression as follows:

$$S(b_0) = K(b_0) + i\omega\, C(b_0) \tag{2-342}$$

The spring coefficient $K(b_0)$ nondimensionalized with $\rho c^2 A_0/z_0$ and the damping coefficient $C(b_0)$ divided by $\rho c A_0$ are shown together with the damping ratio $\zeta(b_0)$ (ratio of the imaginary part of the dynamic-stiffness coefficient to twice the real part) in Fig. 2-62. Note that the ordinate of the damping ratio ζ is plotted logarithmically. As already mentioned, $n = 1$ corresponds to the translational wedge, $n = 2$ to the translational cone, $n = 3$ to the rocking wedge, and $n = 4$ to the rotational cone. The figure demonstrates that the static and high-frequency limits derived previously are joined by smooth curves, in the simplest case for the cone in translation ($n = 2$) by straight lines. The asymptotic value of damping, $\rho c A_0$, is approached from above for $n = 1$, from below for $n = 3$ and $n = 4$. *The damping of the two-dimensional wedge always exceeds that of the three-dimensional cone.* (Compare the curves $n = 1$ and 2 for translation, $n = 3$ and 4 for rotation). The difference is *especially pronounced for low-frequency translational motion*; indeed, *for the wedge* ($n = 1$) *the damping tends to infinity in the static case.* However this "infinite" damping is multiplied by "zero" frequency to obtain the imaginary part of the dynamic-stiffness coefficient (Eq. 2-342), and the product tends to zero. This product is the numerator of the damping ratio $\zeta = \omega C(b_0)/(2K(b_0))$ for an oscillator consisting of a rigid foundation block supported by flexible soil, the natural frequency of the coupled system being ω. The damping ratio shows that two-dimensional modeling significantly exaggerates the radiation damping, usually by more than 100% as specified by the multipliers in the figure. In the formula for the *damping ratio the two-dimensional effects (increased damping in the numerator, reduced stiffness in the denominator) compound one another.* The salient features of the differences between two-dimensional and three-dimensional models are examined further in Section 2.8.

As quite complicated Hankel functions arise for the translational and rocking wedges ($n = 1$, $n = 3$), the practical application of wedges is more complicated than of cones with polynomials in the frequency parameter. Nevertheless, the solution with a wedge is much simpler than the rigorous analysis of a strip on a halfplane.

2.7.2 Aspect Ratio

As for cones, the aspect ratio z_0/b determining the opening angle of the wedges with half-width b (Fig. 2-63) is not chosen arbitrarily. For the translational motions the static-stiffness coefficients of a strip foundation (halfplane) vanish, so these low-frequency limits of the dynamic-stiffness coefficients cannot be used. The wedge's static-stiffness coefficients are also zero (Fig. 2-62, $n = 1$). For the two *translational motions the dynamic-spring coefficients* (real part of the dynamic-stiffness coefficients) *averaged over the low-frequency range of the strip foundation are set equal to the corresponding values of the wedges.* For the horizontal motion the average dynamic-spring coefficient determined from a rigorous halfplane solution equals

$$K = 0.94 \frac{\pi G}{2 - \nu} \tag{2-343}$$

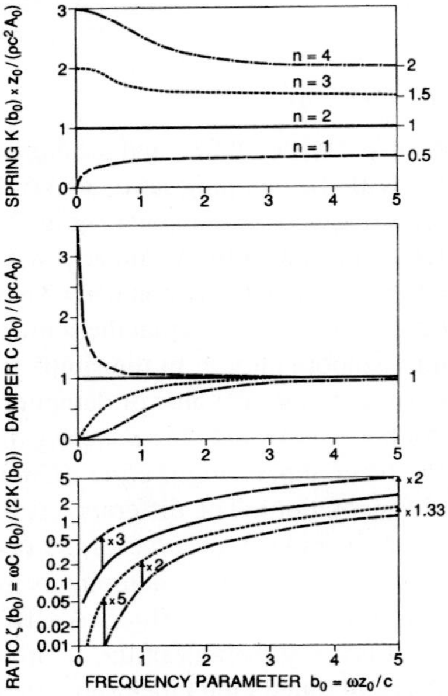

Fig. 2-62 Dynamic-stiffness coefficient and damping ratio for cone-wedge family.

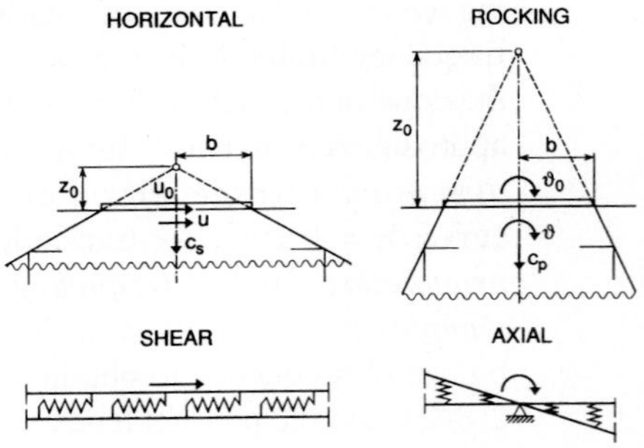

Fig. 2-63 Horizontal and rocking wedges with corresponding apex ratio (opening angle), wave-propagation velocity and distortion.

and for the wedge with $c = c_s$ and $A_0 = 2b$ (see Fig. 2-62, $n = 1$)

$$K = 0.94\,\frac{0.5\rho c_s^2 2b}{z_0} \tag{2-344}$$

Equating the two values yields the aspect ratio for the horizontal motion

$$\frac{z_0}{b} = \frac{2-\nu}{\pi} \tag{2-345}$$

For the *rocking* motion *the static-stiffness coefficient of the strip foundation* (which does not vanish), calculated from a rigorous halfplane solution

$$K_\vartheta = \frac{\pi G b^2}{2(1-\nu)} \tag{2-346}$$

is set equal to the corresponding value of the rocking wedge with $I_0 = 8b^3/12$ (see Fig. 2-62, $n = 3$, $b_0 = 0$)

$$K_\vartheta = \frac{2\rho c^2}{z_0}\,\frac{8b^3}{12} \tag{2-347}$$

For the rocking wedge the appropriate wave-propagation velocity c along the axis is equal to c_p for $\nu \le 1/3$ but restricted to $2c_s$ for $1/3 < \nu \le 1/2$. As discussed in Section 2.1.4 for the cone model, this limit on c in the nearly incompressible case avoids an overestimation of the radiation damping. Equating Eqs. 2-346 and 2-347 leads to

$$\frac{z_0}{b} = \frac{8(1-\nu)}{3\pi}\frac{c^2}{c_s^2} \tag{2-348}$$

The aspect ratios z_0/b for the horizontal and the rocking wedges depend on Poisson's ratio only. As for cones, the wedge with shear waves is squatty and that with dilatational waves slender. For instance for $\nu = 1/3$, $z_0/b = 0.531$ for the horizontal wedge and 2.264 for the rocking wedge. These two wedges are shown together with the occurring distortions in Fig. 2-63.

For horizontal motion with $n = 1$, the dynamic-stiffness coefficient follows for all Poisson's ratios ν from Eq. 2-340 using $c = c_s$ in the expression for b_0 (Eq. 2-341) with the corresponding z_0 (Eq. 2-345). For rocking motion Eq. 2-340 with Eqs. 2-341 and 2-348 applies, with $c = c_p$ in the compressible range ($\nu \le 1/3$) and $c = 2c_s$ for nearly incompressible soil ($1/3 < \nu \le 1/2$). As for cones (Section 2.1.4), an additional *trapped mass moment of inertia* ΔM_ν *arises in the nearly incompressible case*

$$\Delta M_\vartheta = 0.75\left(\nu - \frac{1}{3}\right)\rho\frac{8b^3}{12}b \tag{2-349}$$

ΔM_ϑ modifies the real part of the dynamic-stiffness coefficient. The term $-\omega^2\Delta M_\vartheta = -b_0^2 [(2c_s)^2/z_0^2]\Delta M_\vartheta$ is added to the right-hand side of Eq. 2-340.

2.7.3 Comparison of Dynamic-Stiffness Coefficients for Wedge with Exact Solution

The dynamic-stiffness coefficients of the two-dimensional strip foundation for horizontal and rocking motions are plotted for $\nu = 1/3$ and $\nu = 1/2$ (incompressible case) as a function of the dimensionless frequency $a_0 = \omega b/c_s$ in Figs. 2-64 and 2-65. The spring and damping coefficients are defined in the familiar manner as

$$S(a_0) = K\left[k_h(a_0) + ia_0\, c_h(a_0)\right] \tag{2-350a}$$

$$S_\vartheta(a_0) = K_\vartheta\left[k_r(a_0) + ia_0\, c_r(a_0)\right] \tag{2-350b}$$

For the nondimensionalisation, the average horizontal dynamic-spring coefficient K (Eq. 2-343) and the rocking static-stiffness coefficient K_ϑ (Eq. 2-346) are used. The exact values are taken from Wolf [W4] for $\nu = 1/3$ and from Luco and Hadjian [L1] for $\nu = 1/2$. As in the case of cones (Sections 2.5.2, Fig. 2-45), the exact solution of k_r for $\nu = 1/3$ (Fig. 2-65) exhibits a dip in the intermediate frequency range. This effect, which could be explained by a trapped mass moment of inertia, vanishes again for higher frequencies as the trapped mass "melts" away. Obviously, the simple wedge cannot represent this effect, which is, however, not important, because in the intermediate frequency range c_r (which is multiplied by the dimensionless frequency) dominates. The wedge models also tend to overestimate the damping coefficients c_h and c_r in the lower and intermediate frequency ranges for $\nu = 1/3$. But, in general, the agreement of the results of the wedge models with the exact solutions is satisfactory.

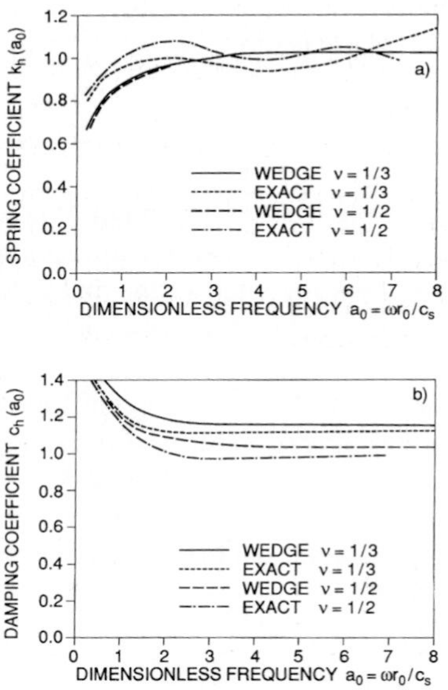

Fig. 2-64 Dynamic-stiffness coefficient for harmonic loading of strip on homogeneous halfplane in horizontal motion for various Poisson's ratios.

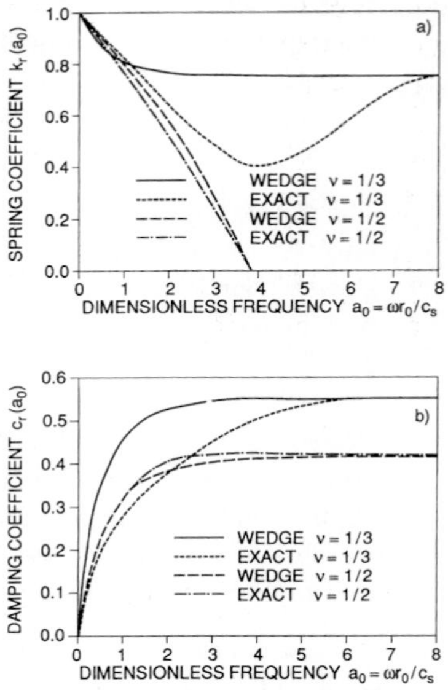

Fig. 2-65 Dynamic-stiffness coefficient for harmonic loading of strip on homogeneous halfplane in rocking motion for various Poisson's ratios.

2.8 INSIGHT ON TWO-DIMENSIONAL VERSUS THREE-DIMENSIONAL FOUNDATION MODELING

2.8.1 Object

To simplify the soil–structure-interaction analysis, three-dimensional problems are often modeled by considering a two-dimensional slice with the same material properties of the soil. This assumption, although convenient, is potentially dangerous. As is shown in Section 2.7.1, the *specific radiation damping* per unit contact area *computed for the two-dimensional case overestimates the actual three-dimensional value*—except for the high-frequency limit where the damping per unit area, density times wave-propagation velocity, is the same. As will be demonstrated in this section, *in addition, the contact area of a two-dimensional model will be larger than that of the three-dimensional case*, which will further increase radiation damping (multiplication of the radiation damping per unit area by this contact area). This latter aspect is verified as follows: The two-dimensional foundation is defined by an equivalent width $2b$ and a depth (length) $2a$ (measured perpendicularly to the two-dimensional plane of analysis), which can be selected somewhat arbitrarily. The width of the two-dimensional slice is not necessarily equal to the width of the three-dimensional problem, but may be modified for calibration purposes. The stiffnesses and mass properties of a two-dimensional slice are by nature referred to a unit depth. They are multiplied by the effective depth to obtain values which are appropriate for the three-dimensional case. The width and depth of the equivalent two-dimensional foundation also determine the area over which the stiffness and mass of the superstructure are distributed. The two-dimensional model can be regarded as a three-dimensional model with the foundation and the structure "repeated" up to infinity in the direction perpendicular to the two-dimensional plane of analysis. In the actual dynamic analysis, an equivalent slice of unit depth of the two-dimensional strip foundation is processed. Assume that a three-dimensional disk foundation with radius r_0 is to be modeled as a two-dimensional strip foundation (plan views in Fig. 2-66). A desirable choice to determine the width and depth of the two-dimensional foundation equates the static-rocking stiffness coefficients and the average horizontal dynamic-spring coefficients (real parts of the dynamic-stiffness coefficients, averaged over the frequency) of the two models. (As the translational static-stiffness coefficient in two dimensions vanishes, it cannot be used for calibration). This results in a width $2b = 1.63r_0$ and a depth $2a = 2.54r_0$, as is shown in Section 2.8.2. The contact area of the strip model, $4.14r_0^2$, is already 1.32 times larger than that of the original disk, $3.14r_0^2$, and is in addition multiplied by the higher specific radiation damping of the two-dimensional case mentioned above. To make matters worse, *two-dimensional modeling always entails an underestimation of the dynamic-spring coefficient* for the translational motions; for static loading the translational spring coefficient in the two-dimensional case vanishes. The *damping ratio* of the two-dimensional case, which is proportional to the ratio of the damping coefficient to the spring coefficient, *will thus be even larger*, resulting, in general, in a non-conservative analysis.

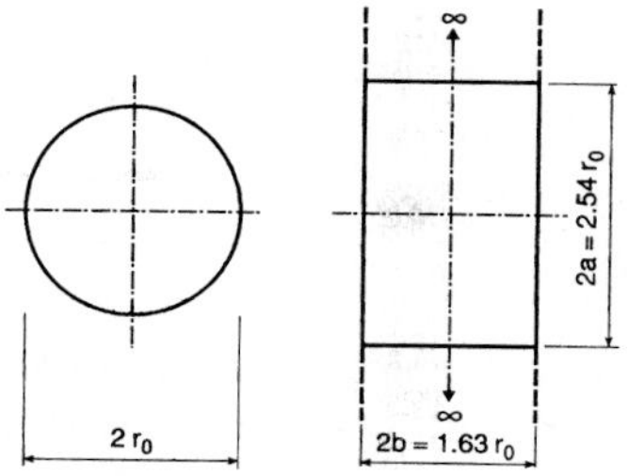

Fig. 2-66 Three-dimensional disk foundation and equivalent two-dimensional strip foundation.

It appears puzzling that the specific radiation damping is larger in the two-dimensional case, although the opportunity for waves to spread is obviously greater in three dimensions than in two. This phenomenon has been termed a paradox [G2]. This common-sense belief based on the geometrical concept of the larger available surface through which waves can potentially propagate under three-dimensional foundations is misleading. As is proven in Section 2.8.5, the essential aspect is the manner in which the amplitudes of the waves decay with distance.

Luco and Hadjian discuss [L1] the feasibility of representing a three-dimensional soil–structure-interaction problem by a two-dimensional model. Using rigorous solutions of three-dimensional elastodynamics to represent the soil, it is demonstrated that it is impossible to match the dynamic-stiffness coefficients for horizontal and rocking motions over a reasonable range of frequencies for the two models.

The object of this section which is based on Wolf and Meek [W15] is to provide conceptual clarity into the damping paradox and the reduced dynamic-spring coefficient of two-dimensional versus three-dimensional foundations without recourse to rigorous elastodynamical solutions, which with their mathematical complexity often obscure physical insight. First, cone–wedge models based on a strength-of-materials approach are addressed in Sections 2.8.2 to 2.8.5. As an alternative when defining the two-dimensional foundation, the impedance of the soil (product of density and wave velocity) can be changed to achieve a much better agreement of the high-frequency limits of the damping coefficients. Second, a strip foundation is also modeled in Section 2.8.6 by an infinite row of point loads. This allows the salient features of two-dimensional translational behavior to be ascertained purely on the basis of considerations of travel time and of decay of wave amplitude. Third, the quite separate but related topic of the transition from square to slender rectangular foundations is examined; such problems are solved via approximate Green's functions in Section 2.8.7.

2.8.2 Equivalent Slice of Two-Dimensional Strip Foundation with Same Material Properties

First, the width $2b$ and depth $2a$ of the equivalent two-dimensional strip foundation are determined to represent adequately the real parts of the dynamic-stiffness coefficients. As mentioned in Section 2.8.1, a reasonable choice equates the static-rocking stiffness coefficients and the average horizontal dynamic-spring coefficients of the disk and the equivalent strip [L1]. This leads to two equations in the unknowns b and a. The static rocking-stiffness coefficients are specified from rigorous solutions as (Eq. 2-346, Table 2-2)

$$K_\vartheta^{2d} = \frac{\pi G b^2 2a}{2(1-\nu)} \tag{2-351a}$$

$$K_\vartheta^{3d} = \frac{8 G r_0^3}{3(1-\nu)} \tag{2-351b}$$

The superscripts $2d$ and $3d$ denote, respectively, the strip and the disk foundations.

Although the two-dimensional strip has no static horizontal stiffness (see Fig. 2-62 for $n = 1$ and $b_0 = 0$), the average horizontal-stiffness coefficient in the low-frequency range is well-defined and is determined from a rigorous solution to be (Eq. 2-343)

$$K^{2d} = 0.94 \frac{\pi G 2a}{2-\nu} \tag{2-352a}$$

By contrast, the static-horizontal stiffness coefficient of the disk exists and is equal to $8Gr_0/(2-\nu)$ (Table 2-2). From Fig. 2-62 for $n = 2$ it would appear that this value remains constant with frequency. However, rigorous results reveal a slight decay in the dynamic-spring coefficient as the frequency increases (Fig. 2-6a).To capture this effect with good accuracy, the average stiffness K^{3d} includes the same factor 0.94 as in Eq. 2-352a

$$K^{3d} = 0.94 \frac{8 G r_0}{2-\nu} \tag{2-352b}$$

Setting Eq. 2-351a equal to Eq. 2-351b and Eq. 2-352a to Eq. 2-352b leads to

$$\frac{b}{r_0} = 0.816 \quad (2\text{-}353a)$$

$$\frac{a}{r_0} = 1.272 \quad (2\text{-}353b)$$

This is illustrated in Fig. 2-66. The high-frequency limits of the radiation damping coefficients are equal to density times appropriate wave velocity ρc multiplied by the contact area for the horizontal motion and the contact moment of inertia for the rocking motion. For compressible soil

$$C^{2d} = \rho c_s 2b2a \quad (2\text{-}354a)$$

$$C^{3d} = \rho\, c_s \pi r_0^2 \quad (2\text{-}354b)$$

$$C_\vartheta^{2d} = \rho\, c_p \frac{8b^3 2a}{12} \quad (2\text{-}355a)$$

$$C_\vartheta^{3d} = \rho\, c_p \frac{\pi r_0^4}{4} \quad (2\text{-}355b)$$

results. Substituting Eq. 2-353, the following ratios are calculated

$$\frac{C^{2d}}{C^{3d}} = \frac{4ab}{\pi r_0^2} = 1.32 \quad (2\text{-}356a)$$

$$\frac{C_\vartheta^{2d}}{C_\vartheta^{3d}} = \frac{16ab^3}{3\pi r_0^4} = 1.18 \quad (2\text{-}356b)$$

For an acceptable two-dimensional model, these ratios should be close to unity. They are, however, significantly larger than one, indicating that the *strip foundation overestimates radiation damping in the high-frequency limit.* Note that the material properties of the soil in the two-dimensional model are the same as those in the three-dimensional case.

2.8.3 Alternative Equivalent Slice of Two-dimensional Strip Foundation Adjusting Impedance

A much better fit of the damping coefficients than in Eq. 2-356 is achieved if, in addition to the width $2b$ and depth $2a$, the ratio q of the impedances (defined as the product of the density and the wave velocity of the slice of the two-dimensional foundation to that of the three-dimensional one) is introduced as a third unknown.

$$q = \frac{\rho^{2d} c_s^{2d}}{\rho c_s} = \frac{\rho^{2d} c_p^{2d}}{\rho c_p} \quad (2\text{-}357)$$

The subscript $3d$ is omitted in the denominator, as these material properties correspond to the actual soil. This extension does not affect the material properties G and ν of the static case. With $\nu^{2d} = \nu$, c_p^{2d}/c_s^{2d} will be equal to c_p/c_s, as reflected in the third term of Eq. 2-357. Using $G^{2d} = \rho^{2d}(c_s^{2d})^2 = G = \rho c_s^2$ and substituting in Eq. 2-357 yields

$$\frac{\rho^{2d}}{\rho} = q^2 \tag{2-358}$$

$$\frac{c_s^{2d}}{c_s} = \frac{1}{q} \tag{2-359}$$

The dynamic material properties density and wave velocity are thus affected. When this alternative procedure to construct an equivalent slice of a two-dimensional model is applied to more complicated cases than the homogeneous halfspace, such as layered sites or a through-soil coupling analysis of adjacent structures with a two-dimensional finite-element model, especially modifying the wave velocity can become problematic.

The dimensions $2b$ and $2a$ of the two-dimensional foundation for this alternative formulation will be the same as described in Section 2.8.2. Equating the static-rocking stiffness coefficients (Eq. 2-351) and the average horizontal dynamic-spring coefficients in the low-frequency range (Eq. 2-352) of the disk and the equivalent strip yields b and a (Eq. 2-353), which is repeated here for easy reference

$$\frac{b}{r_0} = 0.816 \tag{2-360a}$$

$$\frac{a}{r_0} = 1.272 \tag{2-360b}$$

The high-frequency limits of the radiation damping coefficients of the equivalent two-dimensional strip foundation are equal to

$$C^{2d} = \rho^{2d} c_s^{2d}\, 2b2a = q\rho c_s\, 2b2a \tag{2-361}$$

$$C_\vartheta^{2d} = \rho^{2d} c_p^{2d} \frac{8b^3 2a}{12} = q\rho c_p \frac{8b^3 2a}{12} \tag{2-362}$$

Setting Eq. 2-361 equal to C^{3d} (Eq. 2-354b) and Eq. 2-362 to C_ϑ^{3d} (Eq. 2-355b), the corresponding values of the disk, yields

$$q = \frac{\pi r_0^2}{4ab} \tag{2-363a}$$

and

$$q = \frac{3\pi r_0^4}{16ab^3} \tag{2-363b}$$

respectively. Substitution of Eq. 2-360 results in

$$q = 0.756 \tag{2-364a}$$

and

$$q = 0.850 \tag{2-364b}$$

respectively. The average value of Eq. 2-364 is selected

$$q = 0.803 \tag{2-365}$$

This leads to (Eqs. 2-358 and 2-359)

$$\frac{\rho^{2d}}{\rho} = 0.645 \tag{2-366a}$$

$$\frac{c_s^{2d}}{c_s} = 1.245 \tag{2-366b}$$

The following ratios are calculated

$$\frac{C^{2d}}{C^{3d}} = q\frac{4ab}{\pi r_0^2} = 1.061 \tag{2-367a}$$

$$\frac{C_\vartheta^{2d}}{C_\vartheta^{3d}} = q\frac{16ab^3}{3\pi r_0^4} = 0.942 \tag{2-367b}$$

In the high-frequency limit the equivalent slice of the two-dimensional strip foundation overestimates the radiation damping by only 6% for the horizontal motion and underestimates it by the same amount for the rocking motion. This number should be compared with 32% and 18% overestimation of the original equivalent slice of the strip foundation (Eq. 2-356).

2.8.4 Dynamic-Stiffness Coefficients of Disk and Equivalent Slices of Strips

To gain further insight, the dynamic-stiffness coefficients of a three-dimensional disk foundation are compared to those of the two equivalent slices of the two-dimensional strip foundations, keeping the material properties the same and adjusting the impedance. The horizontal and rocking motions are addressed. As is demonstrated in Figs. 2-6 and 2-21 for the cone and in Figs. 2-64 and 2-65 for the wedge, the differences between the results of the simple physical models and the rigorous elastodynamic solutions with the same dimensions are small. This permits cones to be used to model the disk foundation and wedges to represent the strip foundation.

The dynamic-stiffness coefficients in horizontal and rocking motions of a three-dimensional disk of radius r_0 modeled with cones are compared in Fig. 2-67 to those of an equivalent two-dimensional strip foundation with the same material properties represented by wedges. Poisson's ratio ν equals 1/3. For the cone models the results of Figs. 2-6 and 2-21 are used. For the strip foundation, the equivalent width $2b$ and depth $2a$ follow from Eq. 2-353 (Fig. 2-66), which yields the contact area and moment of inertia of the wedge A_0. The aspect ratios z_0/b for the horizontal and rocking wedges are specified in Eqs. 2-345 and 2-348. The equivalent two-dimensional strip foundation with the same material properties (Section 2.8.2) is examined first. The dimensionless frequency parameters b_0 are given in Eq. 2-341 with the corresponding z_0 and with $c = c_s$ for the horizontal wedge and with $c = c_p$ ($c = 2c_s$ for $\nu = 1/3$) for the rocking wedge. The dynamic-stiffness coefficients are calculated from Eq. 2-340 using $n = 1$ for the horizontal wedge and $n = 3$ for the rocking wedge, with the static-stiffness coefficients given by Eq. 2-332 and their high-frequency limits by Eq. 2-339. The spring and damping coefficients are defined in the familiar manner as

$$S(a_0) = K[k(a_0) + ia_0\, c(a_0)] \tag{2-368}$$

in terms of the static-stiffness coefficient of the three-dimensional disk K (Table 2-2, Eq. 2-352b without the factor 0.94, Eq. 2-351b) and $a_0 = \omega r_0/c_s$. From the comparison of the dynamic-stiffness coefficients in Fig. 2-67, it follows that the two-dimensional model leads to a reasonable rocking spring coefficient, but overestimates the rocking damping coefficient in the low and intermediate frequency ranges. For the horizontal motion, the damping ratio ζ, which is proportional to $c(a_0)/k(a_0)$, is much too large in the two-dimensional model. It must be stressed that these discrepancies are caused by replacing the three-dimensional disk foundation by an equivalent slice of a two-dimensional strip foundation (with the same material properties) and not by modeling the disk and the strip foundations with a cone and a wedge. It is thus *impossible to construct an equivalent slice of a two-dimensional model with the same material properties which matches the dynamic-stiffness coefficients of the three-dimensional model.* Because the damping ratio is grossly overestimated, *two-dimensional modeling cannot be recommended for actual engineering applications involving three-dimensional cases.* It is more feasible to take the opposite approach and idealize slender soil–structure interaction problems with a radially symmetric model.

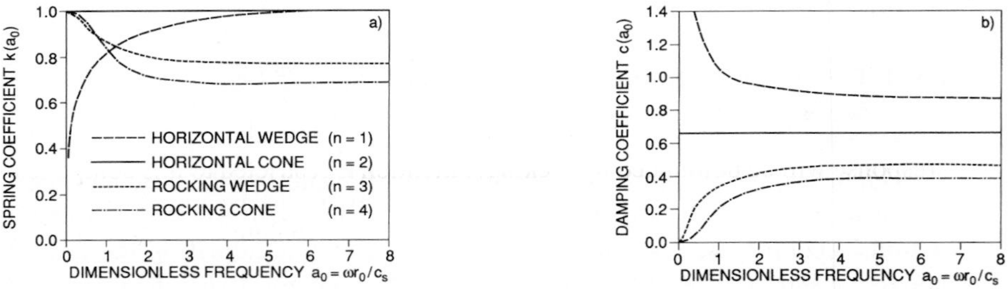

Fig. 2-67 Dynamic-stiffness coefficients for disk modeled as cones and equivalent slice of two-dimensional strip foundation with same material properties modeled as wedges.

The dynamic-stiffness coefficients in horizontal and rocking motions of a three-dimensional disk of radius r_0 modeled with cones are compared in Fig. 2-68 to those of the equivalent two-dimensional strip foundation represented by wedges based on the alternative formulation *adjusting the impedance* (Section 2.8.3). Poisson's ratio ν equals 1/3. Figure 2-68 should be compared with Fig. 2-67 based on the original formulation to construct the equivalent strip foundation with the same material properties as for the disk foundation. The width $2b$ and length $2a$ are specified in Eq. 2-360 (or Eq. 2-353). It follows from Eq. 2-366b that $c_s^{2d} = 1.245c_s$ and with $\nu = 1/3$, $c_p^{2d} = 1.245c_p = 2.490c_s$. The aspect ratios of the wedges specified in Eqs. 2-345 and 2-348 lead to the same values as in the original formulation, as ν does not change and $c_p^{2d}/c_s^{2d} = c_p/c_s = 2$. For the horizontal wedge $z_0 = 0.531b = 0.433r_0$ and for the rocking wedge $z_0 = 2.264b = 1.847r_0$ result. The dimensionless frequency parameter b_0 (Eq. 2-341) is defined for the horizontal wedge as $b_0 = \omega z_0/c_s^{2d}$ $(= 0.348a_0)$ and for the rocking wedge as $b_0 = \omega z_0/c_p^{2d}$ $(= 0.742a_0)$ with $a_0 = \omega r_0/c_s$. As is visible from Fig. 2-68, the agreement of the wedge's damping coefficients for horizontal and rocking motions in the intermediate- and high-frequency ranges is good; in the *low-*

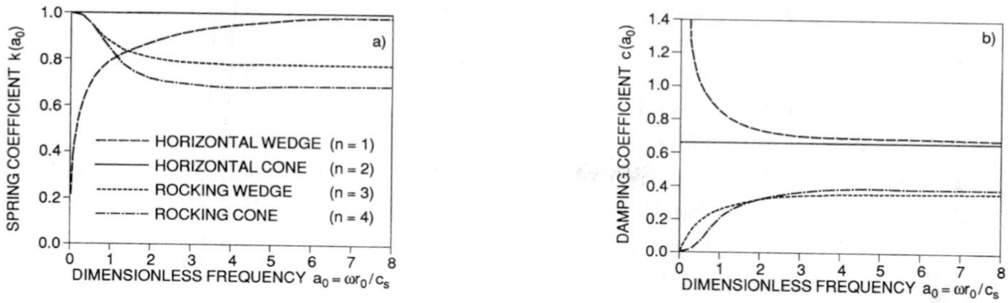

Fig. 2-68 Dynamic-stiffness coefficients for disk modeled as cones and alternative equivalent slice of two-dimensional strip foundation adjusting impedance modeled as wedges.

frequency range the wedge models still overestimate radiation damping, although less than that based on the original formulation (Fig. 2-67).

To illustrate the salient features also numerically, a simplified dynamic system consisting of a rigid structure on the surface of a soil halfspace is addressed as an extreme case, which possibly overemphasizes the effects of two-dimensional versus three-dimensional foundation modeling. The rigid cylindrical structure, which represents a reactor building with radius $r_0 = 20$ m and height 60 m, exhibits an average mass density 700 kg/m^3, leading to a mass = $5.28 \cdot 10^7$ kg and a mass moment of inertia about the base diameter $I_0 = 6.86 \cdot 10^{10}$ kgm^2. The soil with mass density $\rho = 2 \cdot 10^3$ kg/m^3 and Poisson's ratio $\nu = 1/3$ has a shear-wave velocity = 200m/s, resulting in a dilatational wave velocity $c_p = 400$ m/s. Because the structure is taller than it is wide, the soil–structure interaction response will be dominated by rocking. Attention is restricted to this component of motion.

The three-dimensional model is addressed first. The aspect ratio for the rocking cone equals $z_0/r_0 = 2.356$ (Table 2-3B) and yields the apex height $z_0 = 47.12$ m. The static-stiffness coefficient equals $K_\vartheta = 3\,\rho c_p^2 \pi r_0^4/(4z_0) = 2.56 \cdot 10^{12}$ Nm (Eq. 2-332), and the radiation dashpot $C_\vartheta = \rho c_p \pi r_0^4/4 = 1.01 \cdot 10^{11}$ Nms (Eq. 2-339b). The frequency parameter of the cone is defined as $b_0 = \omega \cdot 2.356\, r_0/(2c_s) = 1.178 a_0$. The cone's spring coefficient follows as $k_r(b_0) = 1 - (1/3)b_0^2/(1 + b_0^2)$ and the damping coefficient as $c_r(b_0) = (1/3)b_0^2/(1 + b_0^2)$ (Eq. 2-169). The natural frequency of the three-dimensional dynamic system $\omega_r = (K_\vartheta k_\vartheta(b_0)/I_0)^{0.5}$ is solved by iteration leading to $\omega_r = 5.78$/s (= 0.92 Hz) with $b_0 = 0.68$ and $k_r = 0.895$. The c_r at the same frequency equals 0.12. The values of k_r and c_r can be verified in Fig. 2-62 (after changing the normalisations). The damping ratio at ω_r equals the ratio of the imaginary part of the dynamic-stiffness coefficient to twice its real part, both evaluated at ω_r, $\zeta_r = b_0 c_r/(2k_r) = 0.039$, which can be checked from Fig. 2-62.

The equivalent slice of the two-dimensional strip foundation model with the same material properties is discussed next. The equivalent halfwidth $b = 0.816\, r_0$ is 16.3 m and half-depth $a = 1.272\, r_0$ is 25.4 m (Eq. 2-353). The aspect ratio for the rocking wedge $z_0/b = 2.264$ (Eq. 2-348) yields $z_0 = 36.9$ m. By construction, the static-stiffness coefficient of the wedge is the same as that of the cone $K_\vartheta = 2\rho c_p^2 8b^3 2a/(12z_0) = 2.56 \cdot 10^{12}$ Nm (Eq. 2-332 with $n = 3$). The radiation dashpot $C_\vartheta = \rho c_p 8b^3 2a/12 = 1.19 \cdot 10^{11}$ Nms (Eq. 2-339b) which is as expected 1.18 times larger than that of the cone (Eq. 2-356b). The frequency parameter of the wedge is specified as $b_0 = \omega z_0/c_p = a_0(z_0/r_0)(c_s/c_p) = 0.923\, a_0$. The natural frequency of the equivalent two-dimensional dynamic system is equal to the square root of the ratio of the real part of the dynamic-stiffness coefficient $K(b_0)$ (Eqs. 2-340, 2-342) to the mass moment of inertia I_0, which again has to be solved by iteration. In this frequency range, $k_r(a_0)$ in Fig. 2-67 for the wedge ($n = 3$) is the same as for the cone ($n = 4$). As a result, $\omega_r = 5.78/s$ (= 0.92 Hz) is again obtained with $a_0 = 0.58$ and $b_0 = 0.54$. The spring and damping coefficients $k_r = 0.895$ and $c_r = 0.27$ follow for $a_0 = 0.58$ from Fig. 2-67. The corresponding damping ratio $\zeta_r = a_0 c_r/(2k_r)$ is 0.088 which can also be checked at $b_0 = 0.54$ in Fig. 2-62. (Note that the b_0 of the wedge is smaller than that of the disk.) The damping ratio of the equivalent slice of the two-dimensional strip foundation is thus more than twice that of the three-dimensional foundation.

Suppose now that the coupled soil–structure system is subjected to a harmonically varying rocking moment M, the frequency of which gradually increases, as would be produced by a machine during the start-up phase. The greatest response obviously occurs when the operating frequency corresponds to the natural frequency of the system. If the harmonic exciting moment were applied quasistatically, the corresponding rotation $\vartheta = M/(K_\vartheta k_r)$ in the three- and two-dimensional cases would be the same in this case. Due to resonance, the quasi-static rotation is multiplied by the amplification factor $1/(2\zeta_r) = 12.85$ (three-dimensional) and 5.71 (two-dimensional with same material properties). A non-conservative error of this magnitude is obviously unacceptable and potentially dangerous.

Finally, the dynamic system with the rigid structure is examined based on the alternative formulation of the strip foundation adjusting the impedance. The natural frequency of the equivalent two-dimensional dynamic system is equal to the square root of the ratio of the real part of the dynamic-stiffness coefficient $K(b_0)$ (Eqs. 2-340 and 2-342) to the mass moment of inertia I_0. After iteration, $\omega_r = 5.90/s$ (= 0.94 Hz) results with $b_0 = 0.44$ and $a_0 = 0.59$. $k_r = 0.91$ and $c_r = 0.18$ follow for this a_0 from Fig. 2-68. The corresponding damping ratio $\zeta_r = a_0 c_r/(2k_r)$ equals 0.058, which lies between the values of the three-dimensional disk (0.039) and the original two-dimensional wedge with the same material properties (0.088).

2.8.5 Decay of Waves

In order to study the dynamic behavior of the cone–wedge family in the far field, ($z \rightarrow \infty$), the governing equation of motion (Eq. 2-323) is nondimensionalized. The independent variable $\bar{z} = z\omega/c$ replaces z. Derivatives with respect to z can be expressed by those with respect to $\bar{z}$. For instance, $u(\omega)_{,z} = u(\omega)_{,\bar{z}}\, d\bar{z}/dz = u(\omega)_{,\bar{z}}\, \omega/c$. Substituting in Eq. 2-323 yields

$$u(\omega)_{,\bar{z}\bar{z}} + nu(\omega)_{,\bar{z}} + u(\omega) = 0 \qquad (2\text{-}369)$$

As z and ω appear together as a product in the definition of $\bar{z}$, they play the same role. Thus the high-frequency case $\omega \rightarrow \infty$ corresponds just as well to the far-field case $z \rightarrow \infty$. The asymptotic solution (Eq. 2-335) therefore also describes the behavior "at infinity." As discussed in Section 2.5.2 (Eq. 2-271), conservation of energy at infinity requires that the amplitude decays in inverse proportion to the square root of the area (radiation criterion), in agreement with the exponent $n/2$ in Eq. 2-335. If the exponent $n - 1$ for the near-field static solution (Eq. 2-329) happens to be identical to the exponent $n/2$ for the far field, as is the case only for the translational cone ($n = 2$), the distinction between the near and the far field vanishes; and radiation energy is transferred smoothly for all frequencies. The result is constant frequency-independent damping as shown in Fig. 2-62. In contrast, if $n - 1$ is greater than $n/2$, as is the case for rotation ($n = 3$ or 4), the static displacement will have died off to a small value at the beginning of the far field. For low frequencies very little energy will reach the far-field boundary and penetrate to infinity: The damping begins at zero (Fig. 2-62). With increasing frequency the far-field boundary is pulled in towards the foundation, and the damping gradually increases. Just the opposite occurs for the wedge in translation ($n = 1$). According to the exponent $n - 1 = 0$, the static amplitude arrives undiminished at the beginning of the far field. For low frequencies there is a very pronounced tendency to radiate, which leads to infinite damping in the limit $\omega \to 0$. The essence of radiation damping must be grasped in this fashion. *The heuristic concept of more spreading of waves in three dimensions than in two is misleading.* Indeed, just *the opposite is true: The less the amplitude spreads and diminishes with distance, the greater is the radiation damping.*

2.8.6 Row of Point Loads

The cone–wedge model, while quite simple, requires the solution of a differential equation. The salient features of two-dimensional translational behavior may also be derived purely on the basis of *considerations of travel-time and of decay of the wave amplitude* referred to an infinite row of point loads. The infinite row of identical point loads P separated by the distance d acting on the surface of a halfspace simulates the behavior of a loaded strip foundation (Fig. 2-69). For static and low-frequency dynamic motion the vertical displacement u due to a single point load P decays in inverse proportion to the horizontal radial distance r from the source load. The horizontal velocity of wave propagation in the near field (somewhat less than the Rayleigh-wave velocity, see Fig. 2-53) is denoted as c. In the time domain and for harmonic excitation, the following flexibility relations, defining the Green's functions, apply

$$u(r, t) = \frac{1-\nu}{2\pi G} \frac{P\left(t - \frac{r}{c}\right)}{r} \tag{2-370a}$$

$$u(r, \omega) = \frac{1-\nu}{2\pi G} \frac{P(\omega)}{r} e^{-i\frac{\omega r}{c}} \tag{2-370b}$$

The constant factor $(1 - \nu)/(2\pi G)$ is obtained from the static Boussinesq solution. In the time domain the displacement is a function of the retarded time $t - r/c$. Using superposition,

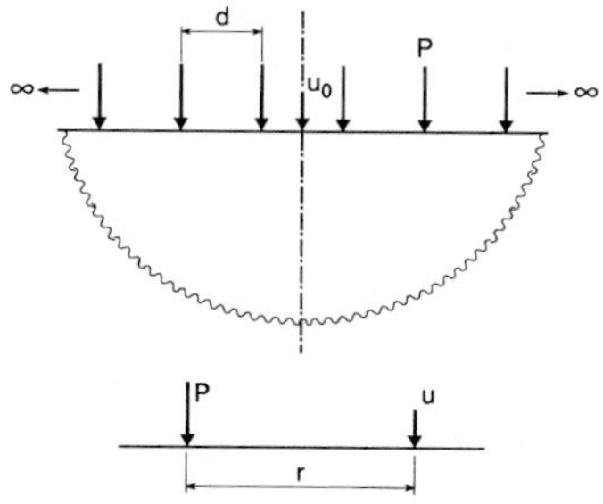

Fig. 2-69 Infinite row of point loads with illustration of Green's function.

the displacement u_0 due to an infinite row of point loads follows as a sum of the contributions described by Eq. 2-370. For harmonic motion and with $r = (j + 0.5)d$

$$u_0(\omega) = \frac{2(1-\nu)}{2\pi G d} P(\omega) \sum_{j=0}^{\infty} \frac{e^{-i\omega(j+0.5)\frac{d}{c}}}{j+0.5} \tag{2-371}$$

results. The leading factor 2 accounts for the contributions of the left and right semi-infinite rows of loads to the displacement amplitude in the center $u_0(\omega)$. The sums for the real and imaginary parts of Eq. 2-371 may be evaluated explicitly [W15]. The resulting expression for the dynamic-stiffness coefficient in the vertical direction $S(b_0) = P(b_0)/u_0(b_0)$ is

$$S(b_0) = \frac{\pi G d}{1-\nu} \left[k_v(b_0) + i b_0 \, c_v(b_0)\right] \tag{2-372}$$

in which the spring and damping coefficients are

$$k_v(b_0) = \frac{\ell n \cot \dfrac{b_0}{4}}{\left(\ell n \cot \dfrac{b_0}{4}\right)^2 + \dfrac{\pi^2}{4}} \tag{2-373a}$$

$$c_v(b_0) = \frac{\dfrac{\pi}{2b_0}}{\left(\ell n \cot \dfrac{b_0}{4}\right)^2 + \dfrac{\pi^2}{4}} \tag{2-373b}$$

The dimensionless frequency parameter b_0 equals $\omega d/c$. The dynamic-stiffness coefficient is plotted in Fig. 2-70. For the low-frequency range $b_0 < 1$ the response is essentially identical to that of the translational wedge ($n = 1$ in Fig. 2-62). The static-stiffness coefficient is zero but the spring coefficient increases abruptly to a more or less constant value. By contrast, the damping coefficient begins from infinity and then diminishes asymptotically. For $b_0 > 1$ the spring coefficient gradually decays, reaching zero when $b_0 = \pi$. This is a resonant condition, in which the distance d between point loads is half a wavelength.

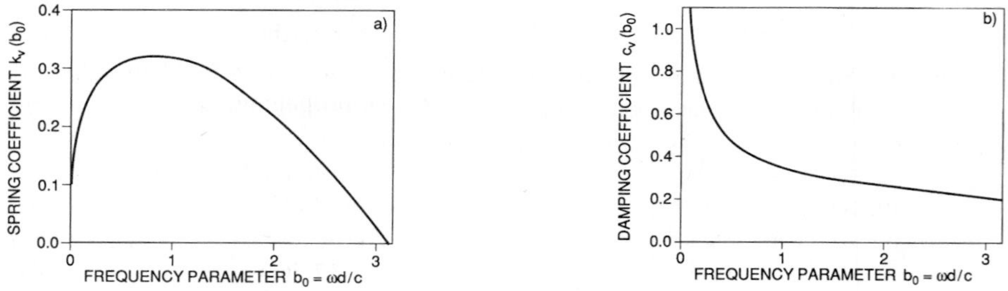

Fig. 2-70 Dynamic-stiffness coefficient for infinite row of point loads.

2.8.7 Square Versus Slender Rectangular Foundations

If the point load is replaced by a small disk with finite radius, travel-time considerations analogous to Eq. 2-370 enable modeling of surface foundations with arbitrary shape. This is discussed in depth in Chapter 5. For better accuracy, the displacement, which decays in inverse proportion to the distance in the near field, is modified to decay in inverse proportion to the square root of the distance in the far field (Rayleigh waves, see also Fig. 2-53).

If a rectangular foundation with width $2b$ and length $2a$ is idealized by an array of small subdisks, the spring and damping coefficients shown in Fig. 2-71 for vertical motion are calculated for $\nu = 1/3$ and for various slenderness ratios a/b. The curves correspond to the conventional representation of the dynamic stiffness in the form of Eq. 2-368, in which a_0 is the dimensionless frequency $\omega b/c_s$. These results computed using the approximate Green's function described above are in excellent agreement with rigorous elastodvnamical solutions, as is demonstrated in Section 5.5.

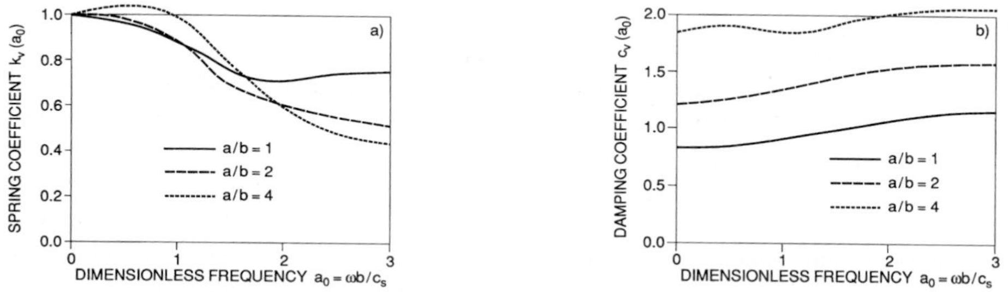

Fig. 2-71 Dynamic-stiffness coefficient for vertical motion of rectangular foundation.

The spring coefficient $k_v(a_0)$ is relatively insensitive to variation of the slenderness ratio a/b. By contrast, it appears at first glance that the damping of slender foundations is several times greater than that of squarish ones. This is, however, a false conclusion, associated with the definition of a_0 in terms of the foundation's shorter half-width b. It must be remembered that the imaginary part of the dynamic-stiffness coefficient which determines the radiation damping is equal to the product of the damping coefficient and the dimensionless frequency. For physical insight it is better to express the damping in terms of the radiation impedance (damping per unit contact area) ρc. The symbol ρ denotes the mass density of the halfspace and c is the appropriate velocity of propagation. In vertical motion dilatational waves emanate from the undersurface of the foundation and thus $c = c_p$. For a foundation with area A_0 and moment of inertia I_0 about the axis of rotation, the vertical and rocking dashpots converge in the high-frequency limit to $C(a_0 = \infty) = \rho c_p A_0$ and $C_\vartheta(a_0 = \infty) = \rho c_p I_0$. These values are confirmed by Figs. 2-72 (vertical case) and 2-73 (rocking case). From Fig. 2-72, a replot of the right portion of Fig. 2-57, it is apparent that for low frequencies, slender foundations are in fact slightly more damped than squarish ones; however, the difference is practically negligible for slenderness ratios $a/b \le 4$. The curves for extremely slender foundations, $a/b = 8$ and 12, imply that for a two-dimensional strip with $a/b = \infty$, $C(a_0 = 0)$ tends to infinity, as discussed in connection with Fig. 2-62. The damping for rocking motion (Fig. 2-73) is always somewhat larger for slender foundations than for squarish

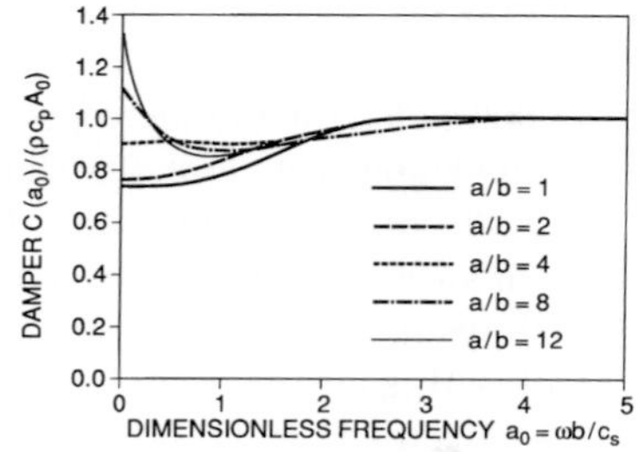

Fig. 2-72 Frequency-dependent damper for vertical motion of rectangular foundation.

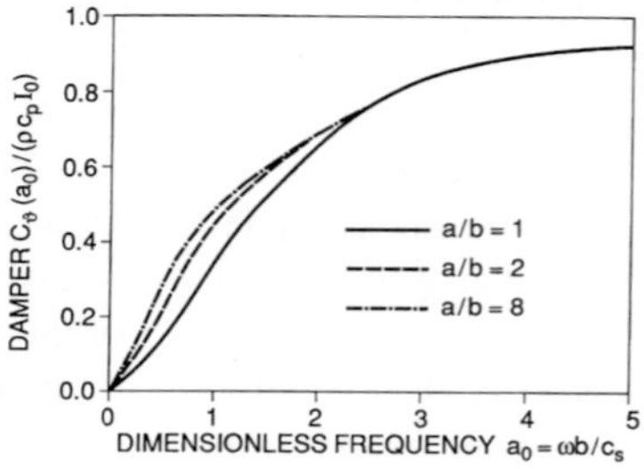

Fig. 2-73 Frequency-dependent damper for rocking motion of rectangular foundation.

ones, but the deviations are less appreciable than in the vertical case; again the difference is relatively unimportant for slenderness ratios $a/b \le 4$.

The static-stiffness coefficient of rectangular foundations is compared to that of a disk with same area A_0 (vertical) or the same moment of inertia I_0 (rotation) in Fig. 2-74. (For a foundation of arbitrary shape, Table 2-2 contains formulas for the static-stiffness coefficients for all motions.) Due to the increased corner-punching effect, the rectangular footing is always stiffer than the equivalent disk; and the stiffness ratio increases with the slenderness ratio. This seems to contradict the fact that the two-dimensional strip is softer than the three-dimensional disk. However, no paradox is involved. The static stiffness in Fig. 2-74 is computed with respect to the total load on the rectangular foundation, not the load per unit length, which is the reference quantity for the two-dimensional strip. If Fig. 2-74 were to be redrawn on a per-unit length basis, the rectangular foundation would appear to be more flexible than the disk.

This suggests that a *rectangular foundation* with $a/b \le 4$ *may be replaced by an equivalent disk* having the same area A_0 (vertical motion) or moment of inertia I_0 (rocking motion). The appealing concept of the equivalent disk is valid, provided that the spring coefficient $k(a_0)$ is corrected by a factor slightly greater than one, as given in Fig. 2-74. This correction factor is not applied to the damping coefficient $c(a_0)$, as can be seen from $C(a_0)$ and $C_\vartheta(a_0)$ plotted in Figs. 2-72 and 2-73.

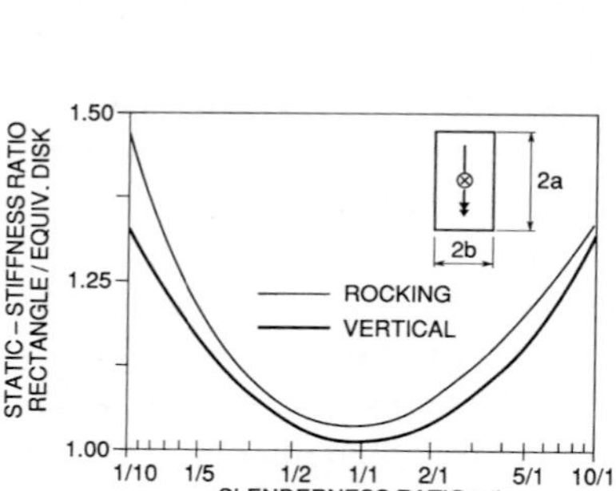

Fig. 2-74 Static-stiffness ratio of rectangular foundation to equivalent disk.

SUMMARY

1. For all components of motion a rigid basemat with area A_0 and (polar) moment of inertia I_0 on the surface of a homogeneous soil halfspace with Poisson's ratio ν, shear-wave velocity c_s, dilatational-wave velocity c_p and density ρ can be modeled with a truncated semi-infinite cone of equivalent radius r_0, apex height z_0, and wave velocity c (Fig. 2-75a). The translational cone model for the displacement u_0 is dynamically equivalent to the spring K- and dashpot C-system (Fig. 2-75b). The rotational cone for the rotation ϑ_0 corresponds exactly to the discrete-element models with one internal degree of freedom ϑ_1 and a small number of spring(s) K_ϑ, dashpot(s) C_ϑ, and in the monkey-tail configuration a mass moment of inertia M_ϑ (Fig. 2-75c). All coefficients are frequency independent. For the vertical and rocking motions in the case of nearly incompressible soil ($1/3 < \nu \le 1/2$), c is limited; and a trapped mass ΔM and a trapped mass moment of inertia ΔM_ϑ assigned to the basemat arise. The properties of the cones and the discrete-element models which are all that is necessary for the modeling of a basemat of arbitrary shape on the surface of the soil (in a general-purpose structural dynamics program working directly in the time domain) are summarized in a nutshell in Table 2-11.

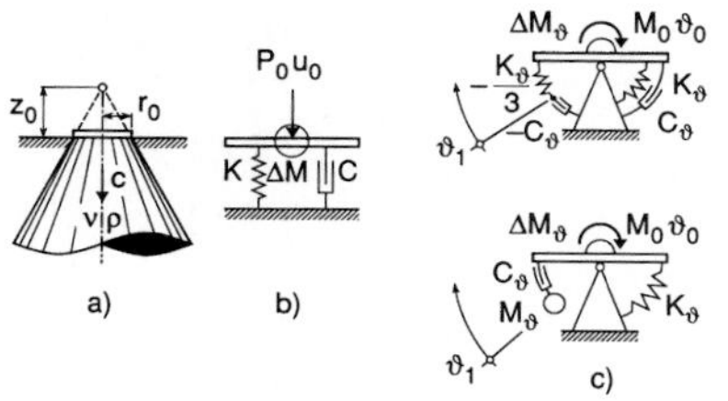

a. Cone.

b. Discrete-element model for translation.

c. Discrete-element model for rotation.

Fig. 2-75 Cone model and equivalent discrete-element model.

2. For the horizontal and torsional cones deforming in shear the appropriate wave velocity c equals c_s. For the vertical and rocking cones deforming axially c equals c_p for $\nu \le 1/3$ and is limited to $2c_s$ for $1/3 < \nu \le 1/2$.
3. The aspect ratio z_0/r_0 determining the opening angle follows from equating the static-stiffness coefficient of the cone to that of the disk on a halfspace. For the translational cone

$$K = \rho c^2 A_0 / z_0$$

and for the disk

$$K_h = \frac{8 G r_0}{2 - \nu} \qquad \textit{horizontal}$$

$$K_v = \frac{4 G r_0}{1 - \nu} \qquad \textit{vertical}$$

For the rotational cone

$$K_\vartheta = 3 \rho c^2 I_0 / z_0$$

TABLE 2-11 BARE ESSENTIALS TO MODEL A FOUNDATION ON SURFACE OF HOMOGENEOUS SOIL HALFSPACE

Motion	Horizontal	Vertical		Rocking		Torsional
Equivalent Radius r_0	$\sqrt{\frac{A_0}{\pi}}$	$\sqrt{\frac{A_0}{\pi}}$		$\sqrt[4]{\frac{4I_0}{\pi}}$		$\sqrt[4]{\frac{2I_0}{\pi}}$
Aspect Ratio $\frac{z_0}{r_0}$	$\frac{\pi}{8}(2-\nu)$	$\frac{\pi}{4}(1-\nu)\left(\frac{c}{c_s}\right)^2$		$\frac{9\pi}{32}(1-\nu)\left(\frac{c}{c_s}\right)^2$		$\frac{9\pi}{32}$
Poisson's Ratio ν	all ν	$\leq\frac{1}{3}$	$\frac{1}{3}<\nu\leq\frac{1}{2}$	$\leq\frac{1}{3}$	$\frac{1}{3}<\nu\leq\frac{1}{2}$	all ν
Wave Velocity c	c_s	c_p	$2c_s$	c_p	$2c_s$	c_s
Trapped Mass ΔM ΔM_ϑ	0	0	$2.4\left(\nu-\frac{1}{3}\right)\rho A_0 r_0$	0	$1.2\left(\nu-\frac{1}{3}\right)\rho I_0 r_0$	0
Discrete-Element Model		$K = \rho c^2 A_0/z_0$ $C = \rho c A_0$		$K_\vartheta = 3\rho c^2 I_0/z_0$ $C_\vartheta = \rho c I_0$ $M_\vartheta = \rho I_0 z_0$		

and for the disk

$$K_r = \frac{8Gr_0^3}{3(1-\nu)} \qquad \textit{rocking}$$

$$K_t = \frac{16}{3}Gr_0^3 \qquad \textit{torsional}$$

This results in

$$\frac{z_0}{r_0} = \frac{\pi}{8}(2-\nu) \qquad \textit{horizontal}$$

$$\frac{z_0}{r_0} = \frac{\pi}{4}(1-\nu)\left(\frac{c}{c_s}\right)^2 \qquad \textit{vertical}$$

$$\frac{z_0}{r_0} = \frac{9\pi}{32}(1-\nu)\left(\frac{c}{c_s}\right)^2 \qquad \textit{rocking}$$

$$\frac{z_0}{r_0} = \frac{9\pi}{32} \qquad \textit{torsional}$$

With the exception of the torsional motion for which it is a constant, z_0/r_0 depends only on ν.

4. The high-frequency behavior is described by a radiation dashpot for the translational cone

$$C = \rho c A_0$$

and for the rotational cone

$$C_\vartheta = \rho c I_0$$

(disregarding the effect of the trapped mass, if any).

5. Cone models are, in principle, doubly asymptotic approximations, exact by construction for the static case and in the high-frequency limit.

6. For nearly incompressible soil ($1/3 < \nu \leq 1/2$) the wave velocity c is limited to $2c_s$ and for the vertical motion a trapped mass

$$\Delta M = \mu\, \rho r_0^3 \qquad \text{with } \mu = 2.4\pi\left(\nu - \frac{1}{3}\right)$$

and for the rocking motion a trapped mass moment of inertia

$$\Delta M_\vartheta = \mu_\vartheta\, \rho r_0^5 \qquad \text{with } \mu_\vartheta = 0.3\pi\left(\nu - \frac{1}{3}\right)$$

are introduced. Assigning ΔM and ΔM_ϑ to the basemat node allows the cone models and their discrete-element models to be constructed in the same way for all ν.

7. For the translational cone, the interaction force $P_0(t)$—displacement $u_0(t)$ relationship of the stiffness formulation equals

$$P_0(t) = Ku_0(t) + C\dot{u}_0(t) + \Delta M\ddot{u}_0(t)$$

The discrete-element model is shown in Fig. 2-75b. For harmonic loading the dynamic-stiffness coefficient of the translational cone is formulated as

$$S(b_0) = K\left[k(b_0) + ib_0\, c(b_0)\right]$$

with the dimensionless spring and damping coefficients

$$k(b_0) = 1 - \frac{\mu}{\pi}\frac{r_0}{z_0}\, b_0^2$$

$$c(b_0) = 1$$

being a function of the dimensionless frequency parameter

$$b_0 = \frac{\omega z_0}{c}$$

In terms of the dimensionless frequency of the disk

$$a_0 = \frac{\omega r_0}{c_s}$$

$$S(a_0) = K\left[k(a_0) + ia_0\, c(a_0)\right]$$

applies with

$$k(a_0) = 1 - \frac{\mu}{\pi}\frac{z_0}{r_0}\frac{c_s^2}{c^2}\, a_0^2$$

$$c(a_0) = \frac{z_0}{r_0}\frac{c_s}{c}$$

For the horizontal motion and the vertical motion with $\nu \le 1/3$, ΔM and thus μ vanish.

8. For the translational cone the displacement $u_0(t)$—interaction force $P_0(t)$ relationship of the flexibility formulation equals for the horizontal motion and the vertical motion for $\nu \le 1/3$

$$u_0(t) = \int_0^t h_1(t-\tau)\frac{P_0(\tau)}{K}\,d\tau$$

where the unit-impulse response function is

$$\begin{aligned} h_1(t) &= \frac{c}{z_0}\,e^{-\frac{c}{z_0}t} && t \ge 0 \\ &= 0 && t < 0 \end{aligned}$$

The convolution integral at time station n can be evaluated efficiently by the recursive procedure

$$u_{0n} = a\,u_{0n-1} + b_0\frac{P_{0n}}{K} + b_1\frac{P_{0n-1}}{K}$$

where the recursive coefficients are specified as (Δt = time step)

$$a = e^{-\frac{c\Delta t}{z_0}}$$

$$b_0 = 1 + \frac{e^{-\frac{c\Delta t}{z_0}} - 1}{\frac{c\Delta t}{z_0}} \qquad b_1 = \frac{-e^{-\frac{c\Delta t}{z_0}} + 1}{\frac{c\Delta t}{z_0}} - e^{-\frac{c\Delta t}{z_0}}$$

The unconditionally stable recursive evaluation which is suitable even for a hand calculation is exact when $P_0(t)$ is piecewise linear.
For the vertical motion when $1/3 < \nu \le 1/2$,

$$u_0(t) = \int_0^t h_4(t-\tau)\frac{P_0(\tau)}{K}\,d\tau$$

applies where

$$\begin{aligned} h_4(t) &= \frac{2c}{z_0}\frac{1}{\sqrt{1-\frac{4}{\gamma}}}\,e^{-\frac{c\gamma}{z_0 2}t}\sinh\frac{c\gamma}{z_0 2}\sqrt{1-\frac{4}{\gamma}}\,t && t \ge 0 \\ &= 0 && t < 0 \end{aligned}$$

with

$$\gamma = \frac{\pi}{\mu}\frac{z_0}{r_0}$$

The recursive evaluation equals

$$u_{0n} = a_1\, u_{0n-1} + a_2\, u_{0n-2} + b_0\, \frac{P_{0n}}{K} + b_1\, \frac{P_{0n-1}}{K} + b_2\, \frac{P_{0n-2}}{K}$$

where

$$a_1 = 2e^{-\frac{c}{z_0}\frac{\gamma}{2}\Delta t} \cos\text{h}\, \frac{c\gamma}{z_0 2}\sqrt{1-\frac{4}{\gamma}}\ \Delta t$$

$$a_2 = -e^{-\frac{c}{z_0}\gamma\,\Delta t}$$

and

$$b_0 = \frac{r(\Delta t)}{\Delta t}$$

$$b_1 = \frac{r(2\Delta t) - (2+a_1)\, r(\Delta t)}{\Delta t}$$

$$b_2 = \frac{r(3\Delta t) - (2+a_1) r(2\Delta t) + (1+2a_1 - a_2) r(\Delta t)}{\Delta t}$$

with $r(t) = r_4(t)$

$$r_4(t) = -\frac{z_0}{c} + t + \frac{1-\frac{2}{\gamma}}{\sqrt{1-\frac{4}{\gamma}}}\frac{z_0}{c}\, e^{-\frac{c}{z_0}\frac{\gamma}{2}t} \sinh \frac{c}{z_0}\frac{\gamma}{2}\sqrt{1-\frac{4}{\gamma}}\ t$$

$$+ \frac{z_0}{c}\, e^{-\frac{c}{z_0}\frac{\gamma}{2}t} \cosh \frac{c}{z_0}\frac{\gamma}{2}\sqrt{1-\frac{4}{\gamma}}\ t \qquad t \geq 0$$

$$= 0 \qquad t < 0$$

For harmonic loading the dynamic-flexibility coefficient equals

$$F(b_0) = S^{-1}(b_0) = \frac{1}{K}\frac{\gamma}{(ib_0)^2 + \gamma\, ib_0 + \gamma}$$

For the horizontal motion and the vertical motion with $\nu \leq 1/3$, $\Delta M = 0$ and thus $\gamma = \infty$

$$F(b_0) = \frac{1}{K}\frac{1}{1 + ib_0}$$

9. For the rotational cone, the interaction moment $M_0(t)$ - rotation $\vartheta_0(t)$ relationship of the stiffness formulation equals

$$M_0(t) = K_\vartheta \vartheta_0(t) + C_\vartheta \dot{\vartheta}_0(t) + \Delta M_\vartheta \ddot{\vartheta}_0(t)$$

$$- \int_0^t h_1(t-\tau)\, C_\vartheta\, \dot{\vartheta}_0(\tau) d\tau$$

with the (same) unit-impulse response function $h_1(t)$ specified in item 8. The convolution integral representing the lingering part of the response M_0^r can be evaluated recursively

$$M_{0n}^r = a\,M_{0n-1}^r + b_0\left(-C_\vartheta\dot{\vartheta}_{0n}\right) + b_1\left(-C_\vartheta\,\dot{\vartheta}_{0n-1}\right)$$

with a, b_0, and b_1 given in item 8. The discrete-element models are shown in Fig. 2-75c. For harmonic loading the dynamic-stiffness coefficient of the rotational cone is formulated as

$$S_\vartheta(b_0) = K_\vartheta\left[k_\vartheta(b_0) + ib_0\,c_\vartheta(b_0)\right]$$

with the dimensionless spring and damping coefficients

$$k_\vartheta(b_0) = 1 - \frac{4\mu_\vartheta}{3\pi}\frac{r_0}{z_0}b_0^2 - \frac{1}{3}\frac{b_0^2}{1+b_0^2}$$

$$c_\vartheta(b_0) = \frac{1}{3}\frac{b_0^2}{1+b_0^2}$$

In terms of a_0

$$S_\vartheta(a_0) = K_\vartheta\left[k_\vartheta(a_0) + ia_0\,c_\vartheta(a_0)\right]$$

applies with

$$k_\vartheta(a_0) = 1 - \frac{4}{3}\frac{\mu_\vartheta}{\pi}\frac{z_0}{r_0}\frac{c_s^2}{c^2}a_0^2 - \frac{1}{3}\frac{a_0^2}{\left(\dfrac{r_0 c}{z_0 c_s}\right)^2 + a_0^2}$$

$$c_\vartheta(a_0) = \frac{z_0\,c_s}{3r_0 c}\frac{a_0^2}{\left(\dfrac{r_0 c}{z_0\,c_s}\right)^2 + a_0^2}$$

For the torsional motion and the rocking motion with $\nu \le 1/3$, ΔM_ϑ and thus μ_ϑ vanish.

10. For the rotational cone the rotation $\vartheta_0(t)$—interaction moment $M_0(t)$ relationship of the flexibility formulation equals for the torsional motion and the rocking motion for $\nu \le 1/3$

$$\vartheta_0(t) = \int_0^t h_2(t-\tau)\frac{M_0(\tau)}{K_\vartheta}d\tau$$

where the unit-impulse response function

$$h_2(t) = \frac{c}{z_0}e^{-\frac{3}{2}\frac{c}{z_0}t}\left(3\cos\frac{\sqrt{3}}{2}\frac{c}{z_0}t - \sqrt{3}\,\sin\frac{\sqrt{3}}{2}\frac{c}{z_0}t\right) \qquad t \ge 0$$

$$= 0 \qquad t < 0$$

The recursive evaluation equals

$$\vartheta_{0n} = a_1\vartheta_{0n-1} + a_2\vartheta_{0n-2} + b_0\frac{M_{0n}}{K_\vartheta} + b_1\frac{M_{0n-1}}{K_\vartheta} + b_2\frac{M_{0n-2}}{K_\vartheta}$$

where

$$a_1 = 2e^{-\frac{3}{2}\frac{c\Delta t}{z_0}}\cos\frac{\sqrt{3}}{2}\frac{c\Delta t}{z_0}$$

$$a_2 = -e^{-3\frac{c\Delta t}{z_0}}$$

and substituting $r(\mathrm{t}) = r_2(t)$

$$r_2(t) = t - \frac{2\sqrt{3}}{3}\frac{z_0}{c}e^{-\frac{3}{2}\frac{ct}{z_0}}\sin\frac{\sqrt{3}}{2}\frac{ct}{z_0} \qquad t \geq 0$$

$$= 0 \qquad t < 0$$

b_0, b_1, and b_2 follow from the equations in item 8. For harmonic loading the dynamic-flexibility coefficient equals

$$F_\vartheta(b_0) = S_\vartheta^{-1}(b_0) = \frac{1}{K_\vartheta}\frac{1 + ib_0}{1 + ib_0 + \left(\frac{1}{3} + \frac{1}{\gamma_\mu}\right)(ib_0)^2 + \frac{1}{\gamma_\mu}(ib_0)^3}$$

with

$$\gamma_\mu = \frac{3\pi}{4\mu_\vartheta}\frac{z_0}{r_0}$$

For the torsional motion and the rocking motion with $\nu \leq 1/3$, $\Delta M_\vartheta = 0$ and thus $\gamma_\mu = \infty$

$$F_\vartheta(b_0) = \frac{1}{K_\vartheta}\frac{3(1+ib_0)}{(ib_0)^2 + 3ib_0 + 3}$$

11. Material damping is introduced in the frequency domain based on the correspondence principle applied to the elastic solution. For noncausal linear-hysteretic damping the shear modulus $G = \rho c_s^2$ and constrained modulus $E_c = \rho c_p^2$ are multiplied by $1 + 2i\zeta_g$ (ζ_g = hysteretic-damping ratio), resulting in the solution for the damped case. The corresponding spring and damping coefficients $k_\zeta(a_0)$, $c_\zeta(a_0)$ are expressed approximately as a function of the elastic values $k(a_0)$, $c(a_0)$

$$k_\zeta(a_0) = k(a_0) - \zeta_g a_0 c(a_0)$$

$$c_\zeta(a_0) = c(a_0) + \frac{2\zeta_g}{a_0}k(a_0)$$

For Voigt viscoelasticity with damping proportional to frequency the factor equals $1 + i\omega 2\zeta_0/\omega_0$, and the correspondence principle is applied directly to the discrete-element model (or lumped-parameter model). Each original spring with coefficient K is augmented by a dashpot with coefficient $C = 2(\zeta_0/\omega_0)K$ in parallel, each original dashpot with coefficient C by a pulley–mass with coefficient $(\zeta_0/\omega_0)\ C$. Masses in the original model remain unchanged.

For frictional (hysteretic) material damping which preserves causality, nonlinear frictional elements replace the augmenting dashpot and pulley–mass. The corresponding forces acting between the nodes with displacements u_0 and u_1 are equal to

$$P = K \mid \overline{u_0 - u_1} \mid \tan\delta \ \text{sgn}(\dot{u}_0 - \dot{u}_1)$$

$$P = C \mid \overline{\dot{u}_0 - \dot{u}_1} \mid 0.5 \tan\delta \ \text{sgn}(\ddot{u}_0 - \ddot{u}_1)$$

with the overbar denoting the short-term memory (current or last peak value) and $\tan\delta = 2\zeta_g$ (δ = friction angle). Incorporation of these frictional elements in the discrete-element model (or lumped-parameter model) permits causal analysis in the time domain taking hysteretic damping independent of frequency into consideration.

12. The agreement of the cone's dynamic-stiffness coefficients with the exact solution is, in general, good. In the higher-frequency range, where the damping coefficients $c(a_0)$, $c_\vartheta(a_0)$ dominate, the cones are very accurate; while in the lower-frequency range the spring coefficients $k(a_0)$, $k_\vartheta(a_0)$ are accurate; but $c(a_0)$, $c_\vartheta(a_0)$ are too large (for $\nu \leq 1/3$), leading to an overestimation of the cone's radiation damping to a certain extent.

13. Reservations against the use of cone models are investigated step by step using comparisons with exact three-dimensional elastodynamic solutions and also by intuitive, physically motivated logical reasoning without solving differential equations. The objections are proven to be unfounded; the aspects omitted by cones are revealed to be of minor physical importance.

Three main reservations are discussed:

Cones are based on strength of materials, not the rigorous theory of the elastic halfspace. It is, however, proven that

- for low-frequency motion the translational cone exhibits a finite damping coefficient, and for the rotational cone the damping coefficient starts from zero. The same properties are also encountered in the exact solution.
- the cones are doubly asymptotic solutions, exact for both very low (static case) and very high frequencies.
- the assumptions that the radius of the cone whose apex is located above the soil's surface increases linearly with depth and that "plane sections remain plane" are (approximately) valid.
- the assumptions for the nearly incompressible soil in vertical and rocking motions used for the cone (limit of the appropriate wave velocity and trapped mass) are confirmed. In particular, for $\nu = 1/2$, $c = 2c_s$ and a trapped mass is present in the exact solution.

The portion of the halfspace outside the cone is neglected. It is demonstrated that

- it is physically correct to assume that the region of the soil halfspace outside the cone is essentially inactive and may be omitted.

- the proportions of the cones are compatible with the proper horizontal wave-propagation velocities which can be determined based on physical reasoning for all motions.

Cone models cannot represent the influence of Rayleigh surface waves. It is proven upon closer examination that this inability is physically correct:

- For low-frequency excitation no Rayleigh waves exist in the portion of the halfspace occupied by the cone.
- For high-frequency excitation the energy transmitted by Rayleigh waves is negligible. In the limit, for $\nu \neq 1/2$, all energy is concentrated in *P*-waves which propagate vertically downward through the prismatic rod of soil directly under the basemat. As ν tends to 1/2, the energy transmitted in the high-frequency vibration approaches zero.

14. In the standard lumped-parameter model (Fig. 2-54), the discrete-element model of the translational cone (Fig. 2-75b) with the same static spring K still applies; but the dashpot C and the mass ΔM (denoted as M), thus two free parameters, are selected to reproduce as closely as possible the actual response of the total dynamic system including the structure regarded as a rigid block. The two coefficients of this standard lumped-parameter model which can also be used for rotational motions are specified in Table 2-9. The use is limited to the frequency range up to $a_0 = 1.5$.

15. In the fundamental lumped-parameter model (Fig. 2.55a) the discrete-element model of the rotational cone (Fig. 2-75c, spring–dashpot model) with the same static direct spring K (the subscript ϑ is omitted) still applies; but the direct dashpot C_ϑ (denoted as C_0) and the other spring (K_1) and dashpot (C_1) connected to the internal node as well as the mass ΔM_ϑ (M_0), thus four free parameters, are selected through curve-fitting to achieve an optimum fit of the dynamic-stiffness coefficient. A monkey-tail model also exists with another direct dashpot being introduced (Fig. 2-55b). The four coefficients of this fundamental lumped-parameter model which can be used for all motions are specified in Table 2-10.

16. A two-dimensional basemat with half-width b on the surface of a homogeneous half-plane (strip foundation) can be modeled by a semi-infinite truncated wedge of half-width b, apex height z_0, and wave velocity c.

As for cones (item 2), c equals c_s for the horizontal wedge deforming in shear; c equals c_p for $\nu \leq 1/3$ and is limited to $2c_s$ for $1/3 < \nu \leq 1/2$ in the vertical and rocking wedges deforming axially.

The aspect ratio z_0/b of the horizontal wedge follows from equating of the dynamic-spring coefficient of the wedge averaged over the frequency to that of the strip foundation.

$$\frac{z_0}{b} = \frac{2-\nu}{\pi} \qquad \textit{horizontal}$$

The static-stiffness coefficients of the wedge and the strip foundation both vanish. For the rocking wedge, its static-stiffness coefficient

$$K_\vartheta = \frac{2\rho c^2}{z_0} \frac{8b^3}{12}$$

is set equal to that of the strip foundation

$$K_\vartheta = \frac{\pi G b^2}{2(1-\nu)}$$

yielding

$$\frac{z_0}{b} = \frac{8(1-\nu)}{3\pi} \frac{c^2}{c_s^2} \qquad \textit{rocking}$$

The high-frequency behavior is described by a radiation dashpot with the same coefficient as for cones—for the translational wedge

$$C = \rho c 2b$$

and for the rocking wedge

$$C_\vartheta = \rho c \frac{8b^3}{12}$$

For nearly incompressible soil ($1/3 < \nu \leq 1/2$) the wave velocity c is limited to $2c_s$, and for the rocking motion a trapped mass moment of inertia

$$\Delta M_\vartheta = 0.75 \left(\nu - \frac{1}{3}\right) \rho \frac{8b^3}{12} b$$

is introduced.
For harmonic loading the dynamic-stiffness coefficient of the wedge equals

$$S(b_0) = \frac{b_0 H^{(2)}_{\frac{n+1}{2}}(b_0)}{H^{(2)}_{\frac{n-1}{2}}(b_0)} \frac{\rho c^2 A_0}{z_0}$$

with

$$b_0 = \frac{\omega z_0}{c}$$

For the translational wedge with $n = 1$ the damping coefficient $c(b_0)$ tends to infinity for the static case and then approaches the high-frequency limit from above. For the rocking wedge with $n = 3$ the damping coefficient $c_\vartheta(b_0)$ starts from zero and tends towards the high-frequency limit from below.

The agreement of the dynamic-stiffness coefficients of the wedge with the exact solution is satisfactory. As for cones, wedge models tend to overestimate radiation damping in the lower-frequency range for $\nu \leq 1/3$.

17. The unified approach of cone-wedge models based on simple approximate strength-of-materials theory allows valuable physical insight to be obtained into the essence of radiation damping and the differences between two-dimensional and three-dimensional solutions. Based on conservation of energy which requires that the wave amplitude decays in inverse propagation to the square root of the area (radiation criterion), it follows that the less the amplitude spreads and diminishes with distance, the greater is the radiation damping. This is the explanation of why radiation damping is much larger in the wedge model and thus in the two-dimensional case than in the cone model used in three dimensions. The heuristic concept of larger surfaces available, through which waves can propagate in three dimensions than in two, is misleading.

To determine the width $2b$ and depth $2a$ of an equivalent two-dimensional strip foundation with the same material properties regarded as a slice through the three-dimensional foundation, the static-rocking stiffness coefficients and the average horizontal dynamic-spring coefficients of the strip and disk foundations are set equal.

$$\frac{b}{r_0} = 0.816$$

$$\frac{a}{r_0} = 1.272$$

An equivalent slice of a two-dimensional strip model with the same material properties cannot be constructed to represent a three-dimensional situation, as the damping ratio is inherently overestimated. Three aspects contribute:

- The contact area of the equivalent two-dimensional model is significantly larger than in the three-dimensional case, which results in too large damping coefficients in the high-frequency limit.
- The damping coefficient per unit area is larger for the translational and rocking motions in two dimensions than in three, as the wave amplitude diminishes less with distance.
- Two-dimensional modeling always entails an underestimation of the dynamic-spring coefficient for the translational motions.

Because the damping ratio is grossly overestimated, two-dimensional modeling of a three-dimensional case with the same material properties cannot be recommended for actual engineering applications. It is more feasible to take the opposite approach and idealize slender soil–structure interaction problems with a radially symmetric model.

An alternative when defining the slice of the two-dimensional strip model exists which changes the impedance ρc of the soil. The density and wave-propagation velocities of the two-dimensional model will differ from those of the actual soil used in the three-dimensional case,

$$\frac{\rho^{2d}}{\rho} = 0.645$$

$$\frac{c^{2d}}{c} = 1.245$$

which can be problematic. A good fit of the damping coefficients in the high-frequency limit is achieved. In the low-frequency range, the two-dimensional model still overestimates radiation damping, although to a lesser extent.

Addressing the transition from square to slender foundations, it follows that the radiation damping for vertical and rocking motions for slender foundations (ratio of length to width smaller than four) is comparable to that of a squarish one and thus to that of an equivalent disk.

3

Foundation on Surface of Soil Layer on Rigid Rock

The homogeneous soil halfspace addressed in Chapter 2 represents one limiting case of a site. The other one consists of a homogeneous soil layer of finite depth resting on rigid rock. To calculate the dynamic response of a massless rigid basemat on the surface of a soil layer for all motions (Fig. 3-1), the truncated semi-infinite cones of the halfspace can be generalized. It will be demonstrated that the layer system can be unfolded into a so-called layered cone. With the help of this *unfolded layered cone* a wave pattern may be postulated which incorporates, in addition to the decay of the amplitudes as the waves propagate away from the basemat, also the reflections at rock interface and on the free surface. Appropriate wave patterns may be written both for translational and rotational motions. From the wave patterns it follows that the elastodynamic effects of the layer are analogous to the familiar acoustic phenomenon of echoes. Echo constants are derived which enable the response of the basemat on the homogeneous halfspace to be converted subsequently to that of the basemat on the layer.

The concept of the unfolded layered cone is examined and echo constants are introduced in Section 3.1. Sections 3.2 and 3.3 address the corresponding flexibility and stiffness formulations, respectively, for the unfolded layered translational and rotational cones.

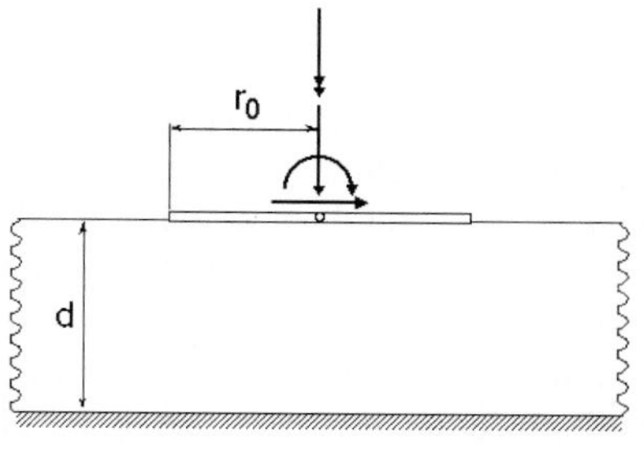

Fig. 3-1 Disk on surface of soil layer resting on rigid rock.

An alternative order of realization is derived in Section 3.4. An illustrative example is discussed in Section 3.5. The formulation of the unfolded layered cone is expanded to the case of a soil layer on flexible rock halfspace in Section 3.6. Lumped-parameter models to represent a basemat on the surface of a soil layer resting on rigid rock are discussed in Section 3.7. The phenomenon of the cutoff frequency, below which there is no radiation damping, is addressed in Section 3.8.

A slightly modified wave pattern can be postulated to construct the dynamic-stiffness matrix of a layer which then allows a disk on a horizontally stratified site with many layers resting on an underlying homogeneous halfspace to be calculated for harmonic loading. This extension is discussed in Appendix D.

The unfolded layered cone is applied to the soil layer on rigid rock in Meek and Wolf [M6], on which this chapter is based. Other aspects are described in other studies [W14, W16, W11, and M4].

3.1 CONSTRUCTION OF UNFOLDED LAYERED CONE MODEL

3.1.1 Layered Translational Cone

Figure 3-2 shows a massless basemat with equivalent radius r_0 of the disk bounded to the surface of an elastic layer with depth d. The substratum is considered to be rigid rock. Material damping is not addressed.

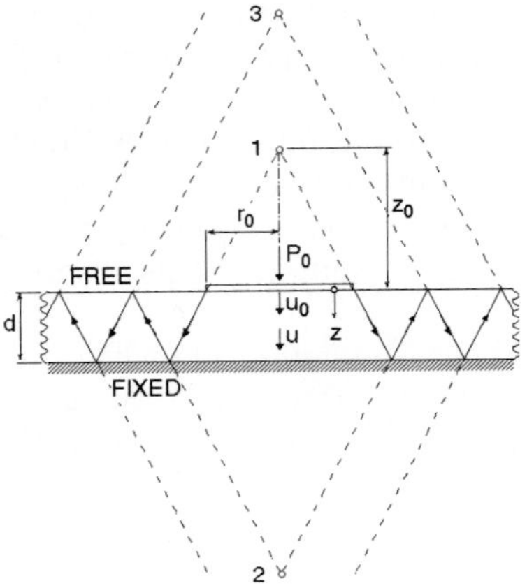

Fig. 3-2 Disk in vertical motion on surface of soil layer with wave pattern in refolded cone.

In the following the vertical motion with axial distortion is addressed. The interaction force P_0 causes dilatational waves to emanate beneath the disk. The waves propagate with velocity c equal to the dilatational wave velocity c_p (and $c = 2c_s$ with c_s = shear-wave velocity for $1/3 <$ Poisson's ratio $\nu \leq 1/2$, see Section 2.1.4), reflecting back and forth at the fixed boundary and the free surface, spreading and decreasing in amplitude (Fig. 3-2). The derivation is also valid for the horizontal motion with shearing distortion, where c equals c_s.

The following nomenclature is introduced for the displacements. The displacement of the (unlayered) cone modeling a disk with load P_0 on a homogeneous halfspace with the material properties of the layer (Fig. 3-3a) is denoted with an overbar as $\bar{u}$ with the value under the disk $\bar{u}_0$. For the disk loaded by the same force P_0 on a layer (Fig. 3-3b), the displacements calculated by superimposing the contributions of all the cones involved (Fig. 3-2) are u and u_0. It will become apparent that the surface motion of the (unlayered) cone representing the homogeneous halfspace $\bar{u}_0$ can be used to generate the motion of the layer u with its surface value u_0. Thus, $\bar{u}_0$ can also be called the *generating function*.

a. Halfspace with generating displacement. **b.** Layer with resulting displacement.

Fig. 3-3 Nomenclature for loaded disk in vertical motion on surface of soil.

The wave pattern in the layer consisting of the superposition of the contributions of the various cones (Fig. 3-2) is now discussed in detail. The force $P_0(t)$ produces dilatational waves propagating downwards from the disk along a cone with apex 1 and height z_0. This initial part of the wave pattern, called the incident wave, and the cone along which it propagates will be the same if the soil is a layer or a halfspace, as the wave generated beneath the disk does not "know" if at a specific depth a fixed interface is encountered or not. The same z_0 thus represents also the apex height of the (unlayered) cone of the homogeneous halfspace (Fig. 3-3a). This logical choice of the same apex height ensures that the static-stiffness coefficient of the layer converges to that of the halfspace for increasing thickness $d \rightarrow \infty$. The generating motion $\bar{u}_0(t)$ can be calculated for a given load $P_0(t)$ using the procedures for a disk on a homogeneous halfspace described in Chapter 2, in particular Section 2.2.2. The function $\bar{u}_0(t)$ vanishes for a negative argument, that is, for $t < 0$. The propagation of $\bar{u}_0(t)$ downwards along the cone is also the same as in the (unlayered) cone addressed in Appendix A. It is shown in Eq. A-6 that the displacement which is a function of $t - z/c$ for propagation in the positive z-direction is inversely proportional to the distance from the apex $z_0 + z$ in the translational cone. Notice that the origin of the coordinate z representing the depth is taken to be on the surface of the layer (Fig. 3-2) and not, as in Chapter 2 and Appendix A, at the apex. This *incident wave* pattern propagating along the cone with apex 1 shown in Fig. 3-4a is thus formulated as

$$\bar{u}(z,t) = \frac{z_0}{z_0 + z}\,\bar{u}_0\left(t - \frac{z}{c}\right) \tag{3-1}$$

At the fixed rock interface the displacement of the incident wave derived by substituting $z = d$ in Eq. 3-1 equals (Fig. 3-4b)

$$\bar{u}(d,t) = \frac{z_0}{z_0 + d}\,\bar{u}_0\left(t - \frac{d}{c}\right) \tag{3-2}$$

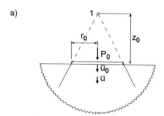

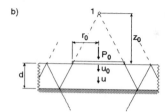

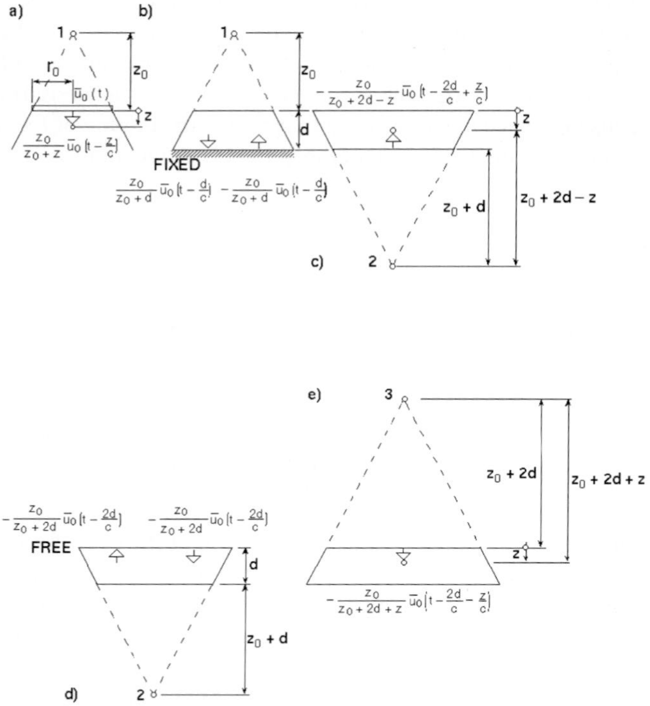

a. Generating wave in halfspace cone.

b. Incident and reflected waves at fixed interface with rigid rock.

c. Upwave in layer from reflection at fixed interface.

d. Incident and reflected waves at free surface.

e. Downwave in layer from reflection at free surface.

Fig. 3-4 Wave pattern and corresponding cones.

As the total displacement vanishes at the fixed boundary, the reflected wave must exhibit the same displacement as in Eq. 3-2 but with the other sign. This reflected wave will propagate upwards along its own cone with apex 2 (Fig. 3-4c). As the cone's opening angle depends only on Poisson's ratio of the soil (Table 2-3A), the opening angle of the cone with apex 2 (and of all cones to be introduced subsequently) will be the same as that of the original (unlayered) cone with apex 1. The apex height of the cone with apex 2 is thus $z_0 + d$. The distance from apex 2 to a point at depth z is $z_0 + 2d - z$; the displacement will again be inversely proportional to this distance. For this *upwave* propagating in the negative z-direction, the displacement will be a function of $t + z/c$. A constant will also arise in the displacements argument which is determined by equating the argument at the fixed boundary to $t - d/c$ (Eq. 3-2). The displacement of the upwave equals

$$-\frac{z_0}{z_0 + 2d - z}\,\bar{u}_0\left(t - \frac{2d}{c} + \frac{z}{c}\right) \tag{3-3}$$

At the free surface of the layer the displacement of the upwave derived by substituting $z = 0$ in Eq. 3-3 equals (Fig. 3-4d)

$$-\frac{z_0}{z_0 + 2d}\,\bar{u}_0\left(t - \frac{2d}{c}\right) \tag{3-4}$$

As is the case at a free boundary of a prismatic bar, the reflected wave in the cone will exhibit the same displacement as in Eq. 3-4. A verification follows below (Eq. 3-6). This reflected wave will propagate downwards along its own cone with apex 3 (Fig. 3-4e). With the geometric relationship shown in this figure and enforcing compatibility of the amplitude and of the argument of the reflected wave's displacement at the free surface, the displacement of the *downwave* is formulated as

$$-\frac{z_0}{z_0 + 2d + z}\bar{u}_0\left(t - \frac{2d}{c} - \frac{z}{c}\right) \tag{3-5}$$

The displacement corresponding to the sum of Eqs. 3-3 and 3-5 must satisfy for $z = 0$ the free-surface boundary condition of vanishing force and thus of zero axial strain. The latter is equal to the partial derivative of the displacement with respect to z. The derivative of Eq. 3-3 at $z = 0$ equals

$$-\frac{z_0}{(z_0 + 2d)^2}\bar{u}_0\left(t - \frac{2d}{c}\right) - \frac{z_0}{z_0 + 2d}\frac{1}{c}\bar{u}_0'\left(t - \frac{2d}{c}\right) \tag{3-6}$$

where $\bar{u}_0'$ denotes differentiation with respect to the total argument of the generating function—$\bar{u}_0' = \partial\bar{u}_0()/\partial()$ with $() = t - 2d/c + z/c$. The derivative of Eq. 3-5 results in the same value but with a sign change. The sum is thus zero.

This process of generating waves will continue. The downwave described by Eq. 3-5 will be reflected at the fixed boundary (Fig. 3-2). The newly created upwave propagating along its own cone will be reflected at the free surface, giving rise to a new downwave. At each fixed and free boundary the apex height of the cone along which the reflected wave propagates increases. *The waves in the layer thus decrease in amplitude and spread resulting in radiation of energy in the horizontal direction.*

The cones shown in Fig. 3-2 and taken apart in Fig. 3-4 can be unfolded into a layered cone. This *unfolded layered cone* shown in Fig. 3-5 contains the same information. As the wave travels downwards through the unfolded layered cone, it encounters every $2d$ either the rock interface or its deeper image. To capture the effect of fixed-boundary reflection, the sign of the displacement is reversed at each crossing of bedrock. Between two subsequent fixed-boundary reflections a free-boundary reflection occurs which doubles the displacement.

The resulting displacement in the layer $u(t,z)$ is equal to the *superposition* of the contributions of all cones; *the displacements of the incident wave* (Eq. 3-1, Fig. 3-4a), of the upwave (Eq. 3-3, Fig. 3-4c), and of the downwave (Eq. 3-5, Fig. 3-4e), *and of all* subsequent *upwaves and downwaves* are summed. This corresponds to refolding Fig. 3-5 into Fig. 3-2. In general the displacement $u(z,t)$ of the layer at depth z and time t may be expressed as a wave pattern.

$$u(z,t) = \underbrace{\frac{z_0}{z_0 + z}\bar{u}_0\left(t - \frac{z}{c}\right)}_{\textit{incident wave}} + \sum_{j=1}^{\infty}(-1)^j\left[\underbrace{\frac{z_0\bar{u}_0\left(t - \frac{2jd}{c} + \frac{z}{c}\right)}{z_0 + 2jd - z}}_{\textit{upwave from rock}} + \underbrace{\frac{z_0\bar{u}_0\left(t - \frac{2jd}{c} - \frac{z}{c}\right)}{z_0 + 2jd + z}}_{\textit{downwave from surface}}\right] \tag{3-7a}$$

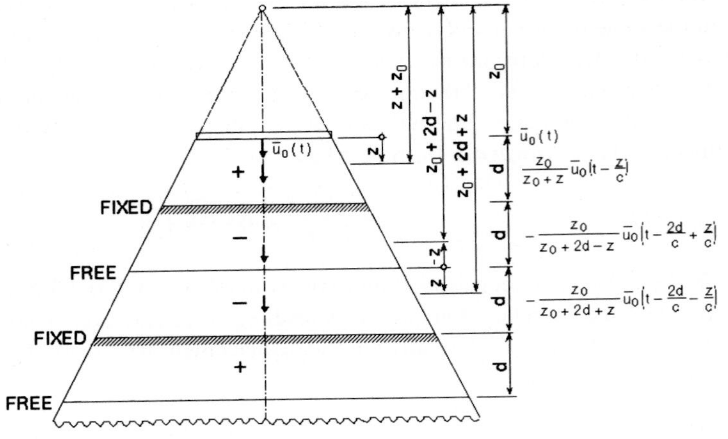

Fig. 3-5 Unfolded layered cone.

Each term of this expression corresponds to a single layer of the unfolded layered cone. As shown in the Fig. 3-5, the various denominators are the respective distances to the apex. The factor $(-1)^j$ takes into account the reflections at bedrock. As $\bar{u}_0$ vanishes for a negative argument, the sum shown in Eq. 3-7a as extending to infinity is limited. The largest j_{max} corresponding to a zero argument equals $j_{max} = (t + z/c)c/(2d)$ for the upwave and $j_{max} = (t - z/c)c/(2d)$ for the downwave. The j_{max} determined from these relationships will, in general, not be an integer. Each sum is terminated at the largest integer which is smaller than or equal to the value of the corresponding j_{max}, denoted as k_u and k_d for the upwave and downwave, respectively. For instance, at $z = d$ and $t = 3.5\ d/c$, for the upwave $j_{max} = 2.25$ which leads to $k_u = 2$, and for the downwave $j_{max} = 1.25$ which yields $k_d = 1$.

$$u(z,t) = \frac{z_0}{z_0 + z}\bar{u}_0\left(t - \frac{z}{c}\right) + \sum_{j=1}^{k_u}(-1)^j \frac{z_0\bar{u}_0\left(t - \frac{2jd}{c} + \frac{z}{c}\right)}{z_0 + 2jd - z} + \sum_{j=1}^{k_d}(-1)^j \frac{z_0\bar{u}_0\left(t - \frac{2jd}{c} - \frac{z}{c}\right)}{z_0 + 2jd + z} \tag{3-7b}$$

As time progresses, k_u and k_d will increase. It should be emphasized that the unfolded layered cone is just a helpful geometric construction to derive Eq. 3-7. It is not a physical model and cannot be represented physically, for example, by finite elements of a tapered bar.

At bedrock $u(z = d,t)$ vanishes, as required by the built-in boundary condition; and at the surface the displacement of the disk $u(z = 0, t) = u_0(t)$

$$u_0(t) = \bar{u}_0(t) + 2\sum_{j=1}^{k}(-1)^j \frac{\bar{u}_0(t - jT)}{1 + j\kappa} \tag{3-8}$$

with $k_u = k_d = k$ and where

$$\kappa = \frac{2d}{z_0} \quad (3\text{-}9a) \qquad\qquad T = \frac{2d}{c} \quad (3\text{-}9b)$$

The value k is equal to the largest integer which is smaller than or equal to $tc/(2d) = t/T$, thus to the number of times at time t that $\bar{u}_0(t)$ has completed the propagation from the free surface (disk) to the rigid rock interface and back to the free surface. The geometric parameter κ is a dimensionless measure of the layer depth. The temporal parameter T corresponds to the travel time from the surface to the rock and back. Everyone is familiar with the acoustic phenomenon of echoes. An explosive-type noise returns at distinct intervals of time with decreasing amplitude. This physical behavior is described mathematically by Eq. 3-8, the time T being the delay between consecutive echoes (echo interval). The introduction of flexibility *echo constants* e_j^F (superscript F anticipating their use in the flexibility formulation) allows Eq. 3-8 to be written concisely as

$$u_0(t) = \sum_{j=0}^{k} e_j^F \, \bar{u}_0(t - jT) \tag{3-10}$$

in which

$$e_0^F = 1 \tag{3-11a}$$

and for $j \geq 1$

$$e_j^F = \frac{2(-1)^j}{1 + j\kappa} \tag{3-11b}$$

Because of the free-surface boundary condition implicit in Eq. 3-8, the reflected waves (sums) do not contribute to the force of the footing. This most important fact, proven in Appendix A3, implies that P_0 and $\bar{u}_0$ (not u_0!) may be computed from one another via the force-displacement relationships of the unlayered cone discussed in Chapter 2. The effect of the layer is captured exclusively by the *echo formula* (Eq. 3-10) which *relates the true surface motion of the layer $u_0(t)$ to present and past values of the generating function $\bar{u}_0(t - jT)$.*

3.1.2 Layered Rotational Cone

The concept discussed in Section 3.1.1 for the translational cone applies also to the rotational cone. Appropriately defined echo constants can again be introduced. The rocking motion with axial distortion is examined. The waves caused by the interaction moment M_0 propagate with c_p (and $= 2c_s$ for $1/3 < \nu \leq 1/2$) along the corresponding cones with reflections occurring repeatedly at the fixed rock interface and at the free surface. The results are also valid for the torsional motion with shearing distortions using $c = c_s$.

The moment M_0 applied to the disk on the homogeneous halfspace modeled with an (unlayered) cone leads to the rocking rotation $\bar{\vartheta}_0$ under the disk which represents the *generating function* (Fig. 3-6a). The rotations under the disk and at a depth of the layer loaded by the same moment acting on the disk calculated by refolding the layered cone are denoted as ϑ_0 and ϑ (Fig. 3-6b).

By contrast to the translational case, the angle of rotation is not simply inversely related to the distance from the apex but decays in inverse proportion to the square and the

a. Halfspace with generating rotation.

b. Layer with resulting rotation.

Fig. 3-6 Nomenclature for loaded disk in rocking motion on surface of soil.

cube of distance. The solution for the unlayered rotational cone, specified in Eq. A-57, is given after introducing a bar by

$$\bar{\vartheta}(z,t) = \left(\frac{z_0}{z_0 + z}\right)^2 \bar{\vartheta}_0\left(t - \frac{z}{c}\right) + \left[\left(\frac{z_0}{z_0 + z}\right)^3 - \left(\frac{z_0}{z_0 + z}\right)^2\right] \bar{\vartheta}_1\left(t - \frac{z}{c}\right) \tag{3-12}$$

At the surface ($z = 0$) the second term vanishes, and only the generating function remains: $\bar{\vartheta}(z = 0,t) = \bar{\vartheta}_0(t)$. The second term involves an auxiliary function $\bar{\vartheta}_1(t)$ (which is equal to the rotation in the internal node of the discrete-element model (see Fig. 2-17)), a convolution (Eq. A-56)

$$\bar{\vartheta}_1(t) = \int_0^t h_1(t - \tau)\bar{\vartheta}_0(\tau)d\tau \tag{3-13}$$

in which the unit-impulse response function is the decaying exponential (Eq. A-24)

$$h_1(t) = \frac{c}{z_0} e^{-\frac{c}{z_0}t} \qquad t \geq 0 \tag{3-14a}$$

$$= 0 \qquad t < 0 \tag{3-14b}$$

The wave pattern for the layered rotational system, verified in Appendix A3 is analogous to Eq. 3-7, except that all factors $z_0/(z_0 + 2jd \mp z)$ are replaced by $[z_0/(z_0 + 2jd \mp z)]^N$ with $N = 2$ or 3. Consequently the denominator $(1 + j\kappa)$ in Eq. 3-8 is replaced by $(1 + j\kappa)^N$. The wave pattern involves both the function $\bar{\vartheta}_0$ (with $N = 2$) and the function $\bar{\vartheta}_1$ (with $N =$ 2 and 3).

Keeping this in mind, the appropriate flexibility *echo constants* for the *layered rotational cone* may be written from Eqs. 3-8 and 3-10 by inspection.

$$\vartheta_0(t) = \sum_{j=0}^{k} \overset{F}{e}_{0j}\, \bar{\vartheta}_0(t - jT) + \sum_{j=0}^{k} \overset{F}{e}_{1j}\, \bar{\vartheta}_1(t - jT) \tag{3-15}$$

with

$$\overset{F}{e}_{00} = 1 \qquad \overset{F}{e}_{10} = 1 - 1 = 0 \tag{3-16a}$$

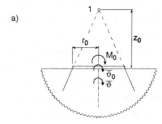

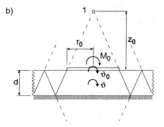

and, for $j \geq 1$

$$e_{0j}^{F} = \frac{2(-1)^j}{(1 + j\kappa)^2} \qquad e_{1j}^{F} = 2(-1)^j \left[\frac{1}{(1 + j\kappa)^3} - \frac{1}{(1 + j\kappa)^2}\right] \tag{3-16b}$$

As in the case of the layered translational cone (see discussion following Eq. 3-7a), the sum in Eq. 3-15 is limited (up to k, the number of echos which have occured up to time t). If $\bar{\vartheta}_0(t)$ is known, it is easy to compute $\bar{\vartheta}_1(t)$ (Eq. 3-13) recursively, as described in Section 2.2.2 (analogous to Eq. 2-83 with Eq. 2-84).

It is worth mentioning that the key expressions to analyze a basemat on a layer, Eqs. 3-10 and 3-15, can also be derived using embedded disks with the corresponding double cones in a full space. This approach is developed in Chapter 4 to calculate embedded foundations. For the basemat on the surface of a layer fixed at its base, one disk is used. To satisfy the free surface and fixed boundary conditions, mirror disks have to be introduced. The Green's functions of the double cones in the full space corresponding to these disks lead to the terms in the *echo formulas*. A thorough treatment is presented in Section 4.1.4.

3.2 FLEXIBILITY FORMULATION

3.2.1 Unit-Impulse Response Function

In the flexibility formulation of a foundation-dynamics problem, the applied interaction force P_0 or moment M_0 is given; and it is desired to calculate the resultant displacement u_0 (Fig. 3-3b) or rotation ϑ_0 (Fig. 3-6b). The solution for a soil layer is obtained by a simple two-step procedure:

1. By applying the procedures described in Section 2.2.2 or 2.3.2, the response of the associated (unlayered) cone modeling the homogeneous halfspace $\bar{u}_0$ (Fig. 3-3a) or $\bar{\vartheta}_0$ (Fig. 3-6a) is determined. It is efficient to apply the recursive formulation (Eqs. 2-83 or 2-178).
2. By using the echo constants, the true surface motion of the layer u_0 or ϑ_0 is evaluated from Eq. 3-10 or Eq. 3-15 using $\bar{u}_0$ or $\bar{\vartheta}_0$. The sums over j are limited (up to k) as for negative arguments $t - jT$, the values $\bar{u}_0(t - jT)$, or $\bar{\vartheta}_0(t - jT)$, $\bar{\vartheta}_1(t - jT)$ vanish.

The accuracy of this approach may be ascertained by considering a simple limiting case of excitation, a Dirac-delta impulse of force or moment, applied to a disk with radius r_0. The depth of the layer is taken to be $d = r_0$ and Poisson's ratio $\nu = 1/3$ ($c_p = 2c_s$). For this combination of parameters the echo constants e_j^F for the horizontal and vertical motions and e_{0j}^F, e_{1j}^F for the rocking motion are listed in Table 3-1. The values of κ (Eq. 3-9a) correspond to the ratios z_0/r_0 given in Table 2-3B. In the expression for T (Eq. 3-9b), $c = c_s$ for the horizontal motion and $c = c_p$ for the vertical and rocking motions. The unit-impulse response functions for the unlayered cone $h_1(t)$ for translation (Eq. 2-75, Fig. 2-8) and $h_2(t)$ (Eq. 2-172, Fig. 2-22) for rocking, which correspond to $\bar{u}_0(t)$ and $\bar{\vartheta}_0(t)$, are multiplied by the echo constants and superposed with time shifts jT according to Eqs. 3-10 and 3-15, 3-13. The results for the layered cone nondimensionalized using the static values of the corresponding halfspace are compared in Figs. 3-7a (horizontal motion), 3-7b (vertical motion), and 3-7c (rocking motion) with "exact" results for the elastic layer [W6]. The values of the abscissa $t/T = 1,2,3, \ldots$ corresponds to the travel time calculated with the appropriate wave velocity from the surface to the rock and back and to multiples thereof. In Fig. 3-7b the unit-impulse response function $h_2(t) = \bar{u}_0(t)$ of the halfspace is also shown. The agreement is quite satisfactory, especially when one takes into account that the exact solutions for the disk on the elastic layer are evaluated by numerical Fourier transformation.

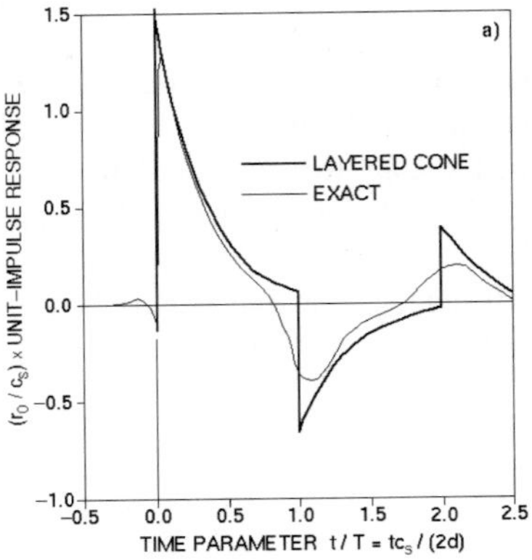

a. Horizontal.

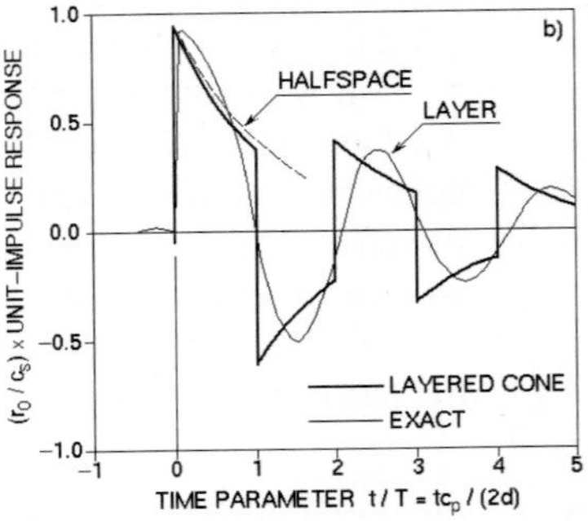

b. Vertical.

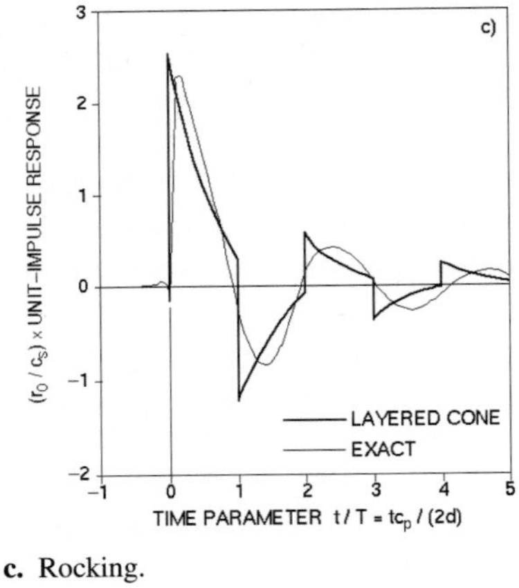

c. Rocking.

Fig. 3-7 Unit-impulse response function of disk on layer.

The implicit low-pass filtering unavoidable in this procedure smears over and thus conceals the jump discontinuities quite probably also present in rigorously exact solutions for the disk on the elastic layer. Note, for example, in Fig. 3-7b that due to smearing the exact results of the layer deviate from those of the halfspace before the earliest possible arrival of reflected waves at $t/T = 1$. This is physically impossible. Prior to $t = T$ the causality principle demands strict equality of the responses of the layer and of the halfspace. Further evidence that the impulse response of the disk on the elastic layer probably exhibits

TABLE 3-1 FLEXIBILITY ECHO CONSTANTS FOR TRANSLATION AND ROTATION

	Horizontal	Vertical	Rocking	
z_0/r_0	0.654	2.094	2.356	
κ	3.058	0.955	0.849	
j	e_j^F	e_j^F	e_{0j}^F	e_{1j}^F
0	1.000	1.000	1.000	0.000
1	−0.493	−1.023	−0.585	0.269
2	0.281	0.687	0.275	−0.173
3	−0.197	−0.517	−0.159	0.114
4	0.151	0.415	0.104	−0.080
5	−0.123	−0.346	−0.073	0.059
6	0.103	0.297	0.054	−0.045
7	−0.089	−0.260	−0.041	0.036
8	0.079	0.231	0.033	−0.029
9	−0.070	−0.208	−0.027	0.024
10	0.063	0.190	0.022	−0.020
11	−0.058	−0.174	−0.019	0.017
12	0.053	0.161	0.016	−0.015
13	−0.049	−0.149	−0.014	0.013
14	0.046	0.139	0.012	−0.011
15	−0.043	−0.131	−0.011	0.010
16	0.040	0.123	0.009	−0.009
17	−0.038	−0.116	−0.008	0.008
18	0.036	0.110	0.008	−0.007
19	−0.034	−0.104	−0.007	0.006
$\sum_{j=0}^{18} e_j^F$	0.692	0.429	0.596	0.161
$\sum_{j=0}^{19} e_j^F$	0.658	0.325	0.589	0.167
Average	0.675	0.377	0.593	0.164
			0.757	

jump discontinuities is furnished by familiar experience with the analogous acoustical case: the echoes from an explosive noise recur with decreasing volume, but without smearing. When calculating the exact values, a small amount of material damping is introduced, which is also partly responsible for the deviation.

The flexibility formulation of the case of nearly incompressible soil is addressed in connection with the stiffness formulation in Section 3.3.4.

3.2.2 Static Case

The accuracy can also be evaluated by addressing the other limit of excitation, the static case. The values $\bar{u}_0$, $\bar{\vartheta}_0$, u_0 and ϑ_0 are constant in this situation. Formulating $\bar{\vartheta}_1$ for a constant $\bar{\vartheta}_0$ (Eq. 3-13 with Eq. 3-14) reveals that $\bar{\vartheta}_1$ converges to the static value $\bar{\vartheta}_0$ as time tends to infinity. This is also visible from the differential equation relating $\bar{\vartheta}_1$ and $\bar{\vartheta}_0$. Equation A-55, rewritten with a bar, is equal to

$$\bar{\vartheta}_1 + \frac{z_0}{c}\dot{\bar{\vartheta}}_1 = \bar{\vartheta}_0 \tag{3-17}$$

As $\dot{\bar{\vartheta}}_1 = 0$, $\bar{\vartheta}_1$ equals $\bar{\vartheta}_0$. The echo formulas, Eqs. 3-10 and 3-15, revert to direct sums of the echo constants e_j^F with the sums extending to infinity

$$u_0 = \bar{u}_0 \sum_{j=0}^{\infty} e_j^F \tag{3-18}$$

$$\vartheta_0 = \bar{\vartheta}_0 \sum_{j=0}^{\infty} e_{0j}^F + \bar{\vartheta}_0 \sum_{j=0}^{\infty} e_{1j}^F \tag{3-19}$$

It is possible to estimate these infinite sums quite accurately by considering only a finite number of terms because the echo constants alternate in sign. The sums are evaluated up to $j = 18$ and 19 at the bottom of Table 3-1, then averaged. The results are

$$u_0 \simeq 0.675\,\bar{u}_0 \quad \text{(horizontal)} \tag{3-20a}$$

$$u_0 \simeq 0.377\,\bar{u}_0 \quad \text{(vertical)} \tag{3-20b}$$

$$\vartheta_0 \simeq 0.757\bar{\vartheta}_0 \quad \text{(rocking)} \tag{3-20c}$$

The inverse of the coefficients in Eq. 3-20 (1.48 for horizontal, 2.65 or vertical and 1.32 for rocking) express how many times stiffer the disk on the layer is than the disk on the homogeneous halfspace. For a rigorous theory-of-elasticity solution [W4], the exact factors are 1.56 (horizontal), 2.56 (vertical), and 1.28 (rocking). If a segment of a homogeneous tapered bar (isolated cone with same opening angle) without horizontal wave spreading were used to calculate these factors, larger values would arise (1.65 for horizontal, 3.09 for vertical, and 1.53 for rocking). The accuracy of the results obtained via echo constants is remarkable (within 5%) for both translational components and for rocking.

The following approximate equations for the static-stiffness coefficients of a disk on the surface of a layer are useful [G3]

$$K_h = \frac{8Gr_0}{2-\nu}\left(1+\frac{1}{2}\frac{r_0}{d}\right) \quad \text{horizontal} \tag{3-21a}$$

$$K_v = \frac{4Gr_0}{1-\nu}\left(1+1.3\frac{r_0}{d}\right) \quad \text{vertical} \tag{3-21b}$$

$$K_r = \frac{8Gr_0^3}{3(1-\nu)}\left(1+\frac{1}{6}\frac{r_0}{d}\right) \qquad \text{rocking} \qquad (3\text{-}21c)$$

$$K_t = \frac{16}{3}Gr_0^3\left(1+\frac{1}{10}\frac{r_0}{d}\right) \qquad \text{torsional} \qquad (3\text{-}21d)$$

It is important to stress that these static-stiffness coefficients of the layer are not used to calibrate the aspect ratios and thus the opening angles of the cones.

3.3 STIFFNESS FORMULATION

3.3.1 Pseudo-Echo Constants for Rotation

In the stiffness formulation of a foundation-dynamics problem, the applied displacement or rotation u_0 or ϑ_0 is given; and it is desired to calculate the resultant interaction force P_0 (Fig. 3-3b) or moment M_0 (Fig. 3-6b). Compared to the flexibility formulation, the roles of input and output are reversed. Again, the solution for the layer is obtained by a simple two-step procedure:

1. The prescribed surface motion u_0 or ϑ_0 is converted to the generating function $\bar{u}_0$ or $\bar{\vartheta}_0$, as described below.
2. Insertion of $\bar{u}_0$ and $\dot{\bar{u}}_0$ into the force-displacement relationship of the (unlayered) cone representing the homogeneous halfspace or of $\bar{\vartheta}_0$ and $\dot{\bar{\vartheta}}_0$ into the moment-rotation relationship yields the desired force P_0 (Fig. 3-3a) or moment M_0 (Fig. 3-6a). This second step is performed using the procedures described in Sections 2.2.1 and 2.3.1.

Turning to the first step, the echo-constant formula relating the generating function (quantity denoted with a bar) $\bar{u}_0$, $\bar{\vartheta}_0$ to the layer's deformation (quantity without a bar) u_0,ϑ_0 has to be inverted—that is, expressing $\bar{u}_0$, $\bar{\vartheta}_0$ as a function of u_0,ϑ_0. For the translational cone, Eq. 3-10 can be inverted straightforwardly, as will be demonstrated. For the rotational cone $\bar{\vartheta}_0$ appears in Eq. 3-12 twice; directly in the first term on the right-hand side and indirectly via the convolution integral $\bar{\vartheta}_1$ (Eq. 3-13). A set of so-called pseudo-echo constants has thus first to be determined before the inversion for the equation of the rotational cone can be performed.

For this purpose a fundamental solution (influence function) is constructed by considering the unit triangular pulse of $\bar{\vartheta}_0$ shown in Fig. 3-8a. The duration of the pulse is $2\Delta t$. A succession of these pulses of varying amplitude exactly represents any piecewise-linear time history $\bar{\vartheta}_0$. The time increment Δt is conveniently defined as an integer fraction of the echo interval: $\Delta t = T$, $T/2$, $T/3$, $T/4$, and so on. To minimize computation, the value of Δt should be chosen as large as possible but small enough for a good piecewise-linear approximation of the input $\bar{\vartheta}_0$. For the rocking motion of the disk on a layer with $d/r_0 = 1$ and $\nu = 1/3$ already addressed in Table 3-1, $\Delta t = T/4$ is selected.

a. Generating functions.

b. Resultant rotation.

Fig. 3-8 Response to triangular pulse.

To calculate $\bar{\vartheta}_1$, the convolution integral of Eq. 3-13 involving h_1 is evaluated recursively. The recursive equation (Eq. 2-83) for time station n is formulated as

$$\bar{\vartheta}_{1n} = a\,\bar{\vartheta}_{1n-1} + b_0\,\bar{\vartheta}_{0n} + b_1\,\bar{\vartheta}_{0n-1} \tag{3-22}$$

where the recursive coefficients a, b_0, b_1 specified in Eq. 2-84 are a function of $c\Delta t/z_0$. For $c\Delta t/z_0 = c[(2d/c)/4]/z_0 = 0.2122$ with $d = r_0$ and $z_0 = 2.356r_0$ (Table 2-3B) $a = 0.80880$, $b_0 = 0.09898$, and $b_1 = 0.09222$ result. Using these, $\bar{\vartheta}_1$ listed in column 4 of Table 3-2 is calculated. The time history of $\bar{\vartheta}_1$ is plotted in Fig. 3-8a. By contrast to the $\bar{\vartheta}_0$-pulse, which is non-zero only over the range $2\Delta t$, the function $\bar{\vartheta}_1$ is of infinite duration (but is obviously negligibly small for $t > 4T$). The triangular pulse $\bar{\vartheta}_0$ generates the surface motion of the layer ϑ_0, samples of which are evaluated in columns 5, 6, and 7 of Table 3-2 and shown in Fig. 3-8b. Column 5 corresponds to the first term on the right-hand side of Eq. 3-15—without the influence of $\bar{\vartheta}_1$. The number of terms in the sums are limited as $\bar{\vartheta}_0$ and $\bar{\vartheta}_1$ vanish for negative arguments. Column 7 lists the complete response determined from Eq. 3-15.

If the sequence of numerical values in column 7 of Table 3-2 is interpreted as a new set of *pseudo-echo constants* $\overset{F}{e}_{rm}$ (r for rotation, m the index for Δt), then the relationship between $\bar{\vartheta}_0$ in column 3 and ϑ_0 in 7 of Table 3-2 may be expressed as

$$\vartheta_0(t) = \sum_{m=0}^{k} \overset{F}{e}_{rm}\,\bar{\vartheta}_0(t-m\Delta t) \tag{3-23}$$

Equation 3-23 is exactly analogous to Eq. 3-10 except that the echo return time T has been replaced by its subdivision Δt. Hence Eq. 3-23 is a *pseudo-echo formula*, derived for a single pulse of $\bar{\vartheta}_0$ but—thanks to the principle of superposition—just as valid for any piecewise-linear variation of $\bar{\vartheta}_0$.

For $\Delta t = T/4$ the evaluation of Eq. 3-23 requires four times as many operations as Eq. 3-10. However, the $\overset{F}{e}_{rm}$ constants decay so rapidly with m that the total numerical effort remains small. Moreover, above $j = 4$ ($m = 16$) the time step Δt may be increased to the full echo internal T.

It should be emphasized that Eq. 3-23 is derived from and thus yields the same result as the original rotational echo formula (Eq. 3-15) in which the contributions of $\bar{\vartheta}_0$ and $\bar{\vartheta}_1$ are evaluated separately. However, the auxiliary function $\bar{\vartheta}_1$ does not appear explicitly in Eq. 3-23. As a result, Eq. 3-23 may be interverted—solved for $\bar{\vartheta}_0(t)$ in terms of $\vartheta_0(t - m\Delta t)$. This turns out to be the decisive prerequisite for the stiffness approach. The pseudo-echo

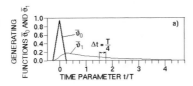

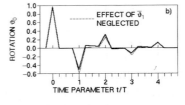

TABLE 3-2 ROTATIONAL RESPONSE TO TRIANGULAR PULSE

j	$\frac{t}{T}$	$\bar{\vartheta}_0(t)$	$\bar{\vartheta}_1(t)$	$\sum_{j=0}^{k} e^F_{0j}\bar{\vartheta}_0(t-jT)$	$\sum_{j=0}^{k} e^F_{1j}\bar{\vartheta}_1(t-jT)$	$\vartheta_0(t)$ e^F_{rm}	m
0	0	1	0.099	1.000	0	1.000	0
	0.25	0	0.172	0	0	0	1
	0.50	0	0.139	0	0	0	2
	0.75	0	0.113	0	0	0	3
1	1.00	0	0.091	−0.585	0.027	−0.558	4
	1.25	0	0.074	0	0.046	0.046	5
	1.50	0	0.060	0	0.037	0.037	6
	1.75	0	0.048	0	0.030	0.030	7
2	2.00	0	0.039	0.275	0.007	0.282	8
	2.25	0	0.032	0	−0.010	−0.010	9
	2.50	0	0.026	0	−0.008	−0.008	10
	2.75	0	0.021	0	−0.007	−0.007	11
3	3.00	0	0.017	−0.159	0.006	−0.153	12
	3.25	0	0.014	0	0.015	0.015	13
	3.50	0	0.011	0	0.013	0.013	14
	3.75	0	0.009	0	0.010	0.010	15
4	4.00	0	0.007	0.104	0.000	0.104	16
	4.25	0	0.006	0	−0.007	−0.007	17
	4.50	0	0.005	0	−0.006	−0.006	18
	4.75	0	0.004	0	−0.005	−0.005	19

constants e^F_{rm} must be ascertained by recalculation of Table 3-2 for the particular situation at hand, but the numerical work is minimal and the results are exact.

3.3.2 Stiffness Echo Constants

The prescribed surface motion of the layer u_0 or ϑ_0 is converted to the generating function $\bar{u}_0$ or $\bar{\vartheta}_0$ by means of a new set of *echo constants* e^K_j or e^K_{rm}. The superscript K denotes the stiffness formulation.

$$\bar{u}_0(t) = \sum_{j=0}^{k} e^K_j\, u_0(t-jT) \tag{3-24}$$

$$\bar{\vartheta}_0(t) = \sum_{m=0}^{k} e^K_{rm}\, \vartheta_0(t-m\Delta t) \tag{3-25}$$

The same formulas are also employed to convert the velocities

$$\dot{\bar{u}}_0(t) = \sum_{j=0}^{k} \overset{K}{e_j}\, \dot{u}_0(t-jT) \tag{3-26}$$

$$\dot{\bar{\vartheta}}_0(t) = \sum_{m=0}^{k} \overset{K}{e_{rm}}\, \dot{\vartheta}_0(t-m\,\Delta t) \tag{3-27}$$

The desired $\overset{K}{e_j}$ or $\overset{K}{e_{rm}}$ may be determined step-by-step from the known $\overset{F}{e_j}$ or $\overset{F}{e_{rm}}$ as follows.

$$\overset{K}{e_0} = 1 \tag{3-28a}$$

and for $j \geq 1$

$$\overset{K}{e_j} = -\sum_{\ell=0}^{j-1} \overset{K}{e_\ell}\, \overset{F}{e_{j-\ell}} \tag{3-28b}$$

or

$$\overset{K}{e_{r0}} = 1 \tag{3-29a}$$

and for $m \geq 1$

$$\overset{K}{e_{rm}} = -\sum_{\ell=0}^{m-1} \overset{K}{e_{r\ell}}\, \overset{F}{e_{rm-\ell}} \tag{3-29b}$$

For instance, Eq. 3-28 can be verified as follows. Formulating the flexibility echo-constants formula (Eq. 3-10) for $u_0(t)$ at times $t, t-T, t-2T, \ldots, t-jT$ including terms up to $\bar{u}_0(t-jT)$ yields

$$\begin{aligned} u_0(t) = {} & \overset{F}{e_0}\, \bar{u}_0(t) + \overset{F}{e_1}\, \bar{u}_0(t-T) + \overset{F}{e_2}\, \bar{u}_0(t-2T) + \ldots + \overset{F}{e_{j-1}}\, \bar{u}_0^F\big(t-(j-1)T \\ & + \overset{F}{e_j}\, \bar{u}_0(t-jT) + \ldots \end{aligned} \tag{3-30a}$$

$$\begin{aligned} u_0(t-T) = {} & \overset{F}{e_0}\, \bar{u}_0(t-T) + \overset{F}{e_1}\, \bar{u}_0(t-2T) + \overset{F}{e_2}\, \bar{u}_0(t-3T) + \ldots + \overset{F}{e_{j-2}}\, \bar{u}_0\big(t-(j-1)T\big) \\ & + \overset{F}{e_{j-1}}\, \bar{u}_0(t-jT) + \ldots \end{aligned} \tag{3-30b}$$

$$\begin{aligned} u_0(t-2T) = {} & \overset{F}{e_0}\, \bar{u}_0(t-2T) + \overset{F}{e_1}\, \bar{u}_0(t-3T) + \overset{F}{e_2}\, \bar{u}_0(t-4T) + \ldots + \overset{F}{e_{j-3}}\, \bar{u}_0\big(t-(j-1)T\big) \\ & + \overset{F}{e_{j-2}}\, \bar{u}_0(t-jT) + \ldots \end{aligned} \tag{3-30c}$$

$$\vdots$$

$$u_0(t-(j-1)T) = \overset{F}{e_0}\, \bar{u}_0(t-(j-1)T) + \overset{F}{e_1}\, \bar{u}_0(t-jT) + \ldots \tag{3-30d}$$

$$u_0(t-jT) = \overset{F}{e_0}\, \bar{u}_0(t-jT) + \ldots \tag{3-30e}$$

The stiffness echo-constants formula (Eq. 3-24) also with terms up to $u_0(t - jT)$ equals

$$\bar{u}_0(t) = \overset{K}{e_0}\, u_0(t) + \overset{K}{e_1}\, u_0(t - T) + \overset{K}{e_2}\, u_0(t - 2T) + \ldots + \overset{K}{e_{j-1}}\, u_0\big(t - (j-1)T\big) + \overset{K}{e_j}\, u_0(t - jT) + \ldots \tag{3-31}$$

Substituting Eq. 3-30 in Eq. 3-31 and equating the coefficients of $\bar{u}_0(t)$, $\bar{u}_0(t-T)$, $\bar{u}_0(t-2T)$, ..., $\bar{u}_0(t-jT)$ results in

$$\overset{K}{e_0}\,\overset{F}{e_0} = 1 \tag{3-32a}$$

$$\overset{K}{e_0}\,\overset{F}{e_1} + \overset{K}{e_1}\,\overset{F}{e_0} = 0 \tag{3-32b}$$

$$\overset{K}{e_0}\,\overset{F}{e_2} + \overset{K}{e_1}\,\overset{F}{e_1} + \overset{K}{e_2}\,\overset{F}{e_0} = 0 \tag{3-32c}$$

$$\overset{K}{e_0}\,\overset{F}{e_j} + \overset{K}{e_1}\,\overset{F}{e_{j-1}} + \overset{K}{e_2}\,\overset{F}{e_{j-2}} + \ldots\, \overset{K}{e_{j-1}}\,\overset{F}{e_1} + \overset{K}{e_j}\,\overset{F}{e_0} = 0 \tag{3-32d}$$

With $\overset{F}{e_0} = 1$ (Eq. 3-11a), it follows from this sequence

$$\overset{K}{e_0} = 1 \tag{3-33a}$$

$$\overset{K}{e_1} = -\overset{K}{e_0}\,\overset{F}{e_1} \tag{3-33b}$$

$$\overset{K}{e_2} = -\overset{K}{e_0}\,\overset{F}{e_2} - \overset{K}{e_1}\,\overset{F}{e_1} \tag{3-33c}$$

$$\overset{K}{e_j} = -\overset{K}{e_0}\,\overset{F}{e_j} - \overset{K}{e_1}\,\overset{F}{e_{j-1}} - \overset{K}{e_2}\,\overset{F}{e_{j-2}} - \ldots - \overset{K}{e_{j-1}}\,\overset{F}{e_1} = -\sum_{\ell=0}^{j-1} \overset{K}{e_l}\,\overset{F}{e_{j-l}} \tag{3-33d}$$

This is the same result as specified in Eq. 3-28. The proof of the relationship for the rotational pseudo-echo constants (Eq. 3-29) is analogous. Again $\overset{F}{e_{r0}} = 1$.

As an example, the disk on the surface of a layer with $d = r_0$ and $\nu = 1/3$ is addressed. Application of this algorithm to columns 2 and 3 of Table 3-1 results in (preliminary) echo constants $\overset{K}{e_j}$ for translation, listed in columns 2 and 5 of Table 3-3 for the horizontal and vertical motions. The corresponding pseudo-echo constants $\overset{K}{e_{rm}}$ for rotation, obtained from column 7 of Table 3-2 are presented in Table 3-4 for rocking.

A comparison of Tables 3-1 and 3-3 for translation or Tables 3-2 and 3-4 for rotation reveals that successive values of the new constants $\overset{K}{e_j}$ or $\overset{K}{e_{rm}}$ die off much more rapidly than the previous constants $\overset{F}{e_j}$ or $\overset{F}{e_{rm}}$. In other words, the stiffness formulation has a "shorter memory" than the flexibility formulation. This fortuitous result suggests that it suffices to consider only the first twenty terms of the infinite sums, Eq. 3-24 or 3-25. For reasons to be discussed, the translational echo constants $\overset{K}{e_j}$ should not simply be truncated at $j = 19$. Better accuracy is obtained by applying a correction $\pm \Delta e_j$, introduced in columns 3 and 6 of Table 3-3 (see Section 3.3.3).

Just as in the flexibility formulation, the static case corresponds to the direct sum of the echo constants $\overset{K}{e_j}$ or $\overset{K}{e_{rm}}$, evaluated at the bottom of Tables 3-3 and 3-4. The stiffness factors which are predicted in the discussion following Eq. 3-20 are confirmed. As already mentioned, the error by comparison to rigorous solutions of the disk on the elastic layer lies within 5%.

TABLE 3-3 STIFFNESS ECHO CONSTANTS FOR TRANSLATION

	Horizontal			Vertical		
j	e_j^K	Δe_j	$E_j^K = e_j^K + \Delta e_j$	e_j^K	Δe_j	$E_j^K = e_j^K + \Delta e_j$
0	1.000	–0.015	0.985	1.000	–0.006	0.994
1	0.493	0.015	0.508	1.023	0.006	1.029
2	–0.038	–0.015	–0.053	0.359	–0.007	0.352
3	0.039	0.015	0.054	0.182	0.007	0.189
4	–0.024	–0.015	–0.039	0.054	–0.006	0.048
5	0.018	0.015	0.033	0.038	0.006	0.044
6	–0.014	–0.015	–0.029	0.004	–0.007	–0.003
7	0.011	0.015	0.026	0.011	0.007	0.018
8	–0.009	–0.015	–0.024	–0.003	–0.006	–0.009
9	0.008	0.015	0.023	0.005	0.006	0.011
10	–0.007	–0.015	–0.022	–0.003	–0.007	–0.010
11	0.006	0.015	0.021	0.003	0.007	0.010
12	–0.005	–0.015	–0.020	–0.003	–0.006	–0.009
13	0.005	0.015	0.020	0.003	0.006	0.009
14	–0.004	–0.015	–0.019	–0.002	–0.007	–0.009
15	0.004	0.015	0.019	0.002	0.007	0.009
16	–0.004	–0.015	–0.019	–0.002	–0.007	–0.009
17	0.003	0.015	0.018	0.002	0.007	0.009
18	–0.003	–0.015	–0.018	–0.002	–0.006	–0.008
19	0.003	0.015	0.018	0.002	0.006	0.008
$\sum_{j=0}^{18} e_j^K$	1.482	0	1.482	2.673	0	2.673
$\sum_{j=0}^{19} (-1)^j e_j^K$	0.302	–0.300	0.002	0.131	–0.130	0.001

3.3.3 Resonant Case for Translation

If the disk is subjected to periodic motion, the frequency of which happens to coincide with a natural frequency of the layer, successive samples $u_0(t)$, $u_0(t - T)$, $u_0(t - 2T)$ and so on are equal in magnitude but alternate in sign. Such behavior corresponds to the sums $(-1)^j e_j^K$ in the last line of columns 2 and 5 of Table 3-3, truncated at $j = 19$. If j were carried out to infinity, the sums would converge to zero. This result is a particular mathematical consequence of the fact that for translational motion the displacement amplitude dies off exactly in inverse proportion to the distance from the apex of the unfolded layered cone.

TABLE 3-4 STIFFNESS ECHO CONSTANTS FOR ROTATION

j	m	Rocking e^K_{rm}
0	0	1.000
	1	0
	2	0
	3	0
1	4	0.558
	5	–0.046
	6	–0.037
	7	–0.030
2	8	0.030
	9	–0.042
	10	–0.032
	11	–0.024
3	12	0.016
	13	–0.019
	14	–0.014
	15	–0.010
4	16	–0.015
	17	–0.007
	18	–0.004
	19	–0.003
	$\sum_{m=0}^{19} e^K_{rm}$	1.321

The infinite sum $\sum_{j=0}^{\infty} (-1)^j e_j^K = 0$ implies that the generating function $\bar{u}_0$ and hence the force P_0 required to sustain the periodic motion vanish. In physical terms this is the phenomenon of perfect undamped resonance, confirmed by the rigorous translational solutions presented later.

To achieve resonance with a finite number of echo constants, the error is divided into $j = 20$ equal portions Δe_j and subtracted from the original echo constants with alternating sign (columns 3 and 6 of Table 3-3). The corrected echo constants E_j^K (columns 4 and 7 of Table 3-3) incorporate the resonant effect; notice that the static stiffness remains unchanged.

For the layered system in rotation, the rotation is not inversely related to the distance from the apex of the unfolded cone but decays in inverse proportion to the higher powers $N = 2$ and 3. Perfect resonance is not observed, and therefore a correction of the pseudo-echo constants e^K_{rm} in Table 3-4 is unnecessary.

3.3.4 Nearly Incompressible Soil

The case of nearly incompressible soil ($1/3 < \nu \leq 1/2$) for vertical and rocking motions is addressed. As discussed in Section 2.1.4, the two special features are the substitution of $c = 2c_s$ for c_p and the inclusion of the trapped mass. The use of $c = 2c_s$ instead of c_p just changes a numerical value and thus has no influence on any method of analysis. The key to understanding what alteration in the analysis the trapped mass may entail is to remember that it is simply assigned to the basemat and does not affect the model of the soil per se.

The trapped mass ΔM and trapped mass moment of inertia ΔM_ϑ affect the calculation of the (unlayered) cone representing the homogeneous halfspace only. In a flexibility formulation the first step determining $\bar{u}_0$ from P_0 and $\bar{\vartheta}_0$ from M_0 and in the stiffness formulation the second step calculating P_0 from $\bar{u}_0$, $\dot{\bar{u}}_0$, $\ddot{\bar{u}}_0$ and M_0 from $\bar{\vartheta}_0$, $\dot{\bar{\vartheta}}_0$, $\ddot{\bar{\vartheta}}_0$ are modified. The same stiffness echo constants e^K_j (Eq. 3-24) and e^K_{rm} (Eq. 3-25) are used for the transformation of the accelerations $\ddot{\bar{u}}_0$ from $\ddot{u}_0$ and $\ddot{\bar{\vartheta}}_0$ from $\ddot{\vartheta}_0$. The trapped mass has no influence on the echo constants.

3.3.5 Dynamic-Stiffness Coefficient

Valuable insight into the accuracy of the layered cone model can be gained by examining the dynamic-stiffness coefficients for harmonic loading. The derivation of the wave pattern of the unfolded layered cone can be performed from the beginning based on the expressions of the displacement amplitudes of the (unlayered) translational cone (Eq. A-8) and of the rotational cone (Eq. A-54). As an alternative, the wave pattern in the frequency domain can be formulated by applying the Fourier transformation to that in the time domain. This approach is followed, whereby the original flexibility echo constants are used.

Layered translational cone. For the translational cone the echo-constant formula relating in the time domain $\bar{u}_0(t - jT)$ to $u_0(t)$ is specified in Eq. 3-10. The Fourier transform of a function $f(t - a)$ equals (time-shifting theorem)

$$\int_0^\infty f(t-a)\, e^{-i\omega t}\, dt = e^{-ia\omega} f(\omega) \tag{3-34}$$

with $f(\omega)$ denoting the Fourier transform of $f(t)$. Applying the Fourier transformation to Eq. 3-10 yields

$$u_0(\omega) = \sum_{j=0}^{\infty} e_j^F\, e^{-ij\omega T}\, \bar{u}_0(\omega) \tag{3-35}$$

Note that the sum extends to infinity. The interaction force-displacement relationship of the (unlayered) cone in the time domain (Eqs. 2-62 and 2-63) can be rewritten as

$$P_0(t) = K\left(\bar{u}_0(t) + \frac{T}{\kappa}\dot{\bar{u}}_0(t)\right) \tag{3-36}$$

with T and κ defined in Eq. 3-9. In the frequency domain

$$P_0(\omega) = K\left(1 + i\frac{\omega T}{\kappa}\right)\bar{u}_0(\omega) \tag{3-37}$$

results. Solving Eq. 3-35 for $\bar{u}_0(\omega)$ and substituting in Eq. 3-37 leads to

$$P_0(\omega) = K\frac{1 + i\dfrac{\omega T}{\kappa}}{\displaystyle\sum_{j=0}^{\infty} e_j^F e^{-ij\omega T}}\, u_0(\omega) \tag{3-38}$$

Substituting the echo constants (Eq. 3-11), the translational dynamic-stiffness coefficient $S(\omega) = P_0(\omega)/u_0(\omega)$ equals

$$S(\omega) = K\frac{1 + i\dfrac{\omega T}{\kappa}}{1 + 2\displaystyle\sum_{j=1}^{\infty}(-1)^j \frac{e^{-ij\omega T}}{1 + j\kappa}} \tag{3-39}$$

Layered rotational cone. For the rotational cone the derivation is analogous. Applying the Fourier transformation to the echo formula (Eq. 3-15) yields

$$\vartheta_0(\omega) = \sum_{j=0}^{\infty} e_{0j}^F e^{-ij\omega T}\bar{\vartheta}_0(\omega) + \sum_{j=0}^{\infty} e_{1j}^F e^{-ij\omega T}\bar{\vartheta}_1(\omega) \tag{3-40}$$

where (Eq. 3-13, 3-17)

$$\bar{\vartheta}_1(\omega) = \frac{1}{i\dfrac{\omega T}{\kappa} + 1}\bar{\vartheta}_0(\omega) \tag{3-41}$$

The coefficient of $\bar{\vartheta}_0(\omega)$ in Eq. 3-41 represents the Fourier transform $H_1(\omega)$ (Eq. A-39) of $h_1(t)$ (Eq. 3-14). The interaction moment-rotation relationship of the (unlayered) cone in the frequency domain (Eqs 2-168, 2-169) can be reformulated as

$$M_0(\omega) = K_\vartheta\left(1 - \frac{1}{3}\frac{(\omega T)^2}{\kappa^2 + (\omega T)^2} + i\frac{\omega T}{3\kappa}\frac{(\omega T)^2}{\kappa^2 + (\omega T)^2}\right)\bar{\vartheta}_0(\omega) \tag{3-42}$$

Solving Eq. 3-40 with Eq. 3-41 for $\bar{\vartheta}_0(\omega)$ and substituting in Eq. 3-42 results after using Eq. 3-16 in the rotational dynamic-stiffness coefficient $S_\vartheta(\omega) = M_0(\omega)/\vartheta_0(\omega)$

$$S_{\vartheta}(\omega) = K_{\vartheta} \frac{1 - \frac{1}{3}\frac{(\omega T)^2}{\kappa^2 + (\omega T)^2} + i\frac{\omega T}{3\kappa}\frac{(\omega T)^2}{\kappa^2 + (\omega T)^2}}{1 + \frac{2}{1 + i\frac{\omega T}{\kappa}}\left(\sum_{j=1}^{\infty} (-1)^j \frac{e^{-ij\omega T}}{(1+j\kappa)^3} + i\frac{\omega T}{\kappa}\sum_{j=1}^{\infty} (-1)^j \frac{e^{-ij\omega T}}{(1+j\kappa)^2}\right)} \tag{3-43}$$

Example. Again, the disk on the surface of a layer with $d = r_0$ and $\nu = 1/3$ is examined. The dynamic-stiffness coefficient $S(a_0)$ (and analogously $S_{\vartheta}(a_0)$) is expressed as

$$S(a_0) = K\left[k(a_0) + i\, a_0\, c(a_0)\right] \tag{3-44}$$

with K denoting the static-stiffness coefficient of the disk on the layer and the dimensionless frequency $a_0 = \omega r_0/c_s$. The spring coefficient $k(a_0)$ and damping coefficient $c(a_0)$ of the layered cone and the exact solution [K1] are plotted in Fig. 3-9a, for the horizontal motion, in Fig. 3-9b for the vertical motion, in Fig. 3-10a for the torsional motion and in Fig. 3-10b for the rocking motion.

For an elastic layer on rigid rock a second dimensionless frequency parameter ωT is physically more meaningful than a_0. Recall that $T = 2d/c$ is the appropriate echo return time, computed with $c = c_s$ for horizontal translation and torsional rotation, and with $c = c_p$ for vertical translation and rocking. The values $\omega T = \pi$, 3π, 5π, and so on correspond to the natural frequencies of the layer. The *fundamental frequency* has special physical significance. As shown in Figs. 3-9 and 3-10 the (radiation) damping coefficients c become appreciable for values of ωT greater than π, whereas for lower frequencies the damping is comparatively small. In view of its role as a *border line between damped and undamped*

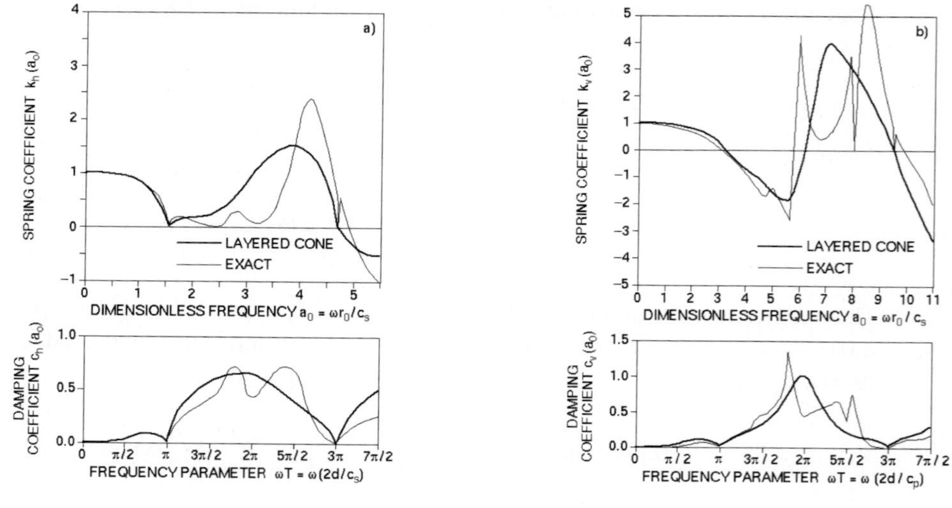

a. Horizontal. **b.** Vertical.

Fig. 3-9 Translational dynamic-stiffness coefficient for harmonic loading of disk on layer.

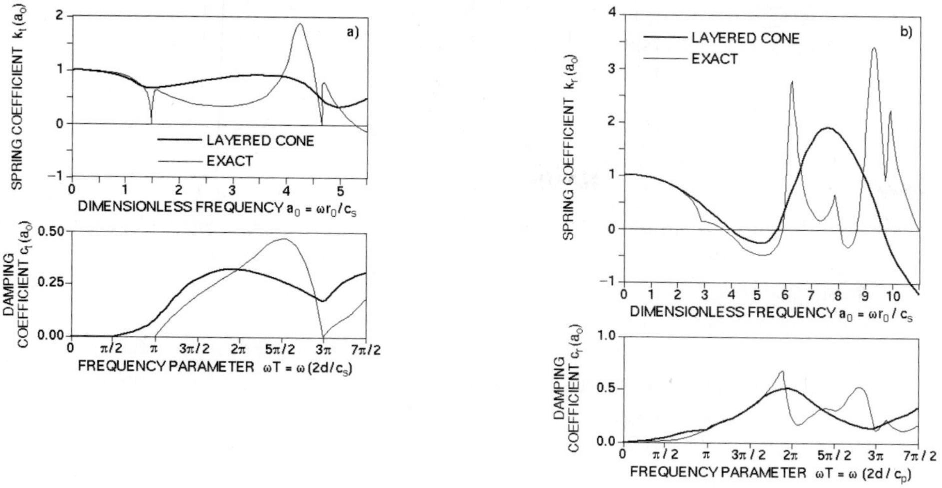

a. Torsional. **b.** Rocking.

Fig. 3-10 Rotational dynamic-stiffness coefficient for harmonic loading of disk on layer.

behavior, the frequency $\omega = \pi/T$ is called the *cutoff frequency*. (This terminology is used in the engineering sense. In the mathematically strict sense of absolutely zero damping, the rigorous cutoff criterion for vertical motion and rocking is actually $\omega T = \pi/2$ for $\nu = 1/3$ ($c_p = 2c_s$). However, the precursor damping up to $\omega T = \pi$, which results from a minimal amount of shear-wave propagation in addition to the predominant axial deformation, is practically negligible.) Note that the layered cone model captures the phenomenon of the cutoff frequency quite satisfactorily, despite a small amount of precursor damping c in the range $\omega T < \pi$.

If the frequency content of the excitation is concentrated below the cutoff frequency, almost no energy is radiated away from the footing. An attached structure will experience very large response when the fundamental frequency of the coupled soil–structure system is less than the cutoff frequency. This potential source of danger is often overlooked in the preliminary design of structures on layered soil. Additional insight on the cutoff frequency is presented in Section 3.8.

Above the cutoff frequency the exact solutions for k and c (thin lines) become increasingly irregular, even erratic. Nonetheless, the smoother results for the layered cone (thick lines) yield a good approximation in the sense of a best fit particularly for the damping coefficient c. It is noteworthy that for translational motion (Fig. 3-9) both k and c are identically zero at $\omega T = \pi$ and 3π. This is precisely the resonance effect incorporated in the corrected translational constants E_j^K (Table 3-3). For rotational motion k_ϑ and c_ϑ become small at $\omega T = \pi$ and 3π but do not vanish; perfect resonance is thus not observed.

High-frequency behavior. Next, the high-frequency limit of the dynamic-stiffness coefficient of a disk on a layer calculated based on layered cones is examined. As the corresponding solution varies strongly with increasing frequency (Figs. 3-9 and 3-10), an average of the limit is determined, calculated between two consecutive zero values of the response of the translational cone. The horizontal and torsional motions are discussed as examples of the translational and rotational cones. The averaged limit equals

$$\tilde{S} = \frac{1}{\pi} \lim_{j \to \infty} \int_{(2j-1)\frac{\pi}{2}}^{(2j+1)\frac{\pi}{2}} S(a_0)\, da_0 = \tilde{K} + i \tilde{a}_0 \tilde{C} \tag{3-45}$$

with the average dimensionless frequency in the investigated interval $\tilde{a}_0 = j\pi$. $S(a_0)$ follows from Eqs. 3-39 and 3-43. The integration is performed numerically. $\tilde{K}$ can be neglected compared to $\tilde{C}$ which is multiplied by $\tilde{a}_0$. For the horizontal motion $\tilde{a}_0 \tilde{C}$ turns out to be equal to $\omega\rho c_s \pi r_0^2$, for the torsional motion to $\omega\rho c_s \pi r_0^4/2$. These are the exact values calculated by multiplying the radiation damping per unit area ρc_s by the area and the polar moment of inertia (Section 2.1.3). It follows that the *layered cone* applied to model a disk on a layer leads to the *exact high-frequency limit of the dynamic-stiffness coefficient.*

3.4 ALTERNATIVE ORDER OF REALIZATION

Both the stiffness and the flexibility formulations for a layer are two-step procedures. One step involves the evaluation of the interaction force-displacement relationship or its inverse for the (unlayered) cone modeling the homogeneous halfspace; this is the topic treated in Chapter 2. The other step, addressed in this chapter, consists of using echo constants to convert the response of the unlayered cone to that of the layered cone representing the layer. These steps may be executed either in the order shown in Fig. 3-11a, which is "natural" for the flexibility formulation, or in the opposite sequence shown in Fig. 3-11b, which is "natural" for the stiffness formulation.

Seen mathematically, *each of the steps* is a linear operator, independent of the other. Therefore, if convenient, they *may be interchanged* without effecting the final results. Specifically, the stiffness formulation may be realized in the "unnatural" sequence in Fig. 3-11a, as explained in the example in Section 3.5. From a conceptual point of view the approach in Fig. 3-11a is much more attractive. The soil is first treated as homogeneous

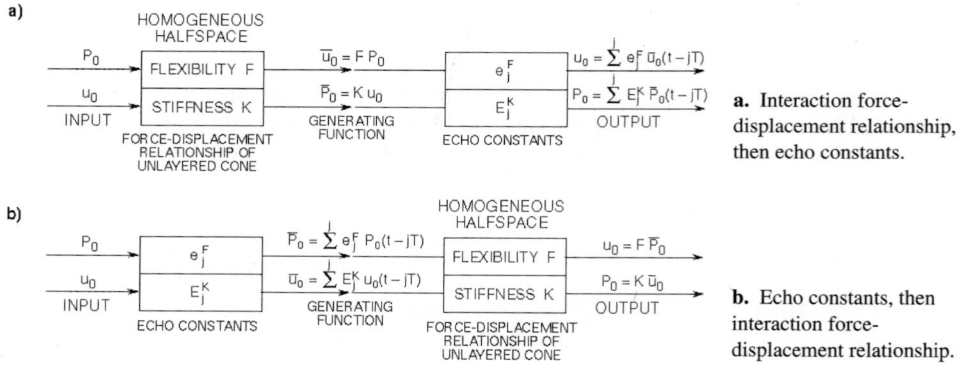

Fig. 3-11 Alternative realization.

(unlayered), then the layer is introduced at the very end. A comparison of the final and next-to-last results reveals the effect of the layer. In addition, the order of calculation in Fig. 3-11a is computationally advantageous for the stiffness formulation; the conversion of velocities according to Eq. 3-26 is avoided.

For notational convenience Fig. 3-11 treats the translational case: interaction force P_0, displacement u_0. It should be apparent that the same considerations apply by analogy to the rotational case: interaction moment M_0, rotation ϑ_0. The following remarks concerning Fig. 3-11, stated in the context of P_0 and u_0, thus also hold for M_0 and ϑ_0.

Notice that regardless of the order or realization, the echo constants e_j^F are always employed in the flexibility formulation. Consequently, the echo constants E_j^K always appear in the stiffness formulation. A further important observation is that the intermediate result, the generating function denoted by a bar, is a displacement in the "natural" sequence (Fig. 3-11a for the flexibility formulation, Fig. 3-11b for the stiffness formulation), but this becomes a force in the "unnatural" sequence (Fig. 3-11a for the stiffness formulation, Fig. 3-11b for the flexibility formulation). The use of interaction forces and moments as generating functions is illustrated by the example.

3.5 ILLUSTRATIVE EXAMPLE

As an illustration of the alternative sequence of calculations in Fig. 3-11a, the stiffness formulation is employed to ascertain the influence of the layer on the vertical and rocking responses of a disk subjected to a rounded triangular pulse of displacement or rotation.

$$u_0(t) = \frac{u_{0\max}}{2}\left(1 - \cos\frac{2\pi t}{T_0}\right) \quad 0 \le t \le T_0 \tag{3-46a}$$

$$\vartheta_0(t) = \frac{\vartheta_{0\max}}{2}\left(1 - \cos\frac{2\pi t}{T_0}\right) \quad 0 \le t \le T_0 \tag{3-46b}$$

The corresponding velocities equal

$$\dot{u}_0(t) = \frac{\pi}{T_0} u_{0\max} \sin\frac{2\pi t}{T_0} \tag{3-47a}$$

$$\dot{\vartheta}_0(t) = \frac{\pi}{T_0} \vartheta_{0\max} \sin\frac{2\pi t}{T_0} \tag{3-47b}$$

The duration of the pulse T_0 is presumed to be twice the echo return time T. The input u_0 and ϑ_0 as well as $\dot{u}_0$ and $\dot{\vartheta}_0$ are plotted in Figs. 3-12a and 3-12b. As before, the layer's depth d is considered to be equal to the disk's radius r_0 and Poisson's ratio $\nu = 1/3$.

For the time increment $\Delta t = T_0/8$ the force and moment histories of the (unlayered) cone modeling the homogeneous halfspace are evaluated in Tables 2-5 and 2-7. These results, which now serve as generating functions for the layered system, are duplicated in columns 4 and 7 of Table 3-5 and denoted by bars. Note that the time increment $\Delta t = T_0/8$ is identical to the value $T/4$ for which the pseudo-echo constants e_{rm}^K in Table 3-4 are determined. This enables direct conversion of the response of the (unlayered) cone to that of the layered cone modeling the layer.

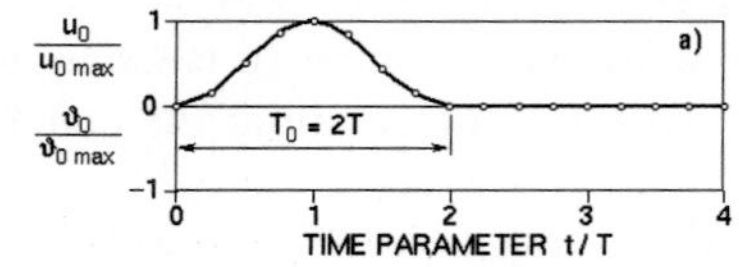

a. Prescribed vertical displacement and rocking pulses.

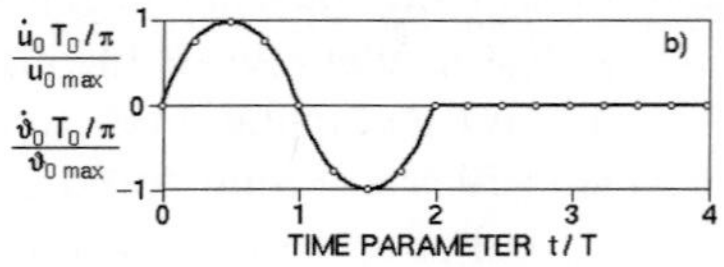

b. Associated vertical and rocking velocities.

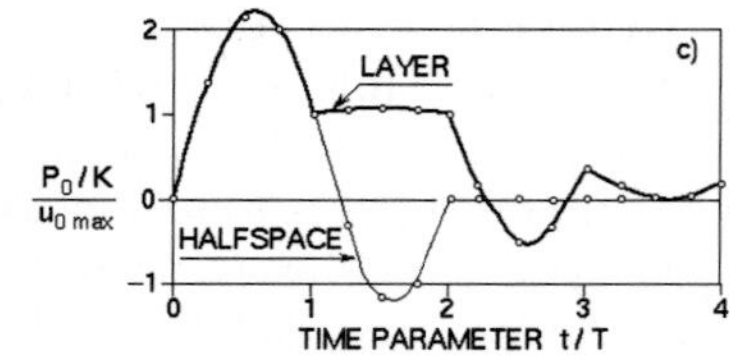

c. Vertical force.

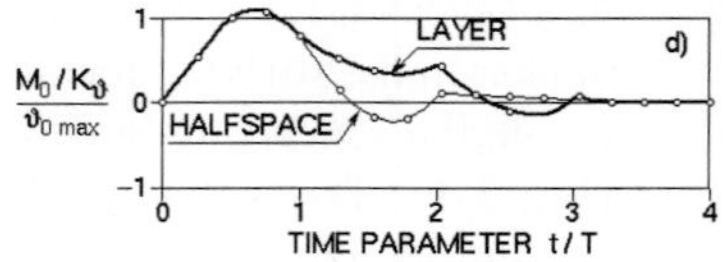

d. Rocking moment.

Fig. 3-12 Example.

TABLE 3-5 EXAMPLE OF STIFFNESS FORMULATION IN "UNNATURAL" ORDER OF REALIZATION

				Vertical			Rocking	
j	m	$\frac{t}{T}$	$\frac{\bar{P}_0}{K u_{0_{max}}}$	18	$\frac{P_0}{K u_{0_{max}}}$	$\frac{\bar{M}_0}{K_\vartheta \vartheta_{0_{max}}}$	e^K_{rm}	$\frac{M_0}{K_\vartheta \vartheta_{0_{max}}}$
0	0	0	0		0	0	0.042	0
	1	0.25	1.3097	0.352	1.3018	0.5395	0.030	0.5395
	2	0.50	2.1449	0	2.1320	0.9807	-0.030	0.9807
	3	0.75	2.0167	0	2.0046	1.0795	-0.037	1.0795
1	4	1.00	1.0000	0	0.9940	0.7897	-0.046	0.7897
	5	1.25	-0.3097	1.029	1.0398	0.2904	0.558	0.5914
	6	1.50	-1.1449	0	1.0691	-0.1182	0	0.4042
	7	1.75	-1.0167	0	1.0646	-0.1907	0	0.3466
2	8	2.00	0	0	1.0290	0.1203	0	0.4588
	9	2.25	0	0.994	0.1423	0.0973	1.000	0.1698
	10	2.50	0		-0.4231	0.0787		-0.0555
	11	2.75	0		-0.3363	0.0636		-0.0979
3	12	3.00	0		0.3520	0.0515		0.0571
	13	3.25	0		0.1385	0.0416		0.0271
	14	3.50	0		0.0024	0.0337		0.0129
	15	3.75	0		0.0233	0.0272		0.0139
4	16	4.00	0		0.1890	0.0220		0.0234

To facilitate hand calculation, it is convenient to write the echo constants E_j^K (column 7 of Table 3-3) and the pseudo-echo constants e_{rm}^K (column 3 of Table 3-4) in inverse order on strips of paper. As indicated in columns 5 and 8 of Table 3-5, the lists of echo constants are moved down the table line-for-line. Adjacent values in columns 4 and 5 (or 7 and 8) are multiplicated and added, resulting in columns 6 and 9, the desired interaction force and moment histories of the layered system. This paper-strip algorithm realizes Eqs. 3-24 and 3-25 exactly, although the variables with and without bars are interchanged.

The different behavior of the layer and the halfspace is evident in Fig. 3-12c (vertical motion) and Fig. 3-12d (rocking). As time progresses, three separate phases of response are observed. Initially, before the earliest arrival of reflected waves at time $t = T$, the disk on the layer experiences exactly the same force or moment as the disk on unbounded soil; this is a logical consequence of the causality principle. Finally, after the prescribed motion has ceased at time $t = T_0 = 2T$, the disk on the layer is subjected to a substantial lingering force or moment not observed (or, for rocking, observed scarcely) for the footing on the halfspace. It is important to realize, however, that this lingering force or moment performs no work on the footing, simply because the footing is at rest.

The intermediate portion of the response, the range $T < t < 2T$, is decisive for the amount of work done on the footing. Notice in Fig. 3-12c and Fig. 3-12d that during this period of time the effect of the layer changes the sign of the force or moment from minus to plus. In the interval $T < t < 2T$ the velocity $\dot{u}_0$ or $\dot{\vartheta}_0$ is negative. Thus the power input to the disk, the product $P_0\,\dot{u}_0$ or $M_0\dot{\vartheta}_0$, which is positive for the (unlayered) cone modeling the halfspace, becomes negative for the layer. The total work done on the disk up to time t, given by the integral

$$W(t) = \int_0^t P_0\,\dot{u}_0\,d\tau \tag{3-48a}$$

for translation or

$$W(t) = \int_0^t M_0\,\dot{\vartheta}_0\,d\tau \tag{3-48b}$$

for rocking, is plotted in Fig. 3-13a (vertical motion) and Fig. 3-13b (rocking). The curves are normalized by regarding the end result $W(2T)$ for the unlayered cone (halfspace) to be 100%. As mentioned above, for the layer the work decreases between time $t = T$ and $2T$ as the disk returns stored energy to the exterior agent that imposes the motion. The final asymptotes $W(2T)$ may be interpreted as the energy dissipated by radiation damping. For the disk on the layer the effective damping is reduced to only 30% (vertical motion) or 49% (rocking) of that for the same disk on the halfspace. The reason for this is that the applied pulse has significant frequency components below the cutoff frequency. As a matter of fact, if the single sine wave of velocity, Fig. 3-12b, were extended periodically, its frequency would be exactly equal to the cutoff frequency. Now, as a thought experiment, suppose that the duration of the pulse were less than half as long: $T_0 < T$. Then the predominant frequency of the sine wave of velocity would be at least double the cutoff frequency. As the footing would in this case come to rest before the arrival of reflected waves at $t = T$, the energy dissipated by radiation damping would be exactly the same for layered and unlayered systems. Stated differently, the end asymptotes of a work diagram analogous to Fig. 3-13 would be attained before $t = T$, in the range where the causality principle dictates exact equivalence of the responses of the layer and the halfspace.

a. Vertical.

b. Rocking.

Fig. 3-13 Work diagram.

3.6 SOIL LAYER ON FLEXIBLE ROCK HALFSPACE

3.6.1 Reflection Coefficient for Translational Cone

In Sections 3.1 to 3.3 the procedure of the unfolded layered cone is developed under the limiting assumption that the underlying rock is rigid, that the soil layer is built-in at its base. In this Section 3.6 this assumption is relaxed, allowing a finite stiffness of the flexible substratum, called "rock" in the following, to be taken into consideration. The unfolded-cone procedure thus enables an approximate analysis of the vibrations of a foundation on the surface of a homogeneous soil layer resting on a flexible rock halfspace. This situation is shown for vertical motion in Fig. 3-14. The translational motion with the corresponding cones is addressed first, followed by the rotational motion in Section 3.6.2. For both types of motion a reflection coefficient at the interface of the layer and the rock based on one-dimensional wave propagation in the cones is derived. It will be demonstrated that this reflection coefficient ($-\alpha$) considering the flexible rock halfspace replaces the corresponding value (-1) for the built-in base in the expression for the echo constants (Eqs. 3-11b and 3-16b). This is the only modification required in the procedure. All flexibility and stiffness formulations for the dynamic analysis remain valid. The modified value of the reflection coefficient is a function of frequency, $-\alpha(\omega)$. The low- and high-frequency limits lead to a frequency-independent $-\alpha$, which allows the dynamic analysis to be performed directly in the familiar time domain. This generalized unfolded layered cone represents a wave pattern whose amplitude decays with distance, and the reflection at the free surface and the reflection and refraction at the layer-rock interface are taken into consideration.

For the translational cone the basic result can be derived in the time domain as follows. Figure 3-14 shows a rigid massless foundation, a disk of radius r_0, on the surface of a homogeneous soil layer of (constant) depth d resting on a flexible rock halfspace. G represents the shear modulus, ν Poisson's ratio, from which the shear- and dilatational-wave velocities c_s and c_p follow with the mass density ρ. Indices L and R are introduced to identify constants associated with the layer and the rock, respectively. Material damping is not

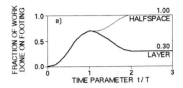

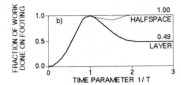

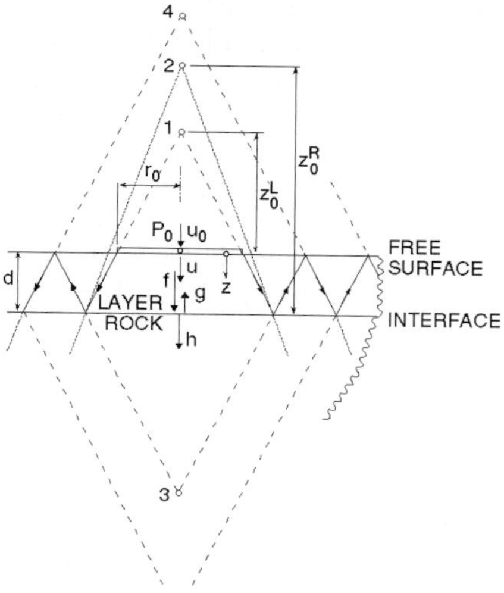

Fig. 3-14 Disk in vertical motion on soil layer resting on flexible rock halfspace with wave pattern in corresponding cones.

addressed. For the predominantly vertical motion caused by a vertical force P_0 acting at the center of the disk presented in Fig. 3-14 the appropriate wave propagation velocity c equals c_p (with the exception of the nearly incompressible case ($1/3 < \nu \leq 1/2$) for which $c = 2c_s$ should be used [Section 2.1.4]). The vertical displacement of the truncated cone is denoted as u with u_0 being its value at the contact area under the disk. The derived equations are also valid for the horizontal motion involving shear waves provided the propagation velocity c is taken to be c_s.

The vertical force $P_0(t)$ produces dilatational waves propagating along a cone of apex height z_0 downwards from the disk. Apex 1 with height z_0^L is specified in such a way as to yield the same static-stiffness coefficients for the truncated semi-infinite cone and the homogeneous halfspace with the material properties of the layer. The ratio z_0^L/r_0 which depends on c_p^L/c_s^L (or ν_L) is specified in Table 2-3A. At the interface of the layer and the rock ($z = d$) the incident wave f will lead to a refracted wave h propagating in the rock in the same direction as the incident wave along its own cone with the apex distance-to-radius ratio $z_0^R/(r_0(z_0^L + d)/z_0^L)$ (apex 2, dotted line). Note that the aspect ratio of the rock's cone is generally different from that of the layer's cone. However, in the special case $\nu_R = \nu_L$, both cones have the same proportions and z_0^R equals $z_0^L + d$. In addition, a reflected wave g is created propagating back through the layer along the indicated cone (apex 3) in the opposite upward direction. The latter wave will reflect back at the free surface and then propagate downwards along the cone (apex 4) shown in Fig. 3-14. Upon reaching the interface of the layer and the rock, a refraction and a reflection again take place. The cones describing the wave propagation in the layer can be unfolded to form a single layered cone shown in Fig. 3-5. The downward wave propagation in the rock is thus also assumed to occur in cones whose apex height depends on the properties of the rock (shown in Fig. 3-14 as dotted lines).

The reflection coefficient $-\alpha$, defined as the ratio of the reflected wave g to the incident wave f, is determined at the first impingement at the interface of the layer and the rock.

The displacement in a cone deforming in translation is inversely proportional to the distance from the apex (Eq. A-6). The displacement in the layer $u_L(z,t)$ after the first impingement at the interface but before the reflection at the free surface equals

$$u_L(z,t) = \frac{z_0^L}{z_0^L + z} f\left(t - \frac{z}{c_L}\right) + \frac{z_0^L}{z_0^L + 2d - z} g\left(t - \frac{2d}{c_L} + \frac{z}{c_L}\right) \tag{3-49}$$

The first term on the right-hand side represents the incident wave and the second the reflected wave. The z is measured from the free surface (Fig. 3-14). The displacement in the rock $u_R(z,t)$ is formulated as

$$u_R(z,t) = \frac{\dfrac{z_0^L z_0^R}{z_0^L + d}}{z_0^R - d + z} h\left(t - \frac{d}{c_L} + \frac{d}{c_R} - \frac{z}{c_R}\right) \tag{3-50}$$

with the numerator chosen for convenience. Notice that the denominators in Eqs. 3-49 and 3-50 are the distances to the apexes of the respective cones. At the interface $z = d$ the arguments of the three functions f, g, and h are the same, $t - d/c_L$. The normal forces N are proportional to

$$N_L(z,t) \propto \rho_L c_L^2 \, u_{L,z}(z,t) \tag{3-51}$$

$$N_R(z,t) \propto \rho_R c_R^2 \, u_{R,z}(z,t) \tag{3-52}$$

Enforcing at the interface ($z = d$) the displacement compatibility

$$u_L(d,t) = u_R(d,t) \tag{3-53}$$

leads to

$$h\left(t - \frac{d}{c_L}\right) = f\left(t - \frac{d}{c_L}\right) + g\left(t - \frac{d}{c_L}\right) \tag{3-54}$$

Using Eq. 3-54 to eliminate h (and h') in the equilibrium equation formulated at the interface

$$N_L(d,t) = N_R(d,t) \tag{3-55}$$

results in a coupled differential equation of first order involving f and g

$$\begin{aligned}(\rho_L c_L + \rho_R c_R)\, g'\left(t - \frac{d}{c_L}\right) + \left(\frac{\rho_L c_L^2}{z_0^L + d} + \frac{\rho_R c_R^2}{z_0^R}\right) g\left(t - \frac{d}{c_L}\right) = \\ (\rho_L c_L - \rho_R c_R)\, f'\left(t - \frac{d}{c_L}\right) + \left(\frac{\rho_L c_L^2}{z_0^L + d} - \frac{\rho_R c_R^2}{z_0^R}\right) f\left(t - \frac{d}{c_L}\right)\end{aligned} \tag{3-56}$$

The single prime symbol (′) denotes differentiation with respect to the argument.

The reflection coefficient $-\alpha$ can be determined for two limits. For the static case with $f' = g' = 0$ Eq. 3-56 yields

$$-\alpha = -\alpha(0) = \frac{g}{f} = \frac{\dfrac{\rho_L c_L^2}{z_0^L + d} - \dfrac{\rho_R c_R^2}{z_0^R}}{\dfrac{\rho_L c_L^2}{z_0^L + d} + \dfrac{\rho_R c_R^2}{z_0^R}} \tag{3-57a}$$

which for $\nu_R = \nu_L$ simplifies to

$$-\alpha = -\alpha(0) = \frac{\rho_L c_L^2 - \rho_R c_R^2}{\rho_L c_L^2 + \rho_R c_R^2} \tag{3-57b}$$

The argument zero of α denotes $\omega = 0$. For the high-frequency limit (argument $\omega = \infty$) with $|f'| \gg |f|$ and $|g'| \gg |g|$

$$-\alpha = -\alpha(\infty) = \frac{g'}{f'} = \frac{\rho_L c_L - \rho_R c_R}{\rho_L c_L + \rho_R c_R} \tag{3-58}$$

results. The latter reflection coefficient corresponds to that encountered in a prismatic bar with a discontinuity in material properties. For the more common case of a more flexible and less denser layer than the rock halfspace ($\rho_L c_L < \rho_R c_R$), $-1 < -\alpha < 0$ applies, for the opposite case $0 < -\alpha < 1$. The two limits $-\alpha = -1$ and $-\alpha = 1$ correspond to the layer on rigid rock and to the layer "on air," respectively. The reflection coefficients specified in Eqs. 3-57 and 3-58 are frequency-independent real numbers which are strongly dependent on the material properties. The analysis can be performed in the time domain. As is illustrated in Section 3.6.4, the low-frequency limit (Eq. 3-57) leads to more accurate results than the high-frequency limit (Eq. 3-58).

To solve the differential equation (Eq. 3-56) exactly, harmonic motion of frequency ω is assumed. With $f'(\omega) = i\omega f(\omega)$ and $g'(\omega) = i\omega g(\omega)$

$$\left[(\rho_L c_L + \rho_R c_R)\, i\omega + \frac{\rho_L c_L^2}{z_0^L + d} + \frac{\rho_R c_R^2}{z_0^R}\right] g(\omega) = \left[(\rho_L c_L - \rho_R c_R)\, i\omega + \frac{\rho_L c_L^2}{z_0^L + d} - \frac{\rho_R c_R^2}{z_0^R}\right] f(\omega) \tag{3-59}$$

results, which leads to the frequency-dependent, complex-valued reflection coefficient

$$-\alpha(\omega) = \frac{g(\omega)}{f(\omega)} = \frac{i\omega(\rho_L c_L - \rho_R c_R) + \dfrac{\rho_L c_L^2}{z_0^L + d} - \dfrac{\rho_R c_R^2}{z_0^R}}{i\omega(\rho_L c_L + \rho_R c_R) + \dfrac{\rho_L c_L^2}{z_0^L + d} + \dfrac{\rho_R c_R^2}{z_0^R}} \tag{3-60}$$

Assuming this reflection coefficient to apply for all impingements at the layer-rock interface, the amplitude ratio of the reflected wave $g_j(\omega)$ after the j–th impingement to the initial incident wave $f_1(\omega)$ follows as

$$\frac{g_j(\omega)}{f_1(\omega)} = (-\alpha(\omega))^j \tag{3-61}$$

f_1 corresponds to the first term on the right-hand side of Eq. 3-49. Equation 3-61 also applies when the frequency-independent, real-valued coefficients of Eqs. 3-57 or 3-58 are used.

The *reflection coefficient* of Eq. 3-60 *is determined by addressing the first impingement of the downwardly propagating (incident) wave in the cone at the interface of the layer and the rock,* as is already mentioned above. A more rigorous procedure can be derived as follows. Each subsequent impingement at the same interface corresponds to a reflection and refraction at the section of the unfolded cone which is located at a distance $2d$ from the previous one. The downwardly propagating incident wave associated with a specific impingement depends on all previous reflections and refractions at the interface. Thus, $-\alpha_k(\omega)$ defined as the ratio of the amplitudes of the reflected wave $g_k(\omega)$ to the incident wave $f_k(\omega)$ for the k-th impingement is specified as

$$-\alpha_k(\omega) = \frac{g_k(\omega)}{f_k(\omega)} = \frac{i\omega(\rho_L c_L - \rho_R c_R) + \dfrac{\rho_L c_L^2}{z_0^L + (2k-1)\,d} - \dfrac{\rho_R c_R^2}{z_0^R + 2(k-1)d}}{i\omega(\rho_L c_L + \rho_R c_R) + \dfrac{\rho_L c_L^2}{z_0^L + (2k-1)\,d} + \dfrac{\rho_R c_R^2}{z_0^R + 2(k-1)d}} \tag{3-62}$$

The frequency-dependent, complex-valued amplitude ratio $g_j(\omega)/f_1(\omega)$ follows from multiplying all coefficients as

$$\frac{g_j(\omega)}{f_1(\omega)} = (-\alpha_1(\omega))\,(-\alpha_2(\omega))\,\ldots\,(-\alpha_k(\omega))\,\ldots\,(-\alpha_j(\omega)) \tag{3-63}$$

Numerical experiments demonstrate that this more rigorous procedure is not necessary. It is sufficient to set $-\alpha_1(\omega) = -\alpha_2(\omega) = \ldots = -\alpha_j(\omega) = -\alpha(\omega)$. That is, the reflection coefficient $-\alpha(\omega)$ of Eq. 3-60 determined from the first reflection-refraction at the interface is assumed to apply for all impingements at the layer-rock interface, as specified in Eq. 3-61.

3.6.2 Reflection Coefficient for Rotational Cone

The rotation in a cone decays with depth as a more complicated function of frequency than the translation does. This is visible when comparing the Green's functions. In the equation for the rotational cone, a frequency-dependent term (last term in Eq. A-98) is present which is missing in the corresponding expression for the translational cone (Eq. A-40). The reflection coefficient $-\alpha$ at the interface of the soil layer and the flexible rock halfspace is thus derived in the frequency domain for the rotational cone. Rocking motion is addressed. The rocking moment M_0 applied to the rigid massless disk causes a rotation ϑ of the cone, with ϑ_0 being its value at the contact area under the disk. For the tensional-compressional motion, the appropriate wave propagation velocity c equals c_p for $\nu \leq 1/3$. The equations to be derived are also valid for the torsional motion involving shear waves, provided the propagation velocity c is taken to be c_s. The wave patterns associated with the various cones discussed in connection with Fig. 3-14 still apply.

The reflection coefficient $-\alpha(\omega)$, defined as the ratio of the reflected-wave amplitude to the incident-wave amplitude, is determined at the first impingement at the interface of the layer and the rock. For a homogeneous halfspace, the solution for the rotational cone with apex height z_0 with waves propagating in the positive z-direction is written as (Eq. A-54)

$$\vartheta(\omega) = \vartheta_0(\omega)\left[\left(\frac{z_0}{z_0+z}\right)^3 + i\frac{\omega}{c}\frac{z_0^3}{(z_0+z)^2}\right]e^{-i\frac{\omega}{c}z} \tag{3-64}$$

The coordinate z is measured from the free surface. The term $z_0 + z$ in the denominator of the second term is the distance from the apex. The rotation amplitude of the rotation in the layer $\vartheta_L(z,\omega)$ after the first impingement at the interface but before the reflection at the free surface equals

$$\begin{aligned}\vartheta_L(z,\omega) = {} & a\left[\left(\frac{z_0^L+d}{z_0^L+z}\right)^3 + i\frac{\omega}{c_L}\frac{\left(z_0^L+d\right)^3}{\left(z_0^L+z\right)^2}\right]e^{-i\frac{\omega}{c_L}z} \\ & + b\left[\left(\frac{z_0^L+d}{z_0^L+2d-z}\right)^3 + i\frac{\omega}{c_L}\frac{\left(z_0^L+d\right)^3}{\left(z_0^L+2d-z\right)^2}\right]e^{-i\frac{\omega}{c_L}(2d-z)}\end{aligned} \tag{3-65}$$

The first term with the integration constant a represents the incident wave and the second with b the reflected wave. The amplitude of the refracted wave in the rock $\vartheta_R(z,\omega)$ is formulated as

$$\vartheta_R(z,\omega) = c\left[\left(\frac{z_0^R}{z_0^R-d+z}\right)^3 + i\frac{\omega}{c_R}\frac{\left(z_0^R\right)^3}{\left(z_0^R-d+z\right)^2}\right]e^{-i\frac{\omega}{c_R}\left(d\frac{c_R}{c_L}-d+z\right)} \tag{3-66}$$

The terms in the denominators in Eqs. 3-65 and 3-66 are the distances to the apexes of the respective cones. Notice that at the interface $z = d$ the arguments of the three exponential functions are the same, $-i\omega d/c_L$. The moments with amplitudes $M(z,\omega)$ are proportional to

$$M_L(z,\omega) \propto \rho_L c_L^2\, \vartheta_{L,z}(z,\omega) \tag{3-67}$$

$$M_R(z,\omega) \propto \rho_R c_R^2\, \vartheta_{R,z}(z,\omega) \tag{3-68}$$

Formulating the compatibility of the rotations at the interface $z = d$

$$\vartheta_L(d,\omega) = \vartheta_R(d,\omega) \tag{3-69}$$

yields

$$a\left[1 + i\frac{\omega}{c_L}\left(z_0^L+d\right)\right] + b\left[1 + i\frac{\omega}{c_L}\left(z_0^L+d\right)\right] = c\left[1 + i\frac{\omega}{c_R}z_0^R\right] \tag{3-70}$$

Enforcing equilibrium at the interface

$$M_L(d,\omega) = M_R(d,\omega) \tag{3-71}$$

results in

$$\begin{aligned} &-a\frac{\rho_L c_L^2}{z_0^L + d}\left[3 + 3i\frac{\omega}{c_L}\left(z_0^L + d\right) + \left(i\frac{\omega}{c_L}\right)^2\left(z_0^L + d\right)^2\right] \\ &+ b\frac{\rho_L c_L^2}{z_0^L + d}\left[3 + 3i\frac{\omega}{c_L}\left(z_0^L + d\right) + \left(i\frac{\omega}{c_L}\right)^2\left(z_0^L + d\right)^2\right] = \\ &-c\frac{\rho_R c_R^2}{z_0^R}\left[3 + 3i\frac{\omega}{c_R}z_0^R + \left(i\frac{\omega}{c_R}\right)^2 z_0^{R2}\right] \end{aligned} \tag{3-72}$$

The reflection coefficient $-\alpha(\omega)$ is defined at the interface as the amplitude ratio of the reflected wave (with constant b) to the incident wave (with constant a). From Eq. 3-65 with $z = d$ it follows that

$$-\alpha(\omega) = \frac{b\left[1 + i\frac{\omega}{c_L}\left(z_0^L + d\right)\right]}{a\left[1 + i\frac{\omega}{c_L}\left(z_0^L + d\right)\right]} = \frac{b}{a} \tag{3-73}$$

Eliminating c from Eqs. 3-70 and 3-72 and then substituting into Eq. 3-73 yields the frequency-dependent reflection coefficient

$$-\alpha(\omega) = \frac{-\dfrac{1 + i\frac{\omega}{c_L}\left(z_0^L + d\right)}{1 + i\frac{\omega}{c_R}z_0^R} + \dfrac{3 + 3i\frac{\omega}{c_L}\left(z_0^L + d\right) + \left(\frac{i\omega}{c_L}\right)^2\left(z_0^L + d\right)^2}{3 + 3i\frac{\omega}{c_R}z_0^R + \left(\frac{i\omega}{c_R}\right)^2 z_0^{R2}} \cdot \dfrac{\rho_L c_L^2 z_0^R}{\rho_R c_R^2\left(z_0^L + d\right)}}{\dfrac{1 + i\frac{\omega}{c_L}\left(z_0^L + d\right)}{1 + i\frac{\omega}{c_R}z_0^R} + \dfrac{3 + 3i\frac{\omega}{c_L}\left(z_0^L + d\right) + \left(\frac{i\omega}{c_L}\right)^2\left(z_0^L + d\right)^2}{3 + 3i\frac{\omega}{c_R}z_0^R + \left(\frac{i\omega}{c_R}\right)^2 z_0^{R2}} \cdot \dfrac{\rho_L c_L^2 z_0^R}{\rho_R c_R^2\left(z_0^L + d\right)}} \tag{3-74}$$

The frequency-dependent reflection coefficient reverts to a constant in two limits. For the low-frequency limit, the static case ($\omega = 0$)

$$-\alpha(0) = \frac{\dfrac{\rho_L c_L^2}{z_0^L + d} - \dfrac{\rho_R c_R^2}{z_0^R}}{\dfrac{\rho_L c_L^2}{z_0^L + d} + \dfrac{\rho_R c_R^2}{z_0^R}} \tag{3-75a}$$

which for $\nu_R = \nu_L$ simplifies to

$$-\alpha(0) = \frac{\rho_L c_L^2 - \rho_R c_R^2}{\rho_L c_L^2 + \rho_R c_R^2} \tag{3-75b}$$

For the high-frequency limit ($\omega = \infty$)

$$-\alpha(\infty) = \frac{\rho_L c_L - \rho_R c_R}{\rho_L c_L + \rho_R c_R} \tag{3-76}$$

follows. Notice that these frequency-independent reflection coefficients for the rotational cone are the same as those for the translational cone (Eqs. 3-57 and 3-58).

As for translational motion, the reflection coefficient $-\alpha(\omega)$ of Eq. 3-74 determined from the first reflection-refraction at the interface is assumed to apply for all impingements at the layer-rock interface.

3.6.3 Flexibility and Stiffness Formulations

All flexibility and stiffness formulations to model a disk on a soil layer resting on rigid rock presented in Sections 3.2 and 3.3 *remain valid for the corresponding case with a flexible rock halfspace*. The only modification consists of *replacing in the expression for the echo constants the reflection coefficient associated with rigid rock, –1, by the corresponding value, $-\alpha$, for the flexible rock.* In a frequency-domain analysis, the frequency-dependent $-\alpha(\omega)$ of Eq. 3-60 for the translational motion and Eq. 3-74 for the rotational motion can be used. In a time-domain analysis, a frequency-independent reflection coefficient must be applied, whereby the static value $-\alpha(0)$ of Eq. 3-57 leads to more accurate results than $\alpha(\infty)$.

Translational cone. The key equations for the translational motion are summarized. Denoting as in Section 3.1.1 the displacement of the initial incident wave as $\bar{u}_0(t)$ (which is also the generating displacement of the homogeneous halfspace with the properties of the layer), the displacement in the layer $u_L(z,t)$ can be expressed as a wave pattern

$$u_L(z,t) = \frac{z_0^L}{z_0^L + z}\,\bar{u}_0\left(t - \frac{z}{c_L}\right) + \sum_{j=1}^{k}(-\alpha)^j\left[\frac{z_0^L}{z_0^L + 2jd - z}\,\bar{u}_0\left(t - \frac{2jd}{c_L} + \frac{z}{c_L}\right) + \frac{z_0^L}{z_0^L + 2jd + z}\,\bar{u}_0\left(t - \frac{2jd}{c_L} - \frac{z}{c_L}\right)\right] \tag{3-77}$$

The first term on the right-hand side corresponds to the initial incident wave, the second term describes the reflected wave after the j-th impingement at the layer-rock interface (propagating upwards), and the third term describes the wave after reflection at the free surface (propagating downwards). The integer k is equal to the largest j for which at least one of the arguments of $\bar{u}_0$ for a specific z and t is positive.

At the surface the displacement is (omitting the subscript L)

$$u_0(t) = u(0,t) = \bar{u}_0(t) + 2\sum_{j=1}^{k}(-\alpha)^j \frac{1}{1 + j\kappa}\,\bar{u}_0(t - jT) \tag{3-78}$$

with the geometric parameter $\kappa = 2d/z_0^L$ and the temporal parameter $T = 2d/c_L$. Introducing the echo constants e_j^F, Eq. 3-78 is written as

$$u_0(t) = \sum_{j=0}^{k} e_j^F \, \bar{u}_0(t-jT) \tag{3-79}$$

with

$$e_0^F = 1 \tag{3-80a}$$

and for $j \geq 1$

$$e_j^F = \frac{2(-\alpha)^j}{1+j\kappa} \tag{3-80b}$$

From a comparison of Eqs. 3-77, 3-78, 3-79 and 3-80 with the corresponding expressions for the case with rigid rock (Eqs. 3-7, 3-8, 3-10 and 3-11), it follows that a complete analogy exists. The factor -1, the reflection coefficient for the case of rigid rock, is replaced by that of flexible rock $-\alpha$. All formulations based on the flexibility and the stiffness approaches thus still apply with this modification.

For a frequency-domain analysis based on the reflection coefficient of the first impingement (Eq. 3-60) the echo constants of Eq. 3-80b are still valid but the real-valued $-\alpha$ is replaced by the complex-valued $-\alpha(\omega)$ resulting in Eq. 3-61. When the reflection coefficient also depends on the number of the impingement (Eq. 3-62), the expression $(-\alpha)^j$ in Eq. 3-80b is replaced by $(-\alpha_1(\omega))(-\alpha_2(\omega)) \dots (-\alpha_j(\omega))$ shown in Eq. 3-63. (As already mentioned, the more rigorous procedure leads to the same results from a practical point of view and thus needs not to be used.) After performing the Fourier transformation of Eq. 3-79 and solving for $\bar{u}_0(\omega)$

$$\bar{u}_0(\omega) = H(\omega) u_0(\omega) \tag{3-81}$$

applies with the translational transfer function H(ω) relating the displacement amplitude of the layer $u_0(\omega)$ to that of the halfspace $\bar{u}_0(\omega)$

$$H(\omega) = \frac{1}{\sum_{j=0}^{\infty} e_j^F e^{-ij\omega T}} \tag{3-82}$$

Finally, the dynamic-stiffness coefficient in the frequency domain $S(\omega)$ of the disk on the soil layer resting on a rock halfspace follows as (see Eq. 3-39):

$$S(\omega) = K \frac{1 + i\frac{\omega T}{\kappa}}{1 + 2\sum_{j=1}^{\infty} \left(-\alpha(\omega)\right)^j \frac{e^{-ij\omega T}}{1+j\kappa}} \tag{3-83}$$

where K denotes the static-stiffness coefficient of the homogeneous halfspace with the properties of the layer. The numerator is the dynamic-stiffness coefficient of the disk on the homogeneous halfspace, which is multiplied by $H(\omega)$ of Eq. 3-82.

Rotational cone. Denoting as in Section 3.1.2 the rotation of the initial incident wave (the so-called generating function) by $\bar{\vartheta}_0(t)$ (which is also the rotation of the homogeneous halfspace with the properties of the layer), the surface rotation directly under the disk at $z = 0$, $\vartheta_0(t)$, is written as (analogous to Eqs. 3-15 and 3-16)

$$\vartheta_0(t) = \sum_{j=0}^{k} \overset{F}{e}_{0j}\, \bar{\vartheta}_0\,(t - jT) + \sum_{j=0}^{k} \overset{F}{e}_{1j}\, \bar{\vartheta}_1\,(t - jT) \tag{3-84}$$

with

$$\overset{F}{e}_{00} = 1 \qquad \overset{F}{e}_{10} = 0 \tag{3-85a}$$

and, for $j \geq 1$

$$\overset{F}{e}_{0j} = \frac{2(-\alpha)^j}{(1+j\kappa)^2} \qquad \overset{F}{e}_{1j} = 2(-\alpha)^j\left[\frac{1}{(1+j\kappa)^3} - \frac{1}{(1+j\kappa)^2}\right] \tag{3-85b}$$

In the frequency domain, in analogy to Eqs. 3-40 and 3-41,

$$\bar{\vartheta}_0(\omega) = H_{\vartheta}(\omega)\, \vartheta_0(\omega) \tag{3-86}$$

applies with the rotational transfer function $H_\vartheta(\omega)$. The rotational dynamic-stiffness coefficient of the disk on the surface of a layer resting on a flexible rock halfspace is

$$S_\vartheta(\omega) = K_\vartheta \frac{1 - \dfrac{1}{3}\dfrac{(\omega T)^2}{\kappa^2 + (\omega T)^2} + i\dfrac{\omega T}{3\kappa}\dfrac{(\omega T)^2}{\kappa^2 + (\omega T)^2}}{1 + \dfrac{2}{1 + i\dfrac{\omega T}{\kappa}}\left(\displaystyle\sum_{j=1}^{\infty} (-\alpha(\omega))^j \frac{e^{-ij\omega T}}{(1 + j\kappa)^3} + i\frac{\omega T}{\kappa}\sum_{j=1}^{\infty} (-\alpha(\omega))^j \frac{e^{-ij\omega T}}{(1 + j\kappa)^2}\right)} \tag{3-87}$$

where K_ϑ denotes the static-stiffness coefficient of the homogeneous halfspace with the properties of the layer. The numerator is the dynamic-stiffness coefficient of the disk on the homogeneous halfspace, which is multiplied by $H_\vartheta(\omega)$. $H_\vartheta(\omega)$ is thus the inverse of the denominator of Eq. 3-87.

An extension to calculate the dynamic-stiffness coefficients of a disk on a stratified site consisting of many layers on an underlying homogeneous halfspace is discussed in Appendix D.

3.6.4 Accuracy Evaluation

Static-stiffness coefficient. The vertical static-stiffness coefficient of a disk on the surface of a soil layer resting on a rock halfspace is calculated first (Fig. 3-14). The Poisson's ratios are fixed $\nu_L = \nu_R = 1/3$ ($c_p/c_s = 2$), and the depth-to-radius ratio d/r_0 and the ratio of the constrained moduli of E_c^L/E_c^R are varied parametrically (see Table 2-1 for the relationship with other material constants).

The ratio $c_p/c_s = 2$ leads to $z_0^L/r_0 = 2.094$ (Table 2-3B). The opening angles of the cones of the layer and the rock are the same. The geometric parameter equals $\kappa = 0.955d/r_0$. The reflection coefficient for the low-frequency limit (Eq. 3-57b) equals

$$-\alpha = -\alpha(0) = \frac{\dfrac{E_c^L}{E_c^R} - 1}{\dfrac{E_c^L}{E_c^R} + 1} \tag{3-88}$$

For the static case u_0 and $\bar{u}_0$ are constant. It follows from Eq. 3-79 that the ratio of the static displacement of the soil layer-rock halfspace system to that of the halfspace with the properties of the soil layer (with the same load acting on the disks), $u_0/\bar{u}_0$, is equal to the sum of all echo constants (Eq. 3-80). The ratio of the corresponding static-stiffness coefficients is thus equal to its inverse. This expression for $S(0)/K$ is also confirmed from Eq. 3-83 evaluated at $\omega = 0$. The sums are evaluated up to $n = 18$ and 19 and then averaged.

The results for $u_0/\bar{u}_0$ are presented in Table 3-6. First, $d/r_0 = 1$ is kept fixed and E_c^L/E_c^R is varied (Table 3-6A). $E_c^L/E_c^R = 0$ corresponds to the case of a soil layer resting on rigid rock. Second, $E_c^L/E_c^R = 0.2$ is kept fixed and d/r_0 is varied as indicated (Table 3-6B). The rigorous solution based on three-dimensional elasticity is denoted as exact. Excellent agreement throughout the vast range of the varied parameters results. The simple procedure involving the sum of the echo constants should prove to be quite helpful for performing static settlement computations.

TABLE 3-6 DIMENSIONLESS VERTICAL STATIC DISPLACEMENT OF DISK ON SOIL LAYER-ROCK HALFSPACE

A. Variable Stiffness Ratio **B.** Variable Depth-to-Radius Ratio

	Stiffness Ratio E_c^L/E_c^R			Depth-to-Radius Ratio d/r_0				
	0	0.2	0.4	0.5	1.0	2.5	5.0	10.0
Cones	0.3757	0.5230	0.6574	0.3952	0.5230	0.7121	0.8275	0.9044
Exact	0.3905	0.5218	0.6473	0.3852	0.5218	0.7437	0.8634	0.9305

Dynamic-stiffness coefficient. The dynamic-stiffness coefficients $S(a_0)$ for all motions of a disk on the surface of a soil layer resting on a flexible rock halfspace are addressed. The dynamic-stiffness coefficient is nondimensionalized as follows

$$S(a_0) = K[k(a_0) + ia_0\, c(a_0)] \tag{3-89}$$

where K is the static-stiffness coefficient of the homogeneous halfspace with the properties of the layer, $k(a_0)$ and (a_0) are the spring and damping coefficients and $a_0 = \omega r_0/c_s^L$. The value of $k(a_0 = 0)$ represents the ratio of the static-stiffness coefficient of the disk on a layer resting on flexible rock to that on the homogeneous halfspace with the properties of the layer.

First, the case of a stiffer and denser layer than the halfspace is examined. This case is appropriate for a new sand fill over soft alluvial virgin soil. It is typical for some sea coast areas. The ratios $d/r_0 = 1$, $G_L/G_R = 5$, $\rho_L/\rho_R = 1.25$, and $\nu_L = \nu_R = 1/3$ are selected, resulting in an impedance ratio $\rho_L c_L/(\rho_R c_R) = 2.5$. The apex ratio z_0^L/r_0 follows for each motion from Table 2-3B; in this case, $z_0^R = z_0^L + d$. For instance, the reflection coefficients of the rotational motions $-\alpha(\omega)$, $-\alpha(0)$, and $-\alpha(\infty)$ are specified in Eqs. 3-74, 3-75b, and 3-76 with

$c = c_p$ for the rocking motion ($a_0 = \omega T$), and $c = c_s$ for the torsional motion ($a_0 = \omega T/2$). Thus, κ and T follow from Eq. 3-9. The dynamic-stiffness coefficient $S_\vartheta(a_0)$ is given by Eq. 3-87. The rigorous results denoted as exact in Figs. 3-15 to 3-18 are taken from Waas [W2].

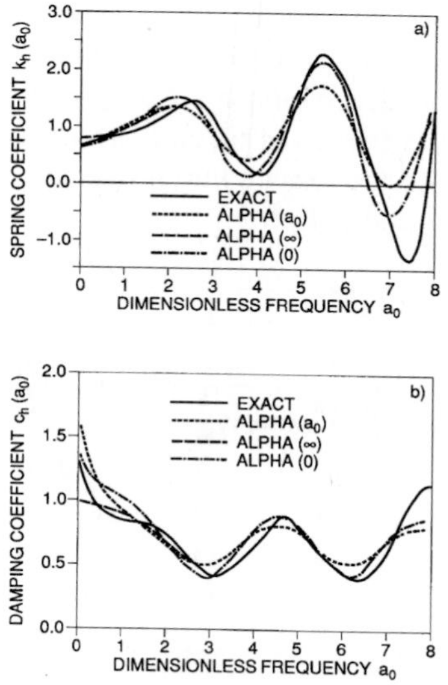

Fig. 3-15 Horizontal dynamic-stiffness coefficient for disk on soil layer stiffer than rock halfspace.

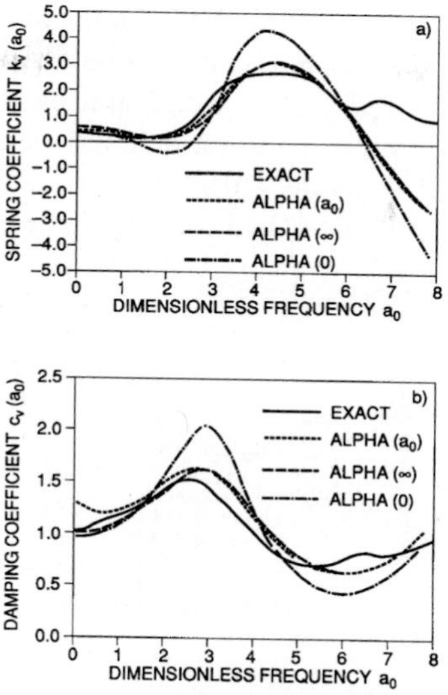

Fig. 3-16 Vertical dynamic-stiffness coefficient for disk on soil layer stiffer than rock halfspace.

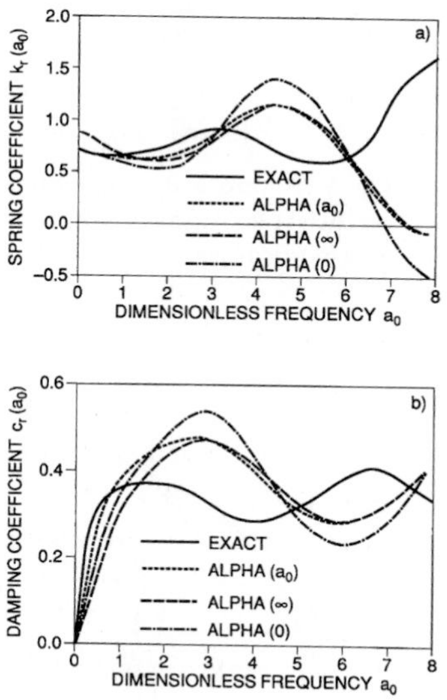

Fig. 3-17 Rocking dynamic-stiffness coefficient for disk on soil layer stiffer than rock halfspace.

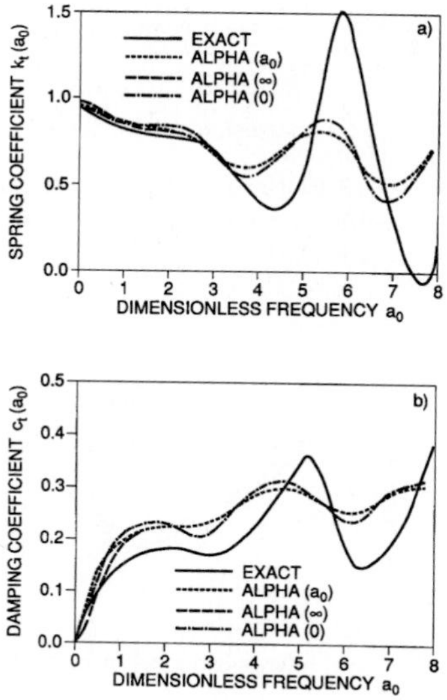

Fig. 3-18 Torsional dynamic-stiffness coefficient for disk on soil layer stiffer than rock halfspace.

The dynamic-stiffness coefficients computed for the three reflection coefficients are compared with the exact solution for the horizontal, vertical, rocking, and torsional motions in Figs. 3-15, 3-16, 3-17, and 3-18. The agreement is good, in certain important frequency ranges even excellent.

Second, the more common case of a more flexible and less denser layer than the halfspace (rock) is addressed. In this example the ratios $d/r_0 = 1$, $G_L/G_R = 0.544$, $\rho_L/\rho_R = 0.85$, and $\nu_L = \nu_R = 0.25$ are selected, leading to an impedance ratio $\rho_L c_L/(\rho_R c_R) = 0.68$. The rigorous solution denoted as exact in the following figures is taken from Luco [L2]. The dynamic-stiffness coefficients are compared with the exact solutions for the horizontal, vertical, and rocking motions in Figs. 3-19, 3-20, and 3-21.

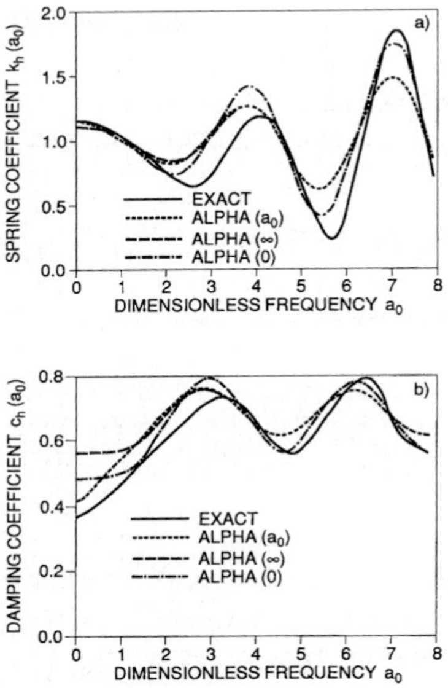

Fig. 3-19 Horizontal dynamic-stiffness coefficient for disk on soil layer more flexible than rock halfspace.

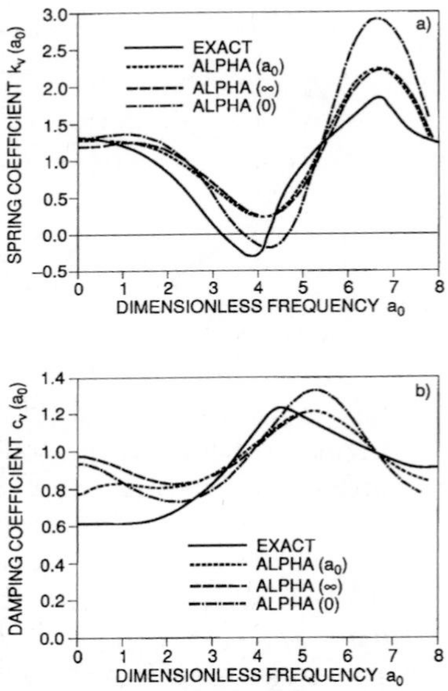

Fig. 3-20 Vertical dynamic-stiffness coefficient for disk on soil layer more flexible than rock halfspace.

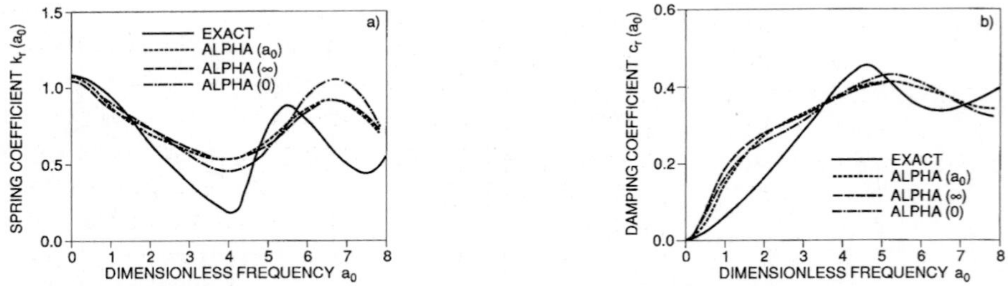

Fig. 3-21 Rocking dynamic-stiffness coefficient for disk on soil layer more flexible than rock halfspace.

It is easy to show numerically that the layered cone model for a disk on a soil layer resting on a flexible rock halfspace leads to the exact high-frequency limit of the dynamic-stiffness coefficient. The same conclusion is thus derived as for the case of a disk on a soil layer resting on rigid rock in Section 3.3.5.

The analysis based on cones leads to an overestimation of the damping coefficients $c(a_0)$ in the low-frequency range for all three motions. To a certain extent this discrepancy is overemphasized, as the imaginary part of the dynamic-stiffness coefficient is divided by the small value of a_0 to calculate $c(a_0)$. The accuracy is better evaluated from the real and imaginary parts of the dynamic-stiffness coefficients ($\text{Re}S(a_0) = k(a_0)$, $\text{Im}S(a_0) = a_0 c(a_0)$) which are actually used in a frequency-domain analysis. The corresponding agreement for all three motions shown in Fig. 3-22 is good.

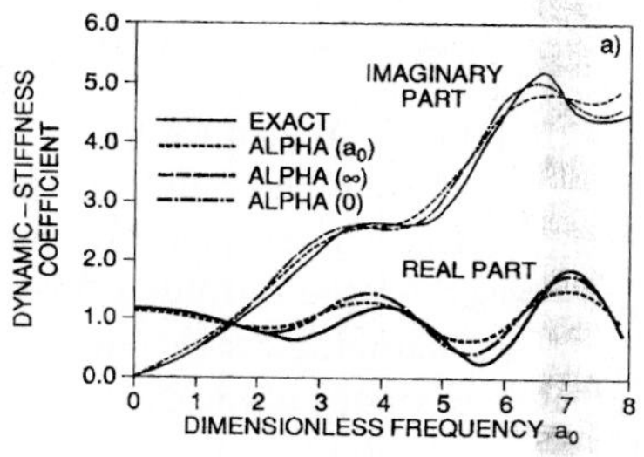

a. Horizontal.

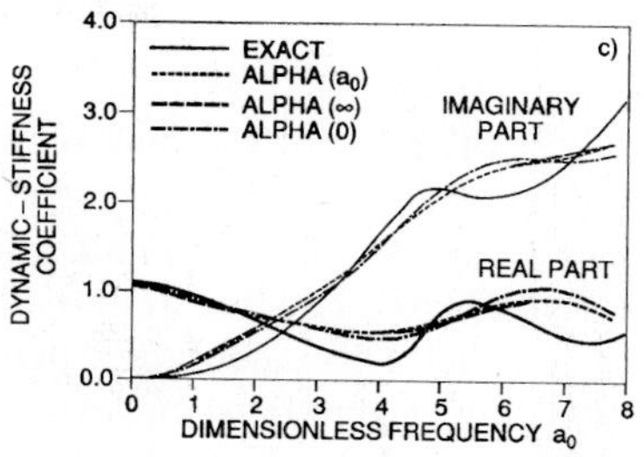

c. Rocking.

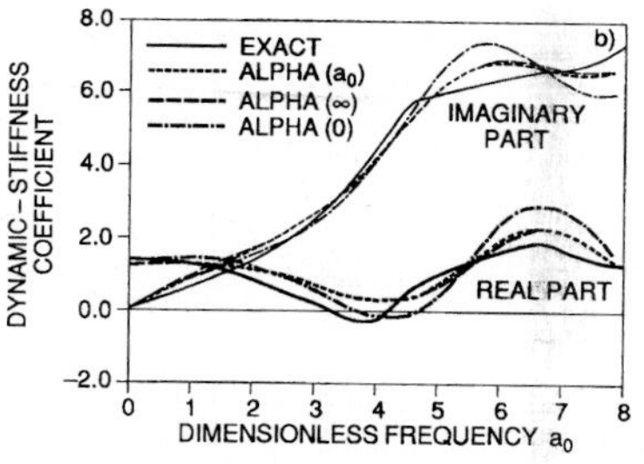

b. Vertical.

Fig. 3-22
Real and imaginary parts of dynamic-stiffness coefficients for disk on soil layer more flexible than rock halfspace.

Transfer function of layer on flexible halfspace to homogeneous half-space. The discrepancy for vertical motion is examined further. As can be seen from Eq. 3-83, the dynamic-stiffness coefficient for the disk on a layer resting on a halfspace is equal to the product of the dynamic-stiffness coefficient for the disk on a halfspace with the properties of the layer and the transfer function. The transfer function $H(a_0)$ (Eq. 3-82) is compared in Fig. 3-23 for the cone models based on the three reflection coefficients with the exact solution determined by dividing the rigorous solution of the dynamic-stiffness

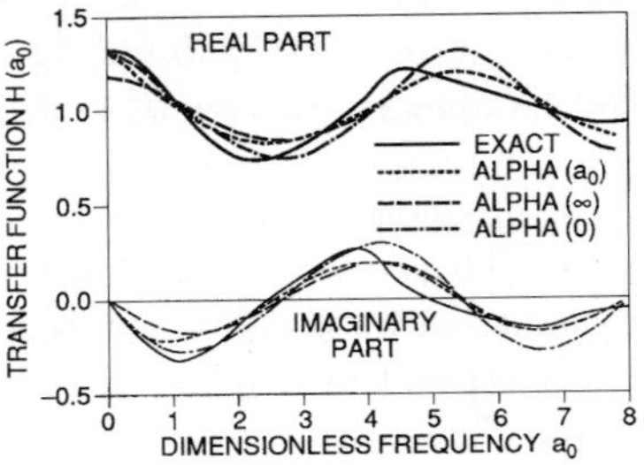

Fig. 3-23 Transfer function of vertical displacement amplitude from layer to halfspace.

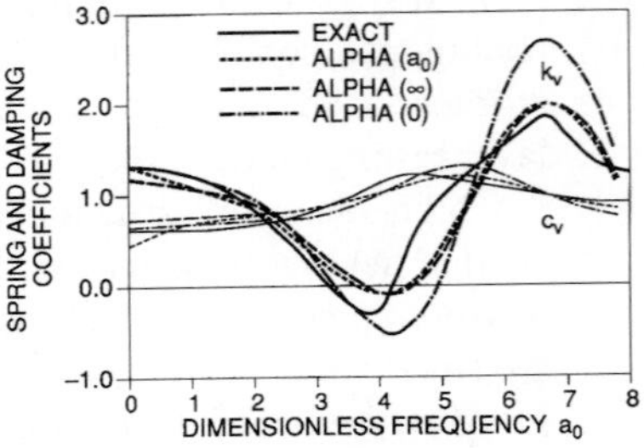

Fig. 3-24 Vertical dynamic-stiffness coefficient of disk on soil layer more flexible than rock halfspace calculated by multiplying exact dynamic-stiffness coefficient of halfspace by transfer function based on cones.

coefficient of the disk on a layer resting on a halfspace by that on a halfspace. The agreement is good throughout the frequency range. The discrepancy thus arises mostly from using the cone model to determine the dynamic-stiffness coefficient for a disk on a halfspace and not from using the transfer function based on the unfolded layered cone model. This is confirmed in Fig. 3-24 where the spring and damping coefficients $k(a_0)$, $c(a_0)$ of the layered system are calculated starting from the exact values of the dynamic-stiffness coefficients of a disk on a halfspace, which are then multiplied by the various transfer functions. The agreement is more than sufficient for practical applications.

Thus, *the dynamic-response of a disk on the surface of a soil layer resting on flexible rock calculated with cones is as accurate as that of a disk on the surface of a homogeneous halfspace.*

3.7 BASIC LUMPED-PARAMETER MODEL

3.7.1 Introductory Example

Besides the refolded layered cone model shown in Fig. 3-2, a lumped-parameter model can be used to represent a basemat on the surface of a soil layer resting on rigid rock. In such a lumped-parameter model besides the foundation node a few additional nodes with degrees of freedom are introduced, which are connected by springs, dashpots, and masses with frequency-independent real coefficients. When a dynamic soil–structure-interaction analysis is performed using a structural dynamics computer program, the lumped-parameter model of the soil layer is attached to the basemat of the structure's model. The total dynamic model is thus processed consistently directly in the time domain. Admittedly, the appreciable number of springs, dashpots, and masses does not allow the lumped-parameter model to be conductive to a simple physical interpretation.

In contrast to the homogeneous soil halfspace where discrete-element models (Figs. 2-5, 2-17) exist which represent exactly the cone models, the *lumped-parameter model is not derived from the corresponding layered cone models*. As described in depth in Appendix B, in this systematic procedure to construct consistent lumped-parameter models *curve-fitting applied to the exact dynamic-stiffness coefficient* results in the coefficients of the lumped-parameter model, whereby a linear system of equations only has to be solved. Calibration with the exact solution is thus performed. The lumped-parameter model's coefficients can be listed in tables for a wide range of parameters, which makes the practical application very attractive, as no curve-fitting has to be done for a specific case.

For the sake of illustration only, a lumped-parameter model is constructed here using the results of the layered cone model for calibration. The corresponding dynamic-stiffness coefficients which are to be reproduced by the lumped-parameter model as closely as possible are specified in Eqs. 3-39 and 3-43 for the translational and rotational layered cones, respectively. Again, the ratio of the depth d to radius r_0 equals 1 and Poisson's ratio $\nu = 1/3$. For each degree of freedom (horizontal, vertical, rocking, torsional), the model shown in Fig. 3-25 is developed. Besides the foundation node with the (generalized) displacement u_0, two internal nodes with u_1, u_2 are used. Of the eight coefficients of the four springs (K_1, K_2, K_3, K_4), of the three dashpots (C_1, C_2, C_3), and of the mass M, actually only six are independent. This is caused by the doubly-asymptotic behavior enforced on the model: Setting the static-stiffness coefficients and the high-frequency damping coefficients of the lumped-parameter model and of the layered cone equal results in two relations which the dynamic-stiffness coefficient of the lumped-parameter model has to satisfy (see Appendix B). The so-called regular part of the dynamic-stiffness coefficient is approximated as a ratio of a second-degree to a third-degree polynomial in ia_0 (Eq. B-4 with $M = 3$). The curve fitting is performed for $0 < a_0 < 5.8$. As an example, the approximate lumped-parameter model's total dynamic-stiffness coefficient for the vertical motion nondimensionalized in the familiar way as described in Eq. 3-44 is compared with the "exact" value to be matched of the layered cone in Fig. 3-26. The fine line is thus the same as the heavy line presented in Fig. 3-9b. Considering that M is only equal to 3, the agreement is good. The partial-fraction expansion (Eq. B-10) consists of one first-order term with two coefficients and one second-order term with four coefficients, resulting in a total of six coefficients. By arranging the discrete-element models of the singular term (Fig. B-1), of the first-order term (Fig. B-2a), and of the second-order term (Fig. B-3b) in parallel, the lumped-parameter model of Fig. 3-25 is constructed. The springs and dashpots connecting the foundation node directly to the rigid support are combined. The result of the construction of the lumped-parameter model is presented for all degrees of freedom in Table 3-7. For $d/r_0 = 1$ and $\nu = 1/3$, the dimensionless coefficients k_i ($i = 1, 2, 3, 4$) of the springs, c_i($i = 1, 2, 3$) of the dashpots, and m of the mass are listed (G = shear modulus), whereby

$$K_i = k_i\, G r_0 \tag{3-90a}$$

$$C_i = c_i\, G \frac{r_0^2}{c_s} \tag{3-90b}$$

$$M = m\, G \frac{r_0^3}{c_s^2} \tag{3-90c}$$

Note that all coefficients are real, but not necessarily positive. For the rocking and torsional cases, Eq. 3-90 corresponds to a dynamic-stiffness coefficient relating the rotation multiplied by r_0 to the moment divided by r_0. To construct a lumped-parameter model representing the relationship between the rotation and the moment, the right-hand side of Eq. 3-90 has to be multiplied by r_0^2. Table 3-7 is provided of the sake of illustration only. As the calibration is performed with the dynamic-stiffness coefficient of the layered cone which is only an approximation, these dimensionless coefficients should not be used in practical applications. Instead, the information in Table 3-8 should be applied which is based on calibration with the exact dynamic-stiffness coefficient.

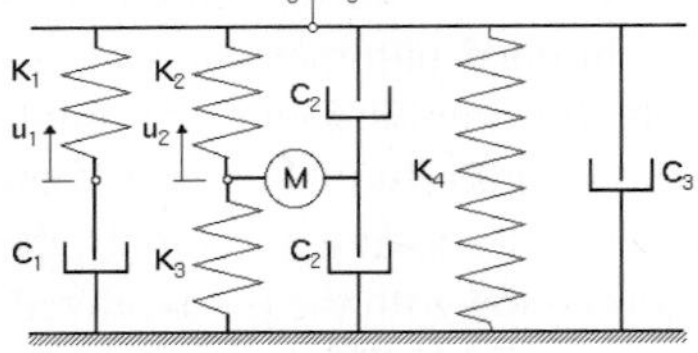

Fig.3-25 Basic lumped-parameter model.

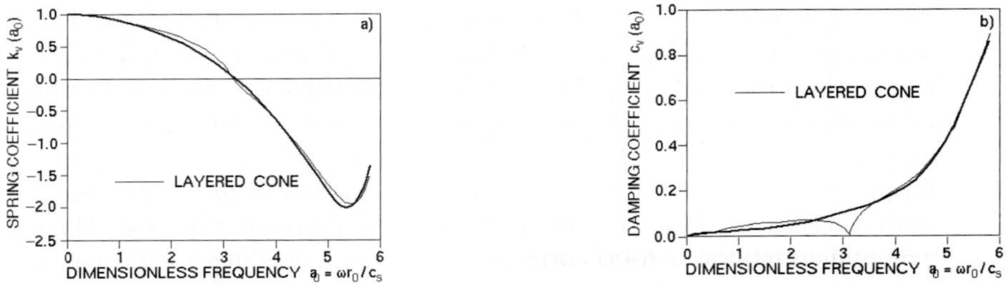

Fig. 3-26 Vertical dynamic-stiffness coefficient for harmonic loading of disk on layer.

TABLE 3-7 DIMENSIONLESS COEFFICIENTS OF LUMPED-PARAMETER MODEL BASED ON LAYERED-CONE RESULTS*

	Horizontal	Vertical	Rocking	Torsional
k_1	−.135504 E+02	−.385021 E+02	−.115333 E+02	−.744350 E+01
k_2	−.889717 E+01	+.520073 E+01	−.424315 E+02	+.147987 E+01
k_3	−.885129 E+02	−.883372 E+03	−.826860 E+02	−.115690 E+02
k_4	+.151865 E+02	+.107474 E+02	+.932713 E+01	+.390900 E+01
c_1	−.463274 E+01	−.574871 E+01	−.191067 E+01	−.972235 E+00
c_2	−.846450 E+01	−.232283 E+02	−.376956 E+01	−.184750 E+01
c_3	+.116060 E+02	+.295083 E+02	+.533956 E+01	+.341750 E+01
m	−.986740 E+01	−.239227 E+02	−.220328 E+01	−.166126 E+01

* Not to be used for practical applications

3.7.2 Achieved Accuracy

To determine the accuracy of the lumped-parameter model to represent a disk with radius r_0 on the surface of a layer of depth d and Poisson's ratio ν, the same case as before ($d = r_0$, $\nu = 1/3$) is examined. The exact values used for calibration are taken from Kausel [K1].

First, it is demonstrated that by choosing a ratio of polynomials of large degree M in ia_0 for the approximation of the (regular part of the) dynamic-stiffness coefficient and that by performing the curve-fitting from zero to the dimensionless frequency a_{0max}, it is possible to reach a very good accuracy throughout this region. $M = 12$ and $a_{0max} = 12.56$ are selected in the following. This will of course lead to a very large number of discrete-element models and thus of springs, dashpots, and masses of the lumped-parameter model, which is too large for many practical applications. For instance for the vertical degree of freedom, all roots of the polynomial in the denominator are complex conjugates, resulting in six discrete-element models for the second-order terms (Fig. B-3b) with a total of six internal degrees of freedom and twenty-four independent coefficients. For the horizontal motion, two and five discrete-element models for the first-order terms (Fig. B-2a) and for the second-order terms (Fig. B-3b), respectively, are obtained with seven internal degrees of freedom and again twenty-four independent coefficients. In Fig. 3-27 the total dynamic-stiffness coefficient of the lumped-parameter model based on Eq. B-12, shown as a heavy line, is compared with the exact value for the horizontal, vertical, rocking, and torsional motions. The spring coefficient $k(a_0)$ and the damping coefficient $c(a_0)$ defined as in Eq. 3-44 are plotted. The results corresponding to the lumped-parameter model capture well the strong frequency dependency of the exact solution. The damping coefficient $c(a_0)$ vanishes (essentially) below the cutoff frequency which is equal to the fundamental frequency in the predominant motion [$c_s/(4d)$ corresponding to $a_0 = \pi/2$ for the horizontal and torsional motions, $c_p/(4d)$ corresponding to $a_0 = \pi$ for the vertical and rocking motions]. In the mathematically strict sense of zero damping, the rigorous criterion for vertical and rocking motions is also equal to the horizontal fundamental frequency ($a_0 = \pi/2$). The precursor damping in the range $\pi/2 < a_0 < \pi$ is from a practical point of view negligible, as it is caused by the minimal amount of shear-wave propagation which can be disregarded compared to the overriding effect of the dominant dilatational wave propagation for vertical and rocking motions.

Second, the same comparison is performed using the lumped-parameter model of Fig. 3-25 applicable for practical cases. $M = 3$ and a_{0max} varying between 4.08 and 5.87 lead for all motions to one real and one pair of complex conjugate roots of the polynomial in the denominator (Eq. B-10), resulting in one discrete-element model for the first-order term (Fig. B-2a) and in one for the second-order term (Fig. B-3b). The agreement shown in Fig. 3-28 is satisfactory, thus demonstrating that a good compromise is reached between the total number of coefficients and the achieved accuracy. As discussed in Section 3.7.1, the lumped-parameter model exactly models the static case and the high-frequency limit (doubly-asymptotic approximation). In general, the accuracy of this lumped-parameter model is similar to that of the layered cone model (Figs. 3-9, 3-10) in the same frequency range.

Next, the unit-impulse response function (dynamic-flexibility coefficient in the time domain) is calculated; the displacement $u_0(\bar{t})$ or rotation $\vartheta_0(\bar{t})$ as a function of the dimensionless time $\bar{t} = tc_s/r_0$ is caused by a unit-impulse load or moment applied at the foundation node at time zero. The lumped-parameter model of Fig. 3-25 ($M = 3$), whose dynamic-stiff-

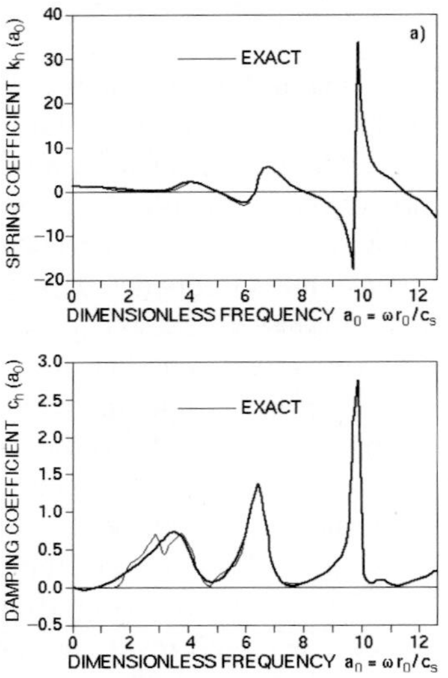

a. Horizontal.

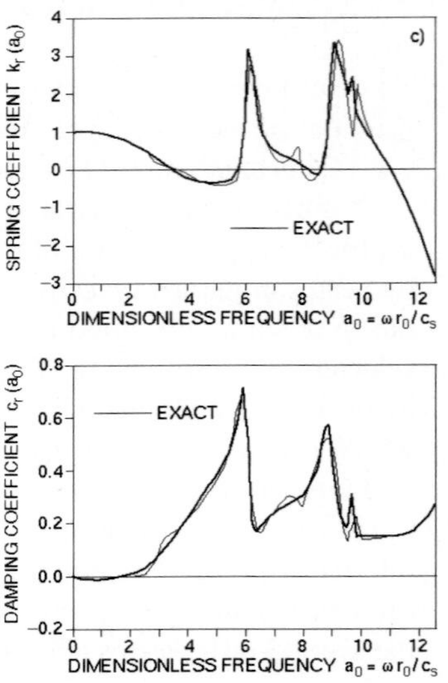

c. Rocking.

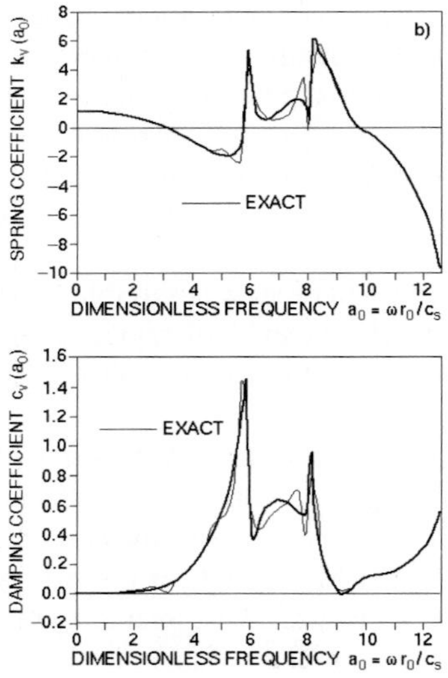

b. Vertical.

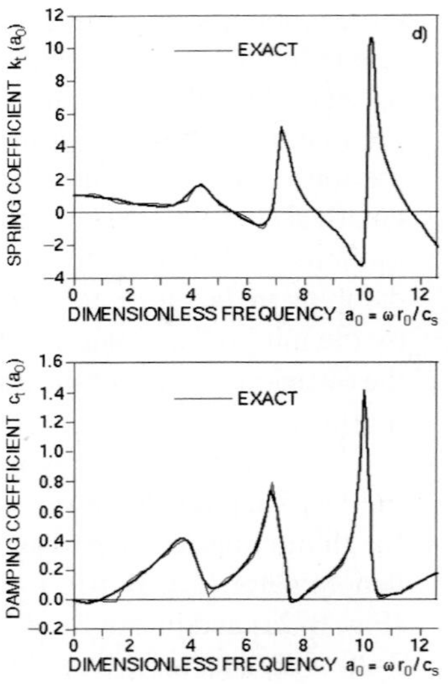

d. Torsional.

Fig. 3-27 Dynamic-stiffness coefficient for harmonic loading of disk on layer (degree of polynomial in denominator = 12).

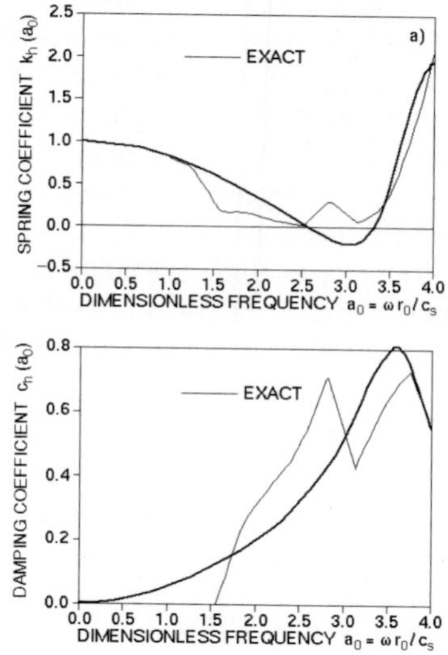

a. Horizontal.

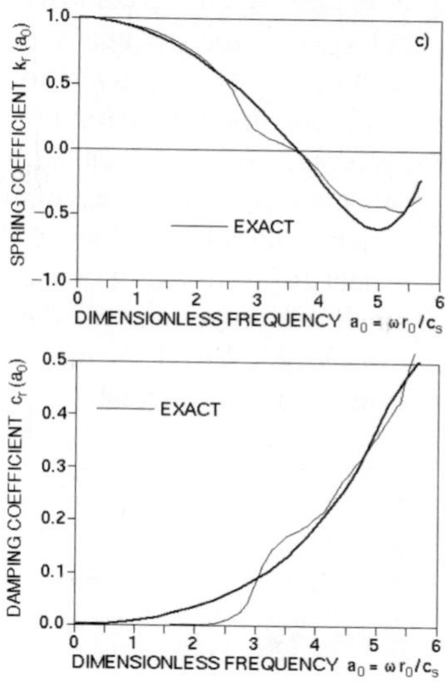

c. Rocking.

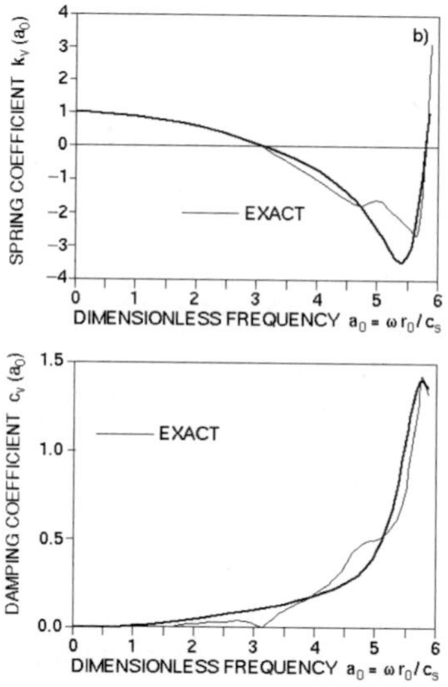

b. Vertical.

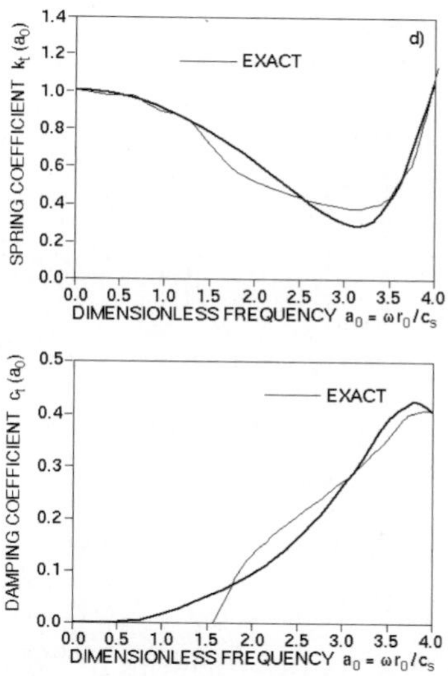

d. Torsional.

Fig. 3-28 Dynamic-stiffness coefficient for harmonic loading of disk on layer (degree of polynomial in denominator = 3).

ness coefficients are shown in Fig. 3-28, is used. This stringent test also evaluates the behavior in the frequency range above a_{0max}. An explicit time integration algorithm (see Section 2.2.4, Eqs. 2-118 to 2-120) with $\Delta\bar{t} = 0.01$ is applied. The unit-impulse response functions nondimensionalized accordingly using the static values of the corresponding halfspace are plotted as solid lines for all motions in Fig. 3-29. As the lumped-parameter models represent the asymptotic values for $a_0 \to \infty$ of the dynamic-stiffness coefficients exactly, the initial values of the unit-impulse response functions at $\bar{t} = 0^+$ are correct. This aspect is discussed in connection with Eq. 2-81. The exact solution which incorporates a small amount of material damping for the horizontal, vertical, and rocking motions is taken from Wolf [W6] (fine line). A comparison with the results based on the layered cone model (Fig. 3-7) is of interest.

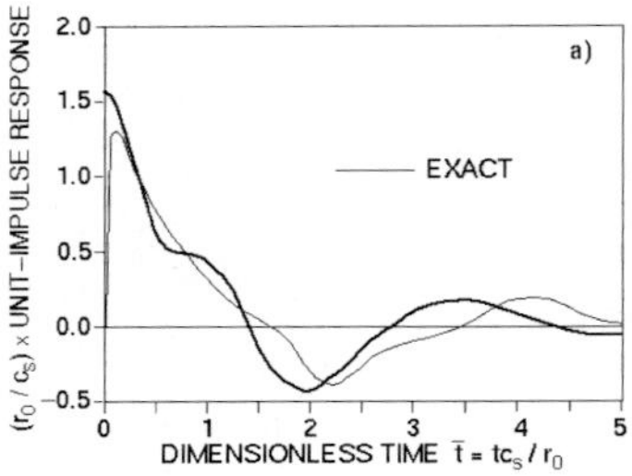

a. Horizontal.

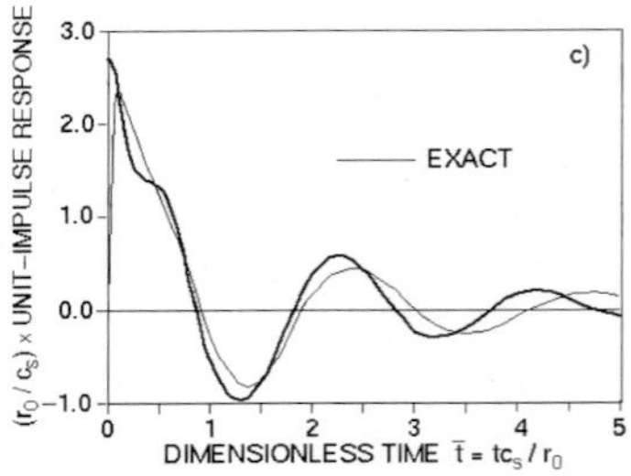

c. Rocking.

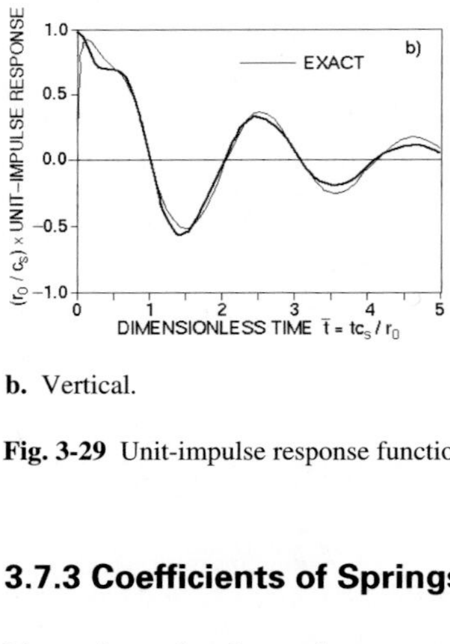

b. Vertical.

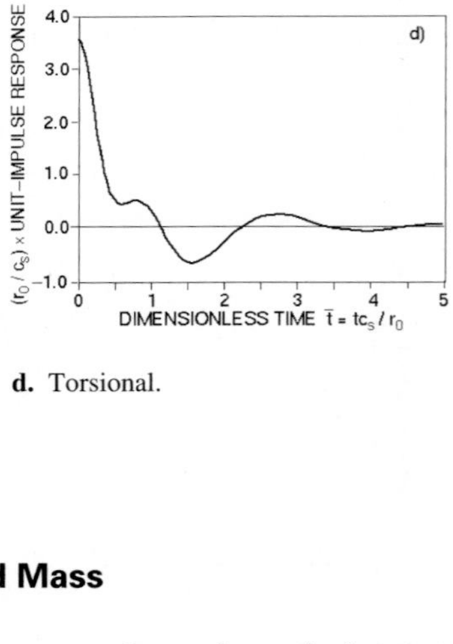

d. Torsional.

Fig. 3-29 Unit-impulse response function of disk on layer.

3.7.3 Coefficients of Springs, Dashpots, and Mass

To analyze the dynamic response of a disk of radius r_0 on the surface of a homogeneous soil layer of depth d, with shear modulus G, Poisson's ratio ν, and shear-wave velocity c_s resting on rigid rock (Fig. 3-1), the stable *lumped-parameter model* of Fig. 3-25 exhibiting sufficient accuracy with only *two* additional *internal degrees of freedom* u_1, u_2 is well suited for practical applications. This lumped-parameter model consists of the *four springs* with coefficients K_1, K_2, K_3, K_4, of the *three dashpots* with coefficients C_1, C_2, C_3, and of the *mass* with the coefficient M. These depend on the dimensionless coefficients k_i ($i = 1, 2, 3, 4$), c_i ($i = 1, 2, 3$), and m as specified in Eq. 3-90. For $r_0/d = 1.00, 0.50, 0.25$, and 0.00 (halfspace) and for $\nu = 0$, 1/3 and 0.45 these coefficients are listed in Table 3-8 for the horizontal, vertical, rocking, and torsional motions. The latter is independent of ν. For the rocking and torsional cases, Eq. 3-90 corresponds to a dynamic-stiffness coefficient relating the rotation multiplied by r_0 to the moment divided by r_0. To construct a lumped-parameter model representing the relationship between the rotation and the moment, the right-hand side of Eq. 3-90 has to be multiplied by r_0^2. The dimensionless coefficients exhibit both signs.

The coefficients of Table 3-7 based on calibration with the dynamic-stiffness coefficients of the layered cone bear comparison with those of Table 3-8.

The effect of material damping in the lumped-parameter model can be considered by either augmenting the elements (Voigt-viscoelastic, Section 2.4.2) or by introducing corresponding frictional-elements (Sections 2.4.4).

3.7.4 Rigid Block on Layer and on Halfspace with Partial Uplift

As an example of a soil–structure-interaction analysis with abrupt nonlinearities occurring, a rigid block with individual footings, which can uplift, resting on the surface of an undamped soil layer (Fig. 3-30) is examined. As the loading consists of an earthquake, the analysis of which is addressed only in Chapter 6, a detailed discussion of the analysis is not possible at this stage. (For a linear system without uplift, see Section 2.4.5, in particular Eq. 2-267 and Fig. 2-39.) The four disks of radius $r_0 = 4$ m are placed at each corner of the rigid body of height $2h = 18$ m, width $2b = 12$ m, mass $m = 12.7 \cdot 10^6$ kg, and mass moment of inertia $I = 6.4 \cdot 10^9$ kgm^2. The layer of depth $d = r_0 = 4$ m is characterized by the shear modulus $G = 1.3 \cdot 10^9$ N/m^2, Poisson's ratio $\nu = 1/3$, and shear-wave velocity $c_s = 750$ m/s. The idealized horizontal earthquake ground acceleration (free field) $\ddot{u}^g(t)$ acting during two seconds is selected as

$$\ddot{u}^g(t) = \frac{g}{10}\left(-3 \sin 8\pi t + 9 \sin 24\pi t\right) \tag{3-91}$$

with $g = 9.81$ m/s^2.

Only the vertical and rocking motions $w(t)$ and $\varphi(t)$ of the block's bottom center are considered in the calculation, with the horizontal motion assumed identical to the ground motion. The lumped-parameter models of the disks representing the vertical motion without any through-soil coupling are shown in Fig. 3-30. Contact is lost when the reaction force becomes negative and is reestablished when penetration occurs (on the left side $u_0^L \geq w - b\varphi$). When the footings are both in contact, compatibility is enforced ($u_0^L = w - b\varphi$), and the total dynamic system has six degrees of freedom (w, φ, u_1^L, u_2^L, u_1^R, u_2^R). When the left footing loses contact, the reaction force remains zero and an additional degree of freedom u_0^L is introduced. An explicit algorithm is well suited for such a nonlinear analysis; as this is based on the predicted displacements and velocities, it can be decided if contact or uplift exists. No iterations arise. The internal forces that contribute to the equilibrium equations are calculated with these predicted values. A time step Δt equal to $0.005\, r_0/c_s$ is chosen. The fundamental frequency of the linear rigid block–soil system associated with rocking equals

TABLE 3-8 DIMENSIONLESS COEFFICIENTS OF BASIC LUMPED-PARAMETER MODEL FOR DISK ON SURFACE OF SOIL LAYER ON RIGID ROCK

Ratio of Radius to Depth r_0/d		Horizontal			Vertical			Rocking			Torsional
		Poisson's Ratio ν									
		0	1/3	0.45	0	1/3	0.45	0	1/3	0.45	
1.00	*k*1	-.109636 E+02	-.125658 E+02	-.107091 E+02	-.185216 E+02	-.312572 E+02	-.585650 E+02	-.538137 E+01	-.127100 E+02	-.125057 E+02	-.920277 E+01
	*k*2	-.199616 E+02	-.100143 E+02	-.277613 E+02	-.689058 E+02	+.564651 E+01	+.533868 E+02	-.118019 E+02	-.127000 E+01	-.102097 E+02	-.488643 E+01
	*k*3	-.596293 E+03	-.236814 E+03	-.837270 E+03	-.803915 E+04	-.297570 E+04	-.972054 E+05	-.370561 E+03	-.106411 E+03	-.114401 E+05	-.762034 E+02
	*k*4	+.262006 E+02	+.172890 E+02	+.350886 E+02	+.781698 E+02	+.101028 E+02	-.297301 E+02	+.152717 E+02	+.665102 E+01	+.171002 E+02	+.104850 E+02
	*c*1	-.423955 E+01	-.391585 E+01	-.443420 E+01	-.564579 E+01	-.620122 E+01	-.533597 E+01	-.152562 E+01	-.168764 E+01	-.159579 E+01	-.209847 E+01
	*c*2	-.144980 E+02	-.969345 E+01	-.164981 E+02	-.573623 E+02	-.372925 E+02	-.162817 E+03	-.511671 E+01	-.464871 E+01	-.205038 E+02	-.424955 E+01
	*c*3	+.176380 E+02	+.128349 E+02	+.196381 E+02	+.618023 E+02	+.435725 E+02	+.173237 E+03	+.622671 E+01	+.621871 E+01	+.231038 E+02	+.581955 E+01
	m	-.444888 E+02	-.177585 E+02	-.804875 E+02	-.355432 E+03	-.896786 E+02	-.759400 E+03	-.136958 E+02	-.294864 E+01	-.748688 E+02	-.501042 E+01
0.50	*k*1	-.101741 E+02	-.756096 E+01	-.103098 E+02	-.869429 E+01	-.178038 E+02	-.211241 E+02	-.558202 E+01	-.315920 E+01	-.544861 E+01	-.584813 E+01
	*k*2	-.711128 E+01	+.221036 E+01	+.353643 E+00	-.211429 E+02	+.869558 E+01	+.237930 E+02	-.260867 E+01	+.429538 E+00	+.544528 E+01	+.779373 E+00
	*k*3	-.376551 E+02	-.183990 E+02	-.386290 E+02	-.301954 E+03	-.648930 E+02	-.574768 E+04	-.120186 E+02	-.563639 E+00	-.495529 E+02	-.267204 E+01
	*k*4	+.114651 E+02	+.370791 E+01	+.631911 E+01	+.266455 E+02	+.167960 E+00	-.104560 E+02	+.553603 E+01	+.268680 E+01	-.714571 E+00	+.448746 E+01
	*c*1	-.563146 E+01	-.169515 E+01	-.312323 E+01	-.635435 E+01	-.300736 E+01	-.885920 E+01	-.180105 E+01	-.217449 E+00	-.226588 E+01	-.873265 E+00
	*c*2	-.853329 E+01	-.484337 E+01	-.900332 E+01	-.118278 E+02	-.967485 E+01	-.348879 E+02	-.209944 E+01	-.485884 E-01	-.872006 E+00	-.664093 E+00
	*c*3	+.116733 E+02	+.798337 E+01	+.121433 E+02	+.162678 E+02	+.159548 E+02	+.453079 E+02	+.320944 E+01	+.161859 E+01	+.347201 E+01	+.223409 E+01
	m	-.125108 E+02	-.142805 E+02	-.222875 E+02	-.438319 E+02	-.104698 E+02	-.156304 E+03	-.166034 E+01	-.550887 E-01	-.105920 E+01	-.758683 E+00
0.25	*k*1	-.500393 E+01	-.569922 E+01	-.635602 E+01	-.650348 E+01	-.866267 E+01	-.939217 E+01	-.197103 E+01	-.131566 E+01	-.185845 E+01	-.317223 E+01
	*k*2	+.117908 E+01	+.113372 E+01	+.126563 E+01	+.212837 E+01	+.360033 E+01	+.591506 E+01	-.908392 E+00	-.159178 E+01	+.140842 E+01	-.110204 E+02
	*k*3	-.531658 E+01	-.627809 E+01	-.861155 E+01	-.111486 E+02	-.206851 E+02	-.294639 E+02	+.320667 E+00	+.889908 E+00	-.198123 E+01	-.272014 E+02
	*k*4	+.330564 E+01	+.414181 E+01	+.444955 E+01	+.290606 E+01	+.353509 E+01	+.239510 E+01	+.280516 E+01	+.228996 E+01	+.228257 E+00	+.133791 E+02
	*c*1	-.753687 E+00	-.123420 E+01	-.118324 E+01	-.147587 E+01	-.301652 E+01	-.652332 E+01	-.111891 E+01	-.117566 E+01	-.166191 E+01	-.175255 E+01
	*c*2	-.320391 E+01	-.343160 E+01	-.476257 E+01	-.545496 E+01	-.633133 E+01	-.153512 E+01	-.192001 E-01	-.120420 E+00	-.608065 E-01	-.576183 E+00
	*c*3	+.634391 E+01	+.657160 E+01	+.790257 E+01	+.989496 E+01	+.126113 E+02	+.119551 E+02	+.131920 E+01	+.199042 E+01	+.266081 E+01	+.214618 E+01
	m	-.197705 E+02	-.277938 E+02	-.353939 E+02	-.202557 E+02	-.262470 E+02	-.217797 E+01	-.496405 E-02	-.234838 E-01	-.166728 E-01	-.408057 E+00
0.00	*k*1	-.135004 E+02	-.388471 E+01	-.517262 E+01	-.196175 E+01	-.741830 E+01	-.174454 E+02	-.177328 E+01	-.371794 E+01	-.398695 E+01	-.347454 E+01
	*k*2	-.953646 E+01	-.159784 E+02	+.239313 E+00	-.586095 E+00	+.149859 E+01	+.318590 E+01	-.825315 E+01	-.530262 E+01	+.488296 E+01	+.161189 E+00
	*k*3	-.152937 E+02	-.214052 E+02	-.491200 E+01	+.418313 E+00	-.108130 E+02	-.145871 E+03	-.960129 E+00	-.456729 E+01	-.157465 E+02	-.175021 E+00
	*k*4	+.100318 E+02	+.139890 E+02	+.491843 E+01	+.253876 E+01	+.426031 E+01	+.401297 E+01	+.363207 E+01	+.648378 E+01	-.222776 E+01	+.329151 E+01
	*c*1	-.108173 E+01	-.406936 E+00	-.431719 E-01	-.540639 E+00	-.308148 E+00	-.287195 E+01	-.105544 E+01	-.150532 E+01	-.158356 E+01	-.257114 E-01
	*c*2	-.164199 E+01	-.441082 E+00	-.433318 E-01	-.316451 E-02	-.760091 E+00	-.496738 E+01	-.396130 E+00	-.400894 E+00	-.408329 E+00	-.525606 E-02
	*c*3	+.478349 E+01	+.358258 E+01	+.318483 E+01	+.470316 E+01	+.704009 E+01	+.153874 E+02	+.150613 E+01	+.197089 E+01	+.300833 E+01	+.157526 E+01
	m	-.207315 E+00	-.331202 E-01	-.126178 E+00	-.110135 E-02	-.348161 E+00	-.240813 E+01	-.245402 E-01	-.633544 E-01	-.125199 E+00	-.126499 E-02

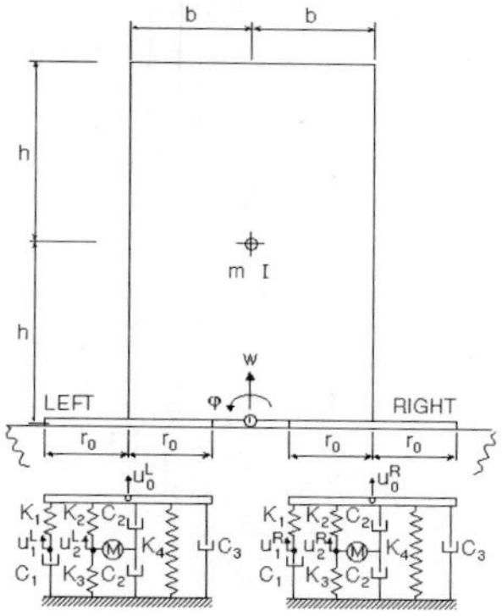

Fig. 3-30 Rigid block on disks with lumped-parameter models.

$$\omega_r = \sqrt{\frac{2b^2 K k(\omega_r)}{I + h^2 m}} \tag{3-92}$$

where K is the vertical static-stiffness coefficient of the disk on the layer and $k(\omega_r)$ describes the frequency-dependency of the dynamic-spring coefficient of the lumped-parameter model. After iteration ω_r corresponds to 6.3 Hz. The corresponding cutoff frequency of the soil layer, which is equal to the fundamental frequency in the vertical direction, is calculated as equal to $c_p/(4d) = 2c_s/(4d) = 93.7$ Hz. As the fundamental frequency of the block–soil system is smaller than the cutoff frequency, no radiation damping and thus a large response will occur.

The time histories of the gaps between the block and the disk [left: $w - b\varphi - u_0^L$] are plotted in Fig. 3-31a. As no radiation damping occurs, the periodic motion after the earthquake has stopped ($t > 2s$) does not decay. Significant uplift occurs; in certain instances even both disks lose contact momentarily. For the sake of comparison the gaps for the same block but resting on a halfspace with the same material properties are shown in Fig. 3.31b. As radiation damping for the halfspace occurs throughout the frequency range, the gaps are

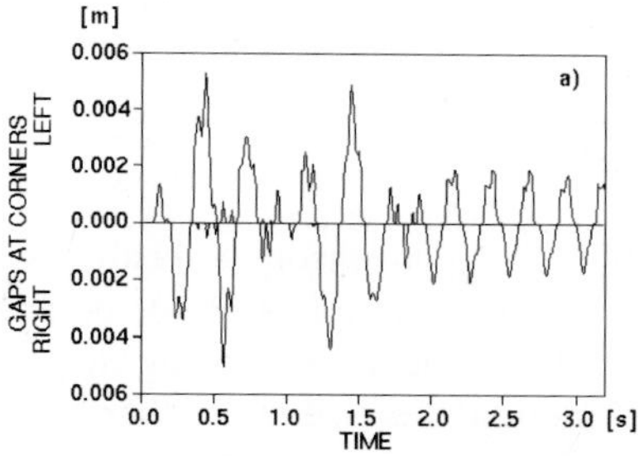

a. Layer.

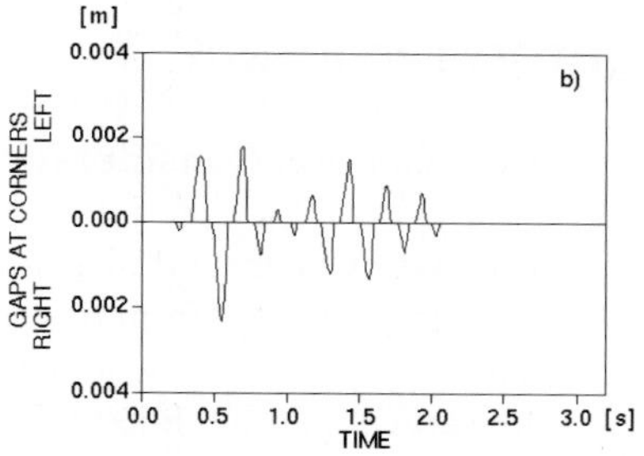

b. Halfspace.

Fig. 3-31 Vertical gap between block and disk.

much smaller and decay rapidly after the earthquake is over. The interaction forces are much larger for the system with the layer (Fig. 3-32a) than for that with the halfspace (Fig. 3-32b). When uplift is prevented—when the system behaves linearly—the reaction forces are smaller (Fig. 3-32c).

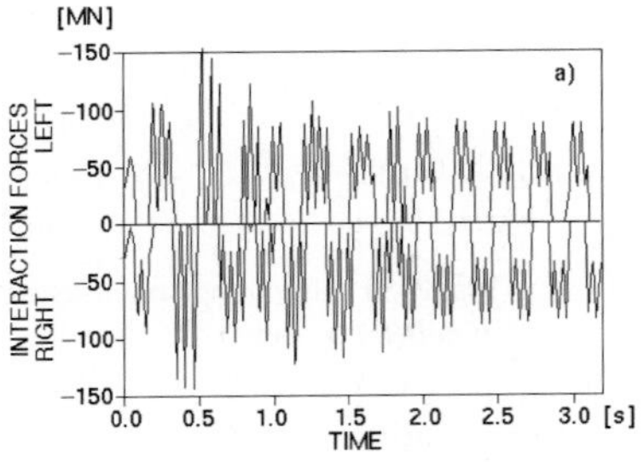

a. Layer with uplift occurring.

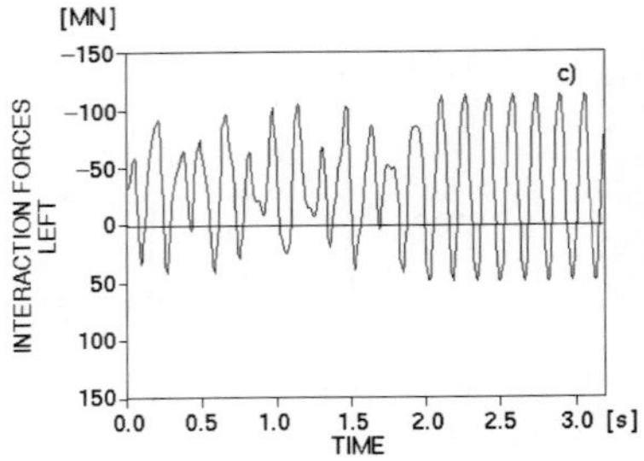

c. Layer with uplift prevented.

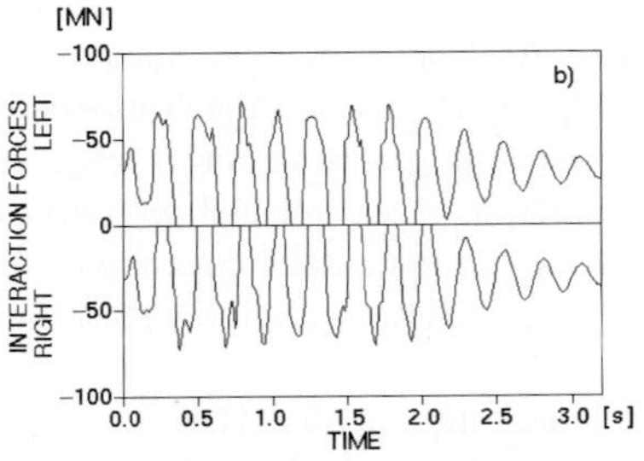

b. Halfspace with uplift occurring.

Fig. 3-32 Interaction forces.

3.8 CUTOFF FREQUENCY

3.8.1 General Considerations

The dynamic behavior of a basemat on the surface of a homogeneous soil halfspace is significantly different from that on the surface of a soil layer on rigid rock. As expected, the static-stiffness coefficient increases for a disk on a layer, the increase being significant for the vertical motion, important for the horizontal motion, diminishing further for rocking, and almost negligible for torsional motion (Eq. 3-21). The unit-impulse response functions of a disk on a layer exhibit vibrational behavior with changes in sign (Fig. 3-7) in contrast to those corresponding to a halfspace where the exponential decay governs (Figs. 2-8, 2-22), indicating that radiation damping is much larger in the latter case. The dynamic-stiffness coefficients of a disk on a layer are strongly frequency dependent with the real parts, the spring coefficients, even changing signs (Figs. 3-9, 3-10). But by far the most important effect, often overlooked by practicing engineers, is that for a soil layer on rigid rock a *cutoff frequency exists, below which radiation of energy does not occur*—the imaginary part of the dynamic-stiffness coefficient, the *damping coefficient, vanishes*. Above the cutoff frequency, in general, the damping coefficient is comparable with that of a disk on a halfspace.

As discussed in connection with Figs. 3-9 and 3-10 the cutoff frequency in the engineering sense depends on the dominant distortion. For the *horizontal and torsional motions* of the disk with shear distortion, the cutoff frequency is equal to the *horizontal fundamental frequency of the layer*, $c_s/(4d)$. For the *vertical and rocking motions* of the disk with axial distortion, the cutoff frequency from a practical point of view can be set equal to the *vertical fundamental frequency of the layer*, $c_p/(4d)$. In the latter case, the effect of the small amount of shear distortion occurring in the range between the two fundamental frequencies on the radiation damping is neglected. As an example, for a layer with Poisson's ratio $\nu = 1/3$, depth $d = 25$ m and shear-wave velocity $c_s = 200$ m/s resulting in a dilatational-wave velocity $c_p = 400$ m/s, the cutoff frequency for the horizontal and torsional motions equals 2 Hz and for the vertical and rocking motions 4 Hz. Structure–soil systems whose natural frequencies are less than the cutoff frequency will be subjected to very large amplitudes of motion during earthquakes and must be dimensioned accordingly.

The two types of physical models to represent the dynamic behavior of a basemat on the surface of a soil layer capture the phenomenon of the cutoff frequency well. For the layered cone model (Fig. 3-2) the echoes of past behavior indeed lead to an approximate representation of the cutoff frequency, as demonstrated in Figs. 3-9 and 3-10 for all motions. As the exact result of the dynamic-stiffness coefficient is used for calibration, the coefficients of the corresponding lumped-parameter model (Fig. 3-25) reflect the effect of the cutoff frequency by construction, as can be verified in Fig. 3-28.

To further demonstrate that the layered cone model can handle the effect of the cutoff frequency, the horizontal motion of a disk on a layer with $d = r_0$ and $\nu = 1/3$ is addressed. The (circular) cutoff frequency ω_c equals $2\pi c_s/(4d) = \pi c_s/(2d)$ which can also be written as π/T with the travel time from the surface to the rock and back, $T = 2d/c_s$. The corresponding dimensionless cutoff frequency a_0 equals $\omega_c r_0/c_s = \pi/2$. For a stiffness formulation the echo constants E_j^K are specified in the fourth column of Table 3-3. The corresponding echo formula (Eq. 3-24) is written as

$$\bar{u}_0(t) = \sum_{j=0}^{19} E_j^K \, u_0(t - jT) \tag{3-93}$$

A harmonic sinusoidal displacement $u_0(t) = u_0 \sin \omega t$ is applied with the associated velocity $\dot{u}_0(t)$. The circular frequency of excitation ω is taken to be slightly below and slightly above the cutoff frequency $\omega_c = \pi/T$. The functions $\bar{u}_0(t)$ and analogously $\dot{\bar{u}}_0(t)$ follow from Eq. 3-93, which then lead to the force $P_0(t)$ based on the interaction force-displacement relationship of the (unlayered) cone. The curves of excitation $u_0(t)$ and response $P_0(t)$ are normalized to unit amplitude in Fig. 3-33. Notice that for $\omega = 0.95\ \omega_c$ the force and displacement are almost exactly in phase: Negligible energy is radiated beneath the cutoff fre-

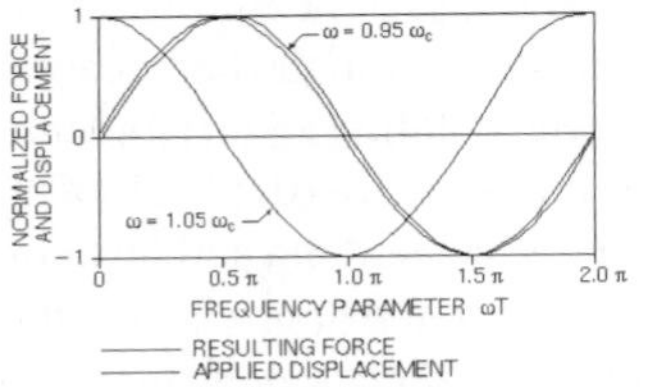

Fig. 3-33 Harmonic response of layered cone above and below cutoff frequency.

quency. By contrast, for $\omega = 1.05\ \omega_c$ a radical change occurs. The force leads the displacement by a phase angle of nearly $\pi/2$ and is thus proportional to the velocity $\dot{u}_0$. This corresponds to a pure dashpot, representing radiation damping.

3.8.2 Practical Significance of Cutoff Frequency

The practical importance of considering the cutoff frequency in design is demonstrated as follows. A typical site of a nuclear power plant [W4] exemplifies the dominant influence of the cutoff frequency on the damping ratio. All calculations are performed with a rigorous boundary-element method. The site consists of four layers of soil with an average shear-wave velocity of 500 m/s resting at a depth of 40 m on bedrock. The shear-wave velocity in the rock is 1000 m/s, just double the value in the soil. Assuming the soil layers to be built in at the interface with the bedrock results in fundamental frequencies of 3.2 Hz and 6.8 Hz in the horizontal and vertical directions. The real and imaginary parts of the dynamic-stiffness coefficient of the reactor building's circular basemat of radius 21 m on the surface of the site are denoted as k and ωc with ω being the frequency. The damping ratios $\omega c/(2k)$ for the horizontal (subscript h) and rocking motions (subscript r) are shown in Fig. 3-34. The fundamental frequency of the coupled structure–soil system is 1.9 Hz, appreciably lower than the (horizontal) cutoff frequency 3.2 Hz. As the shape of the fundamental mode of the reactor building (being a tall structure) consists predominantly of rocking motion, only the material damping ratio of the soil (0.07) is effective. Essentially, no additional energy is dissipated by radiation in the frequency range of the fundamental mode, even though the underlying rock is modeled as a halfspace.

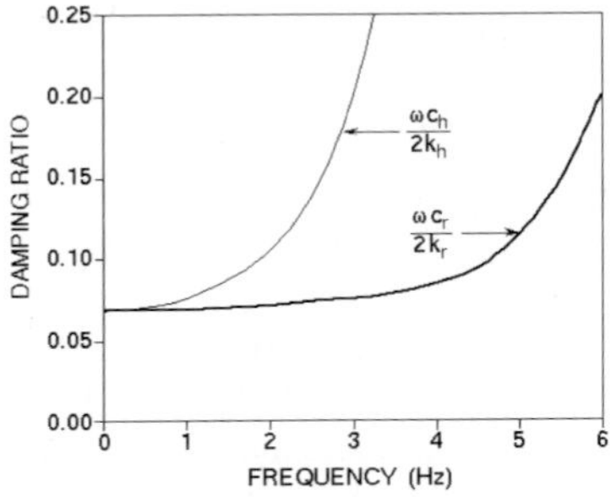

Fig. 3-34 Damping ratios of actual site.

3.8.3 Bar on Elastic Foundation

For translational motion, the layered cone model exhibits an infinite number of natural frequencies where the dynamic-stiffness coefficient vanishes (Fig. 3-9). There exists, however, a very simple model; the bar on elastic foundation (Fig. 3-35a) possesses only a single natural frequency, which is identical to the cutoff frequency. This rod model is much more important than expected, as it represents exactly a single mode of an infinite modal expansion of the layer fixed at its base. The axial motion is denoted as u with A = area, E = modulus of elasticity, and ρ = mass density, from which the rod velocity follows as $c_\ell = \sqrt{E/\rho}$. The spring stiffness per unit length is k_g. This system is analyzed in Section B3.1 by solving its differential equation of motion. In the following, an alternative solution is described, based on wave-propagation techniques. Figure 3-35b shows the left portion of an infinitely long prismatic rod. A longitudinal load p (arising from the spring of the elastic foundation) at some location x induces a stress wave in tension with amplitude $p/2$ which propagates towards infinity and a stress wave in compression with amplitude $p/2$ which propagates in the negative x-direction, reaching the beginning of the rod at time x/c_ℓ later and being reflected at this fixed boundary with double amplitude. The force at the beginning of the rod is therefore given by the sum of the force in the dashpot with coefficient $\rho c_\ell A$ and the reaction forces of all applied longitudinal loads

$$P_0(t) = \rho c_\ell A \dot{u}_0(t) + \int_0^\infty p\left(x, t - \frac{x}{c_\ell}\right) dx \tag{3-94}$$

The distributed load along the rod $p(x, t - x/c_\ell)$ is the product of k_g and the local displacement $u(x, t - x/c_\ell)$

$$P_0(t) = \rho c_\ell A \left(\dot{u}_0(t) + \frac{1}{c_\ell} \frac{k_g}{\rho A} \int_0^\infty u\left(x, t - \frac{x}{c_\ell}\right) dx \right) \tag{3-95}$$

If the rod is regarded as a rigid body vibrating on the elastic foundation, the corresponding natural frequency equals

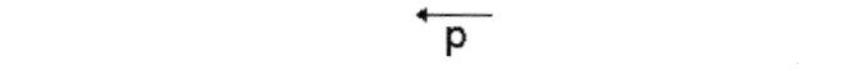

a. Geometry with nomenclature.

b. Wave pattern caused by elastic reaction.

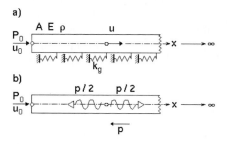

Fig. 3-35 Semi-infinite rod resting on elastic foundation.

$$\omega_c = \sqrt{\frac{k_g}{\rho A}} \tag{3-96}$$

As might be expected, ω_c will turn out to be the single cutoff frequency of the elastic rod on elastic foundation. The Fourier transformation of Eq. 3-95 is evaluated using the time-shifting theorem (Eq. 3-34) leading after substitution of Eq. 3-96 to

$$P_0(\omega) = \rho\, c_\ell A \left(i\omega\, u_0(\omega) + \frac{\omega_c^2}{c_\ell} \int_0^\infty u(x,\omega)\, e^{-i\omega \frac{x}{c_\ell}} dx \right) \tag{3-97}$$

For reasons to be discussed later, there are two sorts of solutions for $u(x,\omega)$ derived in Section B3.1. Below the cutoff frequency the solution is real-valued with decaying amplitude

$$u(x,\omega) = u_0(\omega) e^{-\frac{x}{R}} \tag{3-98}$$

in which R, denoted as the *radius of penetration*, a measure of the rate of decay, is given by

$$R = \frac{\dfrac{c_\ell}{\omega_c}}{\sqrt{1 - \left(\dfrac{\omega}{\omega_c}\right)^2}} \tag{3-99}$$

In the static case $R = c_\ell/\omega_c$. At the cutoff frequency $\omega = \omega_c$ the radius of penetration reaches infinity. Above the cutoff frequency the solution is a complex-valued wave with a constant amplitude

$$u(x,\omega) = u_0(\omega)\, e^{-i\omega \frac{x}{c_a}} \tag{3-100}$$

in which the apparent velocity (phase velocity) is given by

$$c_a = \frac{c_s}{\sqrt{1 - \left(\dfrac{\omega_c}{\omega}\right)^2}} \tag{3-101}$$

At the cutoff frequency $\omega = \omega_c$ the apparent velocity c_a is infinite. As ω increases towards infinity, the apparent velocity decreases asymptotically towards c_ℓ. Substituting for $\omega < \omega_c$ Eq. 3-98 with Eq. 3-99 in Eq. 3-97, or for $\omega > \omega_c$ Eq. 3-100 with Eq. 3-101 in Eq. 3-97, $P_0(\omega)$ is expressed as the dynamic-stiffness coefficient $S(\omega)$ multiplied by $u_0(\omega)$. For instance for $\omega > \omega_c$ the integral in Eq. 3-97 can be evaluated (it vanishes at $x = \infty$) as follows

$$\int_0^\infty u_0(\omega)e^{-i\omega\left(\frac{1}{c_a}+\frac{1}{c_\ell}\right)x}\,dx = u_0(\omega)\lim_{f\to\infty}\int_0^\infty e^{-\left[\frac{1}{f}+i\omega\left(\frac{1}{c_a}+\frac{1}{c_\ell}\right)x\right]}dx$$

$$= u_0(\omega)\lim_{f\to\infty} -\frac{e^{-\frac{x}{f}}e^{-i\omega\left(\frac{1}{c_a}+\frac{1}{c_\ell}\right)x}}{\frac{1}{f}+i\omega\left(\frac{1}{c_a}+\frac{1}{c_\ell}\right)}\Bigg|_0^\infty = \frac{1}{i\omega\left(\frac{1}{c_a}+\frac{1}{c_\ell}\right)}u_0(\omega) \tag{3-102}$$

The dynamic-stiffness coefficient follows as $S(\omega) = P_0(\omega)/u_0(\omega)$ leading to

$$S(\omega) = K\left[k\left(\frac{\omega}{\omega_c}\right)+i\frac{\omega}{\omega_c}\,c\left(\frac{\omega}{\omega_c}\right)\right] \tag{3-103}$$

with the static-stiffness coefficient

$$K = \sqrt{EA\,k_g} \tag{3-104}$$

and the dimensionless frequency

$$\frac{\omega}{\omega_c} = \frac{\omega\, r_0}{c_\ell}\;(= a_0) \tag{3-105}$$

where

$$r_0 = \sqrt{\frac{EA}{k_g}} \tag{3-106}$$

The spring and damping coefficients are equal to

$$k\left(\frac{\omega}{\omega_c}\right) = \sqrt{1-\left(\frac{\omega}{\omega_c}\right)^2} \qquad \text{for } \omega < \omega_c \tag{3-107a}$$

$$c\left(\frac{\omega}{\omega_c}\right) = 0 \tag{3-107b}$$

and

$$k\left(\frac{\omega}{\omega_c}\right) = 0 \qquad \text{for } \omega > \omega_c \tag{3-108a}$$

$$c\left(\frac{\omega}{\omega_c}\right) = \sqrt{1-\left(\frac{\omega_c}{\omega}\right)^2} \tag{3-108b}$$

This is the same result as derived in Section B3.1 (Eq. B-41) and as plotted in Fig. B-5 (solid line). Here ω/ω_c corresponds to a_0. The initial behavior resembles that of the cone model (Fig. 3-9). Because the rod model has only one cutoff frequency, the succession of high-frequency humps in the k and c curves is not observed.

3.8.4 Criterion for Existence of Cutoff Frequency

The rod model shows that if the radius of penetration does not reach infinity, radiation does not take place. The domain is essentially bounded and within a bounded elastic system there is no dissipation of energy. In the absence of radiation, the damping coefficient c is equal to zero (below the cutoff frequency).

For the radius of penetration to reach infinity, it appears that the displacement amplitude must die off in a special way. The appropriate law is derived by a consideration of conservation of radiated energy as the waves pass through successive control surfaces at large radial distances. As shown in Section 2.5.2, the *radiation criterion* (Eq. 2-271) states that the displacement magnitude (amplitude) must decay at infinity $r \to \infty$ in inverse proportion to the square root of the area A of the surface at infinity for radiation of energy to occur

$$|u(r)| \propto \frac{1}{\sqrt{A(r)}} \tag{3-109}$$

In the three-dimensional case the surface at infinity for a layer is a very large cylinder with height d: $A \propto r$ (Fig. 2-42). Thus surface waves (Love waves and generalized Rayleigh waves) of limited depth propagate only when the amplitude decays in proportion to $1/\sqrt{r}$. By contrast, the surface at infinity for the underlying halfspace is a hemisphere with $A \propto r^2$ and propagating body waves must decay as $1/r$. For the rod model investigated in the previous section, the area is constant out to infinity; and therefore the amplitude of propagating waves must also remain constant, as indeed it does (see Eq. 3-100).

These considerations enable a simple determination of whether or not a dynamic system has a cutoff frequency. The static displacement is computed. The *criterion for the existence of a cutoff frequency* states that if *the static displacement decays at infinity more rapidly than the inverse proportion of the square root of the surface at infinity*, $1/\sqrt{A}$, then a cutoff frequency exists. For example, the static displacement of the rod on an elastic foundation is not constant as required by the $1/\sqrt{A}$ law, but decays in proportion to $e^{-x/R}$ (Eq. 3-98).

The most important counterexample of a system without a cutoff frequency is certainly the force-excited disk on the elastic halfspace. According to the classical solution, the static displacement dies off with $1/r$, thus satisfying the $1/\sqrt{A}$ law for the hemispherical surface at infinity. The halfspace radiates energy for all frequencies; its damping coefficient c is essentially constant.

For disks subjected to rotational motion (torsion and rocking) the radiation criterion depends not on the area A, but rather upon the moment of area I of the surface at infinity, which for a cone is proportional to r^4. The $1/\sqrt{A}$ law requires the amplitude of rotation to die off with $1/r^2$. It is shown (Eq. A-54) that for the rotational cone model of the halfspace the rotation is proportional to $1/z^3 + i(\omega/c)/z^2$. In the static case ($\omega = 0$) only the first term remains. The rotation dies off with $1/z^3$ and thus (at very large distances) with $1/r^3$, too rapidly for radiation. On the other hand, for any non-zero frequency the second term dominates at large radial distances, and the radiation criterion is satisfied. Figures 2-20 and 2-21, a plot of the damping coefficients for the torsional and rocking cones, verify that c_ϑ indeed starts from zero but immediately attains positive values.

The cone and bar models indicate that for systems with natural modes of vibration the cutoff frequency is generally equal to the fundamental frequency. At resonance an infinite amount of energy may be stored in the system; the deflection shape is in a sense arbitrary and adjusts itself to correspond to the $1/\sqrt{A}$ law which permits radiation.

SUMMARY

1. For all components of motion, a rigid basemat with equivalent radius r_0 on the surface of a soil layer of depth d resting on rigid rock can be visualized as a folded cone. When unfolded, this layered cone enables a wave pattern to be postulated which incorporates the decay of amplitude as the waves propagate away from the basemat as well as the reflections at rock interface and at the free surface. The computational procedure works exclusively in the familiar time domain. The aspect ratio z_0/r_0 (opening angle) of the unfolded layered cone is the same as that of the truncated semi-infinite (unlayered) cone used to model a disk on a homogeneous halfspace with the same material properties as the layer.
2. From the wave pattern it follows that the translation $u_0(t)$ or rotation $\vartheta_0(t)$ of a basemat on a layer is equal to that of the same basemat on a homogeneous halfspace $\bar{u}_0(t)$ or $\bar{\vartheta}_0(t)$ (generating function), augmented by echoes of the previous response. The appropriate echo constants are derived for the flexibility formulation, then inverted to obtain the echo constants of the stiffness formulation.
The echo formula to be used in a flexibility formulation for translation equals

$$u_0(t) \sum_{j=0}^{k} \overset{F}{e}_j \, \bar{u}_0(t-jT)$$

with the flexibility echo constants

$$\overset{F}{e}_0 = 1$$

$$\overset{F}{e}_j = \frac{2(-1)^j}{1+j\kappa} \qquad j \geq 1$$

and

$$T = \frac{2d}{c} \qquad \kappa = \frac{2d}{z_0}$$

and the appropriate wave velocity c and aspect height z_0. The integer k is equal to the largest index j for which the argument $t - jT$ of $\bar{u}_0$ is positive.
The echo formula to be used in a flexibility formulation for rotation equals

$$\vartheta_0(t) = \sum_{j=0}^{k} \overset{F}{e}_{0j} \, \bar{\vartheta}_0(t-jT) + \sum_{j=0}^{k} \overset{F}{e}_{1j} \, \bar{\vartheta}_1(t-jT)$$

with the flexibility echo constants

$$\overset{F}{e}_{00} = 1$$

$$\overset{F}{e}_{0j} = \frac{2(-1)^j}{(1+j\kappa)^2} \qquad j \geq 1$$

$$\overset{F}{e}_{10} = 0$$

$$\overset{F}{e}_{1j} = 2(-1)^j \left[\frac{1}{(1+j\kappa)^3} - \frac{1}{(1+j\kappa)^2} \right] \qquad j \geq 1$$

and

$$\bar{\vartheta}_1(t) = \int_0^t h_1(t-\tau)\, \bar{\vartheta}_0(\tau)\, d\tau$$

with the unit-impulse response function

$$h_1(t) = \frac{c}{z_0} e^{-\frac{c}{z_0}t} \qquad t \geq 0$$

$$= 0 \qquad t < 0$$

Alternatively, using pseudo-echo constants $\overset{F}{e}_{rm}$ (influence functions) which follow from $\overset{F}{e}_{0j}$ and $\overset{F}{e}_{1j}$ (Table 3-2) the echo formula for rotation equals

$$\vartheta_0(t) = \sum_{m=0}^{k} \overset{F}{e}_{rm}\, \bar{\vartheta}_0(t - m\Delta t)$$

with the time step Δt.

The echo formula to be used in a stiffness formulation for translation equals

$$\bar{u}_0(t) = \sum_{j=0}^{k} \overset{K}{e}_j\, u_0(t - jT)$$

with the stiffness echo constants

$$\overset{K}{e}_0 = 1$$

$$\overset{K}{e}_j = -\sum_{\ell=0}^{j-1} \overset{K}{e}_\ell\, \overset{F}{e}_{j-\ell} \qquad j \geq 1$$

and for rotation

$$\bar{\vartheta}_0(t) = \sum_{m=0}^{k} \overset{K}{e}_{rm}\, \vartheta_0(t - m\Delta t)$$

with the stiffness echo constants

$$e^{K}_{r0} = 1$$

$$e^{K}_{rm} = -\sum_{\ell=0}^{m-1} e^{K}_{r\ell}\, e^{F}_{rm-\ell} \qquad m \geq 1$$

3. In a flexibility formulation, in the first step, the prescribed interaction force $P_0(t)$ or moment $M_0(t)$ is applied to the cone modeling a disk on the associated homogeneous halfspace with the same material properties as the layer, resulting in the surface displacement $\bar{u}_0(t)$ or rotation $\bar{\vartheta}_0(t)$. In the second step, the displacement $u_0(t)$ or rotation $\vartheta_0(t)$ of the basemat on the layer follows from $\bar{u}_0(t)$ or $\bar{\vartheta}_0(t)$ using the corresponding echo formula.
4. In a stiffness formulation, in the first step, the prescribed surface displacement $u_0(t)$ or rotation $\vartheta_0(t)$ is converted to the displacement $\bar{u}_0(t)$ or rotation $\bar{\vartheta}_0(t)$ of a disk on the associated homogeneous halfspace using the corresponding echo formula. In the second step, insertion into the interaction force-displacement relationship of the cone modeling a disk on the associated homogeneous halfspace leads to the interaction force $P_0(t)$ or moment $M_0(t)$ acting on the basemat.
5. In both the flexibility and stiffness formulations, the two steps that are linear operators can be interchanged. For instance, in this alternative order of realization, in the stiffness formulation, the soil is first treated as a homogeneous halfspace, leading to the corresponding interaction force; and second, the effect of the layer is introduced. A comparison of the final result to that after the first step reveals the influence of the layer on the interaction force.
6. The dynamic-stiffness coefficients for harmonic loading based on the layered cone models are equal to for translation

$$S(\omega) = K\frac{1 + i\frac{\omega T}{\kappa}}{1 + 2\sum_{j=1}^{\infty}(-1)^j \frac{e^{-ij\omega T}}{1+j\kappa}}$$

and for rotation

$$S_\vartheta(\omega) = K_\vartheta \frac{1 - \frac{1}{3}\frac{(\omega T)^2}{\kappa^2+(\omega T)^2} + i\frac{\omega T}{3\kappa}\frac{(\omega T)^2}{\kappa^2+(\omega T)^2}}{1 + \frac{2}{1+i\frac{\omega T}{\kappa}}\left(\sum_{j=1}^{\infty}(-1)^j \frac{e^{-ij\omega T}}{(1+j\kappa)^3} + i\frac{\omega T}{\kappa}\sum_{j=1}^{\infty}(-1)^j\frac{e^{-ij\omega T}}{(1+j\kappa)^2}\right)}$$

with the static-stiffness coefficients of the homogeneous halfspace with the material properties of the layer K, K_ϑ.

These dynamic-stiffness coefficients are exact in the high-frequency limit.
7. The unfolded cone used to model a basemat on a soil layer resting on rigid rock summarized above can be generalized to the case of a layer on flexible rock. The only modification consists of replacing in the expressions for the echo constants the reflection coefficient associated with the rigid rock, -1, by the corresponding value $-\alpha$ for

the flexible rock. All flexibility and stiffness formulations for the dynamic analysis remain valid. This layered cone represents a wave pattern whose amplitude decays with distance, and the reflection at the free surface and the reflection and refraction at the layer-rock interface are taken into consideration.

The reflection coefficient $-\alpha$ can be determined based on one-dimensional wave propagation along the cone for the first impingement at the layer-rock interface. In general, the coefficient $-\alpha$ is frequency dependent and can thus only be used in a frequency-domain analysis. However, the high-frequency and the low-frequency limits are frequency independent. For the familiar time domain analysis the latter reflection coefficient, corresponding to the static case, should be used:

translational cone

$$-\alpha(\omega) = \frac{i\omega(\rho_L c_L - \rho_R c_R) + \dfrac{\rho_L c_L^2}{z_0^L + d} - \dfrac{\rho_R c_R^2}{z_0^R}}{i\omega(\rho_L c_L + \rho_R c_R) + \dfrac{\rho_L c_L^2}{z_0^L + d} + \dfrac{\rho_R c_R^2}{z_0^R}}$$

rotational cone

$$-\alpha(\omega) = \frac{-\dfrac{1 + i\frac{\omega}{c_L}\left(z_0^L + d\right)}{1 + i\frac{\omega}{c_R} z_0^R} + \dfrac{3 + 3i\frac{\omega}{c_L}\left(z_0^L + d\right) + \left(\frac{i\omega}{c_L}\right)^2 \left(z_0^L + d\right)^2}{3 + 3i\frac{\omega}{c_R} z_0^R + \left(\frac{i\omega}{c_R}\right)^2 z_0^{R2}} \cdot \dfrac{\rho_L c_L^2 z_0^R}{\rho_R c_R^2 \left(z_0^L + d\right)}}{\dfrac{1 + i\frac{\omega}{c_L}\left(z_0^L + d\right)}{1 + i\frac{\omega}{c_R} z_0^R} + \dfrac{3 + 3i\frac{\omega}{c_L}\left(z_0^L + d\right) + \left(\frac{i\omega}{c_L}\right)^2 \left(z_0^L + d\right)^2}{3 + 3i\frac{\omega}{c_R} z_0^R + \left(\frac{i\omega}{c_R}\right)^2 z_0^{R2}} \cdot \dfrac{\rho_L c_L^2 z_0^R}{\rho_R c_R^2 \left(z_0^L + d\right)}}$$

L and R are indices referring to the layer and the rock, ρ is the density:
high-frequency limit

$$-\alpha(\infty) = \frac{\rho_L c_L - \rho_R c_R}{\rho_L c_L + \rho_R c_R}$$

low-frequency limit

$$-\alpha(0) = \frac{\dfrac{\rho_L c_L^2}{z_0^L + d} - \dfrac{\rho_R c_R^2}{z_0^R}}{\dfrac{\rho_L c_L^2}{z_0^L + d} + \dfrac{\rho_R c_R^2}{z_0^R}}$$

for $\nu^L = \nu^R$

$$-\alpha(0) = \frac{\rho_L c_L^2 - \rho_R c_R^2}{\rho_L c_L^2 + \rho_R c_R^2}$$

8. As an alternative to the layered cone model, the basic lumped-parameter model can be used to represent the dynamic behavior for all components of motion of a basemat on the surface of a soil layer on rigid rock. The stable lumped-parameter model with two additional internal degrees of freedom consists of four springs, three dashpots, and one mass whose real coefficients are specified for various ratios of the radius of the disk r_0 to the depth d of the layer and Poisson's ratio ν for the horizontal, vertical, rocking, and torsional motions in Table 3-8. These coefficients are determined in a consistent procedure by approximating the exact dynamic-stiffness coefficient by a ratio of two polynomials in frequency, which is the dynamic-stiffness coefficient of the lumped-parameter model. The lumped-parameter model can be used to represent the soil in a standard finite-element program for structural dynamics in the time domain.
9. The results obtained with the layered cone model and the basic lumped-parameter model are in remarkable agreement with rigorous solutions. The layered cone model applied in a flexibility formulation reproduces the jump discontinuities present in the unit-impulse response function of a disk on the surface of a layer on rigid rock very accurately. Both the layered cone model and—by construction—the lumped-parameter model capture the phenomenon of the cutoff frequency.
10. A basemat on the surface of a soil layer on rigid rock exhibits a cutoff-frequency, below which no energy is dissipated by radiation; the damping coefficient of the dynamic-stiffness coefficient vanishes. The cutoff frequency f_c in the engineering sense depends on the dominant distortion for the motion of the basemat and is equal to the fundamental frequency of the layer for the same distortion.

$$f_c = \frac{c_s}{4d} \qquad \text{horizontal and torsional}$$

$$f_c = \frac{c_p}{4d} \qquad \text{vertical and rocking}$$

If the fundamental frequency of a coupled structure–soil system lies beneath the cutoff frequency, dynamic excitation will cause a very large response.

In order for waves to penetrate to infinity and transmit energy, the displacement amplitude (magnitude) must decay at infinity in inverse proportion to the square root of the area of the surface at infinity (radiation criterion)

$$|u| \propto \frac{1}{\sqrt{A}}$$

The criterion for the existence of a cutoff frequency states that if the static displacement decays at infinity more rapidly than the inverse proportion of the square root of the area of the surface at infinity, then a cutoff frequency exists.

4

Embedded Foundation and Pile Foundation

The dynamic response of a rigid massless cylindrical foundation embedded in a homogeneous soil halfspace (Fig. 4-1) and in a soil layer resting on rigid or flexible rock for all motions is addressed in this chapter. A pile foundation which can be regarded as an embedded foundation with the embedded domain of large embedment height occupied by the pile is also examined. Material damping of the soil is not addressed explicitly.

The concept of cone models is extended to embedded cylindrical foundations. The building block of the procedure consists of a rigid disk in the embedded domain, which is represented by a double-cone model within the elastic fullspace. Its displacement field defines an approximate Green's function for use in a matrix formulation of structural mechanics based on the familiar force method. This methodology points towards a general strength-of-materials approach to foundation dynamics. The procedure can also be interpreted as an uncomplicated version of the boundary-element method. For the vertical motion Fig. 4-2b shows one such embedded disk with the corresponding double-cone model. It will be demonstrated that a system of equations of the order of the number of embedded disks has to be solved. Although still quite simple when compared to the rigorous procedures of three-dimensional elastodynamics, the analysis of an embedded founda-

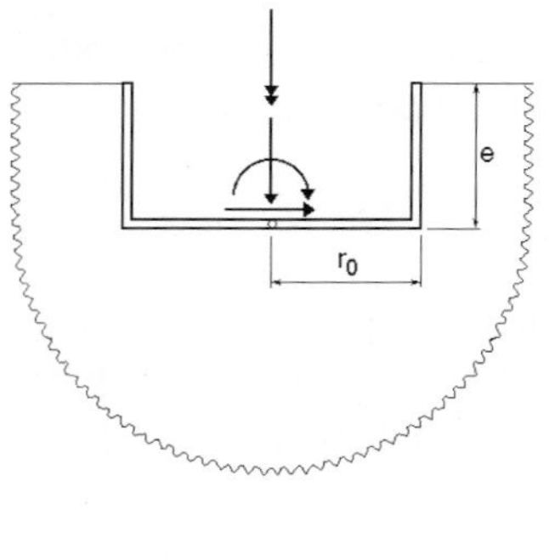

Fig. 4-1 Cylindrical foundation embedded in soil halfspace.

tion is more complicated that the calculation of a surface foundation discussed in Chapters 2 and 3. Hand calculations are not feasible anymore. Unfortunately, it is not possible to extend the concept of a single truncated semi-infinite cone to an embedded foundation. Poor results follow, for example, when for the vertical motion (Fig. 4-2a) the base is modeled with the familiar cone with a vertical axis exhibiting axial distortion and the sidewall by another cone with a horizontal axis subjected to shear distortion. This limitation of the simplest form of the cone model is discussed in Section 4.5.

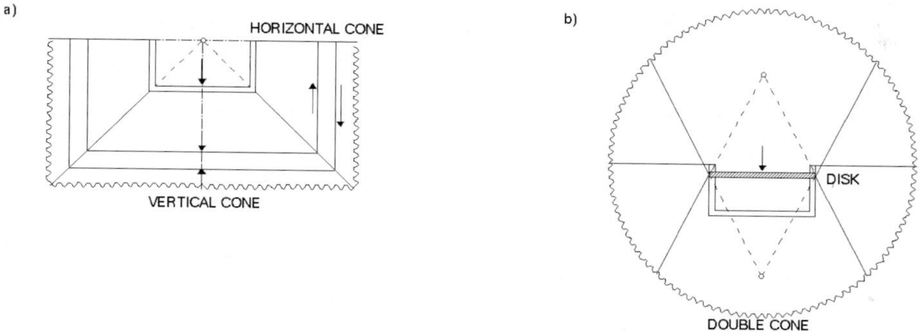

a. Modeled as cone with vertical axis and axial distortion, and cone with horizontal axis and shear distortion.

b. Modeled as embedded disks with corresponding double cone with axial distortion in full space.

Fig. 4-2 Vertical motion of embedded foundation.

Section 4.1 addresses the embedded disk in a fullspace with the double cone model and its Green's function. Using symmetry and anti-symmetry considerations, the free surface of a soil and the fixed boundary of a layer resting on rigid rock can be modeled. The interface between the soil layer and an underlying flexible rock halfspace can be represented too. Section 4.2 discusses the matrix formulation to determine the dynamic-stiffness coefficients for harmonic loading (frequency domain) of a foundation embedded in a halfspace based on disk models. Section 4.3 examines the corresponding formulation to establish the interaction force-displacement relationship in the time domain. Section 4.4 contains the derivation of the case of a foundation embedded in a soil layer resting on rigid or flexible rock based on disk models. Sections 4.6 and 4.7 address the lumped-parameter models to represent the foundation embedded in a halfspace and in a layer. Section 4.8 expands the matrix formulation for harmonic loading to a single pile. Section 4.9 introduces dynamic-interaction factors based on one-dimensional wave propagation and treats the analysis for a pile group including pile–soil–pile interaction. All these sections contain examples to demonstrate the accuracy.

Appendix D describes an extension for harmonic loading to model a foundation embedded in a stratified site with many layers resting on an underlying homogeneous halfspace based on embedded disks with the corresponding double cones.

The concept of using embedded disks with the corresponding double-cone models to analyze foundations embedded in a halfspace is described by Meek and Wolf [M11] in a study on which this chapter is based. Other aspects are discussed elsewhere [W12, W13, and W5].

4.1 EMBEDDED DISK WITH DOUBLE-CONE MODEL AND ITS GREEN'S FUNCTION

4.1.1 One-Sided Cone

A rigid disk of equivalent radius r_0 on the surface of a halfspace can be modeled for all motions with the one-sided cone model shown in Fig. 4-3a. The properties of this cone model, which is discussed in depth in Chapter 2, are summarized in Table 2-3A. In particular, the apex height z_0 is selected so that the static-stiffness coefficient K of the truncated semi-infinite one-sided cone matches that of the disk on a halfspace. The Green's functions in the time domain and for harmonic loading (frequency domain) at a distance a are specified for the translational cone in Eq. A-32 and Eq. A-37 and for the rotational cone in Eq. A-90 and Eq. A-94.

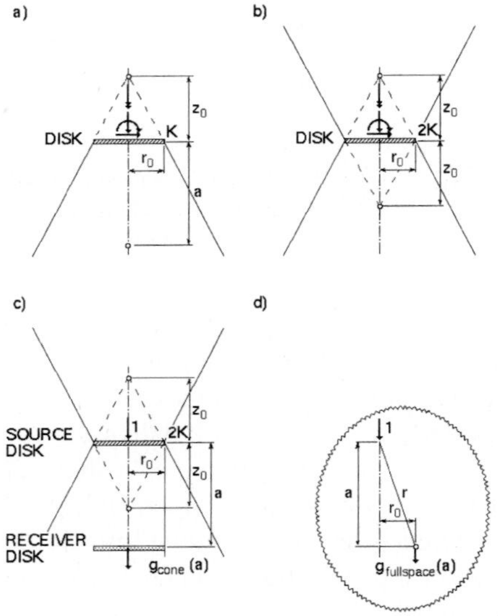

a. Disk on surface of halfspace with one-sided cone model.

b. Disk embedded in fullspace with double-cone model.

c. Green's function based on disk embedded in fullspace with double-cone model.

d. Green's function for vertical motion based on point load in fullspace.

Fig. 4-3 Disk with corresponding cone.

4.1.2 Double Cone

Construction and static-stiffness coefficient. Suppose that, as shown in Fig. 4-3b, a *disk* of radius r_0 is not mounted at the surface of a halfspace, but is instead deeply *embedded within an infinite fullspace*. This situation may be approximated for all motions by a *double-cone model with the same proportions* z_0/r_0 as for the one-sided cone-model in Fig. 4-3a. Addressing the vertical motion, by anti-symmetry the applied load is resisted half in tension (upper cone) and half in compression (lower cone). Stated differently, the stiffness of the surface cone is exactly doubled. The same also applies for the other motions. *The static-stiffness coefficient of the double-cone model is twice that of the* *one-sided cone* and thus is equal to $2K$. The dimensionless spring and damping coefficients $k(a_0)$ and $c(a_0)$ remain unchanged.

Based on the rigorous solution of elastostatics, the exact static-stiffness coefficients of a disk embedded in a fullspace equals

$$K_{exact} = \frac{64(1-\nu)}{7-8\nu} G r_0 \qquad \text{horizontal} \tag{4-1a}$$

$$K_{exact} = \frac{32(1-\nu)}{3-4\nu}) G r_0 \qquad \text{vertical} \tag{4-1b}$$

$$K_{exact} = \frac{64(1-\nu)}{3(3-4\nu)} G r_0^3 \qquad \text{rocking} \tag{4-1c}$$

$$K_{exact} = \frac{32}{3} G r_0^3 \qquad \text{torsional} \tag{4-1d}$$

To investigate the error introduced by using the double-cone model (with the apex ratio z_0/r_0 determined by calibration with the halfspace), the disk's ratio of the static-stiffness coefficient of the double-cone model to that of the exact solution (Eq. 4-1) is calculated for all motions. The static-stiffness coefficients of the double cone are equal to twice the values specified in Table 2-2.

$$\frac{2K}{K_{exact}} = \frac{7-8\nu}{4(2-\nu)(1-\nu)} \qquad \text{horizontal} \tag{4-2a}$$

$$\frac{2K}{K_{exact}} = \frac{3-4\nu}{4(1-\nu)^2} \qquad \text{vertical} \tag{4-2b}$$

$$\frac{2K}{K_{exact}} = \frac{3-4\nu}{4(1-\nu)\nu}{}^{2}) \qquad \text{rocking} \tag{4-2c}$$

$$\frac{2K}{K_{exact}} = 1 \qquad \text{torsional} \tag{4-2d}$$

These ratios for all degrees of freedom equal 1 when $\nu = 0.5$, typical for saturated soil. For $\nu = 0.25$, a lower limit for dry soil, $2K/K_{exact} = 0.888$ for the vertical and rocking motions and equals 0.952 for the horizontal motion.

Green's function. The key aspect in calculating the dynamic behavior of embedded and pile foundations consists in determining the Green's function for a loaded rigid disk embedded in a fullspace. The Green's function in the time domain is defined as the displacement or rotation as a function of time at the receiver point located at a distance a away from the loaded source disk, which is excited at time equals 0 by a unit-impulse force (see Sections A1.4 and A2.4). The Green's function for the displacement $g(a, t - a/c)$ or rotation $g_\vartheta\,(a, t - a/c)$ is a function of a, reflecting the diminishing of the wave after traveling this distance, and of the retarded time $t - a/c$ with c denoting the appropriate wave propagation velocity (Table 2-3A). The quantity a, being a distance, is always positive.

To calculate the dynamic behavior of an *embedded foundation* where the embedment e is of the same order of magnitude as the radius r_0 (Fig. 4-1), the Green's functions of the double cone corresponding to a rigid disk can be used for all motions (Fig. 4-3c). They are equal to the Green's functions of the one-sided cone, doubling the static-stiffness coefficients (Eqs. A-32 and A-37 for the translational cone, Eqs. A-90 and A-94 for the rotational cone).

translational double cone
time domain

$$g\left(a,t-\frac{a}{c}\right)=\frac{1}{2K}\frac{1}{1+\dfrac{a}{z_0}}h_1\left(t-\frac{a}{c}\right) \tag{4-3}$$

harmonic loading (frequency domain)

$$g(a,\omega)=\frac{1}{2K}\frac{1}{1+\dfrac{a}{z_0}}\frac{e^{-i\frac{\omega a}{c}}}{1+i\dfrac{\omega z_0}{c}} \tag{4-4}$$

rotational double cone
time domain

$$g_\vartheta\left(a,t-\frac{a}{c}\right)=\frac{1}{2K_\vartheta}\frac{1}{\left(1+\dfrac{a}{z_0}\right)^2}\left[h_2\left(t-\frac{a}{c}\right)+\left(-1+\frac{1}{1+\dfrac{a}{z_0}}\right)h_3\left(t-\frac{a}{c}\right)\right] \tag{4-5}$$

harmonic loading (frequency domain)

$$g_\vartheta(a,\omega)=\frac{3}{2K_\vartheta}\left(\frac{1}{\left(1+\dfrac{a}{z_0}\right)^3}+i\frac{\omega z_0}{c}\frac{1}{\left(1+\dfrac{a}{z_0}\right)^2}\right)\frac{1}{3+3i\dfrac{\omega z_0}{c}+\left(i\dfrac{\omega z_0}{c}\right)^2}e^{-i\frac{\omega a}{c}} \tag{4-6}$$

The unit-impulse response functions $h_1(t)$, $h_2(t)$, and $h_3(t)$ are specified in Eqs. A-24, A-81, and A-89.

For a *pile foundation* the same Green's functions can be used for the horizontal, rocking, and torsional motions. For these degrees of freedom the active length within which the pile displacements or rotations are appreciable is of the same order of magnitude as the radius r_0 of the pile and thus of the embedment e of a typical embedded foundation. For the vertical motion of a pile (with an embedment e which is one order of magnitude larger than r_0), however, the active length is much larger than in the horizontal direction. Use of the Green's function of the double cone would lead to an over-flexible result for the vertical degree of freedom. This can be verified by addressing the static case. For long piles, the vertical displacement is also needed at large distances from the disk subjected to a vertical load. As a limit, the Green's function of the double cone with axial distortion at an infinite distance a from the disk for the static case follows from Eq. 4-4 with $\omega = 0$ and z_0/r_0 from Table 2-3A as

$$g_{\text{cone}}\,(a\to\infty)=\frac{\pi(1-\nu)^3}{16(1-2\nu)G}\frac{1}{a} \tag{4-7}$$

For this limit the Green's function should be equal to the Kelvin solution $g_{\text{fullspace}}\,(a\to\infty)$, defined as the static displacement in the direction of a unit load acting in an infinite full-space.

$$g_{\text{fullspace}}\,(a\to\infty)=\frac{1}{4\pi Ga} \tag{4-8}$$

From the ratio

$$\frac{g_{\text{cone}}(a \to \infty)}{g_{\text{fullspace}}\ (a \to \infty)} = \frac{\pi^2(1-\nu)^3}{4(1-2\nu)} \tag{4-9}$$

which equals 2.082 for $\nu = 0.25$ and equals 2.665 for $\nu = 0.4$. It follows that the *double-cone model for the vertical degree of freedom is too flexible at large distances from the loaded disk.* The corresponding Green's function has to be modified, as described in the next paragraph for harmonic excitation.

The *modified Green's function for vertical motion*—the axial displacement amplitude at distance a from the loaded disk (Fig. 4-3c)—*is equal to the weighted Green's functions of the double cone and of the fullspace for a point load.*

$$g(a,\omega) = w(a)\, g_{\text{cone}}(a,\omega) + \big(1 - w(a)\big)\, g_{\text{fullspace}}(a,\omega) \tag{4-10}$$

where $g_{\text{cone}}(a,\omega)$ is the Green's function for the double-cone model (Eq. 4-4, Fig. 4-3c) and where $g_{\text{fullspace}}\ (a,\omega)$ is the Green's function for a point load offset by r_0 from the axis of the cone (Fig. 4-3d)

$$g_{\text{fullspace}}\ (a,\omega) = \frac{1}{4\pi G}\left[\psi - \left(\frac{a}{r}\right)^2 \chi\right] \tag{4-11}$$

with

$$\psi = \left(1 - i\frac{1}{a_0} - \frac{1}{a_0^2}\right)\frac{e^{-ia_0}}{r} + \left(i\frac{c_s}{c_p a_0} + \frac{1}{a_0^2}\right)\frac{e^{-i\frac{c_s}{c_p}a_0}}{r} \tag{4-12a}$$

$$\chi = \left(1 - i\frac{3}{a_0} - \frac{3}{a_0^2}\right)\frac{e^{-ia_0}}{r} - \left(\frac{c_s^2}{c_p^2} - i\frac{3c_s}{c_p a_0} - \frac{3}{a_0^2}\right)\frac{e^{-i\frac{c_s}{c_p}a_0}}{r} \tag{4-12b}$$

and

$$r = \sqrt{a^2 + r_0^2} \tag{4-12c}$$

$$a_0 = \frac{\omega r}{c_s} \tag{4-12d}$$

Note that in this definition of a_0 used in specifying the Green's function of the fullspace the characteristic length is the distance r between the source and receiver points (Fig. 4-3d) and not the radius r_0 of the disk. The weighting function $w(a)$ equals

$$w(a) = 1 \qquad \text{for } a \le r_0 \tag{4-13a}$$

$$= e^{-0.8\frac{a-r_0}{r_0}} \qquad \text{for } a > r_0 \tag{4-13b}$$

The farther away the receiver disk is from the source disk (the larger a is), the more important the Green's function of the fullspace becomes.

4.1.3 Embedded Disk with Anti-Symmetric and Symmetric Mirror Disks To Model Free Surface and Fixed Boundary of Halfspace

The concept of the double-cone model representing a disk embedded in a fullspace can be expanded to a disk embedded in a halfspace with either a free boundary surface (standard halfspace with free surface) or a fixed boundary surface.

The halfspace with a free surface is addressed first. The left-hand side of Fig. 4-4a shows a rigid disk as an example in vertical motion, embedded a finite distance e beneath the surface of an elastic halfspace. The rigorous analysis of this situation is a quite complicated three-dimensional problem of elastodynamics. By working with the one-dimensional double-cone model, however, it is easy to find an accurate approximate solution. The key idea is to exploit the anti-symmetry apparent on the right-hand side of Fig. 4-4a. Two identical disks, the *disk and the mirror-image disk*, in the fullspace with their double cones are considered, separated by the distance $2e$ and *excited simultaneously by identical time histories of force*. The middle plane between these two disks coincides with the free surface.The tension waves propagating up from the lower disk arrive at the middle plane at the same time as the corresponding compression waves traveling down from the upper mirror-image disk. The stresses cancel perfectly on the middle plane of the fullspace, thus *satisfying the stress-free boundary condition implicit at the free surface* on the left-hand side of the figure. It is easy to verify that the middle plane is also stress-free for horizontal, rocking, and torsional motions. The Green's function at a distance a for a loaded disk embedded in a halfspace is thus equal to the sum of the Green's functions corresponding to the two loaded disks in the fullspace modeled with double cones. For the disk, the applicable distance is equal to a, and the retarded time equals $t - a/c$ leading to $g(a,t - a/c)$ as specified in Section 4.1.2; for the mirror-image disk, $a' = a + 2e$ and $t - a'/c$ are used resulting, after applying the same equation, in $g(a', t - a'/c)$. For instance, to determine the dynamic-stiffness coefficient for harmonic loading in translational motion of a disk embedded a distance e beneath the free surface of a halfspace, the dynamic-flexibility coefficient of Eq. 4-4 is formulated with $a = 0$ and is added to that with $a = 2e$, and the result is then inverted.

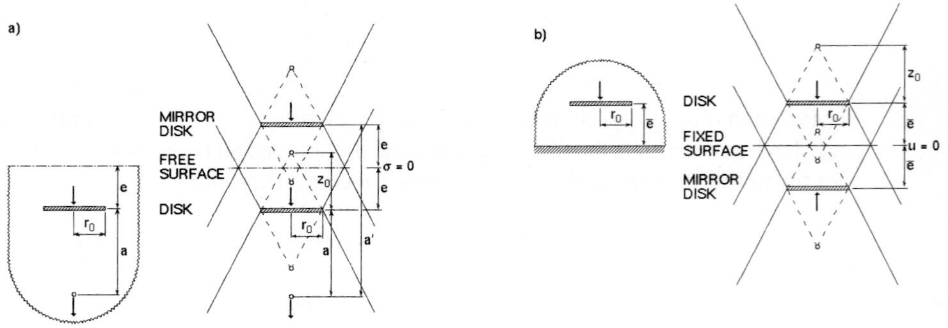

a. Disk embedded in halfspace with free boundary with anti-symmetrically loaded mirror-image disk to represent free surface.

b. Disk embedded in halfspace with fixed boundary with symmetrically loaded mirror-image disk to represent fixed boundary.

Fig. 4-4 Modeling of free and fixed boundaries.

To evaluate the accuracy of this procedure, the vertical static Green's function for $a = 0$—that is, the static-flexibility coefficient of a disk with the embedment ratio e/r_0—for the halfspace is compared to the rigorous result [P2]. In Fig. 4-5a, the vertical static-flexibility coefficients $F(e, 0)$ normalized by the corresponding value for a surface disk $F(0,0) = (1 - \nu)/(4Gr_0)$ are plotted for various values of ν. For small ν (= 0.25) and $e/r_0 > 2$, an observable deviation exists when z_0/r_0 is calculated using the static-stiffness coefficient of the *half*space for calibration. Nevertheless, this procedure is selected for two reasons. First, it guarantees a high accuracy for disks close to the free surface. Second, the calibration based on the halfspace can also be applied for large embedment, as in a pile foundation, when ν is large (Fig. 4-5a). In the following, all analyses with cones are based on the calibration with the halfspace.

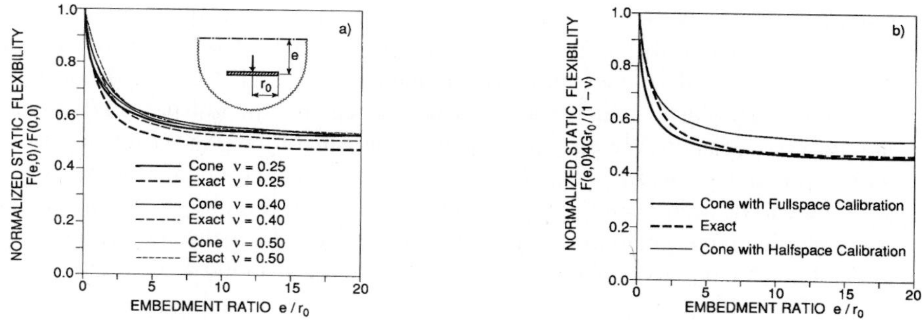

a. Apex height determined through calibration with static-stiffness coefficient of halfspace.

b. Apex height determined through calibration with static-stiffness coefficient of fullspace (Poisson's ratio = 0.25).

Fig. 4-5 Static-flexibility coefficient of rigid disk embedded in halfspace.

As an alternative, z_0/r_0 could be determined by equating the static-stiffness coefficient of the double cone to that of the *full*space. For the vertical motion this leads to (Table 2-3A, Eq. 4-1b)

$$2\frac{\rho c^2 \pi r_0^2}{z_0} = \frac{32(1-\nu)}{3-4\nu}\rho c_s^2 r_0 \tag{4-14}$$

which yields

$$\frac{z_0}{r_0} = \frac{\pi(3-4\nu)}{16(1-\nu)}\frac{c^2}{c_s^2} \tag{4-15a}$$

For $\nu \leq 1/3$, $c = c_p$, which results in

$$\frac{z_0}{r_0} = \frac{\pi(3-4\nu)}{8(1-2\nu)} \tag{4-15b}$$

For $\nu = 0.25$ the static-flexibility coefficient $F(e,0)$ normalized by the exact value $= (1-\nu)/(4Gr_0)$ is shown in Fig. 4-5b for the two calibrations. As expected, the cone with the fullspace calibration performs well for large e/r_0, but the accuracy deteriorates for a shallow embedment. For $e/r_0 = 0$ (for a surface foundation), the error equals 12.5%, as already described in connection with the inverse of Eq. 4-2b. The double-cone model based on the calibration with the fullspace should be used for large e/r_0 and small ν as when representing a tunnel with a large overburden in drained soil or in rock. Such cases are not treated in this book.

Turning to the dynamic case for harmonic loading, the vertical dynamic-flexibility coefficient $F(e,a_0)$ normalized by the static-surface flexibility coefficient $F(0,0)$ is plotted versus the dimensionless frequency $a_0 = \omega r_0/c_s$ for embedment ratios $e/r_0 = 2$ and 5 in Figs. 4-6 and 4-7. The real and imaginary parts are presented. The exact results are again taken from Pak and Gobert [P2].

It can be concluded that the displacement amplitudes for the disk embedded in a halfspace, calculated using a mirror-image disk and with one-dimensional double-cone models to represent the fullspace are quite accurate also for large embedment ratios.

The halfspace with a fixed surface is examined next. The left-hand side of Fig. 4-4b presents a rigid disk, in vertical motion embedded in a halfspace at a distance $\bar{e}$ from the fixed boundary. Again a *mirror-image disk* embedded in the fullspace and modeled with a double cone located at the distance $\bar{e}$ on the other site of the fixed boundary is introduced, which is *loaded by the same time history of the force* as the original disk, but with the *other sign*. As can be seen from the right-hand side of Fig. 4-4b, the geometry and loading are symmetric with respect to the middle plane of the fullspace, which coincides with the *fixed boundary*, yielding a vanishing displacement. The other motions are treated analogously.

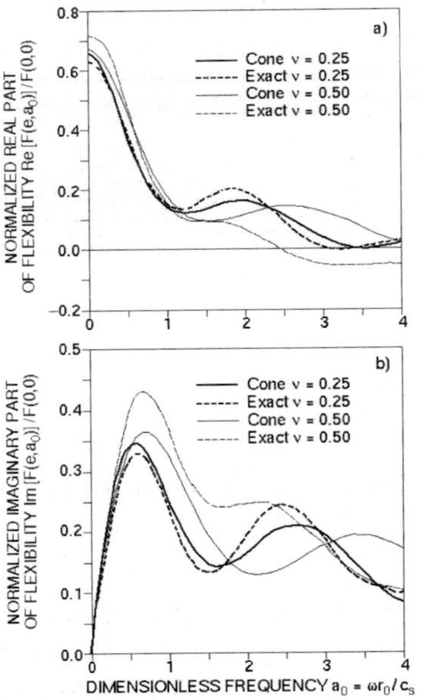

Fig. 4-6 Dynamic-flexibility coefficient for harmonic loading of disk embedded in halfspace for embedment ratio = 2.

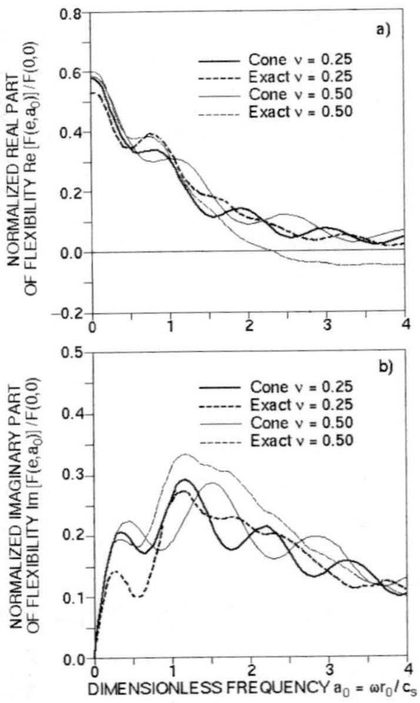

Fig. 4-7 Dynamic flexibility coefficient for harmonic loading of disk embedded in halfspace for embedment ratio = 5.

The Green's function for a loaded disk embedded in a halfspace with a fixed boundary surface again follows as the sum of the Green's functions of the disk and the mirror-image disk.

The case of a disk embedded in a layer with a free surface and a fixed boundary is addressed in the next Section 4.1.4.

4.1.4 Embedded Disk with Anti-Symmetric and Symmetric Mirror Disks To Model Free Surface and Fixed Boundary of Layer

The Green's function in the time domain and thus the wave pattern of a loaded disk embedded in a layer is to be determined. *To model the boundary surfaces of a soil layer resting on rigid rock, the concept of using anti-symmetrically loaded mirror-image disks to represent the free surface and of symmetrically loaded mirror-image disks for the fixed boundary is applied repeatedly.*

First, the case of a disk on the surface of a layer is addressed. Figure 4-8 illustrates the vertical motion, but the concept also works for the horizontal, rocking, and torsional motions. The disk is denoted with the number 1. To enforce the boundary condition on the free surface of the layer, a mirror-image disk embedded in the fullspace with the number 2, separated by an infinitesimal distance, loaded simultaneously with the same unit-impulse load and with the same sign, is introduced as described in Section 4.1.3. The corresponding displacement at depth z occurring from the unit-impulse loads on disks 1 and 2 equals with the distance a between the source disks 1, 2 and the receiver point being z (Fig. 4-8) in Eq. 4-3

$$2\frac{1}{2K}\frac{1}{1+\frac{z}{z_0}}h_1\left(t-\frac{z}{c}\right) \tag{4-16}$$

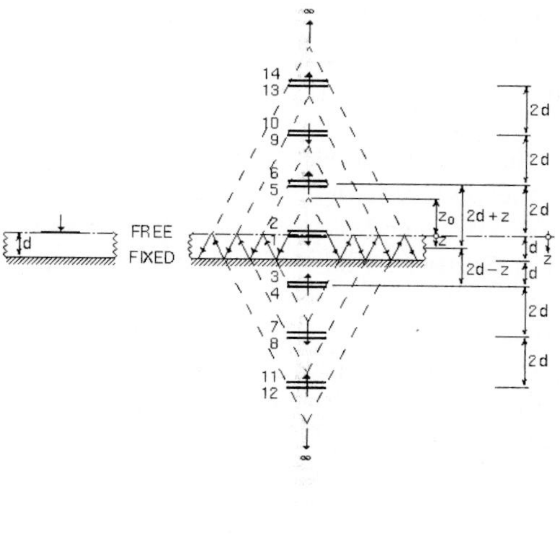

Fig. 4-8 Arrangement of mirror-image disks embedded in fullspace with loads to model disk on surface of soil layer resting on rigid rock.

From the definition of $h_1(t - z/c)$ specified in Eq. 2-76, it follows that for a unit-impulse load $P_0(t)$, $h_1(t - z/c)/K$ can be replaced by the displacement in the halfspace $\bar{u}_0(t - z/c)$, related to the generating function $\bar{u}_0(t)$, yielding

$$\frac{1}{1+\frac{z}{z_0}}\bar{u}_0\left(t-\frac{z}{c}\right) \tag{4-17}$$

which coincides with the incident wave formulated in Eq. 3-1. To satisfy the fixed boundary condition at the base of the layer, the disks 1 and 2 result in introducing the mirror-image disks 3 and 4, loaded simultaneously with the same unit-impulse loads, but acting in the opposite direction, causing the displacement at depth z with $a = 2d - z$ in Eq. 4-3

$$-2\frac{1}{2K}\frac{1}{1+\frac{2d-z}{z_0}}h_1\left(t-\frac{2d-z}{c}\right) = -\frac{1}{1+\frac{2d-z}{z_0}}\bar{u}_0\left(t-\frac{2d-z}{c}\right) \tag{4-18}$$

which coincides with the displacement of the upwave specified in Eq. 3-3. The presence of the loaded disks 3 and 4 embedded in the fullspace violates the free-surface boundary condition. Enforcing this condition, leads to the mirror-image disks 5 and 6, loaded simultaneously with unit-impulse loads acting in the same direction as those acting on disks 3 and 4 resulting with $a = 2d + z$ in Eq. 4-3 in the displacement

$$-2\frac{1}{2K}\frac{1}{1+\frac{2d+z}{z_0}}h_1\left(t-\frac{2d+z}{c}\right) = -\frac{1}{1+\frac{2d+z}{z_0}}\bar{u}_0\left(t-\frac{2d+z}{c}\right) \tag{4-19}$$

which coincides with the displacement of the downwave given in Eq. 3-5. Continuing with this concept results in disks 7, 8, then 9, 10, and so on. In the limit of an infinite number of mirror-image disks, the surface of the fullspace coinciding with the free surface of the layer will be stress free, and a distance d deeper will not displace. It follows that this procedure of working with *anti-symmetrically and symmetrically loaded mirror-image disks* embedded in the fullspace leads to the same formulation derived based on the apparently different assumptions of the *unfolded layered cone* in Section 3.1.1. In particular, the wave pattern specified in Eq. 3-7, and thus the echo constants, follow. *The two formulations are thus identical.*

Next, a disk in a layer with embedment e measured from the free surface is examined (Fig. 4-9). The same concept applies, whereby the paired disks are not separated by an infinitesimal distance but are moved a distance $2e$ out of contact. This array of disks is used to calculate the Green's function; in Section 4.4 this is required to determine the dynamic behavior of a foundation embedded in a layer.

It follows that the procedure based on disks embedded in a fullspace to calculate the dynamic behavior of embedded and pile foundations is a generalization of the unfolded layered cone model for surface foundations on a layer, which itself is an extension of the truncated semi-infinite cone model for surface foundations on a halfspace.

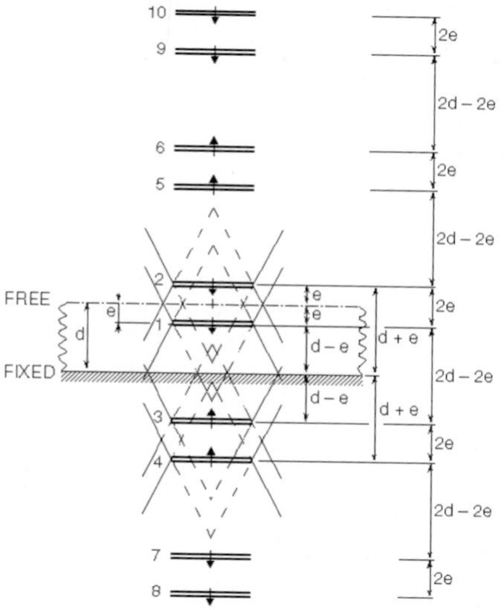

Fig. 4-9 Arrangement of mirror-image disks embedded in fullspace with loads and double cones to model disk embedded in soil layer resting on rigid rock.

4.1.5 Embedded Disk with Mirror Disk to Model Interface with Halfspace

To expand the procedure and at the same time to unify the modeling of the free and fixed boundaries, a disk embedded in a halfspace which has an interface parallel to the disk at a distance $\bar{e}$ with another halfspace with different material properties is addressed (Fig. 4-10). The halfspace containing the disk is denoted with 1 and the other with 2. To calculate the Green's function in halfspace 1, in addition to the disk with the load P, a mirror-image disk placed at the same distance $\bar{e}$ from the interface in the halfspace 2 is introduced. This mirror disk is loaded by $-\alpha P$, where $-\alpha$ is the reflection coefficient between halfspaces 1 and 2. This coefficient is derived based on the first reflection–refraction occurring at the interface for the translational and rotational cones in Sections 3.6.1 and 3.6.2. The two limits of $-\alpha$ for the static and the high-frequency cases are frequency independent, are the same for the two types of cones (Eqs. 3-57, 3-58), and depend strongly or (for the same Poisson's ratios ν in halfspaces 1 and 2) exclusively on the material properties. The values of $-\alpha = -1$ and $-\alpha = 1$ correspond to the fixed boundary and the free surface, respectively. This is also verified in Figs. 4-4b and 4-4a. The disk and the mirror-image disk embedded in their corresponding halfspaces are again modeled with double cones. Poisson's ratio is assumed to be equal for the two halfspaces, yielding the same apex height z_0 for the two cones, as shown in Fig. 4-10. If this is not the case, the geometry becomes more complicated.

At depth z in halfspace 1 with the wave velocity c, the displacement caused by the loaded disk and the mirror-image disk with a frequency-independent reflection coefficient $-\alpha$ equals

$$u(z,t) = \frac{z_0}{z_0+z}\,\bar{u}_0\left(t - \frac{z}{c}\right) - \alpha\,\frac{z_0}{z_0+2\bar{e}-z}\,\bar{u}_0\left(t - \frac{2\bar{e}}{c} + \frac{z}{c}\right) \tag{4-20}$$

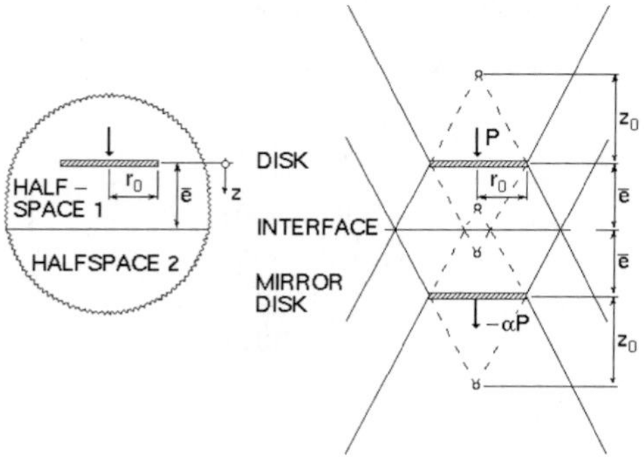

Fig. 4-10 Disk embedded in first halfspace with mirror-image disk with load multiplied by reflection coefficient to model interface with second halfspace.

The $\bar{u}_0\,(t - z/c)$ is the displacement in the halfspace. In the denominators, the distances from the apexes appear. At the interface $z = \bar{e}$, the ratio of the second term to the first term on the right-hand side is equal to the reflection coefficient -α. Equation 4-20 is identical to Eq. 3-49 after introducing the definition of $-\alpha$ specified in Eq. 3-57a.

The concept can be used to model a disk embedded in a soil layer resting on flexible rock (Fig. 4-11). The only difference to the case of rigid rock illustrated in Fig. 4-9 consists of multiplying the loads by $(-\alpha)$ for the mirror-image disks with respect to the interface instead of by -1. For the sake of simplicity in Fig. 4-11, the reflection coefficient $-\alpha$ is assumed to apply for disks 1 and 2. The loads acting on the mirror-image disks with respect to the free surface are unaffected. For the special case of a surface disk ($e = 0$)—disks 1 and 2 are an infinitesimal distance apart—summing the contributions of all disks leads to the same wave pattern as described in Eq. 3-77.

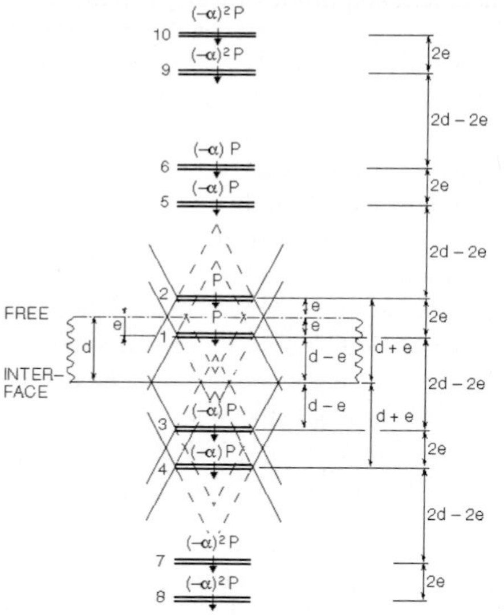

Fig. 4-11 Arrangement of mirror-image disks embedded in fullspace with loads and double cones to model disk embedded in soil layer resting on flexible rock.

4.2 MATRIX FORMULATION FOR HARMONIC LOADING OF FOUNDATION EMBEDDED IN HALFSPACE

4.2.1 Dynamic-Flexibility Matrix of Free Field

Figure 4-12 shows a cylindrical soil region with radius r_0 extending a depth e into the soil halfspace. The soil with mass density ρ, which will later be excavated, is viewed as a sandwich consisting of m rigid disks separated by $m - 1$ soil layers of thickness Δe. In order to formulate this halfspace problem in the fullspace, an additional stack of m mirror-image disks is introduced, as described in Section 4.1.3. For the vertical motion each disk i and its mirror-image disk are subjected to the same vertical force with amplitude $P_i(\omega)$ and experience the same displacement amplitude $u_i(\omega)$ ($i = 1, ..., m$); this guarantees that the middle plane remains stress free.

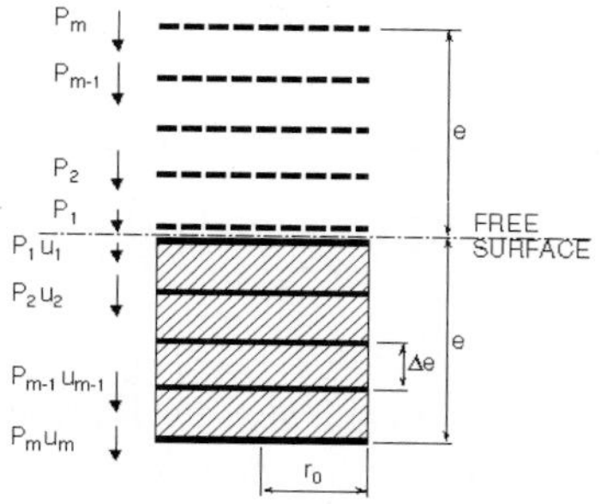

Fig. 4-12 Stack of disks to model embedded cylindrical foundation with anti-symmetrically loaded mirror-image disks.

The m-element vector of displacement amplitudes $\{u(\omega)\}$ is related to the m-element vector of force amplitudes $\{P(\omega)\}$ by an $m \times m$ element dynamic-flexibility matrix $[G(\omega)]$

$$\{u(\omega)\} = [G(\omega)] \{P(\omega)\} \tag{4-21}$$

If instead of amplitudes of forces and displacements, amplitudes of moments and rotations are considered, an analogous flexibility relationship holds

$$\{\vartheta(\omega)\} = [G_\vartheta(\omega)] \{M(\omega)\} \tag{4-22}$$

To avoid unnecessary repetition, the following discussion does not address Eq. 4-22 explicitly. If moments and rotations are viewed in the general sense as forces and displacements, Eq. 4-22 is included implicitly in Eq. 4-21.

An important special property of cone models (not true for rigorous solutions) is that *all four components of disk motion*—horizontal translation, vertical translation, rocking rotation, and torsional rotation—*are initially independent of one another*. The four $m \times m$ matrices $[G(\omega)]_h$, $[G(\omega)]_v$, $[G_\vartheta(\omega)]_r$, and $[G_\vartheta(\omega)]_t$ are uncoupled and may be considered separately. This fact may appear puzzling because of the obvious interdependence between horizontal translation and rocking for an embedded foundation. It turns out, however, that the expected coupling between horizontal and rocking motions has nothing to do with the matrices $[G(\omega)]_h$ and $[G_\vartheta(\omega)]_r$ but may be explained entirely by the kinematics of the disks enforced when formulating the rigid-body motion of the total embedded foundation.

A typical coefficient $g_{ij}(\omega)$ of the dynamic-flexibility matrix specifies the displacement of the i-th disk (receiver) with amplitude $u_i(\omega)$ caused by harmonic force of unit amplitude $P_j(\omega)$ applied to the j-th disk and its mirror image (sources). As illustrated by Fig. 4-13, the receiver disk may lie either between the pair of sources or underneath both of them. In either instance, the distances between the receiver disk and the two source disks are given by the same formulas

$$a = | i - j | \Delta e \tag{4-23a}$$

$$a' = (i + j - 2)\Delta e \tag{4-23b}$$

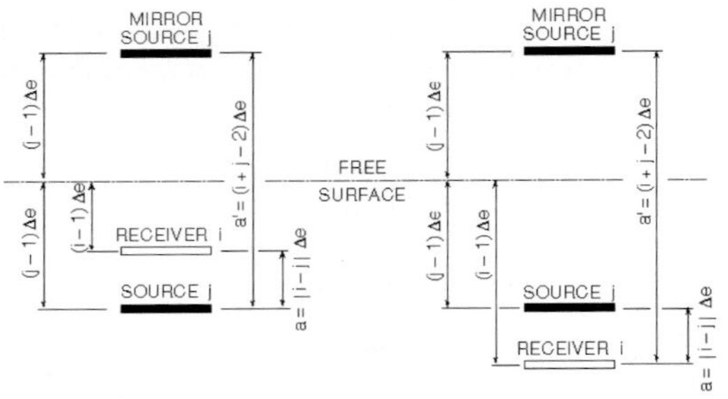

a. Receiver above source.

b. Receiver below source.

Fig. 4-13 Source-receiver relationship.

The absolute-value notation in Eq. 4-23a ensures that distance is always a positive quantity. In Section 4.1.2, closed-form expressions are given for the Green's function $g(a, \omega)$ defined as the displacement amplitude in the double-cone model of the fullspace due to a single-unit source load amplitude acting at distance a away from the receiver location. Summing the contributions of the source and the mirror-image source leads to

$$g_{ij}(\omega) = g_{ij}(a,\omega) + g_{ij}(a',\omega) \tag{4-24}$$

The dynamic-flexibility matrix $[G(\omega)]$ with elements $g_{ij}(\omega)$ formed according to Eq. 4-24 is symmetric.

4.2.2 Dynamic-Stiffness Matrix

Elastodynamical considerations are required solely to set up the dynamic-flexibility matrix $[G(\omega)]$. The rest of the analysis is effected via standard algebraic procedures familiar from the finite-element method. First, Eq. 4-21 is solved for $\{P(\omega)\}$ by inverting the dynamic-flexibility matrix

$$\{P(\omega)\} = [S^f(\omega)]\ \{u(\omega)\} \tag{4-25}$$

in which

$$[S^f(\omega)] = [G(\omega)]^{-1} \tag{4-26}$$

denotes the dynamic-stiffness matrix of the free field (virgin soil before excavation, see Fig. 1-3).

In the free-field soil, the various components of the displacement amplitude vector $\{u(\omega)\}$ are independent of one another. In an embedded foundation, however, *the number of degrees of freedom is restricted to the number of components of permissible rigid-body motion*, written in vector form as $\{u_0(\omega)\}$. For an embedded cylinder it is convenient to choose as the elements of $\{u_0(\omega)\}$ the displacement and rotation amplitudes at the center of the base. The displacement amplitudes of the disks $\{u(\omega)\}$ are related to the rigid-body motion $\{u_0(\omega)\}$ by a kinematic-constraint matrix $[A]$ according to

$$\{u(\omega)\} = [A]\,\{u_0(\omega)\} \tag{4-27}$$

The principle of virtual work implies the existence of a force amplitude vector $\{Q_0(\omega)\}$ associated with $\{u_0(\omega)\}$ such that

$$\{Q_0(\omega)\} = [A]^T\,\{P(\omega)\} \tag{4-28}$$

Equation 4-28 expresses equilibrium with the amplitudes of the resultant $\{Q_0(\omega)\}$. Substituting Eq. 4-27 in Eq. 4-25 and then in Eq. 4-28 results in

$$\{Q_0(\omega)\} = [A]^T[S^f(\omega)][A]\{u_0(\omega)\} \tag{4-29}$$

The matrix triple product in Eq. 4-29 represents the dynamic-stiffness matrix of the shaded volume shown in Fig. 4-12. The *disks and the soil trapped between them* are constrained to execute *a rigid-body motion*. Therefore the *soil may be analytically "excavated"* from Eq. 4-29 simply by subtracting the mass times acceleration of the rigid interior with density ρ. The modified interaction force amplitude vector $\{P_0(\omega)\}$ then becomes

$$\{P_0(\omega)\} = [A]^T\,[S^f(\omega)][A]\{u_0(\omega)\} - [M]\{\ddot{u}_0(\omega)\} \tag{4-30}$$

in which $[M]$ is the rigid-body mass matrix corresponding to the acceleration amplitude vector $\{\ddot{u}_0(\omega)\}$ (see Eq. 4-34). For harmonic motion $\{\ddot{u}_0(\omega)\} = -\omega^2\{u_0(\omega)\}$, substitution of this relationship into Eq. 4-30 leads to the final result

$$\{P_0(\omega)\} = [S^g_{00}(\omega)]\,\{u_0(\omega)\} \tag{4-31}$$

in which $[S^g_{00}(\omega)]$ is the desired dynamic-stiffness matrix of the rigid embedded massless foundation computed as

$$[S^g_{00}(\omega)] = [A]^T\,[S^f(\omega)][A] + \omega^2\,[M] \tag{4-32}$$

The dynamic-stiffness matrix $[S^g_{00}(\omega)]$ is strongly diagonal; only the horizontal and rocking components of motion are coupled. It is advantageous to calculate the various dynamic-stiffness coefficients separately. If this is done, the kinematic matrix $[A]$ degenerates to a vector. This vector (Fig. 4-14) consists either of a column of ones, $\{1\}$, or of an inverted triangle, $\{e\}$, defined as

$$\{e\} = \begin{Bmatrix} e \\ e-\Delta e \\ e-2\Delta e \\ \vdots \\ \Delta e \\ 0 \end{Bmatrix} \tag{4-33}$$

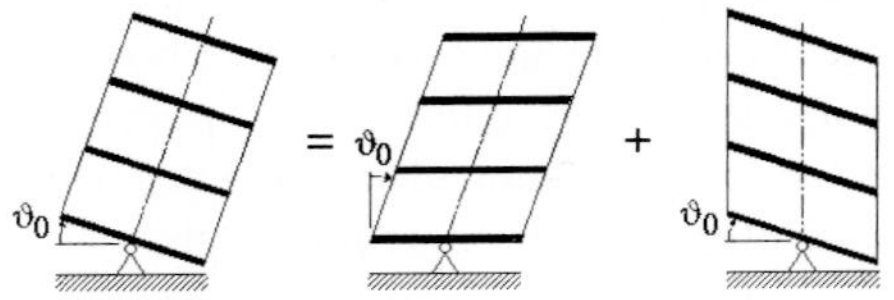

Fig. 4-14 Coupling between rocking and horizontal translation.

The vector $\{1\}$ imposes constant translation or rotation on all the disks, and the vector $\{e\}$ corresponds to a rigid-body rotation about the center of the base. Using these, the various dynamic-stiffness coefficients may be expressed as

$$S_h(\omega) = \{1\}^T [S^f(\omega)]_h \{1\} + \omega^2 m \tag{4-34a}$$

$$S_v(\omega) = \{1\}^T [S^f(\omega)]_v \{1\} + \omega^2 m \tag{4-34b}$$

$$\begin{aligned} S_r(\omega) = {} & \{e\}^T [S^f(\omega)]_h \{e\} + \omega^2 \frac{me^2}{3} \\ & + \{1\}^T [S^f(\omega)]_r \{1\} + \omega^2 \frac{mr_0^2}{4} \end{aligned} \tag{4-34c}$$

$$S_{hr}(\omega) = S_{rh}(\omega) = \{e\}^T [S^f(\omega)]_h \{1\} + \omega^2 \frac{me}{2} \tag{4-34d}$$

$$S_t(\omega) = \{1\}^T [S^f(\omega)]_t \{1\} + \omega^2 \frac{mr_0^2}{2} \tag{4-34e}$$

The various inertial quantities involve the mass $m = A_0 e\rho$ of the excavated soil cylinder. It may be observed that the rocking dynamic-stiffness coefficient $S_r(\omega)$ (Eq. 4-34c) consists of two separate contributions. Rigid-body rotation about the center of the base may be viewed as the sum of a triangular-shaped translation of the disks without rotation (vector $\{e\}$) and a constant rotation of the disks without translation (vector $\{1\}$), as illustrated in Fig. 4-14.

A strong analogy of this procedure to the indirect boundary-element method formulated in the frequency domain exists with the amplitudes of the loads $\{P(\omega)\}$ playing the role of those of the sources. The formulation can straightforwardly be extended to the calculation of the effective foundation input motion, as described in Section 6.3.3.

4.2.3 Example

The dynamic-stiffness coefficients of a rigid cylindrical foundation of radius r_0 and embedment e (Fig. 4-1) embedded in a halfspace with Poisson's ratio $\nu = 0.25$ are calculated. The embedment ratios $e/r_0 = 0.5$, $= 1$, and $= 2$ are processed. The results are determined up to a maximum value of the dimensionless frequency $a_0 = \omega r_0/c_s$ equal to 4.

For good accuracy the number of disks m should be chosen large enough that the slice thickness $\Delta e = e/(m-1)$ does not exceed about one sixth of the shortest wavelength of vertically propagating waves. Setting the wave length $c2\pi/\omega = 2\pi r_0(c/c_s)/a_0$ equal to $6\Delta e = 6e/(m-1)$ yields

$$m - 1 = \frac{3}{\pi} \frac{e}{r_0} \frac{c_s}{c} a_0 \tag{4-35}$$

For the maximum $a_0 = 4$ and, as the shear-wave velocity leads to a larger $m - 1$ than the dilatational wave velocity, with $c = c_s$, $m - 1$ equals 8 for the deepest embedment $e/r_0 = 2$. All analyses are performed with 8 slices (9 disks).

Figure 4-15 shows the static-stiffness factors for all motions—the ratio of the static-stiffness coefficient of the embedded foundation to that of the surface foundation for the same motion—computed using cone models and compared with the rigorous results of Apsel and Luco [A1]. The dashed curves correspond to the empirical formulas [P1], which are applicable for $e/r_0 \leq 2$

horizontal $$K_h = \frac{8Gr_0}{2-\nu}\left(1 + \frac{e}{r_0}\right) \tag{4-36a}$$

vertical $$K_v = \frac{4Gr_0}{1-\nu}\left(1 + 0.54\frac{e}{r_0}\right) \tag{4-36b}$$

rocking $$K_r = \frac{8Gr_0^3}{3(1-\nu)}\left[1 + 2.3\frac{e}{r_0} + 0.58\left(\frac{e}{r_0}\right)^3\right] \tag{4-36c}$$

torsional $$K_t = \frac{16Gr_0^3}{3}\left(1 + 2.67\frac{e}{r_0}\right) \tag{4-36d}$$

coupling $$K_{hr} = \frac{e}{3} K_h \tag{4-36e}$$

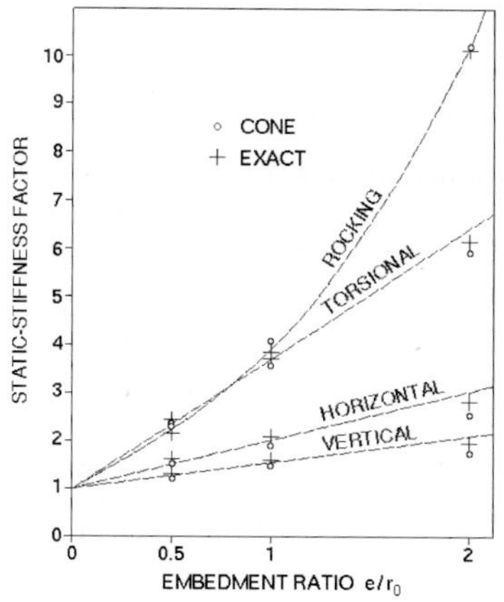

Fig. 4-15 Static-stiffness factors for cylindrical foundation embedded in halfspace.

The accuracy of the cone models is excellent for all components of motion, particularly for the most complicated case, rocking.

The dynamic-stiffness coefficients for harmonic loading, nondimensionalized with the static-stiffness coefficient, are presented for the horizontal, vertical, rocking, and torsional motions in Figs. 4-16, 4-17, 4-18, and 4-19, whereby the real part denoted as the spring coefficient k and the imaginary part divided by a_0 denoted as the damping coefficient c are plotted (see Eq. 3-44 for this decomposition). The deviation from the rigorous results [A1] denoted as exact (shown here for the embedment ratio $e/r_0 = 1$, but comparable for the other embedment ratios) is generally less than 10%. It may be observed that for vertical and horizontal translations (Figs. 4-16 and 4-17) the coefficients k and c are essentially constant and may thus be represented physically by a single spring and a single dashpot with frequency-independent coefficients. The curves of the torsional dynamic-stiffness coefficient of an embedded foundation (Fig. 4-19) have the same shapes as those for a surface disk (Fig. 2-20). The rocking case (Fig. 2-21) is somewhat more complicated. Because part of the stiffness arises from the horizontal component (Fig. 4-14), the rocking coefficients k_r and c_r no longer correspond to pure rotation as for the torsional case. Certain aspects of the horizontal case are observed; for example, the curve for c_r has a non-zero initial value at $a_0 = 0$. For all motions, the damping coefficient c increases with larger embedment ratio e/r_0 throughout the frequency range (although c is affected too by the nondimensionalization with the static-stiffness coefficient K—which also increases when e/r_0 becomes larger).

4.3 MATRIX FORMULATION IN TIME DOMAIN OF FOUNDATION EMBEDDED IN HALFSPACE

4.3.1 Dynamic-Flexibility Matrix of Free Field

The formulation for harmonic loading (frequency domain) of Section 4.2. can be extended to calculate an embedded foundation for a transient excitation directly in the time domain without performing any transformation into the frequency domain. The concept of using a double-cone model to represent the disk within an elastic fullspace enables the spatial discretization in the same manner. In the time domain as well, the approximate Green's functions of the cones provide a link to the boundary-element method and yield the displacement assumption for a strength-of-materials approach to foundation dynamics. As

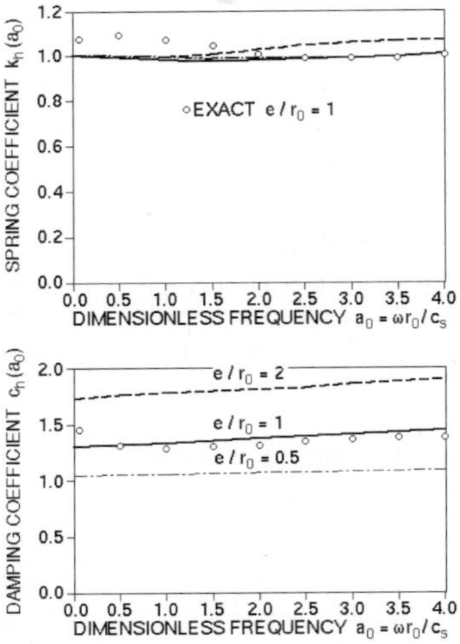

Fig. 4-16 Dynamic-stiffness coefficient for harmonic loading of cylindrical foundation embedded in halfspace in horizontal motion.

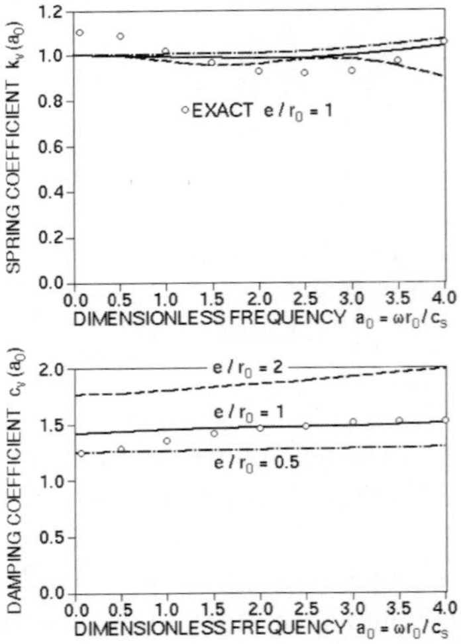

Fig. 4-17 Dynamic-stiffness coefficient for harmonic loading of cylindrical foundation embedded in halfspace in vertical motion.

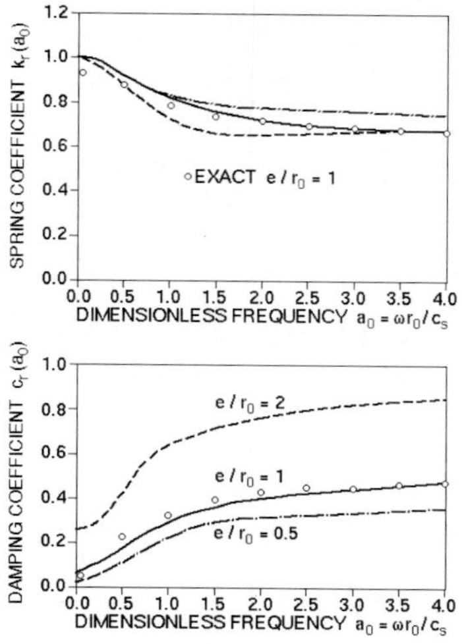

Fig. 4-18 Dynamic-stiffness coefficient for harmonic loading of cylindrical foundation embedded in halfspace in rocking motion.

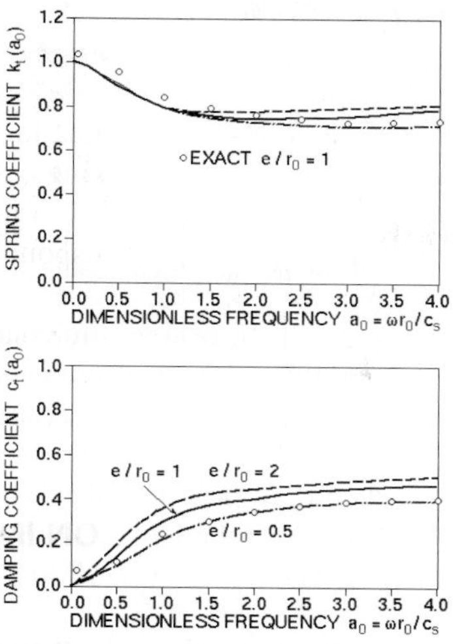

Fig. 4-19 Dynamic-stiffness coefficient for harmonic loading of cylindrical foundation embedded in halfspace in torsional motion.

for surface foundations addressed in Chapter 2, the convolution integrals arising in the formulation are evaluated exactly by recursion, reducing significantly the number of operations and the storage requirements.

The *spatial discretization* for the time-domain analysis is the same as for the frequency-domain calculation. The cylindrical soil region with radius r_0 extending to depth e into the ground and with mass density ρ which will later be excavated is viewed as a sandwich consisting of m rigid disks separated by soil layers of thickness Δe (Fig. 4-20). Each disk is associated with a double-cone model. In order to formulate the halfspace problem in the fullspace (Section 4.1.3), an additional stack of m mirror-image disks is introduced. Each disk and its mirror image are subjected to the same force $P_i(t)$ or moment $M_i(t)$ and experience the same displacement $u_i(t)$ or rotation $\vartheta_i(t)$ ($i = 1, ..., m$), which leads to a stress-free middle plane coinciding with the free surface of the halfspace.

As far as the *temporal discretization* is concerned, the source forces acting on the disks are assumed to vary piecewise linearly between two adjacent time stations (Fig. 4-21). The source forces vanish at time zero.

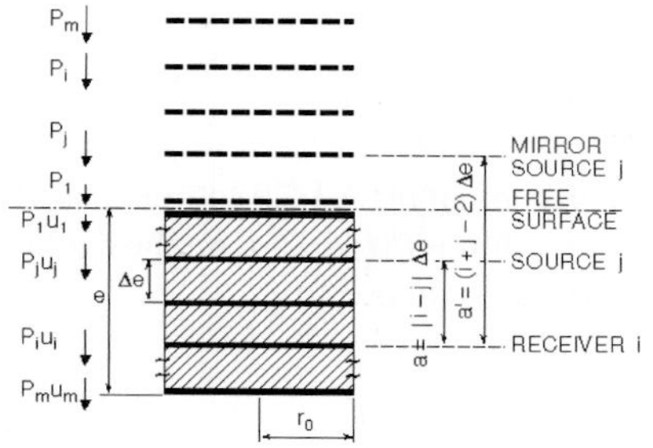

Fig. 4-20 Spatial discretization of embedded cylindrical region with disks and anti-symmetrically loaded mirror-image disks.

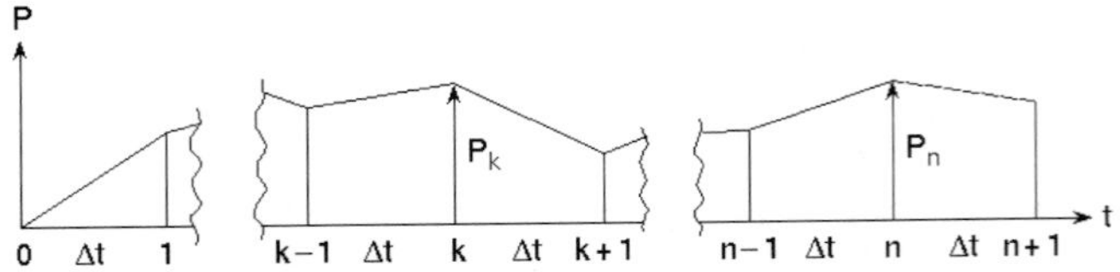

Fig. 4-21 Time discretization of source forces acting on disks.

$\{P\}_k$ denotes the m-element vector of the m source forces acting on all disks including the mirror disks at time station k ($t = k\Delta t$). $\{u\}_n$ represents the m-element vector of the m displacements of the disks at time station n ($t = n\Delta t$). With $\{P(\tau)\}$ denoting the vector of the source forces as a function of time, the contribution of the infinitesimal pulses $\{P(\tau)\}d\tau$ acting at time τ to the displacements $\{du(t)\} = \{du\}_n$ at time $t = n\Delta t \geq \tau$ depends on the matrix of the Green's function evaluated for the time difference $t - \tau$, $[g(t - \tau]$

$$\{du\}_n = [g(t-\tau)]\ \{P(\tau)\}\ d\tau \tag{4-37}$$

Integrating over all pulses form $0 \leq \tau \leq t$ yields

$$\{u\}_n = \int_0^t [g(t-\tau)]\ \{P(\tau)\}\ d\tau \tag{4-38}$$

The matrix $[g(t - \tau)]$ is constructed from the Green's functions for unit impulses specified in Eqs. 4-3 and 4-5. An element g_{ij} of the dynamic-flexibility matrix will contain the contributions of the source j with the distance $a = |i - j|\Delta e$ and of the mirror-image source j with $a' = (i + j - 2)\Delta e$ (Fig. 4-20). For the cone models all four components of each single disk's motion (horizontal, vertical, rocking, torsional) are independent of one another. The corresponding dynamic-flexibility matrices are thus uncoupled. (Of course, when an embedded foundation is simulated as an assemblage of disks, rigid-body constraints introduce a coupling between horizontal and rocking motions).

Recursive evaluation. It is very inefficient to evaluate the convolution integrals directly by numerical quadrature. A much more elegant way is to use *recursion* formulas which express the *displacements at a specific time station as a linear function of the forces at the same time station and of a few past values of the displacements and forces.* These recursive relationships based on a piecewise linear variation of the forces are discussed for the first-order term $h_1(t)$ of the translational cone in Section C2 and applied in Section 2.2.2 and for the second-order terms $h_2(t)$ and $h_3(t)$ of the rotational cone in Section C3 and applied in Section 2.3.2. It is important to emphasize that the recursive evaluation of the convolution integrals does not introduce any approximation.

As an example, the recursive evaluation of one element of Eq. 4-38 is illustrated briefly for the translational motion. The displacement $u(t)$ of the source disk at the n-th time station ($t = n\Delta t$), caused by the piecewise linear force $P(t)$ acting on the same disk, is given formally by the convolution integral (see Eq. 4-3 with $a = 0$).

$$u(t) = \frac{1}{2K}\int_0^t h_1(t-\tau)\ P(\tau)\ d\tau \tag{4-39}$$

In lieu of the convolution integral, discrete samples of $u(t)$, such as $u_n = u(n\Delta t)$, may be evaluated exactly by recursion, whereby

$$u_n = au_{n-1} + b_0 P_n + b_1 P_{n-1} \tag{4-40}$$

with the recursive coefficients specified in Eq. C-24. For a receiver located at a distance a away from the source disk, the source motion will have originated at the earlier retarded time $t - a/c = n\Delta t - a/c = [n - a/(c\Delta t)]\Delta t = n'\Delta t$. The number n' is always less than n, but not necessarily an integer. If n' is not an integer, the appropriate source motion $u_{n'}$ is obtained by interpolating stored past samples of u. The amplitude at the receiver is determined by multiplying $u_{n'}$ by the decay factor $1/(1 + a/z_0)$ (Eq. 4-3)

$$u(a,t) = \frac{u_{n'}}{1 + \dfrac{a}{z_0}} \tag{4-41}$$

The procedure to calculate the rotation $\vartheta(a,t)$ is analogous. The source disk's contributions due to h_2 and h_3 are computed separately by recursion, multiplied by the respective decay factors, and then added.

Time-discretized displacement-interaction force relationship. The evaluation of the convolution integrals leads to the displacement-force relationship

$$\{u\}_n = \sum_{k=1}^{n} [G]_{n-k} \{P\}_k \tag{4-42a}$$

or

$$\{u\}_n = \sum_{k=1}^{n-1} [G]_{n-k} \{P\}_k + [G]_0 \{P\}_n \tag{4-42b}$$

The symmetric dynamic-flexibility matrix $[G]_{n-k}$ denotes the displacements of the m disks at time station n caused by the unit source forces acting on all m disks at time station k (increasing linearly from zero at $k - 1$ to 1 at k, and decreasing linearly from 1 at k to zero at $k + 1$). For $[G]_0$ only the contribution of the forces acting during the n-th time step (from $n - 1$ to n) is considered (and not the "future" contribution of the forces from n to $n + 1$ indicated as a dashed line in Fig. 4-21). It is worth mentioning that the relationship of Eq. 4-42b can be formulated independently for all four motions of the disks.

4.3.2 Interaction Force-Displacement Relationship

The computational procedure for the n-th time step leading from time station n - 1, where all variables are known, to time station n makes use of algebraic manipulations familiar from the finite-element method.

The first term on the right-hand side of Eq. 4-42b can be evaluated, as $\{P\}_k$ ($k = 1, ..., n - 1$) is known; this contribution is denoted as $\{\bar{u}\}_n$.

$$\{\bar{u}\}_n = \sum_{k=1}^{n-1} [G]_{n-k} \{P\}_k \tag{4-43}$$

As mentioned before, the convolution sum should not be evaluated directly; rather, recursion techniques are utilized. Solving Eq. 4-42b for the unknown $\{P\}_n$ leads to

$$\{P\}_n = [S^f]_0 \{u\}_n - [S^f]_0 \{\bar{u}\}_n \tag{4-44}$$

Here $[S^f]_0$ represents the (discretized) instantaneous dynamic-stiffness matrix of the free field in the time domain (inverse of the discretized instantaneous flexibility matrix).

$$[S^f]_0 = [G]_0^{-1} \tag{4-45}$$

In the free field (virgin soil before excavation) the various components of the displacement vector $\{u\}_n$ are independent of one another. For an embedded rigid foundation, however, *the number of degrees of freedom is restricted to the number of components of rigid-body motion* $\{u_0\}_n$. The $\{u_0\}_n$ contains the displacements and rotations at the center of the base of the cylinder. The displacements of the disks $\{u\}_n$ are related to $\{u_0\}_n$ by the kinematic-constraint matrix $[A]$ according to

$$\{u\}_n = [A] \{u_0\}_n \tag{4-46}$$

Virtual-work considerations (equilibrium) lead to a resultant force vector $\{Q_0\}_n$ associated with $\{u_0\}_n$ such that

$$\{Q_0\}_n = [A]^T \{P\}_n \tag{4-47}$$

Substituting Eq. 4-46 in Eq. 4-44 and the result in Eq. 4-47 yields

$$\{Q_0\}_n = [A]^T [S^f]_0 [A] \{u_0\}_n - [A]^T [S^f]_0 \{\bar{u}\}_n \tag{4-48}$$

The second term on the right-hand side is a known load vector. The coefficient matrix $[A]^T [S^f]_0[A]$ represents the (discretized) instantaneous dynamic stiffness in the time domain of the shaded volume in Fig. 4-20 whereby the disks and the soil trapped between them are constrained to execute the rigid-body motion. The forces corresponding to this soil to be *excavated* are equal to the mass matrix $[M]$ times the acceleration vector $\{\ddot{u}_0\}_n$ of the rigid interior with mass ρ. Subtracting these forces from $\{Q_0\}_n$ leads to the interaction forces $\{P_0\}_n$ of the embedded foundation

$$\{P_0\}_n = [A]^T [S^f]_0 [A] \{u_0\}_n - [A]^T [S^f]_0 \{\bar{u}\}_n - [M] \{\ddot{u}_0\}_n \tag{4-49}$$

The matrices appearing in Eq. 4-49 are strongly diagonal. Only the horizontal and rocking components are coupled. It is appropriate to calculate the contributions of the horizontal/rocking, vertical and torsional motions separately. This is linked to the spatial discretization (Section 4.2.2).

Just in passing, it is worth mentioning that the formulation discussed above is nothing else than a special case of the weighted-residual method of the time-domain boundary-integral approach. A piecewise linear variation with time of the source forces is selected. This procedure, which is often also called the indirect boundary-element method, uses the Green's functions of the cones and satisfies compatibility and equilibrium at the end of each time step by using a Dirac impulse as the weighting function. The procedure can also be extended to the calculation of the effective foundation input motion for seismic excitation (Section 6.3.3).

Time-integration scheme. To be able to develop a computational algorithm, a time-integration scheme is needed to express $\{\ddot{u}_0\}_n$ as a function of $\{u_0\}_n$ in Eq. 4-49. The time integration is performed based on the Newmark method with the two parameters β and γ.

$$\{u_0\}_n = \{u_0\}_{n-1} + \Delta t\,\{\dot{u}_0\}_{n-1} + (0.5-\beta)\,\Delta t^2\,\{\ddot{u}_0\}_{n-1} + \beta\Delta t^2\{\ddot{u}_0\}_n \qquad (4\text{-}50a)$$

$$\{\dot{u}_0\}_n = \{\dot{u}_0\}_{n-1} + (1-\gamma)\,\Delta t\,\{\ddot{u}_0\}_{n-1} + \gamma\Delta t\,\{\ddot{u}_0\}_n \qquad (4\text{-}50b)$$

Solving Eq. 4-50a for $\{\ddot{u}_0\}_n$ and substituting in Eq. 4-49 leads to the following system of equations for $\{u_0\}_n$

$$\left([A]^T\,[S^f]_0\,[A] - \frac{1}{\beta\Delta t^2}[M]\right)\{u_0\}_n =$$
$$\{P_0\}_n + [A]^T\,[S^f]_0\,\{\bar{u}\}_n + [M]\left(-\frac{1}{\beta\Delta t^2}\{u_0\}_{n-1} - \frac{1}{\beta\Delta t}\{\dot{u}_0\}_{n-1} - \frac{0.5-\beta}{\beta}\{\ddot{u}_0\}_{n-1}\right) \qquad (4\text{-}51)$$

This discretized system of motion applies to the embedded foundation alone, that is, to a rigid massless empty structure and excavation. Especially for small Δt, the mass $[M]$ of the excavated soil will (because of the minus sign on the left-hand side) lead to an unstable algorithm for the constant-acceleration method with $\gamma = 0.5$ and $\beta = 0.25$. To avoid this instability in an analysis of the embedded foundation alone, a larger γ introducing numerical damping is necessary (see Section 4.3.3). If a structure with its own stiffness and mass is analyzed together with the foundation, which is the normal case encountered in practice, the standard constant-acceleration method can be used.

Alternative modeling of excavated soil. The potential instability in the discretized equation of motion (Eq. 4-51) arises from the subtraction of the forces of the excavated soil which are expressed as the product of the rigid-body mass matrix and the corresponding acceleration vector (Eq. 4-49). As an alternative, the following procedure also based on the boundary-element concept is used for the cylindrical excavated soil. The same disks including their mirror images are used as shown in Fig. 4-20. Each disk is associated with a double-sided semi-infinite cylindrical rod with constant area A_0. Green's functions can be determined straightforwardly for these prismatic rods, formally by setting the apex height $z_0 = \infty$ in the expressions for the cones (Eqs. 4-3 and 4-5). The displacement-force relationship for the infinitely long cylinder discretized in the same points (location of the disks) then follows analogous to Eq. 4-42a. After inversion, the corresponding force-displacement equation analogous to Eq. 4-44 results. To construct the force-displacement relationship of the finite embedded cylinder of length e, that of the cylinder extending from disk m to infinity is subtracted. The corresponding force of this semi-infinite cylinder equals $A_0\rho c\dot{u}_m$. The forces acting on the disks corresponding to the finite embedded cylinder are then subtracted directly from $\{P\}_n$ acting on the disks associated with the cones specified in Eq. 4-44. The further derivation is not affected anymore from this equation onwards. The global force vector calculated by premultiplying these resultant disk forces by $[A]^T$ (Eq. 4-47) leads directly to the final interaction force-displacement relationship valid for the embedded foundation. Unfortunately, this alternative approach does not prevent instabilities from occurring in all cases.

Nearly incompressible soil. In the case of the nearly incompressible soil for $1/3 <$ Poisson's ratio $\nu \leq 1/2$ as discussed in Section 2.1.4, for the vertical and rocking motions besides limiting the wave velocity to $2c_s$, a trapped mass ΔM (Eq. 2-59) and a mass moment of inertia ΔM_ϑ (Eq. 2-60) arise. The intermediate disks of the stack are separated by thin slices of soil. In the course of the analysis, however, the mass of the slices is subtracted out (excavated). As a result, trapped mass need only be included for the undermost disk to represent the cone of soil under the base of the embedded foundation. As for a surface foundation, the trapped mass is conveniently assigned to the basemat and thus processed as part of the structure.

4.3.3 Example

Time-domain solutions for embedded foundations have hardly been reported in the literature. Harmonic excitations with various frequencies acting on the foundation are thus processed in the time domain followed by an example for a transient load applied to an embedded rigid structure.

The embedment ratio e/r_0 equals 1 and Poisson's ratio 0.25. In all calculations, 8 slices with $m = 9$ disks are chosen (Fig. 4-20).

Harmonic loading. In all calculations at least 15 time steps per period are used. After the initial transient starting from zero initial conditions, the steady-state response for harmonic load is achieved. The latter can be represented by a dynamic-stiffness coefficient $S(a_0)$ analogous to a frequency-domain calculation

$$S(a_0) = K\,[k(a_0) + ia_0c(a_0)] \tag{4-52}$$

where K denotes the static-stiffness coefficient of the embedded foundation, $k(a_0)$ and $c(a_0)$ represent the spring and damping coefficients, and $a_0 = \omega r_0/c_s$ is the dimensionless frequency. Figures 4-22 through 4-25 present the results for the horizontal, vertical, rocking, and torsional motions. The solutions based on the cones working in the frequency domain (Section 4.2.3) are shown as solid lines. The static-stiffness coefficients K of this case are used to nondimensionalize the results of all other solutions (Eq. 4-52). The rigorous solutions denoted as exact are reported in Apsel and Luco [A1].

The results based on the cone models in the time domain (directly on Eq. 4-51) are shown as solid dots. For the limit approaching the static case ($a_0 \to 0$) $\gamma = 0.5$ can be selected, but numerical damping must be introduced in the dynamic cases ($0 < a_0 \leq 2$: $\gamma = 0.65$; $2 < a_0 \leq 4$: $\gamma = 0.75$). In all cases, β follows as $0.25\,(\gamma + 0.5)^2$. For the sake of comparison, the transient analysis in the time domain using cones is also performed without excavating the cylinder of soil, for the free field (Eq. 4-48 and not Eq. 4-49 is processed). After having determined the corresponding dynamic-stiffness coefficient, the influence of the excavated part is taken into consideration by subtracting $-\omega^2\,[M]$. In this case no time-integration scheme is needed. These results shown as crosses are more accurate, especially for higher frequencies, than those calculated directly in a transient analysis for the embedded system (solid dots).

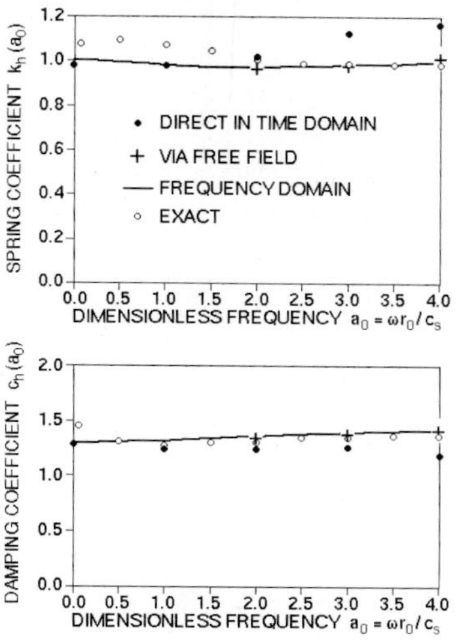

Fig. 4-22 Dynamic-stiffness coefficient of cylindrical foundation with embedment ratio = 1 in horizontal motion.

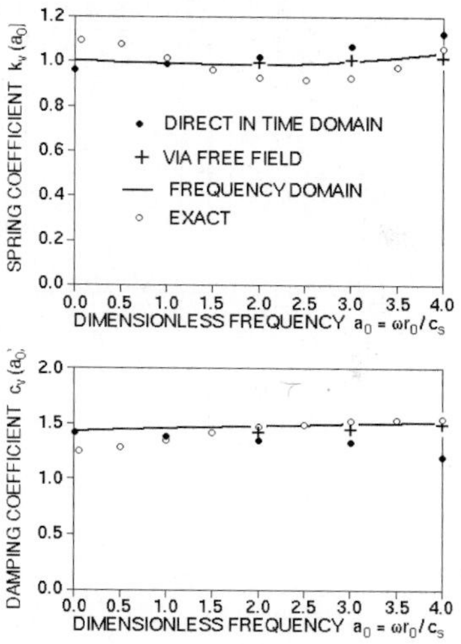

Fig. 4-23 Dynamic-stiffness coefficient of cylindrical foundation with embedment ratio = 1 in vertical motion.

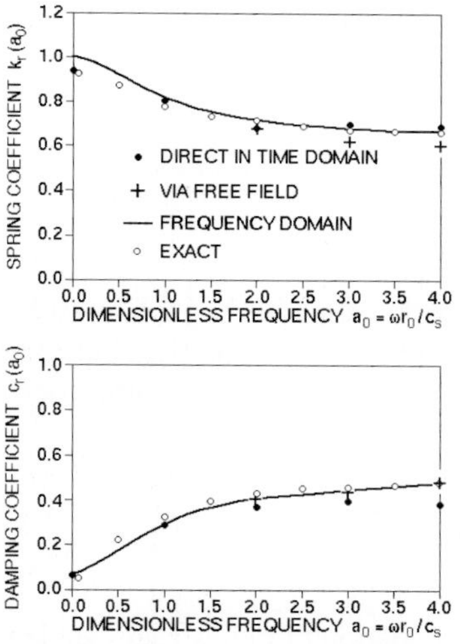

Fig. 4-24 Dynamic-stiffness coefficient of cylindrical foundation with embedment ratio = 1 in rocking motion.

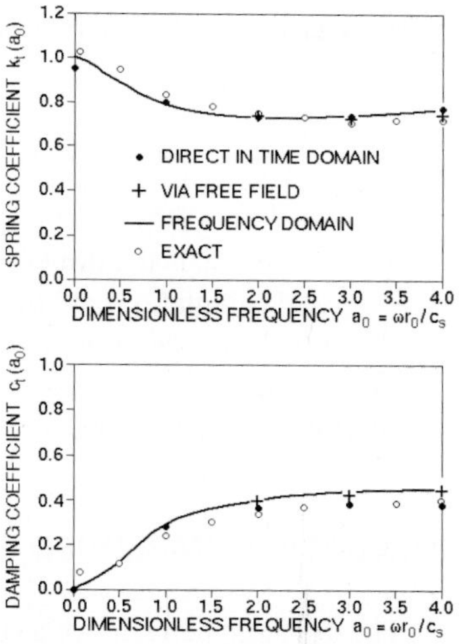

Fig. 4-25 Dynamic-stiffness coefficient of cylindrical foundation with embedment ratio = 1 in torsional motion.

Transient loading. A rigid cylindrical block with height $h = 3r_0$ and radius r_0 is embedded in a homogeneous halfspace with Poisson's ratio = 0.25, with the same mass density as the block and embedment depth $e = r_0$ (Fig. 4-26a). A load $F(t)$ acts horizontally at the base of the block as a rounded triangular pulse

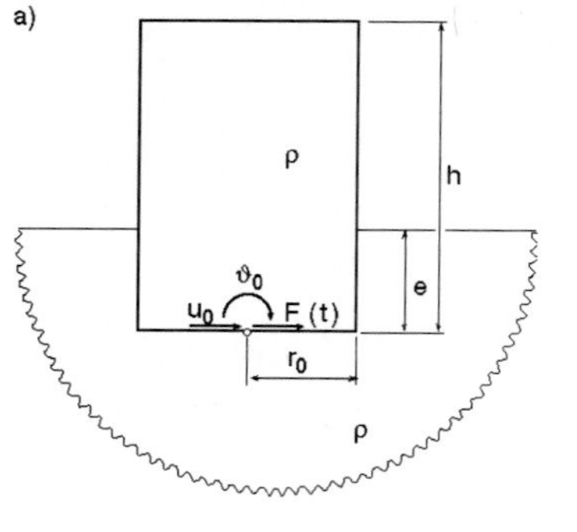

a. Geometry.

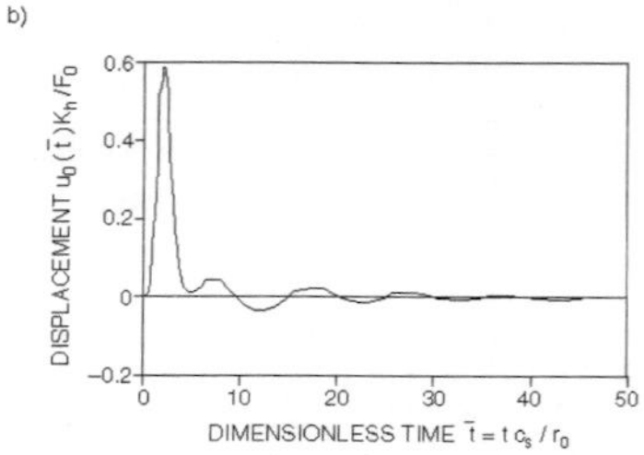

b. Horizontal displacement at base.

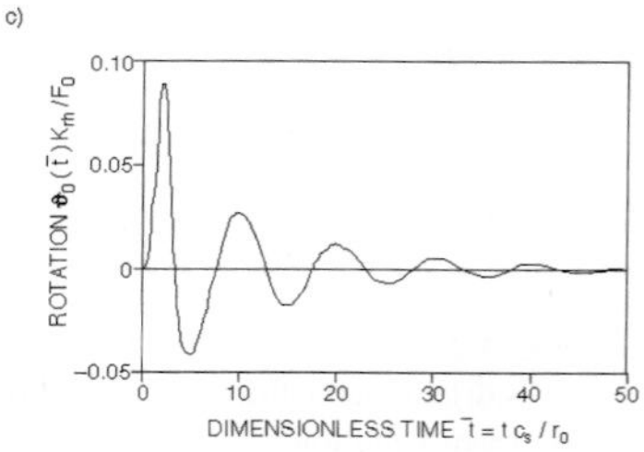

c. Rocking motion.

Fig. 4-26 Rigid cylindrical block embedded in halfspace.

$$F(\bar{t}) = \frac{F_0}{2}\left(1 - \cos 2\pi \frac{\bar{t}}{\bar{T}_0}\right) \qquad 0 < \bar{t} < \bar{T}_0 \tag{4-53}$$

where $\bar{t} = tc_s/r_0$ is the dimensionless time and $\bar{T}_0 = t_0 c_s/r_0 = 2$.

All calculations are performed in the time domain with $\Delta\bar{t} = 0.2$, $\gamma = 0.5$, and $\beta = 0.25$. The time histories of the horizontal displacement at the base $u_0(\bar{t})$ and of the rotation $\vartheta_0(\bar{t})$ are plotted in Figs. 4-26b and 4-26c. K_h and K_{rh} are the static-stiffness coefficients in the horizontal direction and of the coupling term.

4.4 FOUNDATION EMBEDDED IN LAYER

The dynamic behavior of a rigid cylindrical foundation of radius r_0 embedded to depth e in a homogeneous soil layer of depth d resting on rigid rock (Fig. 4-27) is to be analyzed for all motions in the time domain or for harmonic loading (frequency domain). Due to the complicated shape of the surface of this mixed-boundary value problem, a rigorous solution is very difficult to obtain. Based on physical models an approximate solution of high accuracy can be obtained straightforwardly by applying the simple concepts of Sections 4.1 to 4.3.

Again, disks are placed in the embedded region of the soil which will later be excavated (see the lower part of Fig. 4-12). The Green's function for each disk embedded in the fullspace is determined based on the corresponding one-dimensional double-cone model

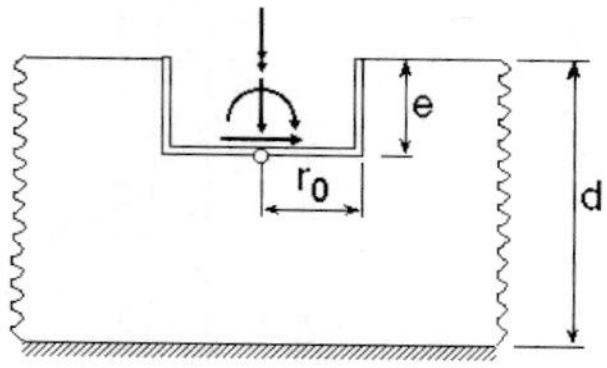

Fig. 4-27 Cylindrical foundation embedded in soil layer resting on rigid rock.

(Fig. 4-3c). To represent the free surface and the fixed surface, anti-symmetrically loaded mirror-image disks (Fig. 4-4a) and symmetrically loaded mirror-image disks (Fig. 4-4b), respectively, are introduced repeatedly without increasing the total number of source forces. For each disk the arrangement shown in Fig. 4-9 will result, which allows the corresponding Green's function to be constructed that satisfies the free-surface and fixed-boundary conditions. The matrix formulation is then the same as described for the frequency and time domains in Sections 4.2 and 4.3, yielding the dynamic-stiffness matrix and the (discretized) interaction force-displacement relationship.

For the sake of illustration, the vertical dynamic-stiffness coefficient for harmonic loading is calculated for an embedment ratio $e/r_0 = 1$ and a radius-to-depth ratio $r_0/d = 1/3$ with Poisson's ratio $\nu = 1/3$ and with a hysteretic damping ratio $\zeta_g = 0.05$. The latter affects the Green's functions through the correspondence principle (Section 2.4.1), modifying the shear modulus $[G(1 + 2i\zeta_g)]$ and the wave velocity $[c(1 + i\zeta_g)]$. Eight disks are placed in the soil cylinder which will later be excavated. To represent approximately the free surface and fixed boundary conditions, over 2,000 mirror-image disks are introduced. The dynamic-flexibility matrix $[G(\omega)]$ of the free field (Eq. 4-21) remains of order 8×8. In Emperador and Dominguez [E2] a rigorous boundary-element solution with a very fine element subdivision, denoted as exact in Fig. 4-28, is presented. The spring coefficient $k(a_0)$ and damping coefficient $c(a_0)$ determined from the dynamic-stiffness coefficient using the exact static-stiffness coefficient for nondimensionalization (Eq. 4-52) are highly accurate throughout the frequency range. Note that due to the material damping the $c(a_0)$ does not vanish below the vertical fundamental frequency $a_0 = 2\pi\, 2c_s/(4d)\, r_0/c_s = \pi/3$.

The following empirically derived equations for the static-stiffness coefficients apply [G3, P1].

$$\text{horizontal } K_h = \frac{8Gr_0}{2-\nu}\left(1+\frac{1}{2}\frac{r_0}{d}\right)\left(1+\frac{e}{r_0}\right)\left(1+\frac{e}{d}\right) \tag{4-54a}$$

$$\text{vertical } K_v = \frac{4Gr_0}{1-\nu}\left(1+1.3\frac{r_0}{d}\right)\left(1+0.54\frac{e}{r_0}\right)\left[1+\left(0.85-0.28\frac{e}{r_0}\right)\frac{\frac{e}{d}}{1-\frac{e}{d}}\right] \tag{4-54b}$$

$$\text{rocking } K_r = \frac{8Gr_0^3}{3(1-\nu)}\left(1+\frac{1}{6}\frac{r_0}{d}\right)\left[1+2.3\frac{e}{r_0}+0.58\left(\frac{e}{r_0}\right)^3\right]\left(1+0.7\frac{e}{d}\right) \tag{4-54c}$$

$$\text{torsional } K_t = \frac{16}{3}Gr_0^3\left(1+\frac{1}{10}\frac{r_0}{d}\right)\left(1+2.67\frac{e}{r_0}\right) \tag{4-54d}$$

$$\text{coupling } K_{hr} = \frac{e}{3}K_h \tag{4-54e}$$

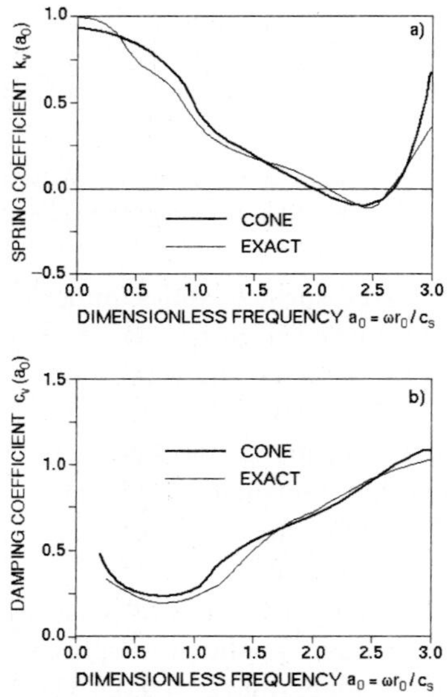

Fig. 4-28 Dynamic-stiffness coefficient for harmonic loading of cylindrical foundation embedded in soil layer resting on rigid rock in vertical motion.

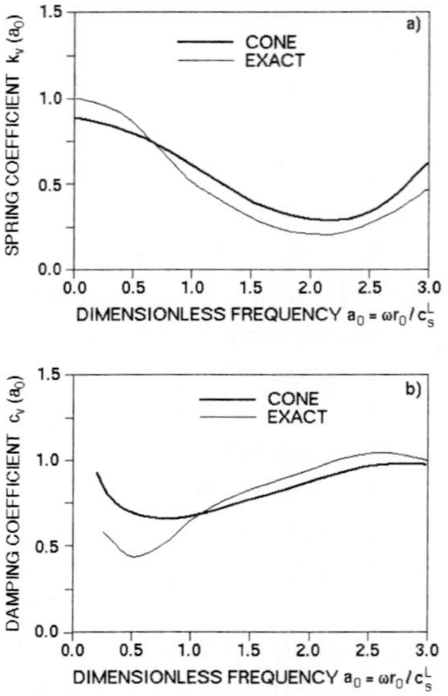

Fig. 4-29 Dynamic-stiffness coefficient for harmonic-loading of cylindrical foundation embedded in soil layer resting on flexible rock in vertical motion.

The first term on the right-hand side is the static-stiffness coefficient of a disk with equivalent radius r_0 on the surface of a halfspace (Table 2-2); the second denotes the stiffening factor moving for a surface foundation from a halfspace to a layer of depth d (Eq. 3-21); the third represents another stiffening factor when a surface foundation is replaced by an embedded foundation with embedment e in a halfspace (Eq. 4-36); and the fourth (not present for the torsional motion) is yet another stiffening factor reflecting embedment in a layer. The applicability of Eq. 4-54 is limited to $e/r_0 \leq 2$ and $e/d \leq 0.5$.

Note that these static-stiffness coefficients are not needed to perform the analysis with the stack of embedded disks.

For the vertical motion of the embedded foundation of Fig. 4.27, Eq. 4-54b leads with $r_0/d = 1/3$, $e/r_0 = 1$ and $e/d = 1/3$ to the product of the three stiffening factors = 2.84 compared to the exact value = 2.69 [E2].

As a generalization, the cylindrical foundation embedded in a damped soil layer with the same geometry, which, however, rests on a flexible rock halfspace, is processed. The Poisson's ratios and mass densities of the rock are equal to those of the soil ($\nu_R = \nu_L$, $\rho_R = \rho_L$), while the wave velocity of the rock is twice that of the soil ($c_R = 2c_L$). The same discretization with 8 disks is used. For the arrangement of the disks shown in Fig. 4-11, the reflection coefficient $-\alpha(\omega)$, which is frequency-dependent (Eq. 3-60), is determined for a surface disk and applied for disks 1 and 2. The vertical dynamic-stiffness coefficient in Fig. 4-29, again nondimensionalized by the exact static-stiffness coefficient, calculated based on disks with the cones agrees well with the solution of Emperador and Dominguez [E2].

An extension to determine the dynamic-stiffness coefficients of a cylindrical foundation embedded in a stratified site consisting of many layers on an underlying homogeneous halfspace is discussed in Appendix D.

4.5 LIMITATION OF SINGLE CONE MODEL

As already mentioned at the beginning of this chapter (Fig. 4-2a), a single truncated semi-infinite cone cannot be used to model the contribution of the sidewall to the dynamic-stiffness coefficient of an embedded foundation. Although simple and, in principle, attractive, this cone with a horizontal axis is not based on a physically sound assumption, as it over-estimates the contribution to the static-stiffness coefficient considerably. This is verified in this section, using the rigorous solution of elastodynamics, which is thus essential to be able to evaluate the performance of any simple physical model. Any approximate method has to be systematically investigated, if necessary modified appropriately, or even abandoned.

It could be suggested that the contributions of the base with radius r_0 and of the cylindrical sidewall with embedment e of the embedded foundation should be calculated separately. The former could be determined as for a surface disk of radius r_0, leading to the same aspect ratio z_0/r_0 and thus opening angle of the cone with a vertical axis. The latter could be calculated again based on a strength-of-materials approach, assuming a "cone" with an apex on the embedded foundation's axis of symmetry, a horizontal axis and a contact area equal to $A_0 = 2\pi r_0 e$. This contribution of the sidewall is illustrated for the vertical motion in Fig. 4-30a, whereby shear distortions occur in the cone. The cone's apex height z_0 determined by geometry equals r_0, yielding an opening angle arctan e/r_0. The area at the radial distance r equals $2\pi(e/r_0)r^2$, the corresponding shear force equals

$$Q = 2\pi \frac{e}{r_0} r^2 \rho c_s^2 u_{,r} \tag{4-55}$$

with u denoting the vertical displacement. Formulating equilibrium of the infinitesimal element taking the inertial load into account

$$-Q+Q+Q_{,r}dr - \rho 2\pi \frac{e}{r_0} r^2 dr\ddot{u} = 0 \tag{4-56}$$

and substituting Eq. 4-55 results in

$$u_{,rr} + \frac{2}{r} u_{,r} - \frac{\ddot{u}}{c_s^2} = 0 \tag{4-57}$$

This is the equation of motion of the translational cone (in shear), shown in Eq. A-3. The corresponding interaction force P_s of the sidewall is calculated for a vertical displacement u_0 and velocity $\dot{u}_0$ of the embedded foundation as (Eqs. 2-62 and 2-63)

$$P_s = K_s u_0 + C_s \dot{u}_0 \tag{4-58}$$

where

$$K_s = \frac{\rho c_s^2 2\pi r_0 e}{r_0} = \rho c_s^2\, 2\pi e \tag{4-59a}$$

$$C_s = \rho c_s\, 2\pi r_0 e \tag{4-59b}$$

The base's contribution equals

$$P_b = K_b u_0 + C_b \dot{u}_0 \tag{4-60}$$

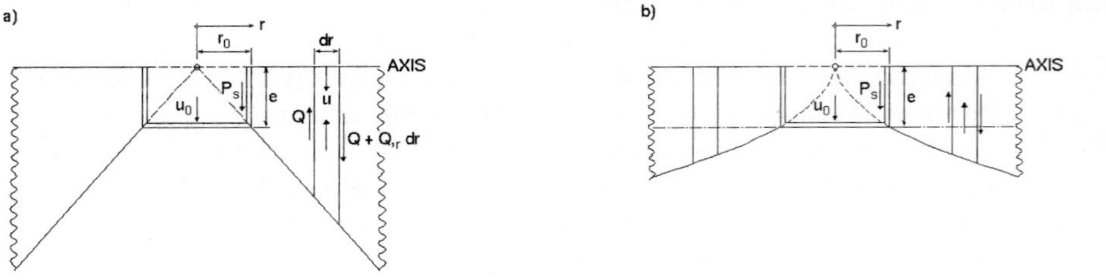

a. Translational cone with opening angle determined by geometry and horizontal axis distorting in shear.

b. Translational cone-wedge with $n = 3/2$ and horizontal axis distorting in shear.

Fig. 4-30 Inappropriate physical models to represent contribution of sidewall to vertical dynamic-stiffness coefficient.

with (for $\nu \leq 1/3$)

$$K_b = \frac{\rho c_p^2 \pi r_0^2}{z_0} \tag{4-61a}$$

$$C_b = \rho c_p \pi r_0^2 \tag{4-61b}$$

Summing the two contributions leads to the interaction force-displacement relationship of the embedded foundation

$$P_0 = K u_0 + C \dot{u}_0 \tag{4-62}$$

with

$$K = K_b + K_s = \frac{\rho c_p^2 \pi r_0^2}{z_0} \left(1 + 2 \frac{e z_0}{r_0^2} \frac{c_s^2}{c_p^2}\right) \tag{4-63a}$$

$$C = C_b + C_s = \rho c_p \pi r_0^2 + \rho c_s 2 \pi r_0 e \tag{4-63b}$$

The dashpot with the coefficient C is the high-frequency limit of the dynamic-stiffness coefficient (ρc times area with $c = c_p$ for perpendicular motion and $c = c_s$ for tangential motion). As the frequency-dependency of the damping coefficient is small (Fig. 4-17), the damping of the embedded foundation is well represented. In contrast, the static-stiffness coefficient K is deficient. For instance for $\nu = 1/3$ ($c_p = 2c_s$), $z_0/r_0 = 2.094$ follows from Table 2-3B yielding from Eq. 4-63a

$$K = \frac{4 G r_0}{1 - \nu}\left(1 + 1.05 \frac{e}{r_0}\right) \tag{4-64}$$

When compared to the accurate formula in Eq. 4-36b with the factor $(1 + 0.54\, e/r_0)$, it follows that the contribution of the sidewall represented by the term $1.05\, e/r_0$ is nearly double what it should be ($0.54\, e/r_0$). Just in passing, it is worth mentioning that if for the apex height of the base's vertical cone the geometric quantity e (Fig. 4-2a) and not z_0 from Table 2-3 based on the static-stiffness coefficient of a surface foundation were used, the discrepancy would be even larger for $e < 2r_0$.

It is instructive to address the horizontal motion, too. Based on an analogous physical model, the sidewall's static-stiffness coefficient can be formulated as

$$K_s = \frac{1}{2}\left(\rho c_s^2\, 2\pi e + \rho c_p^2\, 2\pi e\right) \tag{4-65}$$

In this reasonable assumption shear and dilatational waves are involved in equal parts, as shown in the lower-right part of Fig. 2-49. The total static-stiffness coefficient then follows as

$$K = \frac{\rho c_s^2 \pi r_0^2}{z_0} + K_s = \frac{\rho c_s^2 \pi r_0^2}{z_0}\left(1 + \frac{e z_0}{r_0^2} + \frac{e z_0}{r_0^2}\frac{c_p^2}{c_s^2}\right) \tag{4-66}$$

For $\nu = 1/3$, $z_0/r_0 = 0.654$, yielding

$$K = \frac{8 G r_0}{2-\nu}\left(1 + 3.27\frac{e}{r_0}\right) \tag{4-67}$$

The contribution of the sidewall, $3.27\, e/r_0$, is more than three times too large, as can be verified from Eq. 4-36a with the value e/r_0. The damping coefficient depending on the contact area is again quite accurate.

Returning to the vertical motion, a cylindrical surface corresponding to a cone with vanishing opening angle could be selected to calculate the contribution of the sidewall to the dynamic-stiffness coefficient (dashed-dotted line in Fig. 4-30b). In this case the soil with horizontally propagating shear waves consists of a soil layer of constant depth e with a cylindrical cavity which can be modeled as independent thin layers. This corresponds to $n = 1$ in the cone-wedge family addressed in Section 2.7.1. The static-stiffness coefficient, which is proportional to $n - 1$, vanishes (Eq. 2-332), but the corresponding real part for $a_0 = \infty$; K_∞, specified in Eq. 2-339a, is quite an accurate approximation for the spring coefficient for small ω (Fig. 2-62, curve for $n = 1$) and equals

$$K_s = \frac{\rho c_s^2\, 2\pi r_0 e}{2 r_0} = \rho c_s^2 \pi e \tag{4-68}$$

Using this value, which is half as large as that given in Eq. 4-59a,

$$K = \frac{\rho c_p^2 \pi r_0^2}{z_0}\left(1 + \frac{e z_0}{r_0^2}\frac{c_s^2}{c_p^2}\right) \tag{4-69}$$

results. For $\nu = 1/3$, the term of the sidewall equals $0.57\, e/r_0$, which is quite close to $0.54\, e/r_0$ of Eq. 4-36b.

By applying quite an artificial construction, it is possible to derive an acceptable static-stiffness coefficient (instead of using the real part at $\omega = \infty$, as just discussed), but the dynamic-stiffness coefficient still involves Hankel functions and is thus too complicated to use in practice. The area of the cone-wedge family of strength of materials is assumed to vary as $2\pi(e/r_0)r^{3/2}$ and not as $2\pi(e/r_0)r^2$ as for the cone shown in Fig. 4-30a, and not as $2\pi(e/r_0)r$ as for the cylinder. The corresponding meridian will vary as $r^{1/2}$, as shown as a solid line in Fig. 4-30b, with the apex still lying on the axis of symmetry of the embedded foundation. This corresponds to $n = 3/2$ in the derivation of Section 2.7.1. The sidewall's static-stiffness coefficient of Eq. 2-332 yields

$$K_s = \frac{\rho c_s^2 \, 2\pi r_0 e}{2r_0} = \rho c_s^2 \pi e \tag{4-70}$$

and thus the total static-stiffness coefficient of the embedded foundation is given in Eq. 4-69, which is accurate. However, for $n = 3/2$ the spring coefficient representing the real part of the dynamic-stiffness coefficient will increase for increasing a_0, as can be deducted from Fig. 2-62 by interpolating between the curves for $n = 1$ and 2. This behavior is opposite to the rigorous solution shown in Fig. 4-17. The same contradiction also applies to the behavior of the damping coefficient.

To summarize, although through arbitrary assumptions a good match for the static-stiffness coefficients can be achieved, the agreement of the dynamic-stiffness coefficient of the cone-wedge family (which is, in addition, difficult to calculate) with that of the exact solution, with the exception of the high-frequency limit, is unsatisfactory. It is thus necessary to model the dynamic behavior of an embedded foundation with a stack of disks embedded in the fullspace, as discussed in Sections 4.1 to 4.4.

4.6 FUNDAMENTAL LUMPED-PARAMETER MODEL FOR FOUNDATION EMBEDDED IN HALFSPACE

4.6.1 Coupled Horizontal and Rocking Motions

The very accurate method using disks embedded in a fullspace with the corresponding double cones to model a foundation embedded in a halfspace, described in Sections 4.2 and 4.3, does require using a simple special-purpose computer program. As an alternative the fundamental lumped-parameter model shown in Fig. 2-55 can be applied, which can be easily processed with a general-purpose structural dynamics program working in the time domain. As demonstrated in Section 2.6.2, the accuracy of the dynamic-stiffness coefficients for a surface foundation is high. It is shown in Figs. 4-16 to 4-19 that the frequency dependency of the dynamic-stiffness coefficients for an embedded foundation is similar to that of a surface foundation. It is thus to be expected that the corresponding fundamental lumped-parameter model will perform equally well for an embedded cylindrical foundation.

In this section, only compressible soil with Poisson's ratio $\nu \leq 1/3$ is addressed. For this range of ν the dimensionless coefficient μ_0 for the disk on the surface of a halfspace vanishes for all degrees of freedom (Table 2-10). Thus no mass or mass moment of inertia M_0 connected directly to the foundation node shown in Fig. 2-55 is introduced. The coefficients for the monkey-tail model of Fig. 2-55b are discussed in the following. As described in Section 2.6.2, the coefficients of the spring-dashpot model of Fig. 2-55a can be expressed as a function of those of the monkey-tail model (Eq. 2-305).

It should also be remembered that the fundamental lumped-parameter model corresponds to the consistent model described in Appendix B with a singular term and one first-order term ($M = 1$). The links are provided in Eqs. 2-307 and 2-308. The recursive procedure of Appendix C can thus also be applied.

For the vertical and torsional degrees of freedom (which are independent), the fundamental lumped-parameter models acting in the corresponding directions can be used directly. For a cylindrical foundation of embedment e, a (non-negligible) dynamic-stiffness coefficient, which *couples the horizontal and rocking degrees of freedom* referred to the center 0 of the circular base with equivalent radius r_0 (Fig. 4-31), arises. This necessitates a special treatment of the horizontal and rocking degrees of freedom. As for the horizontal motion the spring and damping coefficients of the dynamic-stiffness coefficient hardly depend on frequency (Fig. 4-16), the corresponding lumped-parameter model is chosen to consist of a spring with the static-stiffness coefficient K_h and a dashpot with the coefficient C_{0h} (to be determined by curve fitting to achieve an optimum fit of the corresponding dynamic-stiffness coefficients) only. The accurate value of Eq. 4-36a is used for K_h. To take the coupling effect into account, this lumped-parameter model corresponding to the horizontal degree of freedom u_0 is *connected eccentrically* to point 0 (Fig. 4-31). The eccentricities are denoted as f_K and f_C. The vertical bar connecting the spring and dashpot to point 0 is rigid. All elements present in the discrete-element model of the rotational cone (Fig. 2-17b) occur in the rocking lumped-parameter model attached in point 0 of the base. A (second) subscript r is used to denote the coefficients corresponding to this degree of freedom. It is important to stress that rocking around point 0 of the base also involves the horizontal lumped-parameter model. The coefficient K_{0r} of the direct spring reflecting rocking of the base is determined in such a way as to lead to the exact (total) rocking static-stiffness coefficient K_r of the embedded foundation

$$K_r = K_{0r} + f_K^2 K_h \tag{4-71}$$

K_r is specified in Eq. 4-36c. By construction, the exact result is thus obtained for the horizontal and rocking motions in the static case.

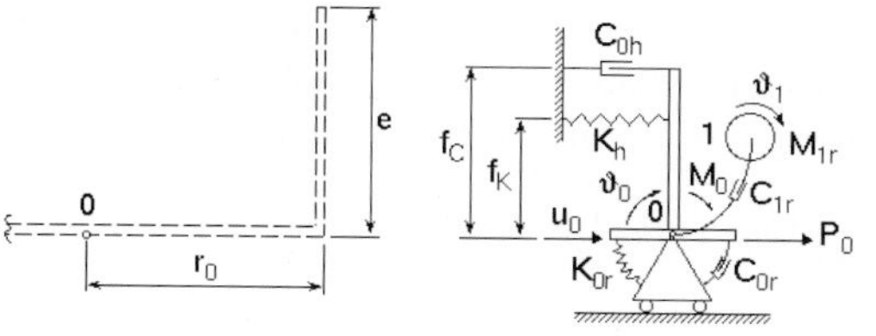

Fig. 4-31 Fundamental lumped-parameter model for foundation embedded in halfspace with coupling of horizontal and rocking motions.

To study the behavior of the coupling term of the horizontal and rocking degrees of freedom, the force-displacement relationships for harmonic load at point 0 are established. The amplitudes of the horizontal load and of the moment are denoted as $P_0(\omega)$ and $M_0(\omega)$; the amplitudes of the horizontal displacement and rotation are $u_0(\omega)$ and $\vartheta_0(\omega)$. Formulating the horizontal- and rotational-equilibrium equations in point 0 and the rotational-equilibrium equation in point 1 and eliminating $\vartheta_1(\omega)$ from these relations leads to

$$P_0(\omega) = K_h\left[1 + i\omega\frac{C_{0h}}{K_h}\right]u_0(\omega) + K_h f_K\left[1 + i\omega\frac{f_C}{f_K}\frac{C_{0h}}{K_h}\right]\vartheta_0(\omega) \tag{4-72a}$$

$$M_0(\omega) = K_h f_K\left[1 + i\omega\frac{f_C}{f_K}\frac{C_{0h}}{K_h}\right]u_0(\omega)$$

$$+ K_{0r}\left[1+\frac{K_h}{K_{0r}}f_K^2 - \frac{\dfrac{\omega^2 M_{1r}}{K_{0r}}}{1+\dfrac{\omega^2 M_{1r}^2}{C_{1r}^2}} + i\omega\left(\frac{M_{1r}}{C_{1r}}\frac{\dfrac{\omega^2 M_{1r}}{K_{0r}}}{1+\dfrac{\omega^2 M_{1r}^2}{C_{1r}^2}} + \frac{C_{0r}}{K_{0r}} + f_C^2\frac{C_{0h}}{K_{0r}}\right)\right]\vartheta_0(\omega) \tag{4-72b}$$

For $\omega \rightarrow 0$, the coefficient of $\vartheta_0(\omega)$ in Eq. 4-72b results in Eq. 4-71.

The dashpots C_{0h}, C_{0r}, C_{1r} and the mass M_{1r} are specified based on the dimensionless coefficients γ_{0h}, γ_{0r}, γ_{1r}, and μ_{1r} with the shear-wave velocity c_s as

$$C_{0h} = \frac{r_0}{c_s}\gamma_{0h}\, K_h \tag{4-73a}$$

$$C_{0r} = \frac{r_0}{c_s}\gamma_{0r}\, K_r \tag{4-73b}$$

$$C_{1r} = \frac{r_0}{c_s}\gamma_{1r}\, K_r \tag{4-73c}$$

$$M_{1r} = \frac{r_0^2}{c_s^2}\mu_{1r}\, K_r \tag{4-73d}$$

Note that the coefficients of the rocking lumped-parameter model are defined with respect to K_r (and not K_{0r}) in Eq. 4-73b–d, although K_{0r} is the coefficient of the direct spring (Fig. 4-31).

The dimensionless coefficients are determined so as to achieve an optimum fit between the dynamic-stiffness coefficients of the (coupled) lumped-parameter model (Eq. 4-72) and the corresponding rigorous values of the embedded foundation. The calibration proceeds as follows. First, the horizontal dynamic-stiffness coefficient $S_h(a_0)$ is addressed. Substituting Eq. 4-73a in the coefficient of $u_0(\omega)$ in Eq. 4-72a leads to

$$S_h(a_0) = K_h\,(1 + ia_0\,\gamma_{0h}) \tag{4-74}$$

with the dimensionless frequency

$$a_0 = \frac{\omega r_0}{c_s} \tag{4-75}$$

As can be seen from the exact result shown in Fig. 4-16, the spring coefficient $k_h(a_0)$ is approximately equal to 1 and the damping coefficient $c_h(a_0)$ is almost constant. The coefficient $c_h(a_0)$ is equal to γ_{0h} which can be expressed as a function of the embedment ratio e/r_0. The results are listed systematically in the next Section 4.6.2. Second, the coupling term of the dynamic-stiffness matrix $S_{hr}(a_0)$ is examined. Substituting Eq. 4-73a in the coefficient of $\vartheta_0(\omega)$ in Eq. 4-72a results in

$$S_{hr}(a_0) = S_{rh}(a_0) = K_h f_K\left(1 + ia_0\frac{f_C}{f_K}\gamma_{0h}\right) \tag{4-76}$$

Curve-fitting of the real part yields an expression for the eccentricity f_K. Calibration of the damping coefficient $c_{hr}(a_0)$, representing the imaginary part, leads to f_C, using the known equation for γ_{0h}. Third, the rocking dynamic-stiffness coefficient leads to the remaining dimensionless coefficients γ_{0r}, γ_{1r} and μ_{1r}. Substituting Eq. 4-73b–d in the coefficient of $\vartheta_0(\omega)$ of Eq. 4-72b and using Eq. 4-71 yields

$$S_r(a_0) = K_r\left[1 - \frac{\mu_{1r}a_0^2}{1 + \frac{\mu_{1r}^2}{\gamma_{1r}^2}a_0^2} + ia_0\left(\frac{\mu_{1r}}{\gamma_{1r}}\frac{\mu_{1r}a_0^2}{1 + \frac{\mu_{1r}^2}{\gamma_{1r}^2}a_0^2} + \gamma_{0r} + f_C^2\gamma_{0h}\frac{K_h}{K_r}\right)\right] \tag{4-77}$$

The last term in Eq. 4-77 is known, and the others are in the same form as in the fundamental lumped-parameter model (Eqs. 2-298 and 2-299).

4.6.2 Coefficients of Spring, Dashpots, and Mass

To analyze the *vertical* and *torsional* dynamic responses of a cylinder with equivalent radius r_0 embedded with depth e in a homogeneous halfspace with shear modulus G, Poisson's ratio ν, and shear-wave velocity c_s, the stable *lumped-parameter model* (monkey-tail arrangement) of Fig. 2.55b with *one internal degree of freedom* u_1 acting in the corresponding directions is well suited for practical applications. This fundamental lumped-parameter model consists of *one spring* with the static-stiffness coefficient K, *two dashpots* with coefficients C_0 and C_1, and *one mass* with the coefficient M_1. The mass M_0 assigned to the foundation node vanishes for compressible soil ($\nu \leq 1/3$). The coefficients depend on the dimensionless coefficients as follows (Eq. 2-294)

$$C_0 = \frac{r_0}{c_s}\gamma_0 K \tag{4-78a}$$

$$C_1 = \frac{r_0}{c_s}\gamma_1 K \tag{4-78b}$$

$$M_1 = \frac{r_0^2}{c_s^2}\mu_1 K \tag{4-78c}$$

For $\nu = 0.25$, γ_0, γ_1, and μ_1 are listed as a function of the embedment ratio e/r_0 together with the accurate equations for the static-stiffness coefficients (Eq. 4-36) in Table 4-1. The lack of reliable data for other Poisson's ratios does not allow these coefficients to be specified as a function of ν. For $e/r_0 = 0$, the coefficients for a disk on the surface for $\nu = 0.25$ (Table 2-10) are recovered.

To analyse the coupled *horizontal* and *rocking* motions, the (coupled) lumped-parameter model of Fig. 4-31 is used. Performing the calibration as described in the previous Section 4.6.1 leads to the results presented in Table 4-1. The horizontal dashpot with the coefficient C_{0h} follows with γ_{0h} (equals γ_0 in Table 4-1) from Eq. 4-73a. The eccentricities of the horizontal spring and dashpot equal

$$f_K = 0.25\,e \tag{4-79a}$$

$$f_C = 0.32\,e + 0.03\,e\left(\frac{e}{r_0}\right)^2 \tag{4-79b}$$

TABLE 4-1 STATIC STIFFNESS AND DIMENSIONLESS COEFFICIENTS OF FUNDAMENTAL LUMPED-PARAMETER MODEL (MONKEY-TAIL ARRANGEMENT) FOR CYLINDER EMBEDDED IN HALFSPACE

	Static Stiffness K	Dimensionless Coefficients of Dashpots γ_0	Dimensionless Coefficients of Dashpots γ_1	Mass μ_1
Horizontal	$\frac{8Gr_0}{2-\nu}\left(1+\frac{e}{r_0}\right)$	$0.68+0.57\sqrt{\frac{e}{r_0}}$	—	—
Vertical	$\frac{4Gr_0}{1-\nu}\left(1+0.54\frac{e}{r_0}\right)$	$0.80+0.35\frac{e}{r_0}$	$0.32-0.01\left(\frac{e}{r_0}\right)^4$	0.38
Rocking	$K_r = \frac{8Gr_0^3}{3(1-\nu)}\left[1+2.3\frac{e}{r_0}+0.58\left(\frac{e}{r_0}\right)^3\right]$ $K_{0r} = K_r - \frac{Gr_0^3}{2(2-\nu)}\left(1+\frac{e}{r_0}\right)\left(\frac{e}{r_0}\right)^2$	$0.15631\frac{e}{r_0}$ $-0.08906\left(\frac{e}{r_0}\right)^2$ $-0.00874\left(\frac{e}{r_0}\right)^3$	$0.40+0.03\left(\frac{e}{r_0}\right)^2$	$0.33+0.10\left(\frac{e}{r_0}\right)^2$
Torsional	$\frac{16Gr_0^3}{3}\left(1+2.67\frac{e}{r_0}\right)$	—	$0.29+0.09\sqrt{\frac{e}{r_0}}$	$0.20+0.25\sqrt{\frac{e}{r_0}}$

The static-stiffness coefficient K_r of the (total) embedded foundation (Eq. 4-36c) is specified in Table 4-1. The rocking lumped-parameter model attached to the base has a static-stiffness coefficient K_{0r} calculated from Eq. 4-71 which is the coefficient of the direct spring. However, to calculate the coefficients of the dashpots C_{0r}, C_{1r}, and of the mass M_{1r}, K_r must be used as specified in Eq. 4-73b–d together with the dimensionless coefficients, determined by calibration again for $\nu = 0.25$ in Table 4-1.

The dynamic-stiffness coefficients for harmonic loading calculated from Eqs. 4-74, 4-77, and so on are plotted as lines for $e/r_0 = 0.5$, 1, and 2 in each of the Figs. 4-32, 4-33, 4-34 and 4-36 for the horizontal, vertical, rocking, and torsional motions. The representation of Eq. 4-52 is used. The rigorous values shown as discrete points are taken from Apsel and Luco [A1]. By construction, k_h equals 1 for all embedment ratios. In Fig. 4-35, the coupling term determined from Eq. 4-76 is examined. k_{hr} equals 1. The agreement is satisfactory.

4.6.3 Hammer Foundation with Partial Uplift of Anvil

As an example of a nonlinear soil–structure-interaction analysis, the vibration of a hammer foundation embedded in soil with an eccentrically mounted anvil is examined. The head impacts against the anvil, which is a massive steel block (Fig. 4-37). This anvil is supported by a viscoelastic suspension (pad) on the foundation block of concrete which is embedded. As a tension-resistant connection is not provided for the pads, the anvil will partially uplift from the block when the dynamic stress (tension) exceeds the static stress (compression). A nonlinear dynamic system thus occurs. Any possible separation of the block from the adjacent soil is disregarded in the following.

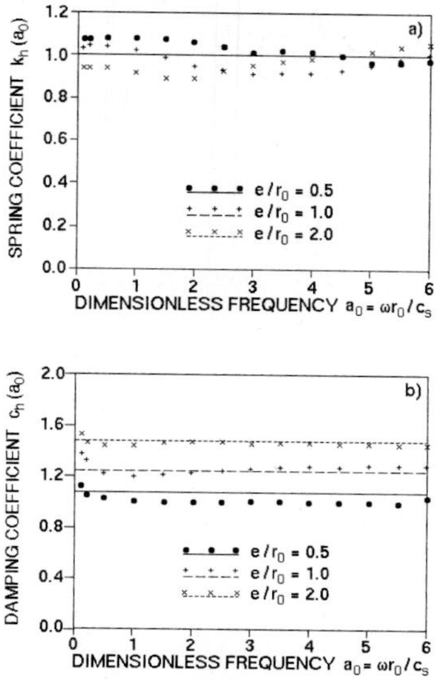

Fig. 4-32 Dynamic-stiffness coefficient for harmonic loading of cylindrical foundation embedded in halfspace in horizontal motion.

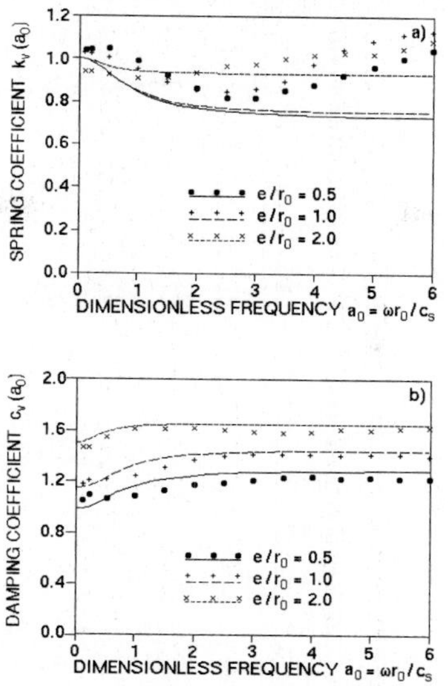

Fig. 4-33 Dynamic-stiffness coefficient for harmonic loading of cylindrical foundation embedded in halfspace in vertical motion.

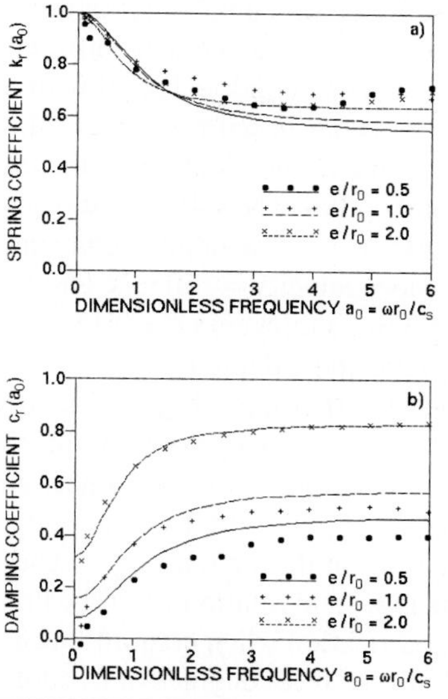

Fig. 4-34 Dynamic-stiffness coefficient for harmonic loading of cylindrical foundation embedded in halfspace in rocking motion.

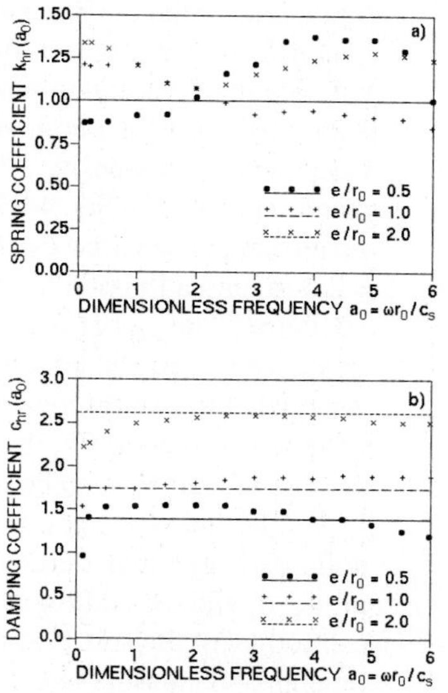

Fig. 4-35 Coupling term of horizontal and rocking motions of dynamic-stiffness matrix for harmonic loading of cylindrical foundation embedded in halfspace.

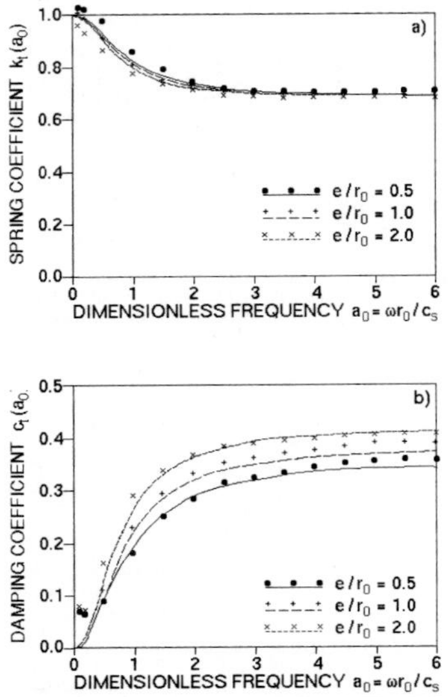

Fig. 4-36 Dynamic-stiffness coefficient for harmonic loading of cylindrical foundation embedded in halfspace in torsional motion.

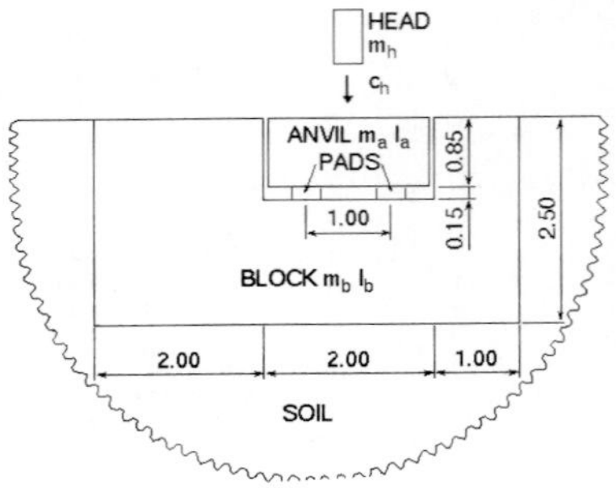

Fig. 4-37 Hammer foundation with inertial block embedded in halfspace.

More specifically, the head with mass $m_h = 1.5 \cdot 10^3$ kg impacts with a velocity c_h = 5 m/s against the anvil, which excites the anvil with an initial velocity. In this conservative simplification, no impact force-time history is used; it is assumed to be infinitely short. The coefficient of restitution (collision) is disregarded. The cylindrical anvil with mass density $7.85 \cdot 10^3$ kg/m^3 is mounted with an eccentricity = 0.5 m with respect to the cylindrical block with mass density $2.5 \cdot 10^3$ kg/m^3. The anvil and the block, with the dimensions in meters specified in Fig. 4-37, are modeled as rigid bodies. The soil consists of an undamped elastic halfspace with a cylindrical excavation (radius = 2.5 m, depth = embedment of block = 2.5 m) and a Poisson's ratio = 0.25, a shear-wave velocity = 150 m/*s*, and a density = $2 \cdot 10^3$ kg/m^3. The pad of the anvil has an area $A = 3$ m^2, a thickness $d = 0.15$ m, a modulus of elasticity $E = 100$ M Pa, a Poisson's ratio $\nu = 0.25$, and a damping ratio $\zeta = 0.05$. The pad is modeled by two discrete elements at a distance of 1.0 m apart (Fig. 4-37). For the analysis with partial anvil uplift, the motion of the anvil, and to a lesser extent of the block, is to be determined, and it is to be compared to the corresponding results of a linear analysis.

The linear dynamic model, which is shown in Fig. 4-38, has eight degrees of freedom: the horizontal, vertical, and rotational motions of the anvil u_a, w_a, ϑ_a and of the block u_b, w_b, ϑ_b and the additional vertical displacement w_1 and rotation ϑ_1 of the lumped-parameter model of the soil. It applies without any modification when no uplift occurs. The coefficients of the dashpots C_0, C_1 and the mass M_1 for the three degrees of freedom of the rigid embedded cylindrical foundation follow from Eqs. 4-73 and 4-78 with the corresponding dimensionless coefficients γ_0, γ_1, μ_1 and the static-stiffness coefficients K specified in Table 4-1. The eccentricities of the lumped-parameter model in the horizontal direction f_K, f_C are calculated from Eq. 4-79. The spring and dashpot coefficients of the two discrete elements of the pad in the horizontal and vertical directions are determined as

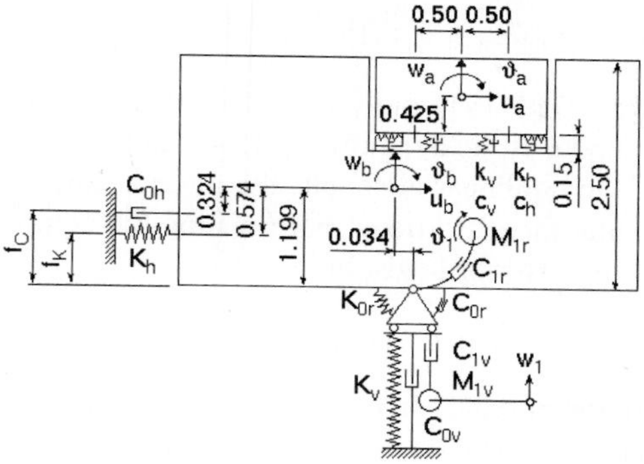

Fig. 4-38 Dynamic model.

$$k_h = \frac{E}{2(1+\nu)}\frac{A}{2d} \tag{4-80a}$$

$$c_h = 2\sqrt{k_h m_a}\zeta \tag{4-80b}$$

$$k_v = \frac{EA}{2d} \tag{4-80c}$$

$$c_v = 2\sqrt{k_v m_a}\zeta \tag{4-80d}$$

where m_a denotes the mass of the anvil. These values apply when no uplift occurs. They are not activated during uplift. Uplift starts when the total force in the vertical direction—calculated as the sum of the dead load and the dynamic part in one of the discrete elements of the pad—becomes zero. The corresponding horizontal force is then also set equal to zero, and an additional degree of freedom is introduced. Contact is regained when the displacement of the anvil relative to the block becomes equal to the length of the unloaded spring (penetration). The number of degrees of freedom is then again equal to that of the linear system.

The dynamic system's response—a free vibration—is triggered by the initial velocity $\dot{w}_a(t = 0^+)$ of the anvil in the vertical direction. All displacements and the other velocities are zero at $t = 0$. Formulating the law of momentum (before and after the impact)

$$m_h c_h = -(m_a + m_h)\,\dot{w}_a(t = 0^+) \tag{4-81}$$

leads to

$$\dot{w}_a(t = 0^+) = -\frac{m_h}{m_a + m_h}c_h \tag{4-82}$$

The following explicit algorithm with a predictor–corrector scheme already examined for the rigid foundation block on a soil halfspace in Section 2.3.4 is used for the time integration. The eight (or nine) degrees of freedom are assembled in the vector $\{u\}$. Starting from the known motion at time $(n-1)\,\Delta t$—that is, $\{u\}_{n-1}$, $\{\dot{u}_{n-1}\}$, $\{\ddot{u}\}_{n-1}$—the displacements and velocities are determined at time $n\Delta t$ as

$$\{u\}_n = \{u\}_{n-1} + \Delta t\,\{\dot{u}\}_{n-1} + \frac{\Delta t^2}{2}\{\ddot{u}\}_{n-1} \tag{4-83a}$$

$$\{\tilde{\dot{u}}\}_n = \{\dot{u}\}_{n-1} + \frac{\Delta t}{2}\{\ddot{u}\}_{n-1} \tag{4-83b}$$

A tilde (~) denotes a predicted value. With $\{u\}_n$ and $\{\tilde{\dot{u}}\}_n$ the internal forces—that is, the linear forces in the springs and dashpots of the dynamic model—can be calculated at time $n\Delta t$. Based on the material law of the nonlinear elements, the internal forces in such elements, the pads, are determined using the distortions. Formulating the equilibrium equations at time $n\Delta t$ results, for vanishing exterior loads, in

$$\{\ddot{u}\}_n = -[M]^{-1}\{F\}_n \tag{4-84}$$

$\{F\}$ is the vector of the resultants of the internal forces. As the mass matrix $[M]$ is diagonal, the accelerations are calculated on an element basis. The corrected velocities at time $n\Delta t$ are equal to

$$\{\dot{u}\}_n = \{\tilde{\dot{u}}\}_n + \frac{\Delta t}{2}\{\ddot{u}\}_n \tag{4-85}$$

which concludes the calculation of the time step. Processing a system with nonlinear elements in an explicit algorithm is thus not more complicated than working with a linear system, as the internal forces are always calculated with the known and predicted values which depend on the motion at the previous time. No iterations arise.

The analysis is performed with $\Delta t = 10^{-3}$ *s*. In Fig. 4-39, the vertical displacement w_a (measured from the static equilibrium position), the horizontal displacement u_a, and the rotation ϑ_a at the center of the anvil are plotted as a function of time. For comparison, the linear results are also indicated. As expected, the partial uplift of the anvil increases the motion significantly.

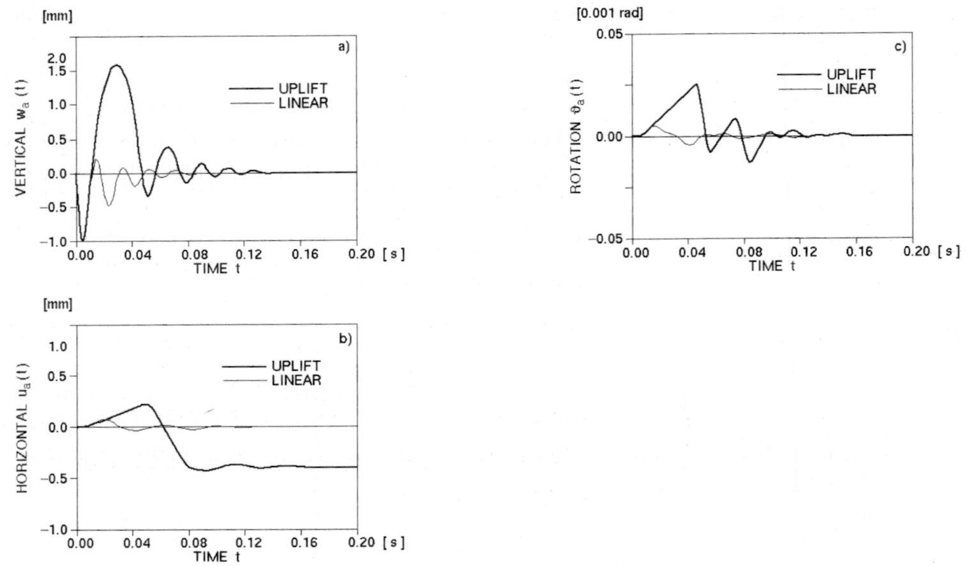

Fig. 4-39 Dynamic response at center of anvil.

4.7 BASIC LUMPED-PARAMETER MODEL FOR FOUNDATION EMBEDDED IN LAYER

4.7.1 Coupled Horizontal and Rocking Motions

For a surface foundation on a homogeneous soil halfspace, the fundamental lumped-parameter model with one internal degree of freedom (Fig. 2-55, Table 2-10) leads to accurate results; whereas for a surface foundation on a soil layer resting on rigid rock, the basic lumped-parameter model with two internal degrees of freedom (Fig. 3-25, Table 3-8) must be used. This is due to the fact that the dynamic-stiffness coefficient for a disk on a layer exhibits a strong frequency dependency. In particular, below the cutoff frequency (fundamental frequency of the layer) the radiation damping is zero; the imaginary part of the dynamic-stiffness coefficient for harmonic loading vanishes. The same behavior also holds for an embedded foundation. For instance, the strong frequency dependency of the vertical dynamic-stiffness coefficient for a cylinder embedded in a layer, somewhat reduced because of material damping, is visible in Fig. 4-28.

The basic lumped-parameter model of Fig. 3-25 with four springs with coefficients K_1, K_2, K_3, K_4, three dashpots with coefficients C_1, C_2, C_3, and one mass M (of which actually only six are independent) is introduced as the fundamental representation. This consistent model with the degree of the polynomial in the denominator $M = 3$ corresponds to one singular term, one first-order term and one second-order term, as described in the systematic procedure in Appendix B. As an alternative to the lumped-parameter model, the interaction forces can be determined based on a recursive evaluation as discussed in Appendix C.

The nomenclature of the foundation is illustrated in Fig. 4-40. Only the horizontal and rocking degrees of freedom u_0 and ϑ_0 which, due to the coupling demand special consideration, are shown in the figure, but the vertical and torsional motions are also studied. The homogeneous undamped soil layer with shear modulus G, Poisson's ratio ν, and shear-wave velocity c_s rests at depth d on rigid rock. The rigid massless cylindrical foundation of radius r_0 and embedment e is laterally in contact with the neighboring soil over the height e_c. This allows a realistic situation to be modeled where after backfill a gap or slippage close to the free surface of the soil has formed.

For the vertical and torsional motions, which are independent, the basic lumped-parameter model acting in the corresponding directions can be applied directly. For the *horizontal* and *rocking* motions *coupling* occurs. This is considered by selecting in addition to the basic lumped-parameter models in the horizontal and rocking directions another one with an eccentricity e acting in the horizontal direction (Fig. 4-41). As the coupling in this physical representation is modeled differently than in that shown in Fig. 4-31, the procedure is described in the following.

It is worth repeating that the dynamic analyst can directly construct the lumped-parameter model based on dimensionless coefficients listed in a table, for instance for various ratios of the radius of the base to the depth of layer r_0/d and contact ratios e_c/e for a given embedment ration e/r_0. The lumped-parameter model can be used as input to a general-purpose structural dynamics program working in the time domain, which even permits the nonlinear behavior of a structure founded on linear soil to be analysed.

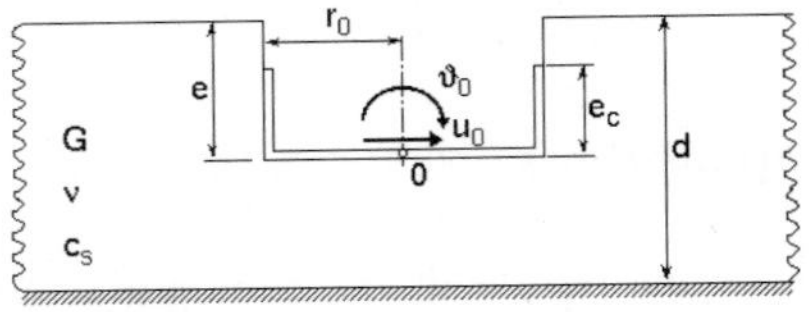

Fig. 4-40 Cylindrical foundation with partial contact over embedment height embedded in soil layer resting on rigid rock.

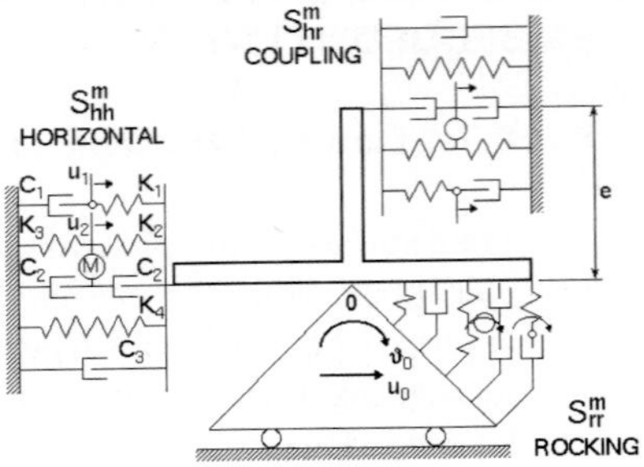

Fig. 4-41 Basic lumped-parameter model for embedded foundation with coupling of horizontal and rocking motions.

The singular part of the dynamic-stiffness coefficient (Section B1.1) is addressed first. In this section only compressible soil (Poisson's ratio $\nu \leq 1/3$) is examined. Neglecting the constant real part (Eq. B-3) and for waves propagating perpendicular to the vibrating surface, which develop for $\omega \rightarrow \infty$ with c_p (dilatational-wave velocity) in the normal and c_s in the tangential directions, the remaining singular part for the various degrees of freedom referred to point 0 of the base is equal to

$$\text{vertical} \quad S_s(a_0) = \pi G r_0 \left(2\frac{e_c}{r_0} + \frac{c_p}{c_s} \right) i a_0 \tag{4-86a}$$

$$\text{torsional} \quad S_s(a_0) = \pi G r_0^3 \left(\frac{1}{2} + 2\frac{e_c}{r_0} \right) i a_0 \tag{4-86b}$$

$$\text{horizontal} \quad S_s(a_0) = \pi G r_0 \left[1 + \frac{e_c}{r_0} \left(1 + \frac{c_p}{c_s} \right) \right] i a_0 \tag{4-86c}$$

$$\text{rocking} \quad S_s(a_0) = \frac{\pi}{3} G r_0^3 \left[3\frac{e_c}{r_0} + \frac{e_c^3}{r_0^3} + \left(\frac{3}{4} + \frac{e_c^3}{r_0^3} \right) \frac{c_p}{c_s} \right] i a_0 \tag{4-86d}$$

$$\text{coupling} \quad S_s(a_0) = + \frac{\pi}{2} G r_0^2 \frac{e_c^2}{r_0^2} \left(1 + \frac{c_p}{c_s} \right) i a_0 \tag{4-86e}$$

with the dimensionless frequency

$$a_0 = \frac{\omega r_0}{c_s} \tag{4-87}$$

For the coupled horizontal and rocking motions, a physical representation involving modeled dynamic-stiffness coefficients must be developed first, denoted with a superscript m. The lumped-parameter model is shown in Fig. 4-41. It consists of three models representing the modeled rocking dynamic-stiffness coefficient $S^m_{rr}(a_0)$, the horizontal one $S^m_{hh}(a_0)$, and the coupling one $S^m_{hr}(a_0)$. The corresponding dynamic-stiffness matrix $[S^m(a_0)]$

with respect to the amplitudes of the horizontal displacement $u_0(a_0)$ and the rotation $\vartheta_0(a_0)$ equals

$$[S^m(a_0)] = \begin{bmatrix} S^m_{hh}(a_0) + S^m_{hr}(a_0) & -e\,S^m_{hr}(a_0) \\ -eS^m_{hr}(a_0) & S^m_{rr}(a_0) + e^2 S^m_{hr}(a_0) \end{bmatrix} \tag{4-88}$$

with e denoting the embedment. Equating $[S^m(a_0)]$ to the corresponding known dynamic-stiffness matrix $[S(a_0)]$ leads to

$$S^m_{hh}(a_0) = S_{hh}(a_0) + \frac{S_{hr}(a_0)}{e} \tag{4-89a}$$

$$S^m_{rr}(a_0) = S_{rr}(a_0) + eS_{hr}(a_0) \tag{4-89b}$$

$$S^m_{hr}(a_0) = -\frac{S_{hr}(a_0)}{e} \tag{4-89c}$$

$S^m_{hh}(a_0)$, $S^m_{rr}(a_0)$, and $S^m_{hr}(a_0)$ are represented by the lumped-parameter model of Fig. 3-25.

As discussed in Section B1.4, the rate of energy transmission $N(a_0)$ must be nonnegative for all frequencies for a stable total dynamic system. Analogous to the scalar case specified in Eq. B-15, $N(a_0)$ is formulated for the matrix case with $u_0(a_0)$ and $\vartheta_0(a_0)$ assembled in $\{u_0(a_0)\}$ as

$$N(a_0) = \frac{c_s}{2r_0} a_0 \{\overset{*}{u}_0(a_0)\}^T Im\,[S(a_0)]\,\{u_0(a_0)\} \tag{4-90}$$

with an asterisk as superscript denoting the complex conjugate value. For $N(a_0) \geq 0$, $\mathrm{Im}[S(a_0)] = a_0[c(a_0)]$, the *damping matrix* $[c(a_0)]$, *must be positive semi-definite* for every a_0.

For the coupled horizontal and rocking degrees of freedom, the damping matrix $[c(a_0)]$ (see Eq. 4-88)

$$[c(a_0)] = \begin{bmatrix} \overset{m}{c}_{hh}(a_0) + \overset{m}{c}_{hr}(a_0) & -e\overset{m}{c}_{hr}(a_0) \\ -e\overset{m}{c}_{hr}(a_0) & \overset{m}{c}_{rr}(a_0) + e^2\overset{m}{c}_{hr}(a_0) \end{bmatrix} \tag{4-91}$$

is addressed. For a positive semi-definite matrix, the eigenvalues must be nonnegative which can be expressed as follows

$$\overset{m}{c}_{hh}(a_0) + \overset{m}{c}_{hr}(a_0) \geq 0 \tag{4-92a}$$

$$\overset{m}{c}_{rr}(a_0) + e^2\overset{m}{c}_{hr}(a_0) \geq 0 \tag{4-92b}$$

$$\begin{aligned} &\left(\overset{m}{c}_{hh}(a_0) + \overset{m}{c}_{hr}(a_0)\right)\left(\overset{m}{c}_{rr}(a_0) + e^2\overset{m}{c}_{hr}(a_0)\right) - e^2\left(\overset{m}{c}_{hr}(a_0)\right)^2 = \\ &\overset{m}{c}_{hh}(a_0)\,\overset{m}{c}_{rr}(a_0) + \overset{m}{c}_{rr}(a_0)\,\overset{m}{c}_{hr}(a_0) + e^2\overset{m}{c}_{hh}(a_0)\,\overset{m}{c}_{hr}(a_0) \geq 0 \end{aligned} \tag{4-92c}$$

For all investigated ratios of the radius of the foundation to the depth of the layer r_0/d and contact ratios e_c/e, these conditions are satisfied.

4.7.2 Achieved Accuracy

The results of Tassoulas [T1] are used for calibration. As this analysis is performed with some material damping, the influence of the material damping is subsequently eliminated approximately. The resulting dynamic-stiffness coefficients are accurate and are denoted as "exact" in the figures to be discussed. The damping coefficient $c(a_0)$ below the cutoff frequency remains small, but does not vanish.

For the vertical and torsional degrees of freedom the curve fitting is performed from zero to $a_0 = 2.5$, for the coupled horizontal and rocking motions up to $a_0 = 1$. This covers the frequency range of seismic excitation and of most machine vibrations. For all cases presented, the embedment ratio e/r_0 equals 1 and Poisson's ratio $\nu = 1/3$. For a ratio of the radius of the foundation to the depth of the layer $r_0/d = 0.5$ and a contact ratio $e_c/e = 1$, Fig. 4-42 presents the spring coefficients $k(a_0)$ and the damping coefficients $c(a_0)$ defined in the standard manner (see Eq. 4-52) corresponding to the basic lumped-parameter model of Fig. 3-25 for the vertical and torsional degrees of freedom, and to the elements of the matrix of Eq. 4-88 describing the lumped-parameter model of Fig. 4-41 referred to point 0 for the other degrees of freedom. Figures 4-42a, 4-42c, and 4-42d thus correspond to $S^m_{hh}(a_0)$ + $S^m_{hr}(a_0)$, $S^m_{rr}(a_0) + e^2 S^m_{hr}(a_0)$ and - $e S^m_{hr}(a_0)$, respectively. The agreement with the exact values is good; in particular, the small values of $c(a_0)$ below the cutoff frequency are well represented. The contact ratio e_c/e is varied for the vertical degree of freedom in Fig. 4-43. As the results for $e_c/e = 0.5$ and $= 0$ show when compared to $e_c/c = 1$ (Fig. 4-42b), a loss of contact has a marked effect on $c(a_0)$ not always resulting in a reduction! Finally, the ratio of the radius of the foundation to the depth of the layer r_0/d is varied in Fig. 4-44. As expected, $c(a_0)$ is larger for $r_0/d = 1/3$ than for $r_0/d = 1/2$ (Fig. 4-42b).

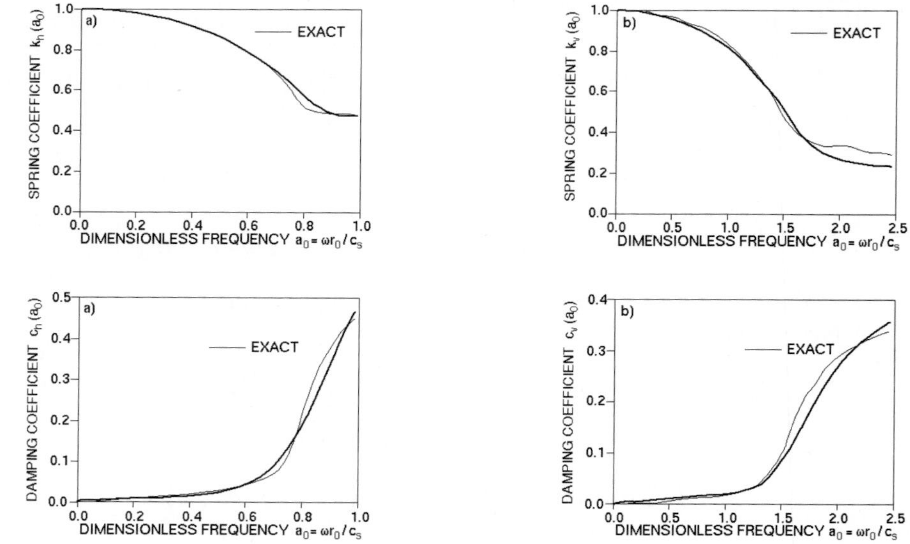

a. Horizontal **b.** Vertical.

Fig. 4-42 Dynamic-stiffness coefficient for harmonic loading of cylindrical foundation embedded in layer (embedment ratio = 1) for ratio of radius of foundation to depth of layer = 0.5 and contact ratio = 1.

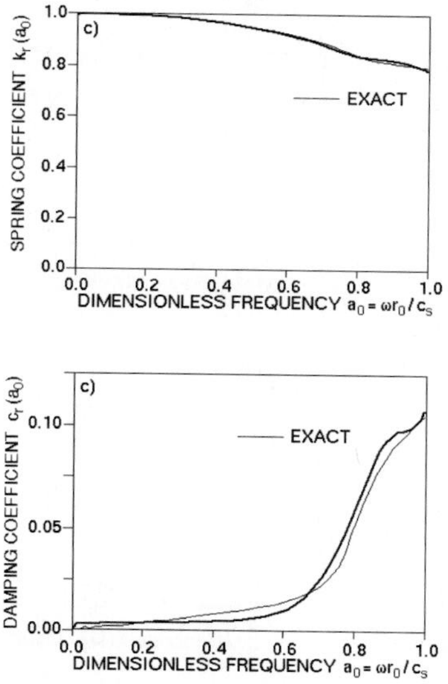

c. Rocking.

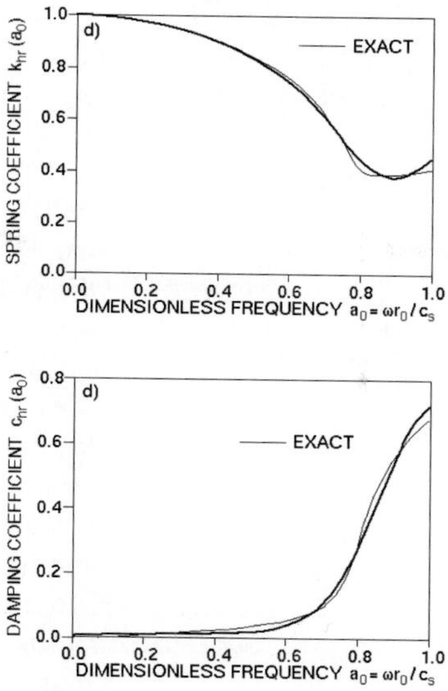

d. Coupling.

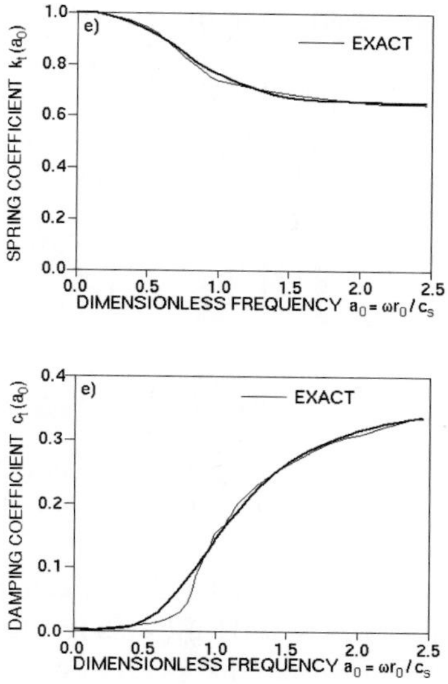

e. Torsional.

Fig. 4-42 (continued)

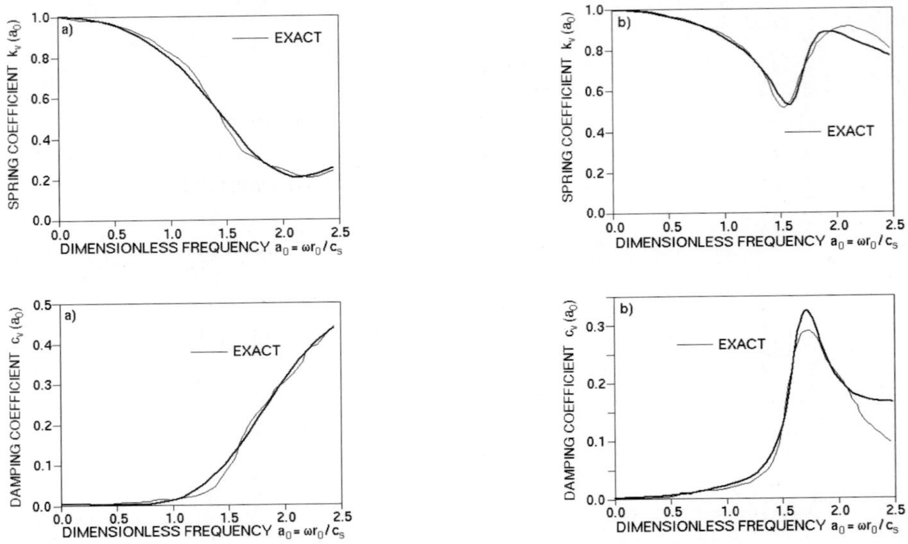

a. Contact ratio = 0.5.

b. Contact ratio = 0.

Fig. 4-43 Vertical dynamic-stiffness coefficient for harmonic loading of cylindrical foundation embedded in layer (embedment ratio = 1) for ratio of radius of foundation to depth of layer = 0.5.

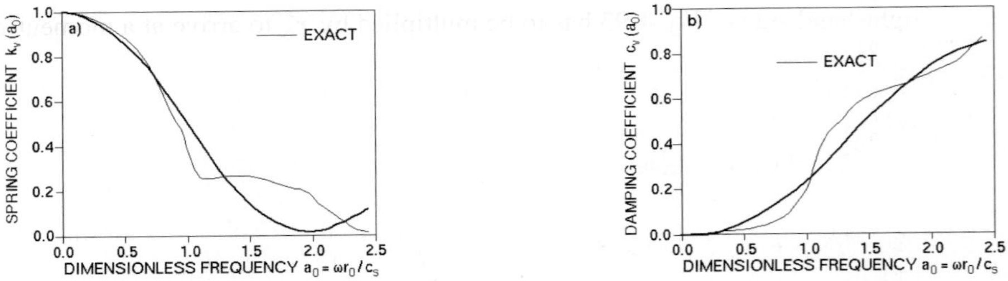

Fig. 4-44 Vertical dynamic-stiffness coefficient for harmonic motion of cylindrical foundation embedded in layer (embedment ratio = 1) for ratio of radius of foundation to depth of layer = 1/3 and contact ratio = 1.

If a better agreement up to a larger frequency is required, the degree of the polynomial in the denominator M must be chosen larger (Section B1.2). This results in a larger number of springs, dashpots, and masses as well as of internal degrees of freedom than present in Figs. 3-25 and 4-41. In Section 3.7.2, excellent agreement is demonstrated with $M = 12$ up to $a_0 = 12.56$ for all degrees of freedom of a surface disk on a soil layer resting on rigid rock.

4.7.3 Coefficients of Springs, Dashpots, and Mass

The rigid massless cylindrical foundation (Fig. 4-40) embedded in a soil layer of shear modulus G, Poisson's ratio $\nu = 1/3$, shear-wave velocity c_s, and depth d resting on rigid rock is addressed. The foundation of radius r_0 and embedment e is laterally in contact with the neighboring soil over the height e_c. The embedment ratio e/r_0 equals 1. The stable basic lumped-parameter model of Fig. 3-25 with only two internal degrees of freedom u_1, u_2 consists of four springs with coefficients K_1, K_2, K_3 K_4; three dashpots with coefficients C_1, C_2, C_3; and a mass with the coefficient M. This basic lumped-parameter model is used directly to model the vertical and the torsional degrees of freedom. For the coupled horizontal and rocking degrees of freedom, three basic lumped-parameter models are arranged as shown in Fig. 4-41. Note that the basic lumped-parameter model denoted with the word coupling is placed at a vertical eccentricity e (equals embedment) from the center of the base 0. For the basic lumped-parameter model involving translations (vertical, horizontal, coupling), the coefficients of the springs, dashpots, and the mass depend on the dimensionless coefficients k_i ($i = 1, 2, 3, 4$), $c_i = (i = 1, 2, 3)$, and m.

$$K_i = k_i G r_0 \tag{4-93a}$$

$$C_i = c_i G \frac{r_0^2}{c_s} \tag{4-93b}$$

$$M = mG \frac{r_0^3}{c_s^2} \tag{4-93c}$$

For the basic lumped-parameter model representing rotations (torsional and rocking), the right-hand side of Eq. 4-93 has to be multiplied by r_0^2 to arrive at a moment-rotation relationship.

For the two ratios of the radius of the foundation to the depth of the layer $r_0/d = 1/2$, $= 1/3$ and for the three contact ratios $e_c/e = 1.0$ (full lateral contact), $= 0.5$, $= 0$ (no lateral contact), the eight dimensionless coefficients k_i, c_i, m of the basic lumped-parameter models for the vertical and torsional degrees of freedom (Fig. 3-25) and for the coupled horizontal and rocking motions (Fig. 4-41) are specified in Table 4-2. With the information in the columns denoted as Horizontal, Rocking, Coupling, the coupled lumped-parameter model of Fig. 4-41 can be constructed directly. For the case of no lateral contact ($e_c/e = 0$) the coupling terms turn out to be very small and can thus be neglected. Note that certain dimensionless coefficients are negative in Table 4-2.

Material damping can be incorporated in the lumped-parameter model by either augmenting the discrete-elements (Voigt viscoelasticity, Section 2.4.2) or by introducing the corresponding frictional elements (Section 2.4.4).

4.7.4 Hammer Foundation with Partial Uplift of Anvil

As an example of a nonlinear soil–structure-interaction analysis, the vibration of a hammer foundation embedded in a soil layer resting on rigid rock with an eccentrically mounted anvil is examined (Fig. 4-45a). All dimensions are specified in meters. The arrangement is the same as defined and discussed in Section 4.6.3, the only difference being that the supporting soil consists of a layer and not a halfspace.

TABLE 4-2 DIMENSIONLESS COEFFICIENTS OF BASIC LUMPED-PARAMETER MODEL FOR CYLINDER EMBEDDED IN SOIL LAYER ON RIGID ROCK (EMBEDMENT RATIO $e/r_0 = 1$)

Radius to Depth r_0/d	Contact Ratio e_c/e		Vertical	Horizontal	Rocking	Coupling	Torsional
1/2	1.0	*k*1	-.203759 E+02	-.124401 E+02	-.125229 E+02	-.618776 E+01	-.139252 E+02
		*k*2	+.339543 E+01	+.286199 E+01	-.583152 E+00	+.202777 E+01	-.275441 E+01
		*k*3	-.617014 E+01	-.208541 E+02	-.814822 E-01	-.141784 E+02	+.178780 E+01
		*k*4	+.166202 E+02	+.794575 E+01	+.130945 E+02	+.337083 E+01	+.161164 E+02
		*c*1	-.918456 E+01	-.590158 E+01	-.315268 E+01	-.333135 E+01	-.774712 E-02
		*c*2	-.596381 E+00	-.516028 E+01	-.885823 E-01	-.340080 E+01	-.736101 E+00
		*c*3	+.131164 E+02	+.130103 E+02	+.322858 E+01	+.811310 E+01	+.858610 E+01
		m	-.987169 E+00	-.163126 E+02	-.680666 E+00	-.146553 E+02	-.962102 E+00
	0.5	*k*1	-.190169 E+02	-.123585 E+02	-.918010 E+01	-.311508 E+01	-.150459 E+02
		*k*2	+.102770 E+02	+.382788 E+01	+.934512 E+00	+.786487 E+00	+.149201 E+01
		*k*3	-.256293 E+02	-.116229 E+02	-.466308 E+01	-.869559 E+01	-.230599 E+01
		*k*4	+.480379 E+01	+.697738 E+01	+.821627 E+01	+.184030 E+01	+.132374 E+02
		*c*1	-.803919 E+00	-.129978 E+01	-.212247 E+01	-.715314 E+00	-.513171 E+00
		*c*2	-.378972 E+01	-.357027 E+01	-.316747 E+00	-.208337 E+01	-.403901 E+00
		*c*3	+.131677 E+02	+.102413 E+02	+.266675 E+01	+.326137 E+01	+.511390 E+01
		m	-.364874 E+01	-.820645 E+01	-.342125 E+01	-.888905 E+01	-.515523 E+00
	0.0	*k*1	-.199866 E+02	-.113528 E+02	-.801960 E+01	—	-.820959 E+01
		*k*2	+.324059 E+01	+.187819 E+01	+.103933 E+01	—	+.236828 E+00
		*k*3	-.138239 E+03	-.141228 E+02	-.800817 E+01	—	-.295213 E+00
		*k*4	+.151110 E+02	+.837372 E+01	+.584466 E+01	—	+.794727 E+01
		*c*1	-.577181 E+01	-.169786 E+01	-.101867 E+01	—	-.288545 E+00
		*c*2	-.891247 E+01	-.396633 E+01	-.157192 E+01	—	-.308176 E-01
		*c*3	+.151425 E+02	+.710633 E+01	+.313092 E+01	—	+.160082 E+01
		m	-.485815 E+02	-.142894 E+02	-.217586 E+01	—	-.372596 E-01
1/3	1.0	*k*1	-.215677 E+02	-.800686 E+01	-.112339 E+02	-.531331 E+01	-.158881 E+02
		*k*2	+.995664 E+01	+.248098 E+01	+.271244 E+01	+.128879 E+01	-.216892 E+01
		*k*3	-.299529 E+02	-.530555 E+01	-.112792 E+02	-.117090 E+02	+.122884 E+01
		*k*4	+.122789 E+01	+.460883 E+01	+.830774 E+01	+.314281 E+01	+.175253 E+02
		*c*1	-.214856 E+01	-.638370 E-01	-.185381 E+01	-.345899 E+01	-.770582 E+00
		*c*2	-.703468 E+01	-.234186 E+01	-.147482 E+01	-.442673 E+01	-.114118 E+01
		*c*3	+.195563 E+02	+.101919 E+02	+.461482 E+01	+.913903 E+01	+.899118 E+01
		m	-.476605 E+01	-.598035 E+01	-.101760 E+02	-.222249 E+02	-.244900 E+01
	0.5	*k*1	-.263609 E+02	-.105510 E+02	-.812675 E+01	-.258694 E+01	-.164865 E+02
		*k*2	+.106994 E+02	+.323771 E+01	+.327590 E+01	+.487010 E+00	+.162631 E+01
		*k*3	-.415582 E+02	-.101866 E+02	-.183711 E+02	-.708382 E+01	-.359665 E+01
		*k*4	+.391023 E+00	+.579774 E+01	+.434718 E+01	+.155304 E+01	+.138158 E+02
		*c*1	-.734715 E-02	-.691681 E-01	-.831614 E+00	-.752538 E+00	-.786309 E+00
		*c*2	-.101472 E+02	-.475156 E+01	-.272228 E+01	-.265221 E+01	-.129218 E+01
		*c*3	+.195330 E+02	+.114226 E+02	+.507228 E+01	+.383021 E+01	+.600218 E+01
		m	-.674277 E+01	-.148975 E+02	-.147137 E+02	-.128622 E+02	-.159889 E+01
	0.0	*k*1	-.147108 E+02	-.922525 E+01	-.736535 E+01	—	-.790274 E+01
		*k*2	+.600489 E+01	+.187933 E+01	-.907967 E+00	—	+.176502 E-01
		*k*3	-.355109 E+02	-.788239 E+01	-.157724 E+03	—	-.179897 E-01
		*k*4	+.527313 E+01	+.637232 E+01	+.684877 E+01	—	+.788488 E+01
		*c*1	-.203850 E+01	-.425306 E-01	-.168579 E+01	—	-.128670 E+00
		*c*2	-.830045 E+01	-.368700 E+01	-.114538 E+02	—	-.263292 E-03
		*c*3	+.145304 E+02	+.682700 E+01	+.130128 E+02	—	+.157125 E+01
		m	-.200705 E+02	-.139626 E+02	-.920928 E+02	—	-.331649 E-03

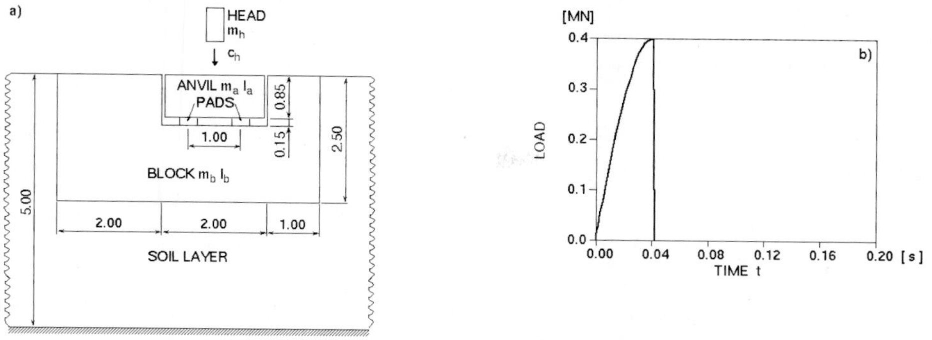

a. General layout. **b.** Impact-load time history.

Fig. 4-45 Hammer foundation with inertial block embedded in soil layer resting on rigid rock.

The dynamic model, which is shown in Fig. 4.46 has 14 degrees of freedom: the horizontal, vertical, and rocking motions of the anvil u_a, w_a, ϑ_a and of the block u_b, w_b, ϑ_b; and for each of the four basic lumped-parameter models two internal degrees of freedom u_{1v}, u_{2v}, u_{1h}, u_{2h}, ϑ_{1r}, ϑ_{2r}, u_{1c}, u_{2c}. The coefficients of the springs, dashpots, and masses for the vertical, horizontal, rocking, and coupling models follow from the first rows of Table 4-2 ($r_0/d = 0.5$, $e_c/e = 1$).

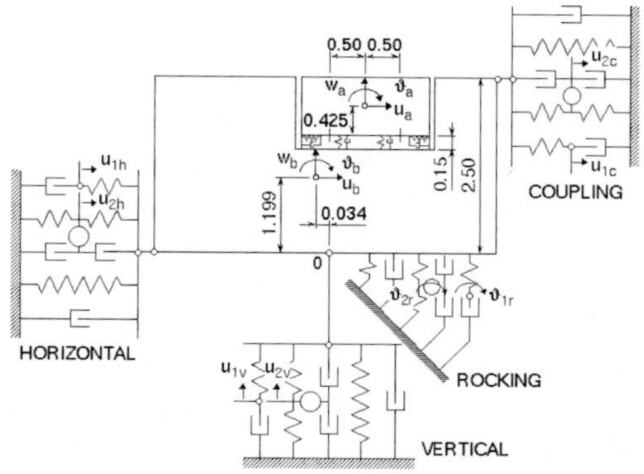

Fig. 4-46 Dynamic model.

The dynamic analysis is performed with an explicit time integration using a predictor–corrector scheme with a time step $\Delta t = 3.3 \cdot 10^{-4} s$. The vertical displacement w_a of the anvil's center is plotted in Fig. 4-47. The linear result is also indicated. As expected, the partial uplift of the anvil increases the motion significantly. For comparison, the response of this hammer foundation embedded in a homogeneous halfspace with the same soil properties, taken from Section 4.6.3 (Fig. 4-39a), is also shown. The response of the foundation with the soil layer is similar to that with the soil halfspace.

As a second possibility to excite the dynamic system, the impact load of Fig. 4-45b is applied to the anvil at rest. The vertical displacement of the anvil is presented in Fig. 4-48. For comparison, the soil is also modeled as a homogeneous halfspace with a cylindri-

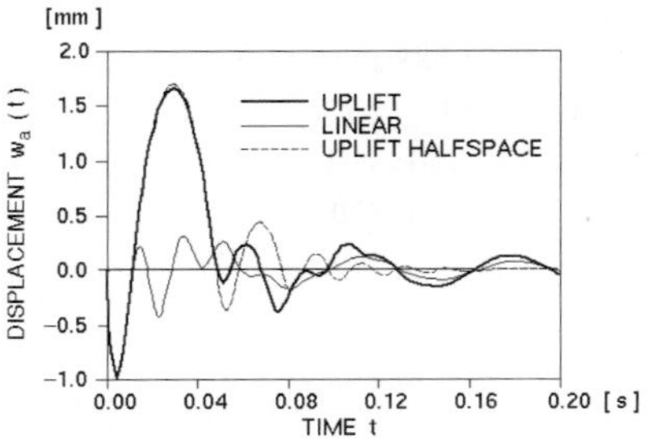

Fig. 4-47 Vertical displacement at center of anvil for load transmitted as initial velocity of anvil.

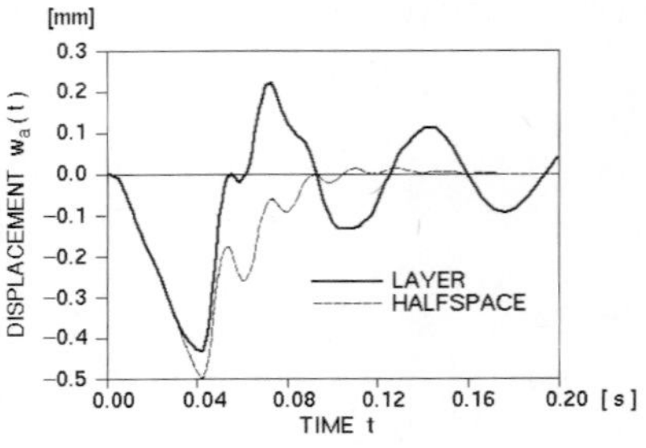

Fig. 4-48 Vertical displacement at center of anvil for impact load applied to anvil.

cal excavation based on the fundamental lumped-parameter model of Section 4.6. As the dominant frequencies of the impact load lie mostly below the cutoff frequency of the layer, where the radiation damping vanishes, the response of the system with the soil layer is significantly larger and decays slower than that with the soil halfspace after the initial phase.

4.8 SINGLE PILE

4.8.1 Analogy to Embedded Foundation

The dynamic analysis of a single pile is *analogous* to that of an embedded foundation. The building block of the procedure consists of a rigid disk of radius equal to that of the pile embedded in the soil which is represented by a double-cone model within the elastic fullspace. For the horizontal, rocking, and torsional motions the corresponding Green's functions for harmonic loading and in the time domain are applied directly. For the vertical motion (where the active length within which the displacement along the pile is appreciable compared to the radius), a weighted Green's function of the double cone and of the fullspace for a point load is constructed (Section 4.1.2). Anti-symmetrically and symmetrically loaded mirror disks are introduced to model the free surface and the fixed boundary of the soil. The soil region within the pile is regarded as a sandwich of rigid disks separated by soil layers, which are later analytically excavated and replaced by the pile material. The analysis can be performed for all motions in the frequency and time domains for a pile embedded in a halfspace and in a soil layer resting on rigid rock.

As an example, the matrix formulation for harmonic loading of a pile embedded in a halfspace is discussed in the next Section 4.8.2.

4.8.2 Matrix Formulation for Harmonic Loading of Pile Embedded in Halfspace

Dynamic-flexibility matrix of free field. A floating pile of diameter $2r_0$ and length ℓ in the soil's halfspace is addressed. The vertical degree of freedom is used for illustration. The horizontal and rocking motions follow analogously. The soil cylinder which will later be replaced by the pile is viewed as a sandwich consisting of m rigid disks equally spaced with soil between them (Fig. 4-49). In order to formulate this halfspace problem in the fullspace, an additional stack of m mirror-image disks is introduced. Each disk and its mirror image are subjected to the same force with amplitude $P_i(\omega)$ acting in the same direction and experience the same displacement with amplitude $u_i(\omega)$ ($i = 1, \ldots, m$); this guarantees that the middle plane coinciding with the free surface remains stress free.

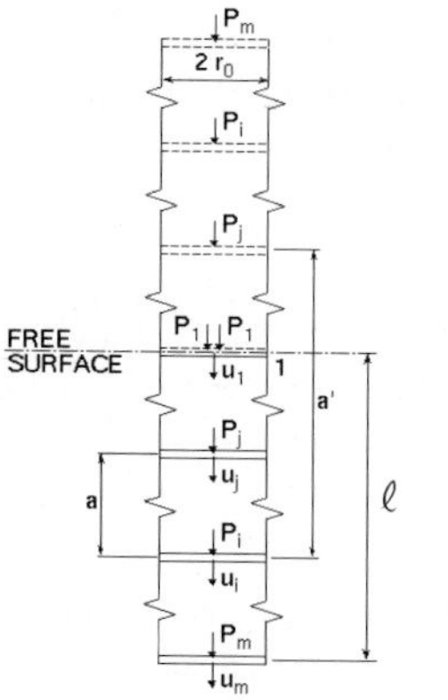

Fig. 4-49 Stack of disks to model pile embedded in halfspace with anti-symmetrically loaded mirror-image disks.

The displacement amplitudes $\{u(\omega)\}$ are related to the force amplitudes $\{P(\omega)\}$ by an $m \times m$ element dynamic-flexibility matrix $[G(\omega)]$

$$\{u(\omega)\} = [G(\omega)]\ \{P(\omega)\} \tag{4-94}$$

A typical coefficient $g_{ij}(\omega)$ of this dynamic-flexibility matrix specifies the displacement amplitude of the i-th disk (receiver) $u_i(\omega)$ caused by the harmonic forces of unit amplitude $P_j(\omega)$ applied to the j-th disk at distance a, $g_{ij}(a,\omega)$ and to its mirror image disk at distance a', $g_{ij}(a',\omega)$

$$g_{ij}(\omega) = g_{ij}(a,\omega) + g_{ij}(a',\omega) \tag{4-95}$$

The modified Green's function for vertical harmonic motion is specified in Eq. 4-10 with Eq. 4-4 and Eqs. 4-11, 4-12, and 4-13. The dynamic-flexibility matrix of the free field $[G(\omega)]$ is symmetric.

Dynamic-stiffness matrix. Equation 4-94 is solved for $\{P(\omega)\}$ by inverting the dynamic-flexibility matrix

$$\{P(\omega)\} = [S^f(\omega)]\ \{u(\omega)\} \tag{4-96}$$

in which $[S^f(\omega)] = [G(\omega)]^{-1}$ denotes the dynamic-stiffness matrix of the free field (virgin halfspace), discretized in the nodes corresponding to the disks. Replacing the cylindrical soil region by the pile results in a dynamic-stiffness matrix $[\Delta S(\omega)]$ with the same discretization and the difference of the properties of the pile and the soil

$$[\Delta S(\omega)] = [\Delta K] - \omega^2\ [\Delta M] \tag{4-97}$$

where $[\Delta K]$ and $[\Delta M]$ are the standard pile's static-stiffness and mass matrices calculated based on rod (beam) theory after subtracting the stiffness and mass of the (excavated) soil cylinder. Assembling $[S^f(\omega)]$ and $[\Delta S(\omega)]$ leads to the force-displacement relationship of the pile foundation

$$\{Q(\omega)\} = \left([S^f(\omega)] + [\Delta S(\omega)]\right) \{u(\omega)\} \tag{4-98}$$

The sum $[S^f(\omega)] + [\Delta S(\omega)]$ represents the dynamic-stiffness matrix of the pile embedded in soil discretized in all nodes correponding to the disks. For a vertical load with amplitude $Q_1(\omega)$ applied at the head of the pile, $\{Q(\omega)\} = [Q_1(\omega), 0, 0, \ldots 0]^T$, $\{u(\omega)\}$ follows from Eq. 4-98 as

$$\{u(\omega)\} = \left([S^f(\omega)] + [\Delta S(\omega)]\right)^{-1} \{Q(\omega)\} \tag{4-99}$$

The reciprocal of the first element in the coefficient matrix is equal to the vertical dynamic-stiffness coefficient $S(\omega)$ of the single pile

$$Q_1(\omega) = S(\omega)\ u_1(\omega) \tag{4-100}$$

It is worth mentioning that the horizontal and rocking degrees of freedom of a single pile are coupled. For the free field, the dynamic-stiffness matrices of horizontal motion (calculated with the translational double cones) and for rocking motion (calculated with the rotational double cones) corresponding to Eq. 4-96 are independent. The expected coupling is introduced through beam theory applied to the pile elements (Eq. 4-97).

The derivation presented above does not only lead to the dynamic-stiffness matrix of a single pile but also permits the displacements and pile forces as a function of depth to be calculated.

For nearly incompressible soil (1/3 < Poisson's ratio $\nu \leq 1/2$), again the same procedure for the vertical and rocking motions applies as discussed for an embedded foundation at the end of Section 4.3.2. It follows that the trapped mass need only be included for the undermost disk. Obviously, for a long, slender pile the mass trapped under the pile tip is extremely small by comparison to the mass of the pile itself and may be neglected with no appreciable loss of accuracy.

4.8.3 Example of Single Floating Pile

The dynamic-stiffness coefficients of a single (stiff) pile with slenderness ratio $\ell/(2r_0) = 15$, modulus of elasticity ratio $E_p/E = 1000$, mass density ratio $\rho_p/\rho = 1.43$ in an elastic half-space with $\nu = 0.4$ is calculated (subscript p denotes the pile). The hysteretic damping ratio ζ_g equals 0.05. This material damping affects the Green's function for harmonic motion. The static-stiffness coefficient K is multiplied by $1 + 2i\zeta_g$ and the wave propagation velocities by $1 + i\zeta_g$ (Section 2.4.1). The highest dimensionless frequency of interest $a_0 = \omega 2r_0/c_s$ equals 1. For 10 points per wave length, in analogy to Eq. 4-35

$$m - 1 = \frac{5}{\pi}\frac{l}{r_0}\frac{a_0}{2} \tag{4-101}$$

is formulated. This results for $a_0 = 1$ to $m - 1 = 23.9$; 25 equally spaced disks ($m = 25$) are selected.

To estimate the static-stiffness coefficients of a pile in a homogeneous halfspace, the accurate equations [G3] can be used.

$$\text{horizontal} \qquad K_h = 2r_0\, E \left(\frac{E_p}{E}\right)^{0.21} \qquad (4\text{-}102a)$$

$$\text{vertical} \qquad K_v = 3.8\, r_0\, E \left(\frac{\ell}{2r_0}\right)^{0.67} \qquad (4\text{-}102b)$$

$$\text{rocking} \qquad K_r = 1.2\, r_0^3\, E \left(\frac{E_p}{E}\right)^{0.75} \qquad (4\text{-}102c)$$

$$\text{coupling} \qquad K_{hr} = 0.88\, r_0^2\, E \left(\frac{E_p}{E}\right)^{0.5} \qquad (4\text{-}102d)$$

For the vertical direction, Eq. 4-102b actually applies to a floating pile in a layer with depth $= 2\ell$.

The static-stiffness coefficients of a single pile K are compared with the solutions presented in Eq. 4-102, which are shown in parentheses. The cone model results for the vertical motion in 17.5 Er_0 (23.1), for the horizontal in 7.4 Er_0 (8.5), for the rocking in 211.8 Er_0^3 (213.4), and the coupling term equals 25.2 Er_0^2 (27.8).

The dynamic-stiffness coefficients are decomposed in the standard manner as described in Eq. 4-52. The spring coefficient $k(a_0)$ and damping coefficent $c(a_0)$ are plotted as a function of a_0 for the vertical, horizontal, and rocking degrees of freedom in Fig. 4-50. The agreement with the rigorous solution of Kaynia and Kausel [K2], denoted as exact in the figure, is good. To make the deviations more visible, $k(a_0)$ is plotted starting from the value 0.6, not 0.

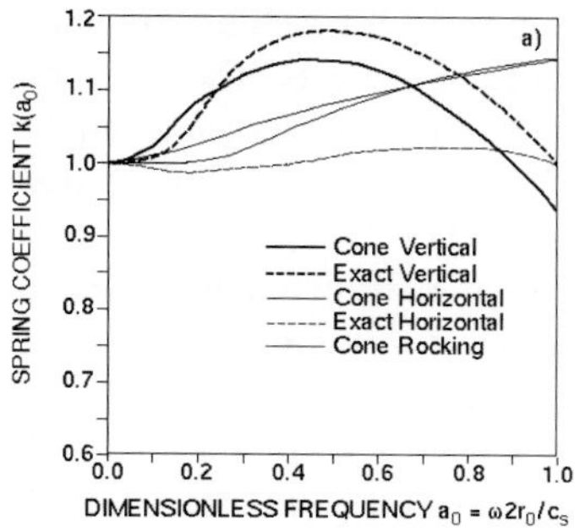

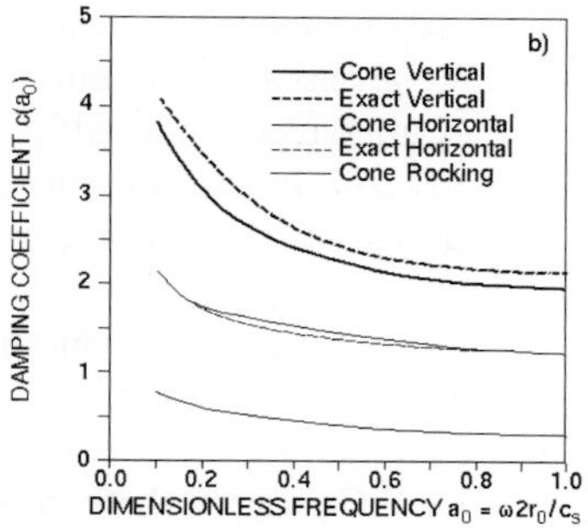

Fig. 4-50 Dynamic-stiffness coefficient for harmonic loading of single pile embedded in halfspace.

4.9 PILE GROUP

4.9.1 Group Behavior

To analyze the dynamic behavior of a group of piles with their heads connected by a rigid cap, the basemat, and with vertical axes, the interaction of the individual piles via the soil has to be taken into consideration. For instance, to calculate the dynamic-stiffness coefficients for harmonic loading of a pile group it is not permissible just to add the contributions of the individual piles calculated in Section 4.8. For static loads, the displacement of a specific pile increases due to the displacement fields created by the other piles of the group. As a result, the overall displacement of the pile group will be larger than that of a single pile subjected to the average load. The sum of the static-stiffness coefficients of the individual piles is thus larger than the static-stiffness coefficient of the pile group, the increase being significant for a large number of piles with small axial distances. For dynamic loads, *pile–soil–pile interaction* also takes place caused by waves that are emitted from the shaft of each pile and then propagate to "strike" neighboring piles with resulting refractions and reflections. The mechanism is much more complicated than in the static case. The magnitudes of the dynamic-stiffness coefficient of a pile group for a specific frequency can be *smaller or larger* than that of the sum of the contributions of the individual piles. The group's dynamic-stiffness coefficients are also very *strongly frequency dependent*, as will be demonstrated.

The rigorous analysis of a pile group based on the boundary-element method, which solves the elastodynamical equations with the complicated boundary and interface conditions, leads to an enormous computational effort. A real need exists to develop a simple physically motivated procedure with sufficient accuracy that considers dynamic pile–soil–pile interaction. The proposed method uses the dynamic behavior, such as the dynamic-stiffness coefficients, of a single pile, which can be calculated based on rigid disks with corresponding double cones as described in Section 4.8. To capture the interference of the essentially *cylindrical wave fields* originating along each pile shaft and propagating radially outwards, a displacement pattern in the horizontal plane is introduced, which is based on a sound physical approximation. This allows dynamic-interaction factors describing the effect of one pile—the source, on another pile—the receiver, to be determined. From the study of only two piles at a time, it is then possible to determine the dynamic response of a pile group using familiar matrix methods of structural analysis. The method leads to very accurate results for stiff piles embedded in soft soil and can be less accurate for (relatively) stiff soil. In addition, the pile spacing should not be too small. The procedure formulated for harmonic loading is described in several studies [D1, M1].

As an example, the formulation of a floating pile group in a homogeneous halfspace is developed for harmonic loading in this section. The method can, however, also be applied to a floating pile group in a soil layer resting on rigid rock.

4.9.2 Dynamic-Interaction Factor

When determining the *dynamic-interaction factors*, only two piles, the loaded source pile and the unloaded receiver pile, are addressed. The presence of the other piles is disregarded; the corresponding reflections and refractions are not taken into consideration.

First, the dynamic-interaction factor in the *vertical* direction is addressed (Fig. 4-51a), which will be used to calculate the vertical and rocking responses of a pile group. It is assumed that only *cylindrical SV-waves* (with a vertical particle motion) are emitted from the harmonically loaded source pile and propagate with the shear-wave velocity c_s in the radial horizontal direction r towards the location of the receiver pile. It is also postulated that all cylindrical waves emanate *simultaneously* from all points along the pile length.

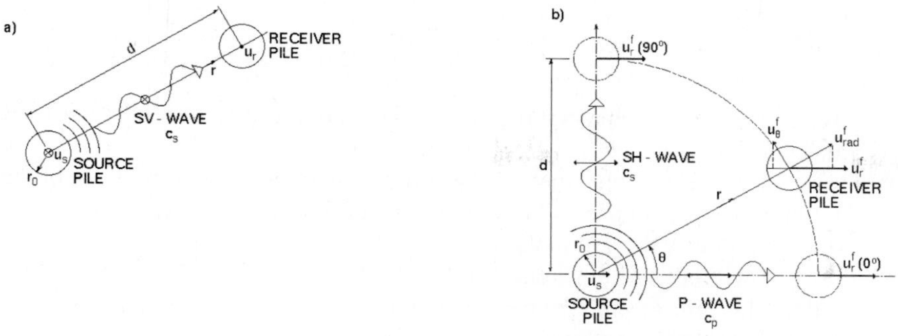

a. Vertical. **b.** Lateral (horizontal).

Fig. 4-51 Cylindrical waves emitted from shaft of loaded source pile and propagating towards receiver pile.

They spread out in phase and form a cylindrical wave front. This does not mean, however, that the wave amplitude along the wave front (r = constant) is constant with depth; in general, it will decrease with depth z characterized by $A(z)$. The asymptotic solution for cylindrical waves is used. For $r \to \infty$ the area of the cylindrical surface is proportional to r; the radiation criterion (Eq. 2-271) states that the variation of the amplitude in the radial direction will thus be proportional to $1/\sqrt{r}$. For the large-argument limit, the solution determining the phase angle for undamped soil will be equal to $e^{-i\omega r/c_s}$(see Eq. 2-333). Introducing material damping with the hysteretic damping ratio ζ_g yields (Section 2.4.1, Eq. 2-231a)

$$e^{-i\omega\frac{r}{c_s(1+i\zeta_g)}} \simeq e^{-\zeta_g\omega\frac{r}{c_s}} e^{-i\omega\frac{r}{c_s}} \tag{4-103}$$

The vertical displacement amplitude thus equals

$$u(r,\omega) = A(z)\frac{1}{\sqrt{r}} e^{-\zeta_g\omega\frac{r}{c_s}} e^{-i\omega\frac{r}{c_s}} \tag{4-104}$$

The displacement amplitude of the soil at the location of the axis of the receiver pile (subscript r) follows with $r = d$ (axis-to-axis distance between the two piles) as

$$u_r(\omega) = A(z)\frac{1}{\sqrt{d}} e^{-\zeta_g\omega\frac{d}{c_s}} e^{-i\omega\frac{d}{c_s}} \tag{4-105}$$

that of the source pile (subscript s) can be written as (the phase angle and the material damping effect between the axis and the perimeter vanish)

$$u_s(\omega) = A(z)\frac{1}{\sqrt{r_0}} \tag{4-106}$$

The dynamic-interaction factor in the vertical direction is defined as the *ratio of the (additional) displacement amplitude at the location of the receiver pile $u_r(\omega)$ to the displacement amplitude of the loaded source pile $u_s(\omega)$.*

$$\alpha_v(\omega) = \frac{u_r(\omega)}{u_s(\omega)} \tag{4-107}$$

Substituting Eqs. 4-105 and 4-106 in Eq. 4-107 yields

$$\alpha_v(\omega) = \sqrt{\frac{r_0}{d}}\, e^{-\zeta_g \omega \frac{d}{c_s}}\, e^{-i\omega \frac{d}{c_s}} \tag{4-108}$$

Here, $\alpha_v(\omega)$ is a frequency-dependent complex quantity depending strongly on the distance d.

It should be noted that for the vertical direction the displacement of the receiver pile (at its head) is assumed to be equal to that of the soil (free field); the presence of the pile does not modify the soil motion (Eq. 4-105).

It will become apparent that the simple expression $\alpha_v(\omega)$ is all that is needed to calculate the pile group's dynamic-stiffness coefficient in vertical motion, besides the dynamic-stiffness coefficient of the single pile.

Second, the dynamic-interaction factor in the *lateral* direction $\alpha_h(\omega)$ is examined (Fig. 4-51b), which will be used to determine the horizontal and torsional responses of a pile group. A more detailed analysis than for $\alpha_v(\omega)$ is necessary. In contrast to the vertical direction, in the more flexible lateral direction of the pile the motion $\{u_r(\omega)\}$ of the receiver pile in all nodes corresponding to the disks (Fig. 4-49) will differ significantly from that of the soil in the free field $\{u_r^f(\omega)\}$ determined in the same nodes (location). The superscript f is used to denote quantities referred to the free field.

The dynamic-interaction factor $\alpha_h^f(\omega)$ is determined first. Besides being a function of the distance d between the two piles, $\alpha_h^f(\omega)$ will depend also on the angle θ between the horizontally applied load and the line connecting the two piles (Fig. 4-51b). It is, however, only necessary to calculate the dynamic-interaction factors referred to the free field of the soil at the same distance d at $\theta = 0°$, $\alpha_h^f(0°, \omega)$, and at 90°, $\alpha_h^f(90°, \omega)$. Two sorts of waves are postulated to exist (see also lower-right part of Fig. 2-49). The location at 0° is affected by *cylindrical P-waves* which propagate with the dilatational-wave velocity c_p yielding in analogy to Eq. 4-108

$$\alpha_h^f(0°,\omega) = \frac{u_r^f(0°,\omega)}{u_s(\omega)} = \sqrt{\frac{r_0}{d}}\, e^{-\zeta_g \omega \frac{d}{c_p}}\, e^{-i\omega \frac{d}{c_p}} \tag{4-109}$$

For the nearly incompressible soil discussed in Section 2.1.4 ($1/3 < \nu \le 1/2$), c_p equals $2c_s$. For the location at 90°, *cylindrical SH-waves* (with a horizontal particle motion) propagating with c_s are involved. This results in

$$\alpha_h^f(90°,\omega) = \frac{u_r^f(90°,\omega)}{u_s(\omega)} = \sqrt{\frac{r_0}{d}}\, e^{-\zeta_g \omega \frac{d}{c_s}}\, e^{-i\omega \frac{d}{c_s}} \tag{4-110}$$

which is the same expression as for the dynamic-interaction factor in the vertical direction $\alpha_v(\omega)$ specified in Eq. 4-108. The radial and circumferential displacement amplitudes vary with the first harmonic of θ as

$$u_{\text{rad}}^f(\omega) = \cos\theta\, u_r^f(0°,\omega) \tag{4-111a}$$

$$u_\theta^f(\omega) = -\sin\theta\, u_r^f(90°,\omega) \tag{4-111b}$$

The amplitude of the displacement parallel to the applied source load (Fig. 4-51b) equals

$$u_r^f(\omega) = \cos\theta\, u_{\text{rad}}^f(\omega) - \sin\theta\, u_\theta^f(\omega) \tag{4-112}$$

The dynamic-interaction factor referred to the free field is defined as

$$\alpha_h^f(\omega) = \frac{u_r^f(\omega)}{u_s(\omega)} \tag{4-113}$$

Substituting Eqs. 4-112, 4-111, 4-110 and 4-109 yields

$$\alpha_h^f(\omega) = \cos^2\theta\alpha_h^f(0°,\omega) + \sin^2\theta\alpha_h^f(90°,\omega) \tag{4-114}$$

The transformation from $\alpha_h^f(\omega)$ to the dynamic-interaction factor of the (receiver) pile $\alpha_h(\omega)$ referred to its head is performed as follows. Consistent with the assumption that all waves emanate simultaneously from the disks along the source pile, the free-field motion of the soil at the location of the disks of the receiver pile (but without the presence of the latter) equals

$$\{u_r^f(\omega)\} = \alpha_h^f(\omega)\,\{u_s(\omega)\} \tag{4-115}$$

with the horizontal motion of the source pile $\{u_s(\omega)\}$ specified as $\{u(\omega)\}$ in Eq. 4-99. (The load is applied only at the first disk located at the pile's head). Using the concept of substructuring with replacement as in the derivation of the basic equation of soil–structure-interaction analysis (explained in Section 6.1.1), the horizontal motion of the receiver pile equals (analogous to Eq. 6-15)

$$\{u_r(\omega)\} = \left([S^f(\omega)] + [\Delta S(\omega)]\right)^{-1} [S^f\{\omega)]\,\{u_r^f(\omega)\} \tag{4-116}$$

$[S^f(\omega)]$ and $[\Delta S(\omega)]$ are defined in Eqs. 4-96 and 4-97. Equation 4-116 leads to the displacement amplitude at the pile's head (node 1 in Fig. 4-49), denoted as $u_r(\omega)$. The final lateral dynamic-interaction factor follows as

$$\alpha_h(\omega) = \frac{u_r(\omega)}{u_s(\omega)} \tag{4-117}$$

with $u_s(\omega)$ being equal to the amplitude of the displacement at the source pile's head (corresponds to $u_1(\omega)$ in Eq. 4-100).

As an approximation which still leads to an acceptable accuracy, the motion of the receiver pile's head can be assumed to be equal to the free-field motion of the soil at the same location. In this case the lateral dynamic-interaction factor is given in Eq. 4-114 with Eqs. 4-109 and 4-110 applying.

4.9.3 Matrix Formulation for Dynamic-Stiffness Matrix

Based on the dynamic-stiffness coefficients of the single pile embedded in soil $S(\omega)$ for all motions (Section 4.8.2) and the dynamic-interaction factors in the vertical direction $\alpha_v(\omega)$ and in the lateral direction $\alpha_h(\omega)$, the dynamic-stiffness matrix for harmonic loading of the pile group can be determined using the standard matrix formulation of structural analysis.

The procedure is discussed for the vertical degree of freedom. Figure 4-52 shows the plan view of a general arrangement of n piles with the axial force amplitudes at the heads $P_j(\omega)$ ($j = 1, ..., m$) and the axis-to-axis distances d_{ij} between piles i and j. For the sake of simplicity, it is assumed that all piles are the same. Using the definition of the dynamic-interaction factor, the vertical displacement amplitude at the head of pile i (receiver) $u_{ij}(\omega)$ caused by the axial load applied to the pile j (source) resulting in its own vertical displacement amplitude $u_{jj}(\omega)$ equals

$$u_{ij}(\omega) = \alpha_v(d_{ij},\omega)\ u_{jj}(\omega) \tag{4-118}$$

where

$$P_j(\omega) = S(\omega)\ u_{jj}(\omega) \tag{4-119}$$

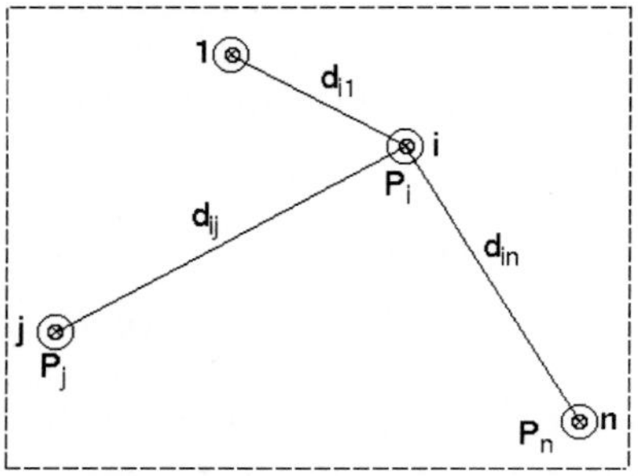

Fig. 4-52 Plan view of arrangement of piles in group.

The $\alpha_v(d_{ij},\omega)$ is specific in Eq. 4-108. Using superposition, the total displacement amplitude of pile i from the loads applied to all piles is formulated as

$$u_i(\omega) = \sum_{j=1}^{i-1} \alpha_v(d_{ij},\omega)\ u_{jj}(\omega) + u_{ii}(\omega) + \sum_{j=i+1}^{n} \alpha_v(d_{ij},\omega)\ u_{jj}(\omega) \tag{4-120}$$

Substituting Eq. 4-119 results in

$$u_i(\omega) = \frac{1}{S(\omega)} \left(\sum_{j=1}^{i-1} \alpha_v(d_{ij},\omega)\ P_j(\omega) + P_i(\omega) + \sum_{j=i+1}^{n} \alpha_v(d_{ij},\omega)\ P_j(\omega) \right) \tag{4-121}$$

$$i = 1, \ldots, n$$

Introducing the n-element vectors of amplitudes $\{u(\omega)\}$ and $\{P(\omega)\}$, Eq. 4-121 is written as

$$\{u(\omega)\} = [G(\omega)]\ \{P(\omega)\} \tag{4-122}$$

with the dynamic-interaction factor $\alpha_v(d_{ij},\omega)$ assembled in the $n \times n$ element dynamic-flexibility matrix $[G(\omega)]$, which is also inversely proportional to the dynamic-stiffness coefficient of the single pile $S(\omega)$.

Formulating compatibility of the vertical displacement at the head of pile i yields

$$u_i(\omega) = u_0(\omega) \quad i=1, \ldots, n \tag{4-123}$$

with the amplitude of the rigid basemat $u_0(\omega)$. With the kinematic constraint matrix $[A]$ consisting for the vertical degree of freedom of a column of ones, Eq. 4-123 is written as

$$\{u(\omega)\} = [A]\, u_0(\omega) \tag{4-124}$$

Formulating equilibrium, the resulting force with the amplitude $P_0(\omega)$ on the basemat equals

$$P_0(\omega) = \sum_{i=1}^{n} P_i(\omega) \tag{4-125}$$

or

$$P_0(\omega) = [A]^T \{P(\omega)\} \tag{4-126}$$

Eliminating $\{P(\omega)\}$ and $\{u(\omega)\}$ from Eqs. 4-122, 4-124, and 4-126 leads to

$$P_0(\omega) = S_v(\omega)\, u_0(\omega) \tag{4-127}$$

in which $S_v(\omega)$ is the desired dynamic-stiffness coefficient in the vertical direction of the pile group computed as

$$S_v(\omega) = [A]^T\, [G(\omega)]^{-1}\, [A] \tag{4-128}$$

The other degrees of freedom are processed analogously. For the sake of illustration, the rocking motion (with coupling to the horizontal motion) for the symmetric arrangement with 3 × 3 piles with the spacing s is briefly discussed (Fig. 4-53). Considering symmetry and anti-symmetry, rocking of the rigid basemat by an angle with the amplitude $\vartheta_0(\omega)$ is resisted by the three pairs of axial forces acting at the heads of the piles with amplitudes $P_1(\omega)$, $P_2(\omega)$, and by nine moments with amplitude $M(\omega)$. The shear forces activated due to the coupling of the individual piles are not shown in Fig. 4-53. It is assumed that no interaction through the soil takes place due to the rocking of the individual pile. For an analysis by hand without using a computer program, the following equations are processed

$$M(\omega) = S_\vartheta(\omega)\, \vartheta_0(\omega) \tag{4-129}$$

$$\begin{aligned} u_1(\omega) = {} & \left[1+\alpha_v(2s,\omega) - \alpha_v(2s,\omega) - \alpha_v\left(2\sqrt{2}s,\omega\right)\right] \frac{P_1(\omega)}{S(\omega)} \\ & + \left[\alpha_v(s,\omega) - \alpha_v\left(\sqrt{5}s,\omega\right)\right] \frac{P_2(\omega)}{S(\omega)} \end{aligned} \tag{4-130a}$$

$$\begin{aligned} u_2(\omega) = {} & \left[2\alpha_v(s,\omega) - 2\alpha_v\left(\sqrt{5}s,\omega\right)\right] \frac{P_1(\omega)}{S(\omega)} \\ & + \left[1 - \alpha_v(2s,\omega)\right] \frac{P_2(\omega)}{S(\omega)} \end{aligned} \tag{4-130b}$$

$$u_1(\omega) = s\,\vartheta_0(\omega) \tag{4-131a}$$

$$u_2(\omega) = s\,\vartheta_0(\omega) \tag{4-131b}$$

$$M_0(\omega) = 4\, s P_1(\omega) + 2 s P_2(\omega) + 9M(\omega) \tag{4-132}$$

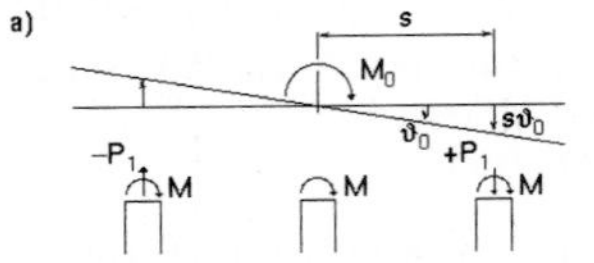

a. Elevation.

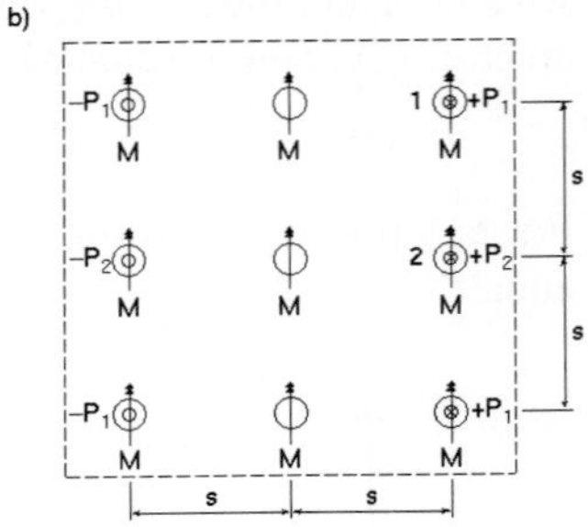

b. Plan view.

Fig. 4-53 Activated pile forces and moments for rocking motion contributing to dynamic-stiffness coefficient on diagonal.

The dynamic-stiffness coefficients of the single pile embedded in soil are denoted for the vertical motion as $S(\omega)$ and for the rocking motion as $S_\vartheta(\omega)$. Eliminating $u_1(\omega)$, $u_2(\omega)$, $P_1(\omega)$, and $P_2(\omega)$ and substituting in Eq. 4-132 leads to the relationship

$$M_0(\omega) = S_r(\omega)\ \vartheta_0(\omega) \tag{4-133}$$

with the rocking dynamic-stiffness coefficient $S_r(\omega)$, the term on the diagonal of the matrix, of the pile group.

The procedure allows not only the dynamic-stiffness matrix of a pile group to be calculated, but also the amplitudes of the displacements, forces, and moments along the axis of each pile.

4.9.4 Example of Floating 3 × 3 Pile Group

The floating 3 × 3 rigidly capped pile group in a homogeneous soil halfspace shown in Fig. 4-54 is addressed. The properties of the single pile and the soil are described at the beginning of Section 4.8.3. Two ratios of the pile spacing to the diameter, $s/(2r_0) = 5$ and 10, are examined.

The dynamic-stiffness coefficients for the vertical, rocking, and horizontal motions as a function of $a_0 = \omega 2r_0/c_s$ are presented in Figs. 4-55, 4-56, and 4-57. For the horizontal degree of freedom the dynamic-interaction factor α_h referred to the receiver pile (Eq. 4-117) is used. The dynamic-stiffness coefficients are decomposed in the standard manner

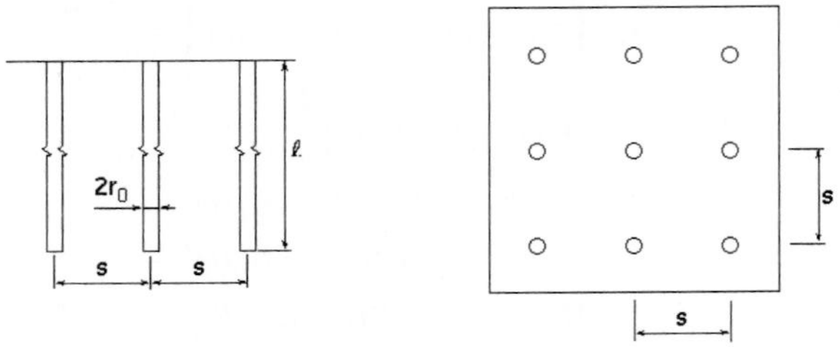

Fig. 4-54 Elevation and plan view of 3 × 3 pile group.

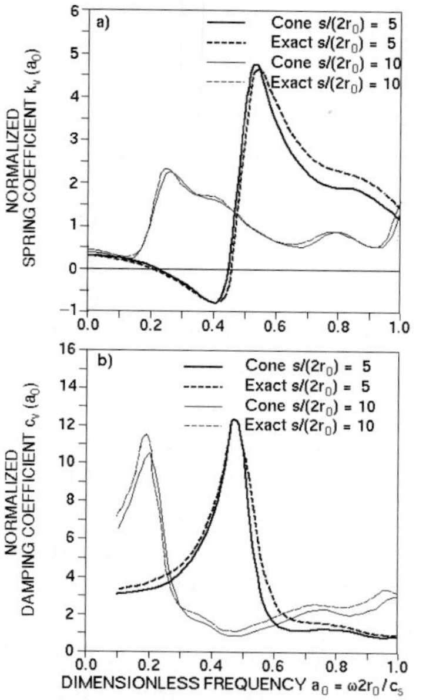

Fig. 4-55 Dynamic-stiffness coefficient for harmonic loading of pile group in vertical motion.

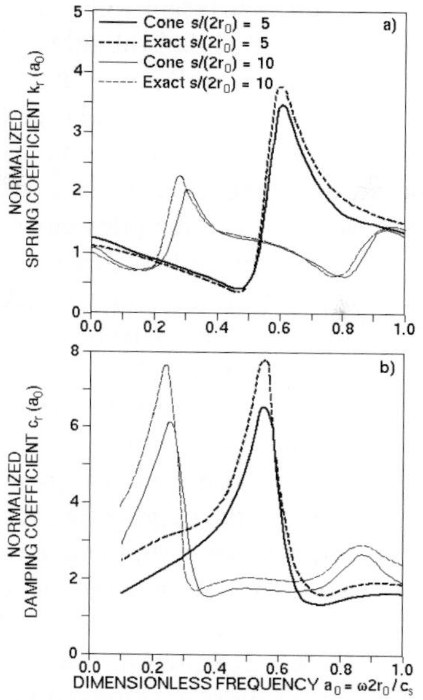

Fig. 4-56 Dynamic-stiffness coefficient for harmonic loading of pile group in rocking motion.

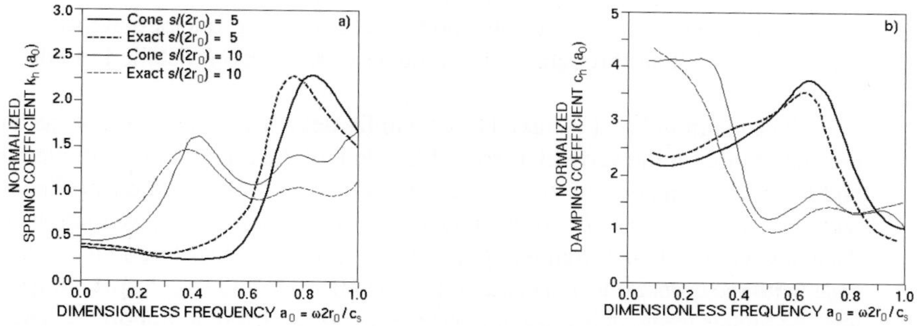

Fig. 4-57 Dynamic-stiffness coefficient for harmonic loading of pile group in horizontal motion.

as described in Eq. 4-52. For the two translational motions, the normalization is performed with the sum of the corresponding static-stiffness coefficients of the single piles ($9K$); for the rocking motion it is performed with the sum of the vertical static-stiffness coefficient of a single pile multiplied by the square of the leverarm—the contribution of the single pile axial stiffness to the rocking static-stiffness coefficient of the pile group ($6s^2K$). The rigorous solution of Kaynia and Kausel [K2], denoted as exact, is also shown.

As for the two translational motions, the normalized static-spring coefficients $k(a_0 = 0)$ in Figs. 4.55a and 4.57a are significantly smaller than one; it is confirmed that the static behavior of the group is more flexible than that of the sum of the single piles. For all three motions the frequency dependency of the dynamic-stiffness coefficients is very strong and is quite different from that of the single pile shown in Fig. 4-50, which corresponds to neglecting pile–soil–pile interaction. The strong oscillatory behavior results in $k(a_0)$ attaining values well above unity for certain frequencies.

The vertical and rocking motions, where, to calculate $\alpha_v(\omega)$, the postulated cylindrical waves originating on the shafts of piles propagate with c_s, are discussed further [D1]. For a large wave length λ compared to the spacing s of the piles (low-frequency range), the soil between the piles moves as a rigid body. This trapped mass leads to a downward-parabolic tendency of $k(a_0)$, as discussed in connection with Eq. 2-281. The limit occurs at $\lambda = 3s$, resulting with $\lambda = c_s 2\pi/\omega$ in $a_0 = (2\pi/3)\, 2r_0/s$. This results for $s/(2r_0) = 5$ in $a_0 = 0.4$ and for $s/(2r_0) = 10$ in $a_0 = 0.2$, which corresponds to the turning points (valleys) of $k_v(a_0)$ in Fig. 4-55a and of $k_r(a_0)$ in Fig. 4-56a. For the frequency range above this limit, strong wave-interference effects occur. The a_0 of the predominant peaks in the spring coefficients can also be identified. For the vertical motion, the corresponding λ will be equal to $2s$. In this case, the cylindrical waves originating with a certain phase from the source pile will arrive at the neighboring receiver pile with exactly the opposite phase, yielding a negative displacement amplitude $u_r(\omega)$ compared to the positive $u_s(\omega)$ due to the source pile's own load. The dynamic-interaction factor $\alpha_v(\omega)$ (Eq. 4-107) will be negative, resulting in a large force which must be applied to enforce displacement compatibility. The corresponding a_0 for $\lambda = 2s$ equals $a_0 = \pi 2r_0/s$, yielding $a_0 = 0.6$ for $s/(2r_0) = 5$ and $a_0 = 0.3$ for $s/(2r_0) = 10$. The peaks of k_v at these dimensionless frequencies are clearly visible in Fig. 4-55a. It is also possible to identify the a_0 corresponding to the next peak for the larger spacing $s/(2r_0) = 10$ which occurs for $\lambda = 2s/3$, leading to $a_0 = 3\pi 2r_0/s = 0.9$. For the rocking motion, the peaks of k_r appear at the same dimensionless frequencies (Fig. 4-56a), the maximum wave interference occuring between piles 1 and 2 on the same side of the rocking axis (Fig. 4-53). The agreement of the results of the cone model using dynamic-interaction factors with the exact solution is excellent throughout the frequency range. Even detailed trends are accurately captured.

The magnitudes of the axial force amplitudes $N(a_0)$ at the heads of the corner pile and the center pile for the vertical degree of freedom of the pile group are plotted in Figs. 4-58a and 4-58b as a function of the dimensionless frequency. The force amplitudes are normalized with the average static force of the pile group. Again, strong oscillations with a_0 occur. As expected, in the low frequency range the axial force distribution is similar to the static case, with the corner pile and center pile carrying more and less, respectively, than the average. At higher frequencies, wave interference effects become important. The center pile is particularly sensitive to the cylindrical waves arriving in phase from the four edge piles at distance s and from the four corner piles at distance $\sqrt{2}\, s$. As explained above, when the wave length λ is twice the distance; in the range between $2s$ and $2\sqrt{2}\, s$, the center pile will tend to move in the opposite direction of the other piles: This means, that to satisfy compatibility, the load acting on the center pile must increase. The corresponding range for a_0 is bounded by $\pi 2r_0/s$ and $(\pi/\sqrt{2})\, 2r_0/s$. The force amplitude of the center pile is indeed large for $0.4 < a_0 < 0.6$ when $s/(2r_0) = 5$ and for $0.2 < a_0 < 0.3$ when $s/(2r_0) = 10$ (Fig. 4-58b). $|N(a_0)|$ of the corner pile (Fig. 4.58a) shows of course the opposite behavior. The average force amplitude of the center pile can be almost twice and that of the corner pile nearly half the average value in this frequency range, a dramatic reversal of the static case.

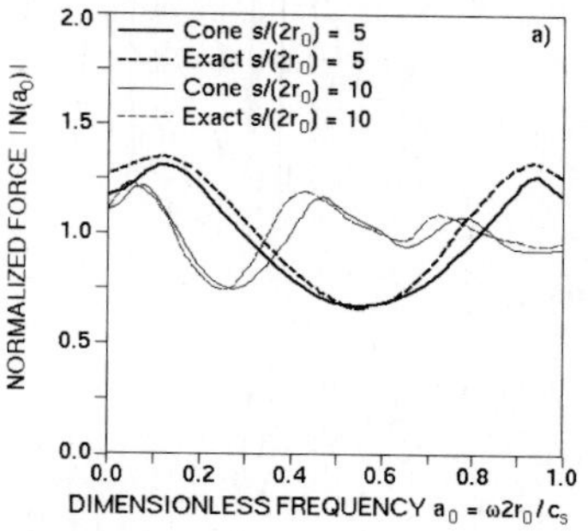

a. At head of corner pile.

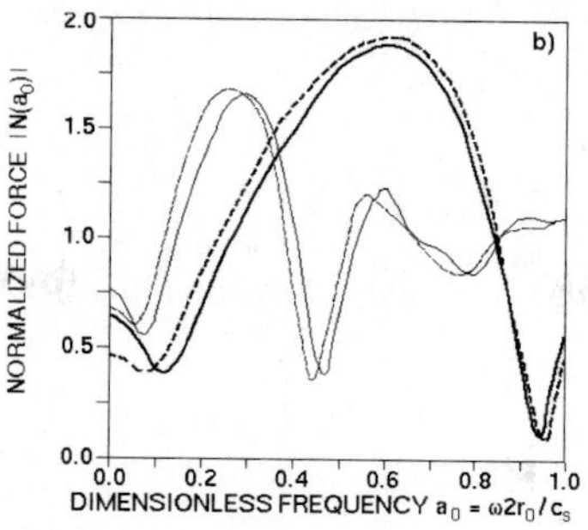

b. At head of center pile.

Fig. 4-58 Axial force amplitude for harmonic loading of pile group in vertical motion.

SUMMARY

1. The building block to analyze an embedded foundation for all components of motion consists of a rigid disk of radius r_0 embedded in a full-space, which is modeled with a double-cone model. The aspect ratio z_0/r_0 which determines the opening angle follows, as for the one-sided cone used to model a surface disk, from equating the static-stiffness coefficient of the cone to that of a disk on a halfspace. The only change consists of doubling the static-stiffness coefficients K and K_ϑ. The double cone's displacement field defines approximate Green's functions for use in a matrix formulation of structural mechanics.
2. The following Green's functions apply for embedded foundations with the distance a measured from the source disk to the receiver point on the axis and c denoting the appropriate wave velocity.

translational double cone

$$g\left(a, t-\frac{a}{c}\right) = \frac{1}{2K}\frac{1}{1+\dfrac{a}{z_0}}h_1\left(t-\frac{a}{c}\right)$$

with the unit-impulse response function

$$\begin{aligned} h_1(t) &= \frac{c}{z_0}e^{-\frac{c}{z_0}t} & t \geq 0 \\ &= 0 & t < 0 \end{aligned}$$

$$g(a,\omega) = \frac{1}{2K}\frac{1}{1+\dfrac{a}{z_0}}\frac{e^{-i\frac{\omega a}{c}}}{1+i\dfrac{\omega z_0}{c}}$$

rotational double cone

$$g_\vartheta\left(a, t-\frac{a}{c}\right)=\frac{1}{2K_\vartheta}\frac{1}{\left(1+\frac{a}{z_0}\right)^2}\left[h_2\left(t-\frac{a}{c}\right)+\left(-1+\frac{1}{1+\frac{a}{z_0}}\right)h_3\left(t-\frac{a}{c}\right)\right]$$

with the unit-impulse response functions

$$h_2(t) = \frac{c}{z_0}e^{-\frac{3}{2}\frac{c}{z_0}t}\left(3\cos\frac{\sqrt{3}}{2}\frac{c}{z_0}t-\sqrt{3}\sin\frac{\sqrt{3}}{2}\frac{c}{z_0}t\right) \qquad t\geq 0$$

$$= 0 \qquad t<0$$

$$h_3(t) = 2\sqrt{3}\frac{c}{z_0}e^{-\frac{3}{2}\frac{c}{z_0}t}\sin\frac{\sqrt{3}}{2}\frac{c}{z_0}t \qquad t\geq 0$$

$$= 0 \qquad t<0$$

$$g_\vartheta(a,\omega)=\frac{3}{2K_\vartheta}\left(\frac{1}{\left(1+\frac{a}{z_0}\right)^3}+i\frac{\omega z_0}{c}\frac{1}{\left(1+\frac{a}{z_0}\right)^2}\right)\frac{1}{3+3i\frac{\omega z_0}{c}+\left(i\frac{\omega z_0}{c}\right)^2}e^{-i\frac{\omega a}{c}}$$

3. For vertical motion of a pile the weighted Green's function of the double cone $g_{\text{cone}}(a,\omega)$ as specified under item 2 and of the fullspace for a point load $g_{\text{fullspace}}(a,\omega)$ is used

$$g(a,\omega) = w(a)\, g_{\text{cone}}(a,\omega) + (1-w(a))g_{\text{fullspace}}(a,\omega)$$

with the weighting function

$$w(a) = 1 \qquad a\leq r_0$$

$$= e^{-0.8\frac{a-r_0}{r_0}} \qquad a>r_0$$

and where with the shear modulus G

$$g_{\text{fullspace}}(a,\omega)=\frac{1}{4\pi G}\left[\psi-\left(\frac{a}{r}\right)^2\chi\right]$$

and with

$$\psi=\left(1-i\frac{1}{a_0}-\frac{1}{a_0^2}\right)\frac{e^{-ia_0}}{r}+\left(i\frac{c_s}{c_p a_0}+\frac{1}{a_0^2}\right)\frac{e^{-i\frac{c_s}{c_p}a_0}}{r}$$

$$\chi=\left(1-i\frac{3}{a_0}-\frac{3}{a_0^2}\right)\frac{e^{-ia_0}}{r}-\left(\frac{c_s^2}{c_p^2}-i\frac{3c_s}{c_p a_0}-\frac{3}{a_0^2}\right)\frac{e^{-i\frac{c_s}{c_p}a_0}}{r}$$

$$r=\sqrt{a^2+r_0^2}$$

$$a_0=\frac{\omega r}{c_s}$$

4. To model the free surface of a soil, a mirror-image disk (embedded in the fullspace and also represented by a double cone) loaded by the same time history of force as the original disk and with the same sign is introduced. The same procedure can be used to model a fixed boundary, but the force on the mirror-image disk acts in the opposite direction. To model a disk embedded in a soil layer resting on rigid rock, the concept of anti-symmetrically loaded mirror-image disks to represent the free surface and of symmetrically loaded mirror-image disks for the fixed boundary is applied repeatedly. For a surface disk on the soil layer resting on rigid rock, this concept leads to the same formulation as that based on the unfolded layered cone of Chapter 3.
5. Modeling the soil region which will later be excavated by a sandwich of rigid disks separated by soil, the dynamic-stiffness matrix for harmonic loading $[S^g_{00}(\omega)]$ of an embedded cylindrical foundation with respect to the rigid-body displacement amplitudes $\{u_0(\omega)\}$ and corresponding force amplitudes $\{P_0(\omega)\}$ is formulated with standard matrix methods of structural analysis as

$$\{P_0(\omega)\} = [S^g_{00}(\omega)]\,\{u_0(\omega)\}$$

where

$$[S^g_{00}(\omega)] = [A]^T\,[S^f(\omega)]\,[A] + \omega^2\,[M]$$

$[A]$ is the kinematic-constraint matrix of the rigid foundation with $\{u(\omega)\}$ denoting the displacement amplitudes of the disks

$$\{u(\omega)\} = [A]\,\{u_0(\omega)\}$$

$[M]$ is the rigid-body mass matrix of the excavated soil, and $[S^f(\omega)]$ is the dynamic-stiffness matrix of the free field

$$[S^f(\omega)] = [G(\omega)]^{-1}$$

where the dynamic-flexibility matrix of the disks embedded in the soil equals

$$\{u(\omega)\} = [G(\omega)]\,\{P(\omega\}\}$$

with the corresponding force amplitudes $\{P(\omega)\}$.
6. In the time domain, the corresponding interaction force-displacement relationship of an embedded foundation at time $n\Delta t$ equals

$$\{P_0\}_n = [A]^T\,[S^f]_0[A]\,\{u_0\}_n - [A]^T\,[S^f]_0\,\{\bar{u}\}_n - [M]\,\{\ddot{u}_0\}_n$$

$[S^f]_0$ is the instantaneous dynamic-stiffness matrix of the free field

$$[S^f]_0 = [G]_0^{-1}$$

where the displacement-force relationship of the disks equals

$$\{u\}_n = \{\bar{u}\}_n + [G]_0\,\{P\}_n$$

with

$$\{\bar{u}\}_n = \sum_{k=1}^{n-1} [G]_{n-k}\,\{P\}_k$$

$[G]_{n-k}$ is the dynamic-flexibility matrix in the time domain (displacements at time n caused by unit forces acting at time k).

7. The static-stiffness coefficients of a cylindrical foundation of radius r_0 embedded with height e in a soil layer of depth d resting on rigid rock can be calculated from the following accurate approximations

$$K_h = \frac{8Gr_0}{2-\nu}\left(1+\frac{1}{2}\frac{r_0}{d}\right)\left(1+\frac{e}{r_0}\right)\left(1+\frac{e}{d}\right) \qquad \text{horizontal}$$

$$K_v = \frac{4Gr_0}{1-\nu}\left(1+1.3\frac{r_0}{d}\right)\left(1+0.54\frac{e}{r_0}\right)\left[1+\left(0.85-0.28\frac{e}{r_0}\right)\frac{\frac{e}{d}}{1-\frac{e}{d}}\right] \qquad \text{vertical}$$

$$K_r = \frac{8Gr_0^3}{3(1-\nu)}\left(1+\frac{1}{6}\frac{r_0}{d}\right)\left[1+2.3\frac{e}{r_0}+0.58\left(\frac{e}{r_0}\right)^3\right]\left(1+0.7\frac{e}{d}\right) \qquad \text{rocking}$$

$$K_t = \frac{16}{3}Gr_0^3\left(1+\frac{1}{10}\frac{r_0}{d}\right)\left(1+2.67\frac{e}{r_0}\right) \qquad \text{torsional}$$

$$K_{hr} = \frac{e}{3}K_h$$

The equations also apply to a cylindrical foundation embedded in a halfspace with $d = \infty$ and to a surface foundation with $e = 0$.

8. As an alternative, a cylindrical rigid foundation embedded in a halfspace can also be modeled with the simple fundamental lumped-parameter model. It introduces at most for each degree of freedom one additional internal degree of freedom and consists of the direct spring with the static-stiffness coefficient, two dashpots and one mass, whose coefficients are specified as a function of the embedment ratio e/r_0 in Table 4-1. The coupling between the horizontal and rocking motions is achieved by connecting the horizontal lumped-parameter model with an eccentricity to the base.
9. To represent a cylindrical rigid foundation embedded in a soil layer resting on rigid rock, a lumped-parameter model for all degrees of freedom is described. For the vertical and torsional motions the basic lumped-parameter model with four springs, three dashpots, and a mass introducing two internal degrees of freedom is directly applicable. For the coupled horizontal and rocking motions, a physical representation exists, consisting of three basic lumped-parameter models, one of which is attached with an eccentricity to take the coupling into consideration. For various ratios of the radius of the foundation to the depth of the layer and lateral contact ratios, the frequency-independent real coefficients for all springs, dashpots and masses are specified in Table 4-2.
10. The formulation based on disks embedded in a fullspace with their corresponding double cones and anti-symmetrically and symmetrically loaded mirror-image disks to model a free surface and a fixed boundary can straightforwardly be applied to the dynamic analysis of a single pile. For the vertical motion the weighted Green's function of item 3 must be used. The cylindrical soil region between the disks is not just analytically excavated as for an embedded foundation, but replaced by the difference of the material properties of the pile and the soil.

11. For a pile group, pile–soil–pile interaction can be considered with one dynamic-interaction factor in the vertical direction and one in the lateral direction $\alpha(\omega)$, defined as the amplitude ratio of the displacement at the head of a receiver pile $u_r(\omega)$ to the corresponding displacement of the loaded source pile $u_s(\omega)$ under its own dynamic load.

$$\alpha(\omega) = \frac{u_r(\omega)}{u_s(\omega)}$$

Cylindrical waves are assumed to emanate simultaneously from all points along the source pile shaft and thus to spread out in phase propagating horizontally towards the receiver pile. Cylindrical *SV*-waves are created in the case addressed to determine the vertical dynamic-interaction factor

$$\alpha_v(\omega) = \sqrt{\frac{r_0}{d}}\, e^{-\zeta_g \omega \frac{d}{c_s}}\, e^{-i\omega \frac{d}{c_s}}$$

with the radius r_0, pile distance d and material damping ratio ζ_g. A combination of cylindrical *P*- and *SH*-waves occurs analogously when the lateral dynamic-interaction factor is derived

$$\alpha_h^f(\omega) = \cos^2\theta\ \alpha_h^f(0°, \omega) + \sin^2\theta \alpha_h^f\ (90°,\omega)$$

with the angle θ between the horizontal load and the line connecting the source and the receiver piles.

$$\alpha_h^f(0°,\omega) = \sqrt{\frac{r_0}{d}}\, e^{-\zeta_g \omega \frac{d}{c_p}}\, e^{-i\omega \frac{d}{c_p}}$$

$$\alpha_h^f\ (90°,\omega) = \alpha_v(\omega)$$

The dynamic-interaction factor referred to the free field of the soil (at the location of the receiver pile) $\alpha_h^f(\omega)$ has to be transformed to the corresponding factor referred to the head of the receiver pile itself $\alpha_h(\omega)$ based on the concept of substructuring with replacement. Based on the dynamic-stiffness coefficients of the single pile and the dynamic-interaction factors, the dynamic-stiffness matrix of a pile group follows using the standard matrix formulation of structural analysis.

12. The results obtained with the embedded disks and the lumped-parameter models for embedded foundations and single piles are accurate. This means, for instance, that the greatly simplified solution based on disks embedded in a fullspace modeled with double cones and subsequent analytical excavation captures the essential features of the very complicated boundary-value problem of an embedded foundation. The strong oscillations with frequency of the dynamic-stiffness coefficients of a floating group of stiff piles is well reproduced using the dynamic-interaction factors. This demonstrates that the key physical issue of the very complex pile–soil–pile interaction is the interference of the cylindrical waves originating along the shaft of each pile.

5

Simple Vertical Dynamic Green's Function

To analyze the dynamic behavior of a foundation of arbitrary shape with a rigid basemat on the surface of a homogeneous halfspace, an equivalent radius of a circular disk can be determined with the same area as the basemat for translation or with the same moment of inertia for rotation. For each degree of freedom, the disk foundation is represented by its own cone model. This procedure discussed in Chapter 2 is accurate and can, for instance, be used for a rectangular foundation up to a slenderness ratio of length to width equal to four (Section 2.8.7). The equivalent-radius concept cannot be applied to very slender rectangular or, for instance, L-forms.

To analyze the vertical and rocking motions of such irregular foundations on a homogeneous halfspace, a strength-of-materials formulation based on an approximate Green's function (fundamental solution) is presented. The approximate Green's function for harmonic motion is derived in three steps. First, the fundamental solution for a point load is ascertained by logical, nonmathematical physical reasoning, supplemented by a few constants taken from rigorous solutions. Next, a group of point loads arranged in a circle yields the approximate Green's function for a ring. Finally, the *approximate Green's function for a disk* is constructed from concentric rings. Then a *foundation of any desired shape may be modeled by an assemblage of such subdisks.* Also, more complicated cases such as a flexible basemat and the through-soil coupling of adjacent surface foundation can be analyzed. In principle, this procedure could be expanded to lateral motion, leading to the horizontal and torsional dynamic-stiffness coefficients of foundations of arbitrary shape. This extension, however, lies outside the scope of this text.

Besides providing a practical tool to calculate the vertical and rocking dynamic-stiffness coefficients for harmonic loading of an irregular basemat, the procedure leads to physical insight into the dynamic behavior of a halfspace. The latter is essential, as it forms the basis to demonstrate that a cone can indeed represent a disk on a halfspace (Section 2.5). In particular, the role of Rayleigh waves can be well understood based on the approximate Green's function.

The simplified Green's function describes a displacement pattern in the horizontal plane outside the basemat on the free surface. Discretizing the basemat into subdisks leads to a larger computational effort than does determining an equivalent disk for the cone model. From this point of view, the procedure is the same as used in the boundary-element method, enabling the practically oriented engineer to understand the key concepts behind this rigorous procedure.

This chapter is based on work by Meek and Wolf [M9].

5.1. GREEN'S FUNCTION OF POINT LOAD ON SURFACE OF HALFSPACE

5.1.1 Static Case

To facilitate insight, the appropriate Green's function is deduced via logical reasoning. A vertical point load P indents the surface of an elastic halfspace. If the soil is homogeneous, all points of a circle with radius r centered on the applied load on the free surface experience the same vertical displacement $u(r)$. As the radius increases, the displacement decreases in proportion to $1/r^n$, the exponent n being initially unknown. Doubling the load P must double the displacement, and doubling the soil's stiffness (shear modulus G) must halve the displacement; hence the formula for u must take the form

$$u(r) = \frac{\alpha}{Gr^n} P \tag{5-1a}$$

with α a constant dependent on Poisson's ratio ν. In order that the right-hand side of Eq. 5-1a possess the proper dimensions of length, the exponent n must be 1. The expression of Eq. 5-1a has been derived to this point solely by logical reasoning, without recourse to the theory of elasticity. The rigorous elasticity solution of Boussinesq reveals the numerical value of the constant α to be $(1 - \nu)/(2\pi)$, so that the exact static Green's function is

$$u(r) = \frac{1-\nu}{2\pi Gr} P \tag{5-1b}$$

5.1.2 Dynamic Case

Salient features of the dynamic Green's function for harmonic loading may also be inferred by logical reasoning. It is well known that a vertical vibrating source with amplitude $P(\omega)$ produces surface waves (Rayleigh waves) that propagate with horizontal velocity c_R. The Rayleigh-wave velocity c_R is slightly less than the shear-wave velocity c_s; for example, $c_R = 0.9325\ c_s$ for soil with $\nu = 1/3$. Like an ocean wave, the Rayleigh wave dies off very rapidly with depth; in effect, it is confined to a thin layer near the surface. As a result, successive wave fronts are cylinders, not hemispheres. For cylindrical wave fronts, the radiation criterion enforcing conservation of radiated energy requires that the displacement amplitude decays with $1/\sqrt{r}$ at large distances from the source (Eq. 2-271). This is the behavior in the so-called *far field*. The question arises of where the far field begins. The location must be specified in terms of a characteristic radius r_f (subscript f for the far field). The only physical quantities available to define the radius r_f are the propagation velocity c_R and the circular frequency ω. Since the quotient c_R/ω has the proper units of length, the far field boundary is located at

$$r_f = \beta \frac{2\pi c_R}{\omega} = \beta\lambda_R \tag{5-2}$$

in which β is a constant, initially unknown, and $\lambda_R = 2\pi c_R/\omega$ is the wave length of the Rayleigh wave.

Next, the near field in the vicinity of the source is addressed. In the static case ($\omega = 0$) the near field extends according to Eq. 5-2 to infinity; and from Eq. 5-1b, it follows that the displacement amplitude decays in proportion to $1/r$ in this interior region. As the frequency ω increases, the far field boundary r_f moves in toward the source; the near field shrinks. In the near field, where Rayleigh waves have not yet formed, the propagation velocity is not quite identical to c_R. To a first approximation the near-field propagation velocity along the surface is taken to be γc_R, with γ a constant, initially unknown.

The displacement magnitude $|u(\omega)|$, which is proportional to $1/r$ in the near field and to $1/\sqrt{r}$ in the far field, may be summarized as follows:

$$\text{Near field } r \le r_f \qquad |u(\omega)| = \frac{1-\nu}{2\pi G r} P(\omega) \tag{5-3a}$$

$$\text{Far field } r > r_f \qquad |u(\omega)| = \frac{1-\nu}{2\pi G \sqrt{r_f r}} P(\omega) \tag{5-3b}$$

These expressions are equal at the far field boundary $r = r_f$.

The ripples which would be observed in an aerial photo of the vibrating halfspace are sinusoidal waves, most conveniently represented by multiplying the displacement magnitude $|u(\omega)|$ by a complex exponential function $e^{-i\varphi(\omega)}$

$$u(r,\omega) = |u(\omega)| e^{-i\varphi(\omega)} \tag{5-4}$$

For each wavelength λ away from the source, the phase angle $\varphi(\omega)$ completes a revolution of 2π radians, so that

$$\varphi(\omega) = \frac{2\pi r}{\lambda} \tag{5-5}$$

Substituting Eq. 5-5 in Eq. 5-4 yields an exponential function with the argument proportional to i times ω times r with a minus sign. This corresponds to an outward-propagating wave (see Eq. A-8).

The wave length λ must be computed using the appropriate velocity of propagation. In the near field $\lambda = 2\pi\gamma c_R/\omega$ and thus

$$\varphi(\omega) = \frac{\omega r}{\gamma c_R} \tag{5-6a}$$

In the far field at large distance $r \gg r_f$ the velocity of propagation changes over to the Rayleigh-wave velocity c_R and thus $\lambda = \lambda_R$. Note, that this changeover does not occur at $r = r_f$. The phase angle is

$$\varphi(\omega) = \Delta\varphi + \frac{\omega r}{c_R} \tag{5-6b}$$

The phase shift $\Delta\varphi$, required to match Eqs. 5-6a and 5-6b at their point of intersection, is a further constant, initially unknown. The expression of Eqs. 5-6a and 5-6b defines an approximate bilinear phase relationship. Notice that if the phase shift $\Delta\varphi$ is positive, the (positive) constant γ must be less than one; otherwise, the straight lines (Eqs. 5-6a and 5-6b) do not intersect.

To augment the logical reasoning, numerical values for the constants β, γ, and $\Delta\varphi$ are required. These may be ascertained with the help of the exact solution for $\nu = 1/4$ shown in Fig. 5-1, taken from Rücker [R2]. The double-logarithmic plot of the magnitude (Fig. 5-1a) includes the factor r in the ordinate. As a result, the horizontal asymptote in the left-hand portion of the curve represents the near-field decay law $\propto 1/r$. The right-hand asymptote, a straight line with slope 1/2, corresponds to the far-field decay law $\propto 1/\sqrt{r}$. The analytical expression of the limiting far-field solution, derived in Miller and Pursey [M12], may be written more conveniently in the form suggested by Eq. 5-3b as

$$u(r,\omega) = \frac{1-\nu}{2\pi G}\,\frac{P(\omega)e^{-i(\pi/4+\omega r/c_R)}}{\sqrt{\beta^*\lambda_R r}} \tag{5-7}$$

with β^* a function of Poisson's ratio ν and the velocities c_s (shear waves), c_p (dilatational waves), and c_R (Rayleigh waves). The formula for the dimensionless constant β^* follows from Rücker [R2] (Eqs. A3 and A7):

$$\beta^* = \left(\frac{1-\nu}{2\pi}\right)^2 \frac{\left\{8\frac{c_s}{c_R} - \left[48 - 32\left(\frac{c_s}{c_p}\right)^2\right]\left(\frac{c_s}{c_R}\right)^3 + 48\left[1-\left(\frac{c_s}{c_p}\right)^2\right]\left(\frac{c_s}{c_R}\right)^5\right\}^2}{\left[2\left(\frac{c_s}{c_R}\right)^2 - 1\right]^4\left[\left(\frac{c_s}{c_R}\right)^2 - \left(\frac{c_s}{c_p}\right)^2\right]} \tag{5-8}$$

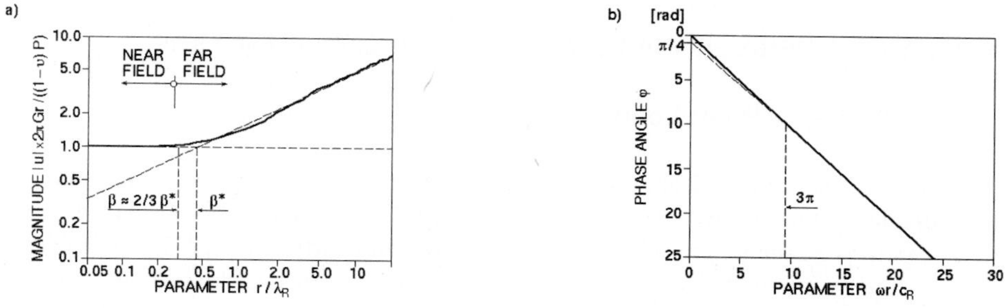

a. Magnitude. **b.** Phase angle.

Fig. 5-1 Exact Green's function for vertical point load.

This expression is evaluated numerically in Table 5-1 for typical values of Poisson's ratio ν. In Fig. 5-1a the near- and far-field asymptotes intersect at the point $\beta^*\lambda_R$. From Table 5-1 $\beta^* = 0.313$ for $\nu = 0$ and $\beta^* = 0.534$ for $\nu = 1/2$; hence, for larger values of Poisson's ratio, the near field becomes larger.

TABLE 5-1 LOCATION OF FAR-FIELD BOUNDARY

	Value of Poisson's Ratio ν			
	0	1/4	1/3	1/2
Velocity Ratio c_s/c_p	$1/\sqrt{2}$	$1/\sqrt{3}$	1/2	0
Velocity Ratio c_s/c_R	1.1441	1.0877	1.0724	1.0468
Parameter β^*	0.3131	0.4233	0.4652	0.5340

By comparison of Eq. 5-7 with Eq. 5-3b, it would appear that r_f should be $\beta^*\lambda_R$. However, numerical experiments have shown that the far field effectively begins at a radius r_f somewhat less than $\beta^*\lambda_R$, at the location where the curve of the exact solution starts to deviate from the near field asymptote. According to Fig. 5-1a, it is appropriate in Eq. 5-2 to select $\beta \approx (2/3)\beta^*$. For the halfspace with $\nu = 1/3$ considered herein, the choice $\beta = 0.3$ in Eq. 5-2 yields good results in Eq. 5-3b. As a rule of thumb for physical insight, the far-field boundary is located at one-third a wavelength from the source.

A comparison of Eqs. 5-7 and 5-6b reveals that the phase shift $\Delta\varphi$ is equal to $\pi/4$. The exact solution for the phase angle $\varphi(\omega)$ is plotted in Fig. 5-1b, which also shows that the straight lines described by Eqs. 5-6a and 5-6b intersect when $\omega r/c_R \approx 3\pi$—that is, 1.5 wave lengths λ_R away from the source. This location is five times farther out than the radius r_f which is appropriate as the far-field boundary for magnitude. Corresponding to the intersection at $\omega r/c_R = 3\pi$, the constant γ must equal 12/13.

For the halfspace with $\nu = 1/3$, the selection of the constants $\beta = 0.3$, $\gamma = 12/13$, and $\Delta\varphi = \pi/4$ completes the specification of the approximate dynamic Green's function, which converges to the exact static Green's function in the limit $\omega \to 0$.

5.2 SOLUTIONS DERIVED VIA POINT-LOAD GREEN'S FUNCTION

5.2.1 Ring of Point Loads

If evenly spaced vertical point loads, all acting in phase, are arranged in a circle, the vertical displacement at any desired location except directly under a load may be computed by superposing the various Green's functions. For any particular load the radius r is simply the distance from the source to the location of interest, denoted as the receiver. Due to the decay law $\propto 1/r$, the displacement becomes infinite whenever the receiver coincides with a source ($r = 0$). This singular behavior is a main reason why exact solutions of elasticity problems are mathematically so complicated. Fortunately, there is an engineering approach to avoid the singularity: One just considers a large number of loads and computes the displacement on the perimeter of the ring between any two of them. The inverse of this dynamic-flexibility coefficient is equal to the dynamic-stiffness coefficient.

For a ring of 60 loads at radius r_0 the vertical dynamic-stiffness coefficient for $\nu = 1/3$ is plotted in Fig. 5-2. The results are presented in the customary form described, for instance, in Eq. 4-52. The frequency dependence is expressed by the spring and damping coefficients $k_v(a_0)$ and $c_v(a_0)$, which are functions of the dimensionless frequency $a_0 =$

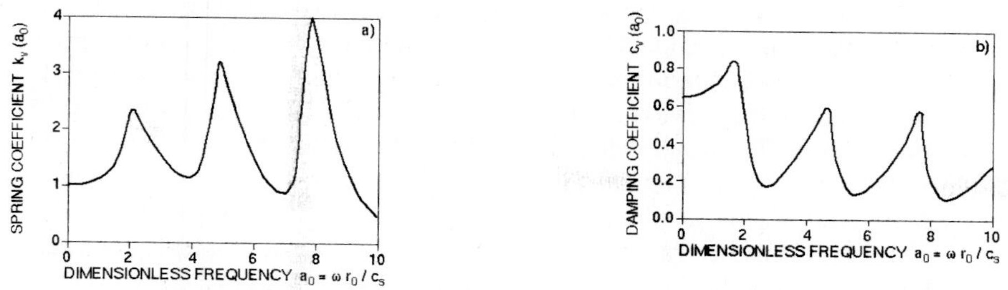

Fig. 5-2 Dynamic-stiffness coefficient for harmonic loading of ring on homogeneous halfspace in vertical motion.

$\omega r_0/c_s$. By convention, the parameter a_0 is always defined with respect to the shear-wave velocity c_s, even though the Rayleigh-wave velocity c_R would appear to be more appropriate in this case. The peaks and valleys in the curves are resonance effects; the interior region of the ring is analogous to a drumhead (membrane) with many natural frequencies.

5.2.2 Rigid Disk in Vertical Motion

To simulate a rigid disk in vertical motion, 20 concentric rings of radius up to r_0 are superposed. Each ring j consists of 60 equal loads, but the load amplitudes $P_j(a_0)$ are different for different rings. The rigidity constraint requires that all rings have the same vertical displacement amplitude $u_0(a_0)$. A solution of simultaneous equations yields the appropriate $P_j(a_0)$-values for the various rings, which, when summed, add up to the total load amplitude $P(a_0)$. This matrix formulation is analogous to that described in Section 5.4. The spring and damping coefficients of the dynamic-stiffness coefficients for $\nu = 1/3$ calculated using the approximate Green's function, Fig. 5-3, are in excellent agreement with the exact solution of Veletsos and Verbic [V3]. For low frequencies the interaction force-displacement relationship converges to

$$P(a_0) = K(1+0.74ia_0)\, u_0(a_0) \tag{5-9}$$

in which K is the vertical static-stiffness coefficient $4Gr_0/(1-\nu)$, evaluated for $\nu = 1/3$.

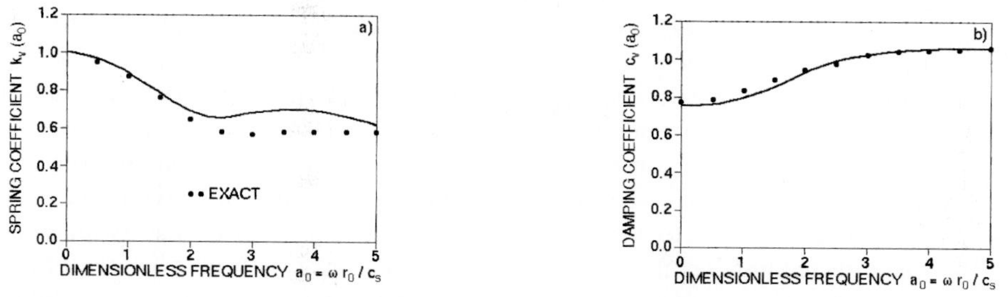

Fig. 5-3 Dynamic-stiffness coefficient for harmonic loading of disk on homogeneous halfspace in vertical motion.

After the $P_j(a_0)$-values are known, it is easy to calculate the displacement amplitude $u(r, a_0)$ at points outside the disk by summing the contributions of the separate rings; the result, expressed in terms of magnitude and phase angle, is shown in Fig. 5-4. Within drawing accuracy, the curves computed for various values of a_0 in the range 0 to 1.5 are indistinguishable. The magnitude plot may be viewed as a cross section of the static settlement valley. The uniform static displacement under the disk is $u_0 = (1 - \nu)P/(4Gr_0)$. From Eq. 5-1b it follows that a point load P would produce a displacement $u(r_0)$ smaller by the factor $2/\pi$. This difference (dashed curve) in Fig. 5-4a vanishes rapidly. For values $r/r_0 > 2$ the displacement of the disk is essentially identical to that due to a central point load, in agreement with the principle of St-Venant. The phase angle for the disk is also practically the same as that of the point load, except that the origin if shifted from the center to the location $2r_0/\pi$ where the displacement amplitudes of the disk and the point load coincide.

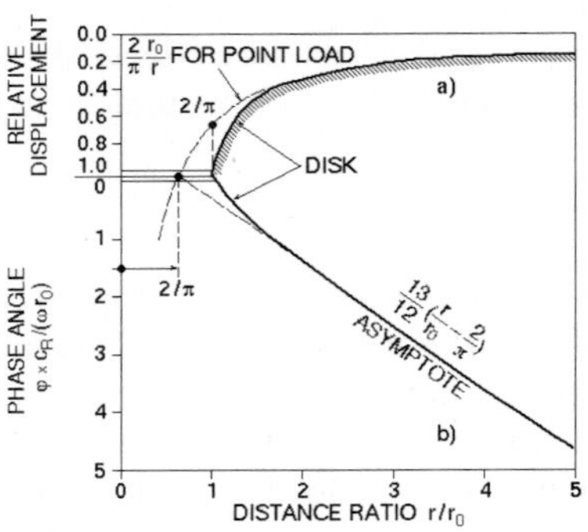

a. Magnitude.

b. Phase angle.

Fig. 5-4 Approximate Green's function for disk in vertical motion.

5.3 APPROXIMATE GREEN'S FUNCTION OF DISK

5.3.1 Illustrative Example

For small values of a_0 the approximate Green's function for the disk is defined by Eq. 5-9 in conjunction with Fig. 5-4. If the load amplitude $P(a_0)$ is given, Eq. 5-9 may be solved for the disk's displacement amplitude $u_0(a_0)$. Then the curves in Fig. 5-4 yield the magnitude and phase angle of exterior displacements $u(r,a_0)$ in the near field. In the far field the decay law $\propto 1/\sqrt{r}$ applies, as specified by Eq. 5-3b. Before summarizing in the next Section 5.3.2 all equations used to construct the approximate Green's function, the steps of calculation are illustrated for the case of the near field by the example pictured in Fig. 5-5. Two identical disks separated by the distance b are excited by a pair of loads with amplitude $P(a_0)$, one directed upward, the other downward. This situation is an example for the interaction of neighboring foundations; at the same time it is a primitive model for the rocking case. The displacement amplitude of the left disk due to its own downward load is

$$u_{11}(a_0) = \frac{P(a_0)}{K(1 + 0.74 i a_0)} \tag{5-10}$$

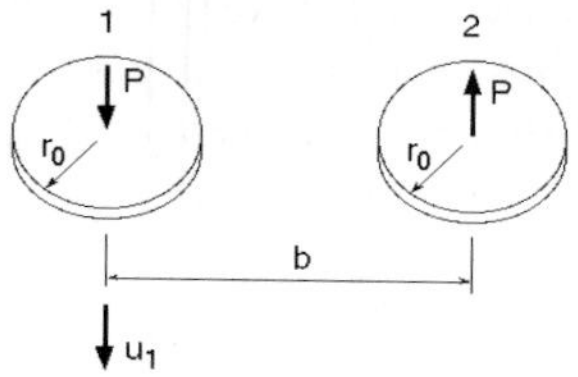

Fig. 5-5 Two disks loaded antisymmetrically as simple model for rocking motion.

The additional displacement amplitude of the left disk due to the upward load on the right disk is

$$u_{12}(a_0) = \frac{-P(a_0)}{K(1+0.74ia_0)}\frac{2}{\pi}\frac{r_0}{b}\, e^{-i\frac{13}{12}\frac{\omega}{c_R}\left(b-\frac{2r_0}{\pi}\right)} \tag{5-11}$$

The sum of Eqs. 5-10 and 5-11 yields the total displacement amplitude of the left disk

$$u_1(a_0) = u_{11}(a_0) + u_{12}(a_0) = \frac{P(a_0)\left[1 - \frac{2}{\pi}\frac{r_0}{b}\, e^{-0.74ia_0\left(\frac{\pi b}{2r_0} - 1\right)}\right]}{K(1+0.74ia_0)} \tag{5-12}$$

in which the phase angle has been rewritten in terms of $a_0 = \omega r_0/c_s$. Notice that the factors 12/13 (= γ) and $2/\pi$ (= displacement ratio in Fig. 5-4) combine with the velocity ratios $c_R/c_s = 0.9325$ to yield the same constant 0.74 appearing in the denominator. As might be expected, this is not a chance result. The reason becomes obvious in the low-frequency limit. For small values of a_0 [or equivalently for small phase angles $\varphi(a_0)$], $e^{-i\varphi(a_0)}$ tends to $1 - i\varphi(a_0)$ and Eq. 5-12 simplifies to

$$u_1(a_0) \approx \frac{P}{K}\frac{(1+0.74ia_0)}{(1+0.74ia_0)}\left(1-\frac{2}{\pi}\frac{r_0}{b}\right) \tag{5-13}$$

The complex-valued terms $1 + 0.74ia_0$ in the numerator and denominator cancel. The displacement and force are in phase with one another for small a_0, not just for $a_0 = 0$. Consequently, the damping coefficient $c(a_0)$ for rocking vanishes for a_0 approaching zero. This fact, observed in rigorous solutions and cone models for the rocking case, is discussed in Section 2.5.2. It has an interesting physical explanation. Whereas the static displacement due to a single load decays with $1/r$, the static displacement of a couple, namely

$$u = \frac{(1-\nu)\,P}{2\pi G}\left(\frac{1}{r} - \frac{1}{r+b}\right) \approx \frac{(1-\nu)Pb}{2\pi G r^2} \quad \text{for} \quad \frac{b}{r} \ll 1 \tag{5-14}$$

decays with $1/r^2$, too rapidly for waves to penetrate to infinity (see also discussion in connection with Eq. 2-272).

5.3.2 Summary of Equations

For small values of the dimensionless frequency $a_0 = \omega r_0/c_s$, the displacement of a rigid disk with amplitude $u_0(a_0)$ produced by a vertical load with amplitude $P(a_0)$ is

$$u_0(a_0) \approx \frac{P(a_0)}{K(1+0.74ia_0)} \tag{5-15}$$

The numerical value 0.74 corresponds to an elastic halfspace with Poisson's ratio $\nu = 1/3$, for which the static stiffness, given by $K = 4Gr_0/(1 - \nu)$ in the general case, becomes $K = 6Gr_0$. On the surface of the soil outside the disk ($r > r_0$), the vertical displacement amplitude may be expressed as

$$u(r,a_0) = u_0(a_0)Ae^{-i\varphi(a_0)} \tag{5-16}$$

in which A is an amplitude-reduction factor and $\varphi(a_0)$ is a phase angle.

Amplitude-reduction factor. According to Fig. 5-4a, the amplitude-reduction factor in the near field tends asymptotically to the dashed curve

$$A = \frac{2}{\pi}\frac{r_0}{r} \qquad \text{for } r \le r_f \tag{5-17}$$

The decay law for magnitude becomes proportional to $1/\sqrt{r}$ in the far field

$$A = \frac{2}{\pi}\frac{r_0}{\sqrt{r_f r}} \qquad \text{for } r > r_f \tag{5-18}$$

The Eqs. 5-17 and 5-18 yield the same value of A at the far field boundary $r = r_f$, which for $\nu = 1/3$ is located about 0.3 Rayleigh wavelengths away from the center of the disk

$$r_f \approx 0.3\lambda_R = 0.3\frac{2\pi c_R}{\omega} = \frac{1.76}{a_0}r_0 \tag{5-19}$$

The expression in terms of a_0 follows from the fact that for $\nu = 1/3$, $c_R = 0.9325c_s$.

If a large foundation is idealized as an assemblage of smaller subdisks, adjacent subdisks are separated by $r = 1.77r_0$. (As subdisks are defined to have the same area as subsquares, they overlap slightly; as a result $r = 1.77\ r_0$ instead of $2r_0$.) For $r \ge 1.77r_0$ the Eqs. 5-17 and 5-18, which represent the dashed asymptote in Fig. 5-4a involve practically no error. If even more accuracy is desired, and especially if smaller values $r < 1.77r_0$ are of interest, the distance r in Eqs. 5-17 and 5-18 may be replaced by an effective distance $\tilde{r}$:

$$\tilde{r} = r - \frac{\left(1 - \frac{2}{\pi}\right)r_0}{\left(\frac{r}{r_0}\right)^2} \tag{5-20}$$

If A is calculated with $\tilde{r}$ instead of r, essentially exact agreement with the solid curve in Fig. 5-4a is obtained. In particular, $A = 1$ at the edge of the disk, not $2/\pi$ as given by Eq. 5-17.

Phase angle. As shown in Fig. 5-4b, the origin of the phase angle $\varphi(a_0)$ is located not at the center of the disk, but rather at the point $2r_0/\pi$. The phase angle $\varphi(\omega)$ is defined by the relationships

$$\varphi(\omega) = \frac{\omega\left(r - \frac{2r_0}{\pi}\right)}{\frac{12}{13}c_R} \qquad \text{for} \qquad \frac{\omega\left(r - \frac{2r_0}{\pi}\right)}{c_R} \le 3\pi \tag{5-21a}$$

$$\varphi(\omega) = \frac{\omega\left(r - \frac{2r_0}{\pi}\right)}{c_R} + \frac{\pi}{4} \quad \text{for} \quad \frac{\omega\left(r - \frac{2r_0}{\pi}\right)}{c_R} > 3\pi \tag{5-22a}$$

which may be expressed in terms of a_0 as

$$\varphi(a_0) = 1.16 a_0 \left(\frac{r}{r_0} - \frac{2}{\pi}\right) \quad \text{for} \quad r - \frac{2r_0}{\pi} \leq 5 r_f \tag{5-21b}$$

$$\varphi(a_0) = \frac{\pi}{4} + 1.07 a_0 \left(\frac{r}{r_0} - \frac{2}{\pi}\right) \quad \text{for} \quad r - \frac{2r_0}{\pi} > 5 r_f \tag{5-22b}$$

5.4 MATRIX FORMULATION OF SURFACE FOUNDATION WITH ARBITRARY SHAPE MODELED WITH SUBDISKS

The matrix formulation to determine the surface foundation's dynamic-stiffness matrix for harmonic loading in the vertical and rocking motions is analogous to that of a pile group (Section 4.9.3). A surface foundation of any desired shape may be idealized as an assemblage of n smaller subdisks with radius Δr. The approximate Green's function for a subdisk is equal to that of a disk with radius $r_0 = \Delta r$. The dimensionless frequency of the subdisk, $a_0^{\Delta r} = \omega \Delta r / c_s$, remains small by comparison to that of the large irregular foundation, $a_0 = \omega b / c_s$ [with b a typical dimension, for rectangular basemats the halfwidth of the shorter side, provided the discretization is fine enough ($\Delta r \ll b$)]. In order for the linearized dynamic-stiffness relationship (Eq. 5-9) to remain accurate, $a_0^{\Delta r}$ for the subdisk should not exceed 1/2. Introducing the n-element vectors of amplitudes of vertical displacements $\{u(a_0)\}$ and of vertical forces $\{P(a_0)\}$, the displacement-force relationship for the subdisks formulated at their centers equals

$$\{u(a_0)\} = [G(a_0)]\,\{P(a_0)\} \tag{5-23}$$

The dynamic-flexibility matrix $[G(a_0)]$ is constructed based on the approximate Green's function summarized in Section 5.3.2.

With the rigid basemat's amplitudes of the vertical displacement $u_0(a_0)$ and rotation $\vartheta_0(a_0)$ assembled in $\{u_0(a_0)\}$, the compatibility condition yields

$$\{u(a_0)\} = [A]\,\{u_0(a_0)\} \tag{5-24}$$

The first column of the kinematic constraint matrix $[A]$ consists of ones; the second column contains the distances of the subdisk's centers to the axis of rocking. Formulating equilibrium with $\{P_0(a_0)\}$ containing the amplitudes of the vertical resultant and moment yields

$$\{P_0(a_0)\} = [A]^T\,\{P(a_0)\} \tag{5-25}$$

Substituting Eq. 5-24 in Eq. 5-23 which is solved for $\{P(a_0)\}$ and then substituted in Eq. 5-25 results in

$$\{P_0(a_0)\} = [S(a_0)]\,\{u_0(a_0)\} \tag{5-26}$$

with the dynamic-stiffness matrix

$$[S(a_0)] = [A]^T [G(a_0)]^{-1} [A] \tag{5-27}$$

As this procedure is based on an approximate Green's function, it must be checked that the resulting damping matrix of $[S(a_0)]$ is positive semi-definite for all a_0 to guarantee a non-negative rate of energy transmission (Sections 4.7.1 and B1.4).

The matrix $[G(a_0)]^{-1}$ represents the dynamic-stiffness matrix with respect to the sub-disks in the vertical direction. This permits a flexible basemat to be modeled.

5.5 EXAMPLE OF SQUARE SURFACE FOUNDATION

To illustrate the use of subdisks, the dynamic-stiffness coefficients of a square rigid basemat of length $2b$ on a halfspace with $\nu = 1/3$ is computed for vertical and rocking motions. One quadrant is discretized with 7 x 7 subdisks. The other three quadrants are treated by symmetry considerations. Per wavelength 8 subdisks are selected for good accuracy. Setting the wavelength $c_s 2\pi/\omega = 2\pi b/a_0$ equal to $8b/7$ leads to $a_0 = 5.5$, the maximum dimensionless frequency which can be represented accurately.

The dynamic-stiffness coefficients, decomposed in the standard manner, are plotted in Fig. 5-6. The accuracy by comparison to rigorous solutions denoted as exact presented in Wolf [W18] is excellent particularly for the damping coefficient. The frequency dependence of $k(a_0)$ and $c(a_0)$ is captured, even though the k and c values of the subdisk (Eq. 5-9) are constant. Notice that the k and c curves for the square (Fig. 5-6) and the disk (Fig. 5-3) are practically identical. This implies that if a large disk were to be constructed from smaller subdisks, the results would match; in other words, the approximate Green's function clones (reproduces) itself.

Further results for rectangular foundations are presented in Section 2.8.7 (Figs. 2-71 to 2-74).

5.6 THROUGH-SOIL COUPLING OF SURFACE FOUNDATIONS

The through-soil coupling in the vertical direction of two square rigid basemats of length $2b$ and distance $d = 2b$ on a halfspace with $\nu = 1/3$ is addressed (Fig. 5-7). Each basemat is

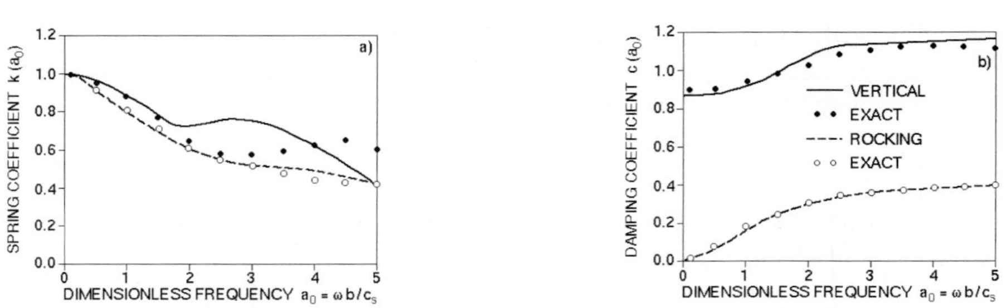

Fig. 5-6 Dynamic-stiffness coefficients for harmonic loading of square foundation on homogeneous halfspace in vertical and rocking motions.

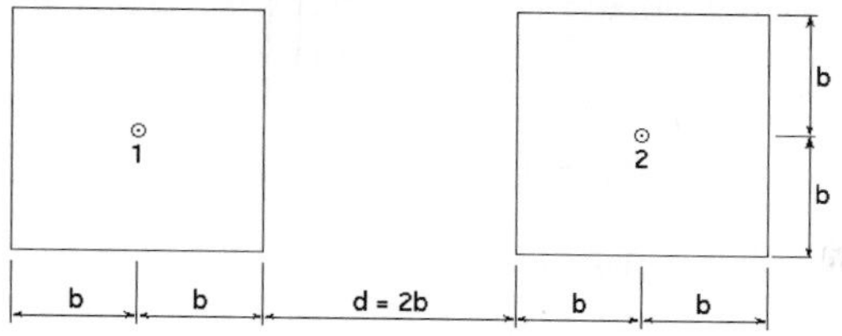

Fig. 5-7 Plan view of two-foundation model.

discretized into 10 x 10 subdisks. The dynamic-stiffness coefficient $S_{12}(a_0)$ representing the through-soil coupling of the two basemats in their centers of gravity points 1 and 2 in the vertical direction is normalized as

$$S_{12}(a_0) = Gb[k_{v12}(a_0) + ia_0c_{v12}(a_0)] \tag{5-28}$$

with $a_0 = \omega b/c_s$. A good agreement (Fig. 5-8) exists with the rigorous solution of Wong and Luco [W19] denoted as exact, which is based on a boundary-element procedure for welded contact and with a small damping ratio $\zeta_g = 0.02$.

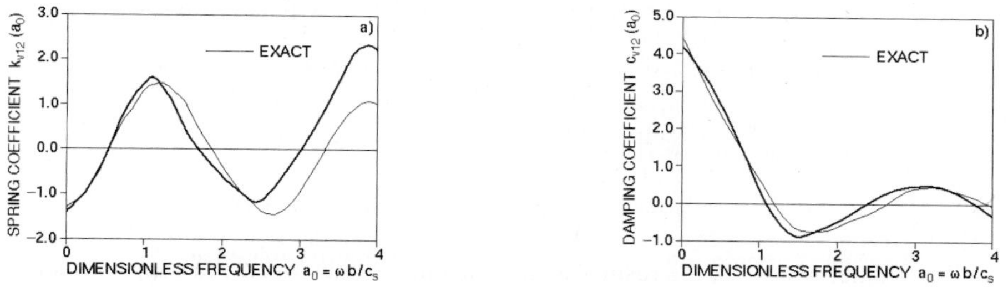

Fig. 5-8 Vertical through-soil coupling dynamic-stiffness coefficient for harmonic loading of two-foundation model.

SUMMARY

1. Based on physical reasoning with some constants determined from a rigorous elastodynamic solution, the Green's function of a vertical point load with amplitude $P(\omega)$ on the surface of a halfspace for harmonic loading is approximated as

$$u(r,\omega) = |u(\omega)|e^{-i\varphi(\omega)}$$

The vertical displacement magnitude equals

$$|u(\omega)| = \frac{1-\nu}{2\pi Gr}P(\omega) \qquad \text{for} \qquad r \le r_f$$

$$|u(\omega)| = \frac{1-\nu}{2\pi G\sqrt{r_f r}}P(\omega) \qquad \text{for} \qquad r > r_f$$

with the shear modulus G and the radial distance r between the source and the receiver. The far-field boundary is located for Poisson's ratio $\nu = 1/3$ at

$$r_f = 0.3\lambda_R = 0.3\,\frac{2\pi c_R}{\omega}$$

with the wave length λ_R and the propagation velocity $c_R = 0.9325c_s$ of the Rayleigh wave.

The phase angle $\varphi(\omega)$ equals

$$\varphi(\omega) = \frac{\omega r}{\frac{12}{13}c_R} \qquad \text{for} \qquad \frac{\omega r}{c_R} \le 3\pi$$

$$\varphi(\omega) = \frac{\pi}{4} + \frac{\omega r}{c_R} \qquad \text{for} \qquad \frac{\omega r}{c_R} > 3\pi$$

2. After using a circle of point loads to represent a ring, a group of concentric rings is employed to construct the approximate Green's function of a rigid disk. For small values of the dimensionless frequency $a_0 = \omega r_0/c_s$, the amplitude of the vertical displacement $u_0(a_0)$ produced by a vertical load with amplitude $P(a_0)$ equals

$$u_0(a_0) = \frac{P(a_0)}{K(1 + 0.74\, i a_0)}$$

with the static-stiffness coefficient

$$K = 6Gr_0 \qquad \text{for} \qquad \nu = 1/3$$

The Green's function outside the disk is approximated as

$$u(r,a_0) = u_0(a_0)Ae^{-i\varphi(a_0)}$$

The amplitude-reduction factor A equals

$$A = \frac{2}{\pi}\frac{r_0}{r} \qquad \text{for} \qquad r \le r_f$$

$$A = \frac{2}{\pi}\frac{r_0}{\sqrt{r_f r}} \qquad \text{for} \qquad r > r_f$$

with

$$r_f = 0.3\,\lambda_R = \frac{1.76}{a_0}\,r_0$$

The phase angle $\varphi(a_0)$ equals

$$\varphi(a_0) = 1.16\,a_0\left(\frac{r}{r_0} - \frac{2}{\pi}\right) \qquad \text{for} \qquad r - \frac{2r_0}{\pi} \le 5\,r_f$$

$$\varphi(a_0) = \frac{\pi}{4} + 1.07\,a_0\left(\frac{r}{r_0} - \frac{2}{\pi}\right) \qquad \text{for} \qquad r - \frac{2r_0}{\pi} > 5\,r_f$$

3. Arbitrarily shaped surface foundations are modeled by an assemblage of subdisks. The familiar matrix formulation of structural mechanics leads to the dynamic-stiffness coefficients for vertical and rocking motions. The dynamic-stiffness matrix with respect to the displacement amplitudes of the rigid basemat $\{u_0(a_0)\}$ equals

$$[S(a_0)] = [A]^T\,[G(a_0)]^{-1}\,[A]$$

The vertical displacement amplitudes of the subdisks $\{u(a_0)\}$ are equal to

$$\{u(a_0)\} = [A]\,\{u_0(a_0)\}$$

with the kinematic constraint matrix $[A]$. With the vertical force amplitudes acting on the subdisks $\{P(a_0)\}$

$$\{u(a_0)\} = [G(a_0)]\,\{P(a_0)\}$$

applies, where the dynamic-flexibility matrix $[G(a_0)]$ is constructed based on the approximate Green's function of the subdisk specified in item 2.

6

Seismic Excitation

The use of the simple physical models to represent the soil for problems excited by loads acting on the structure arising from machine vibrations is well documented in the previous chapters. The case of seismic excitation addressed in the following is much more puzzling, as it is not immediately clear in what form and at what location the earthquake input should be applied. It will be demonstrated that as for the rigorous elastodynamical solution, the *seismic excitation can be converted to an equivalent load* acting on the dynamic system. This is to be expected. After all, if the soil is viewed generally as a linear "black box," a non-ambiguous solution procedure must be valid, regardless whether the unknown contents of the black box are a simple physical model or a complicated rigorous solution. Various possibilities exist. It is thus as easy to process an earthquake as it is a load acting directly on the structure in a dynamic soil–structure-interaction analysis where the soil is represented by a simple physical model.

As in the previous chapters, emphasis is placed on the substructure method of analysis. In Section 6.1 the basic equation of motion of the dynamic system consisting of the structure and the soil for seismic excitation is derived. Both frequency- and time-domain analyses are addressed. Various forms of the equivalent loading are discussed. The first step of any seismic analysis is the calculation of the free-field response of the site, which is examined in Section 6.2. From the free-field motion the so-called effective foundation input motion is determined, which is discussed in Section 6.3. For surface foundations horizontally propagating waves can be assumed and for embedded foundations, vertically propagating waves. This effective foundation input motion leads to the load acting on the dynamic system. In Section 6.4 the seismic analysis in the time domain by hand with a pocket calculator of a surface foundation modeled with cones (soil halfspace) or unfolded layered cones (soil layer) is addressed. Use is made of the recursive evaluation of the convolution integrals and of echo constants. Section 6.5 examines the various possibilities of performing the seismic analysis where the soil is represented by a lumped-parameter model in the time domain. Finally, Section 6.6 establishes the basic equation of motion for the direct method of analysis. The possibility of using a local transmitting boundary based on cones is mentioned.

6.1 BASIC EQUATION OF MOTION OF SUBSTRUCTURE METHOD

6.1.1 Formulation in Total Displacements for Harmonic Excitation

To formulate the basic equation of motion, the structure with a rigid base consisting of the basemat and the adjacent walls embedded in soil shown in Fig. 6-1 is examined for seismic excitation. It is customary to perform this derivation for harmonic excitation and then transform the final result to the time domain.

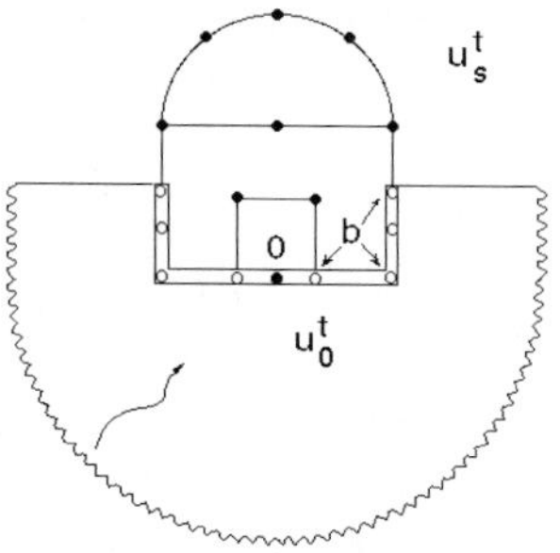

Fig. 6-1 Structure–soil system with rigid base.

The dynamic system consists of two substructures, the actual structure and the soil with excavation. Subscripts are used to denote the nodes of the discretized system. The node on the rigid structure–soil interface (base) is denoted by 0, the remaining nodes of the structure by s. To differentiate between the various subsystems, superscripts are used when necessary. The structure is indicated by s (when used with a property matrix), the soil with excavation by g (for ground). Other reference soil systems are introduced later, as are other nodes.

The equations of motion of the structure are formulated in total displacement amplitudes $\{u^t(\omega)\}$ as

$$\begin{bmatrix} [S_{ss}(\omega)] & [S_{s0}(\omega)] \\ [S_{0s}(\omega)] & [S^s_{00}(\omega)] \end{bmatrix} \begin{Bmatrix} \{u^t_s(\omega)\} \\ \{u^t_0(\omega)\} \end{Bmatrix} = \begin{Bmatrix} \{0\} \\ -\{P_0(\omega)\} \end{Bmatrix} \tag{6-1}$$

The word total (superscript t) expresses that the motion is referred to an origin that does not move. The dynamic-stiffness matrix $[S^s(\omega)]$ of the structure, which is a bounded system, is calculated as

$$[S^s(\omega)] = [K]\,(1+2i\zeta) - \omega^2[M] \tag{6-2}$$

where $[K]$ and $[M]$ are the static-stiffness and mass matrices, respectively, and ζ denotes the hysteretic structural damping ratio. The vector $\{u^t(\omega)\}$ of order equal to the number of dynamic degrees of freedom of the total discretized system can be decomposed into the subvectors $\{u_s^t(\omega)\}$ and $\{u_0^t(\omega)\}$. The latter denotes the amplitudes of the rigid-body motion (three translations and three rotations) of the base. $[S^s(\omega)]$ is decomposed accordingly. To avoid using unnecessary symbols, the superscript s (for structure) is used only when confusion would otherwise arise. Finally, $\{P_0(\omega)\}$ are the amplitudes of the interaction forces of the other substructure, the soil system ground. Notice that in this formulation in total displacements the nodes not in contact with the soil are not loaded (vector $\{0\}$ on the right-hand side of Eq. 6-1).

To express $\{P_0(\omega)\}$, the unbounded soil system ground with excavation and rigid massless structure–soil interface (Fig. 6-2) is addressed. $[S_{00}^g(\omega)]$ denotes its dynamic-stiffness matrix for harmonic loading and $\{u_0^g(\omega)\}$ the displacement amplitudes of the soil system ground caused by the earthquake. How $\{u_0^g(\omega)\}$ can be expressed as a function of the seismic free-field motion is discussed later. For the motion $\{u_0^g(\omega)\}$ the interaction forces acting at the node 0 vanish because for this loading state the rigid base shown in Fig. 6-2 is a free surface. The interaction forces of the soil will thus depend on the motion relative to $\{u_0^g(\omega)\}$ and can be expressed as

$$\{P_0(\omega)\} = [S_{00}^g(\omega)]\left(\{u_0^t(\omega)\} - \{u_0^g(\omega)\}\right) \tag{6-3}$$

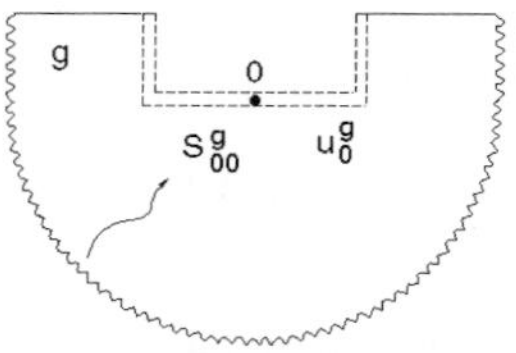

Fig. 6-2 Soil system ground with excavation and rigid structure–soil interface.

Substituting Eq. 6-3 in Eq. 6-1 yields the *basic equation of motion of the structure–soil system* with a rigid base expressed in *total displacement* amplitudes

$$\begin{bmatrix} [S_{ss}(\omega)] & [S_{s0}(\omega)] \\ [S_{0s}(\omega)] & [S_{00}^s(\omega)] + [S_{00}^g(\omega)] \end{bmatrix} \begin{Bmatrix} \{u_s^t(\omega)\} \\ \{u_0^t(\omega)\} \end{Bmatrix} = \begin{Bmatrix} \{0\} \\ [S_{00}^g(\omega)]\{u_0^g(\omega)\} \end{Bmatrix} \tag{6-4}$$

In this formulation, the earthquake excitation is characterized by $\{u_0^g(\omega)\}$, the seismic rigid-body motion of the node of the soil reference system ground (taking the excavation into account). It is worth mentioning that $\{u_0^g(\omega)\}$ does not occur in the structure–soil system. As $\{u_0^g(\omega)\}$ appears on the right-hand side of Eq. 6-4 determining the equivalent load, it is denoted as the *effective foundation input motion*. By setting $\{u_0^t(\omega)\} = 0$ in Eq. 6-4, it can be deduced that the right-hand side $\{P_0^g(\omega)\} = [S_{00}^g(\omega)]\{u_0^g(\omega)\}$ represents the amplitudes of the loads (three forces and three moments) exerted on the rigid base in node 0 by the seismic motion when the base is kept fixed. They are called *driving loads* (forces).

To express the effective foundation input motion $\{u_0^g(\omega)\}$ as a function of the free-field motion, nodes with subscript b (for base) on the base of the structure (Fig. 6-1) are introduced. These nodes can also be identified in the free field of the soil (left-hand side of Fig. 6-3). For this reference system, free field of the soil is denoted with a superscript f;

$[S^f_{bb}(\omega)]$ and $\{u^f_b(\omega)\}$ are the dynamic-stiffness matrix and the vector of the displacement amplitudes. The dynamic-stiffness matrix of the soil reference system ground, without enforcing the rigid-body motion and discretized in the nodes b, is denoted as $[S^g_{bb}(\omega)]$ (right-hand side of Fig. 6-3). The dynamic-stiffness matrix $[S^e_{bb}(\omega)]$ (superscript e for excavation) of the excavated soil, a bounded domain, follows analogously from Eq. 6-2 using the properties of the soil.

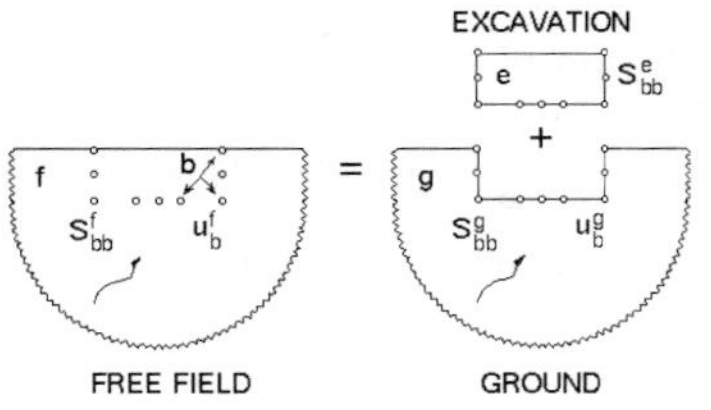

Fig. 6-3 Dynamic-stiffness matrix and earthquake excitation referred to soil's reference systems free field and ground.

The free-field system results when the excavated part of the soil is added to the soil system ground with excavation (Fig. 6-3). This also holds for the assembly process of the dynamic-stiffness matrices

$$[S^e_{bb}(\omega)] + [S^g_{bb}(\omega)] = [S^f_{bb}(\omega)] \tag{6-5}$$

The equations of motion of the excavated part of the soil are written as

$$[S^e_{bb}(\omega)]\{u^t_b(\omega)\} = -\{P_b(\omega)\} \tag{6-6}$$

This equation is analogous to Eq. 6-1 when the "structure" consists of the excavated part of the soil only and the rigid-body constraint is not enforced along the base. The interaction forces with amplitudes $\{P_b(\omega)\}$ of the soil system ground follow as in Eq. 6-3

$$\{P_b(\omega)\} = [S^g_{bb}(\omega)]\left(\{u^t_b(\omega)\} - \{u^g_b(\omega)\}\right) \tag{6-7}$$

Substituting Eq. 6-7 in Eq. 6-6 yields

$$\left([S^e_{bb}(\omega)] + [S^g_{bb}(\omega)]\right)\{u^t_b(\omega)\} = [S^g_{bb}(\omega)]\{u^g_b(\omega)\} \tag{6-8}$$

For this free-field system under seismic excitation (Fig. 6-3), $\{u^t_b(\omega)\} = \{u^f_b(\omega)\}$. Using in addition Eq. 6-5 leads to

$$[S^f_{bb}(\omega)]\{u^f_b(\omega)\} = [S^g_{bb}(\omega)]\{u^g_b(\omega)\} \tag{6-9}$$

This equality of forces is quite a remarkable result in its own right. Although for the substructure of the soil with excavation (system ground) the line with the nodes b (where the motion is equal to $\{u^g_b(\omega)\}$) is a free surface as shown in Fig. 6-3, the forces with amplitudes $\{P^g_b(\omega)\} = [S^g_{bb}(\omega)]\{u^g_b(\omega)\}$ are not zero. (The influence of the exterior boundary located at "infinity" with an applied earthquake motion also has to be taken into account when calculating the forces in nodes b.) Equation 6-9 expresses the motion of the soil with excavation $\{u^g_b(\omega)\}$ as a function of that of the free field $\{u^f_b(\omega)\}$, which does not depend on the excavation (with the exception of the location of the nodes, in which it is to be calculated).

As the soil–structure interface is subjected to rigid-body motion (Fig. 6-2),

$$\{u_b^g(\omega)\} = [A]\,\{u_0^g(\omega)\} \tag{6-10}$$

applies. $[A]$ represents the kinematic-constraint matrix with geometric quantities only. The principle of virtual work expressing equilibrium results in

$$\{P_0^g(\omega)\} = [A]^T\,\{P_b^g(\omega)\} \tag{6-11}$$

Substituting Eq. 6-10 in Eq. 6-9 and premultiplying by $[A]^T$ yields

$$[A]^T[S_{bb}^f(\omega)]\,\{u_b^f(\omega)\} = [A]^T[S_{bb}^g(\omega)]\,[A]\,\{u_0^g(\omega)\} = [S_{00}^g(\omega)]\,\{u_0^g(\omega)\} \tag{6-12}$$

where the dynamic-stiffness matrix of the soil system ground with the rigid-body constraint enforced equals

$$[S_{00}^g(\omega)] = [A]^T[S_{bb}^g(\omega)]\,[A] \tag{6-13}$$

The right-hand side of Eq. 6-12 can be written as $\{P_0^g(\omega)\} = [S_{00}^g(\omega)]\,\{u_0^g(\omega)\}$. Substituted in Eq. 6-4, the basic equation of motion of the structure–soil system is rewritten as

$$\begin{bmatrix} [S_{ss}(\omega)] & [S_{s0}(\omega)] \\ [S_{0s}(\omega)] & [S_{00}^s(\omega)] + [S_{00}^g(\omega)] \end{bmatrix} \begin{Bmatrix} \{u_s^t(\omega)\} \\ \{u_0^t(\omega)\} \end{Bmatrix} = \begin{Bmatrix} \{0\} \\ [A]^T[S_{bb}^f(\omega)]\,\{u_b^f(\omega)\} \end{Bmatrix} \tag{6-14}$$

In this formulation in total displacement amplitudes, the *driving load* vector is expressed as the *product of the dynamic-stiffness matrix of the free-field* $[S_{bb}^f(\omega)]$ (discretized in the nodes at which the structure is subsequently inserted) and of the *free-field motion* $\{u_b^f(\omega)\}$ in the same nodes. It should be emphasized that in the system f for which $[S_{bb}^f(\omega)]$ is calculated, the soil is not excavated (Fig. 6.-3). The vector $\{u_b^f(\omega)\}$ has only to be calculated in those nodes b which subsequently will lie on the structure–soil interface.

The effective foundation input motion follows from Eq. 6-12 as

$$\{u_0^g(\omega)\} = [S_{00}^g(\omega)]^{-1}\,[A]^T\,[S_{bb}^f(\omega)]\{u_b^f(\omega)\} \tag{6-15}$$

The basic equation of motion in total displacements specified in Eq. 6-4 can be physically interpreted as illustrated in Fig. 6-4. The discretized structure $[S^s(\omega)]$ is supported on a generalized spring characterized by $[S_{00}^g(\omega)]$. The far end of the generalized spring not connected to the structure is excited by $\{u_0^g(\omega)\}$, which is calculated from $\{u_b^f(\omega)\}$ and $[S_{bb}^f(\omega)]$ using the kinematic-constraint matrix $[A]$. It is important to note that the support motion $\{u_0^g(\omega)\}$ to be applied at the end of the generalized spring is calculated from the free-field response $\{u_b^f(\omega)\}$ at the structure–soil interface and not at some (fictitious) boundary at a depth where the underlying medium can be regarded as very stiff. For instance, consider a structure supported on the surface of a layer of soil resting on rigid rock. The generalized spring represents the stiffness and damping of the basemat on the surface of the soil layer built-in at the interface with the rock. The applied support motion depends on the free-field motion at the free surface of the layer and not at its base. For a better understanding of the foregoing, it is helpful to think of the generalized spring as having a length approaching zero.

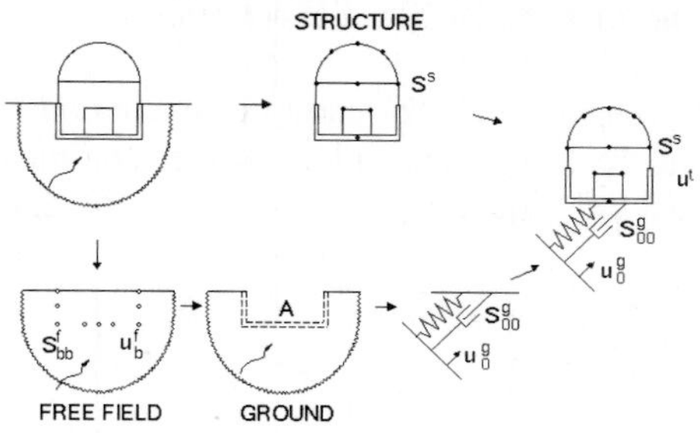

Fig. 6-4 Physical interpretation of basic equation of motion in total displacements with effective foundation input motion.

As explained in Section 6.5 when the soil is represented by a lumped-parameter model with masses, the physical interpretation of Fig. 6-4 must be modified.

The derivation allows the basic equation of motion to be formulated also for a flexible base with nodes b (Fig. 6-1)

$$\begin{bmatrix} [S_{ss}(\omega)] & [S_{sb}(\omega)] \\ [S_{bs}(\omega)] & [S^s_{bb}(\omega)] + [S^g_{bb}(\omega)] \end{bmatrix} \begin{Bmatrix} \{u^t_s(\omega)\} \\ \{u^t_b(\omega)\} \end{Bmatrix} = \begin{Bmatrix} \{0\} \\ [S^g_{bb}(\omega)]\,\{u^g_b(\omega)\} \end{Bmatrix} = \begin{Bmatrix} \{0\} \\ [S^f_{bb}(\omega)]\,\{u^f_b(\omega)\} \end{Bmatrix} \tag{6-16}$$

with the effective foundation input motion

$$\{u^g_b(\omega)\} = [S^g_{bb}(\omega)]^{-1}\,[S^f_{bb}(\omega)]\,\{u^f_b(\omega)\} \tag{6-17}$$

The term on the right-hand side of Eq. 6-16 can be further developed. Substituting Eq. 6-5 yields

$$[S^f_{bb}(\omega)]\,\{u^f_b(\omega)\} = [S^g_{bb}(\omega)]\,\{u^f_b(\omega)\} + \{P^f_b(\omega)\} \tag{6-18}$$

where the force amplitudes of the excavated region caused by the free-field motion are

$$\{P^f_b(\omega)\} = [S^e_{bb}(\omega)]\,\{u^f_b(\omega)\} \tag{6-19}$$

In this formulation the *driving loads* can be calculated using the same dynamic-stiffness matrix of the soil $[S^g_{bb}(\omega)]$ as appearing on the left-hand side of the equations, in addition to the *free-field motion* $\{u^f_b(\omega)\}$ and the *corresponding force* amplitudes $\{P^f_b(\omega)\}$.

The derivation of the basic equation of motion is based on substructuring with replacement. By adding the excavated part of the soil (system e) to the irregular system g a regular system f is formed on which the load depends.

6.1.2 Formulation in Total Displacements in Time Domain

To transform the basic equation of motion specified for harmonic excitation in Eq. 6-4 to the time domain, it is appropriate to address the contributions of the (nonlinear) structure and the linear soil separately. The structure's equation of motion is specified as

$$\begin{bmatrix} [M_{ss}] & [M_{s0}] \\ [M_{0s}] & [M_{00}] \end{bmatrix} \begin{Bmatrix} \{\ddot{u}_s^t(t)\} \\ \{\ddot{u}_0^t(t)\} \end{Bmatrix} + \begin{Bmatrix} \{F_s(t)\} \\ \{F_0(t)\} \end{Bmatrix} = \begin{Bmatrix} \{0\} \\ -\{P_0(t)\} \end{Bmatrix} \tag{6-20}$$

$[M]$ is the mass matrix and $\{F(t)\}$ the vector of the (nonlinear) internal forces of the structure at time t. For a linear structure $\{F(t)\}$ can be expressed as a function of the static-stiffness matrix $[K]$ and the damping matrix $[C]$. This results in

$$\begin{aligned} &\begin{bmatrix} [M_{ss}] & [M_{s0}] \\ [M_{0s}] & [M_{00}] \end{bmatrix} \begin{Bmatrix} \{\ddot{u}_s^t(t)\} \\ \{\ddot{u}_0^t(t)\} \end{Bmatrix} + \begin{bmatrix} [C_{ss}] & [C_{s0}] \\ [C_{0s}] & [C_{00}] \end{bmatrix} \begin{Bmatrix} \{\dot{u}_s^t(t)\} \\ \{\dot{u}_0^t(t)\} \end{Bmatrix} + \\ &\begin{bmatrix} [K_{ss}] & [K_{s0}] \\ [K_{0s}] & [K_{00}] \end{bmatrix} \begin{Bmatrix} \{u_s^t(t)\} \\ \{u_0^t(t)\} \end{Bmatrix} = \begin{Bmatrix} \{0\} \\ -\{P_0(t)\} \end{Bmatrix} \end{aligned} \tag{6-21}$$

The interaction forces of the soil that result in the soil's contribution to the basic equation of motion of the coupled system are specified for harmonic excitation in Eq. 6-3. Also in the time domain, $\{P_0(t)\}$ will depend on the motion of the base relative to that of the soil system ground (effective foundation input motion), $\{u_0^t(t)\} - \{u_0^g(t)\}$. As discussed for the scalar case in Appendix C, the product of the transfer function and the input amplitude in the frequency domain (Eq. C-3) corresponds to a convolution integral of the corresponding unit-impulse response function and the input in the time domain (Eq. C-1). The transfer function and the unit-impulse response function form a Fourier transform pair (Eq. C-4). As mentioned in the discussion following Eq. C-1, the regular part of the interaction force is obtained in a stiffness formulation, corresponding to the lingering part. The singular part defined in Eq. B-3 corresponds to the instantaneous response in the time domain calculated with a spring and a dashpot. These various aspects result in

$$\begin{aligned} \{P_0(t)\} = {} & [K_\infty]\left(\{u_0^t(t)\} - \{u_0^g(t)\}\right) + [C]\left(\{\dot{u}_0^t(t)\} - \{\dot{u}_0^g(t)\}\right) \\ & + \int_0^t [S_{r,00}^g(t-\tau)]\left(\{u_0^t(\tau)\} - \{u_0^g(\tau)\}\right) d\tau \end{aligned} \tag{6-22}$$

$[K\infty]$ and $[C]$ are the spring and dashpot matrices corresponding to the singular term, which follows from the limit of $[S_{00}^{g}(\omega)]$ for $\omega \to \infty$. Its regular part is equal to

$$[S_{r,00}^{g}(\omega)] = [S_{00}^{g}(\omega)] - [K_{\infty}] - i\omega[C] \tag{6-23}$$

The unit-impulse response function matrix follows formally from

$$[S_{r,00}^{g}(t)] = \frac{1}{2\pi} \int_{-\infty}^{+\infty} [S_{r,00}^{g}(\omega)] e^{i\omega t}\, d\omega \tag{6-24}$$

Substituting Eq. 6-22 in Eq. 6-20 results in the equation of motion, whereby on the left- and right-hand sides convolution integrals appear.

$$\begin{bmatrix} [M_{ss}] & [M_{s0}] \\ [M_{0s}] & [M_{00}] \end{bmatrix} \begin{Bmatrix} \{\ddot{u}_s^t(t)\} \\ \{\ddot{u}_0^t(t)\} \end{Bmatrix} + \begin{Bmatrix} \{F_s(t)\} \\ \{F_0(t)\} \end{Bmatrix} + \begin{Bmatrix} \{0\} \\ [K_{\infty}]\{u_0^t(t)\} + [C]\{\dot{u}_0^t(t)\} + \int_0^t [S_{r,00}^{g}(t-\tau)]\{u_0^t(\tau)\}d\tau \end{Bmatrix} = \begin{Bmatrix} \{0\} \\ [K_{\infty}]\{u_0^g(t)\} + [C]\{\dot{u}_0^g(t)\} + \int_0^t [S_{r,00}^{g}(t-\tau)]\{u_0^g(\tau)\}d\tau \end{Bmatrix} \tag{6-25}$$

The right-hand side, depending on the effective foundation input motion $\{u_0^g(t)\}$, can be evaluated before the actual interaction analysis is performed. This results in the effective (driving) load as a function of time applied to the dynamic system described by the left-hand side (structure supported on generalized spring).

The rather formal derivation presented above does not have to be followed in an actual application. It is sufficient to replace in the soil's interaction force-displacement relationship in the time domain the motion by the total motion $\{u_0^t(t)\}$ minus the effective foundation input motion $\{u_0^g(t)\}$, as discussed for surface foundations on a soil halfspace and on a soil layer in Section 6.4 and for embedded foundations in Section 6.3. The convolution integrals appearing on the left- and right-hand sides of the equations can be evaluated recursively (Appendix C).

Many slightly different formulations exist, involving the rotational cone velocities and not displacements in the convolution. A powerful alternative consists of using the lumped-parameter model developed in Appendix B, as examined in Section 6.5. The right-hand side of the equations expressing the driving load as a function of the effective foundation input motion is also determined based on the same lumped-parameter model as used to represent the generalized spring on the left-hand side. It is sufficient to apply the effective foundation input motion $\{u_0^g(t)\}$ to the lumped-parameter model at the foundation node and to determine the reaction force at the same location that is equal to the driving load $\{P_0^g(t)\}$. The latter is then applied at the foundation node to the total dynamic system consisting of the model of the structure and the lumped-parameter model of the soil.

6.1.3 Kinematic and Inertial Interactions

In the equations of motion formulated in total displacements (Eqs. 6-4 and 6-25), the effective driving load for earthquake excitation is applied only at the base, on the structure–soil interface. The structural engineer is, however, accustomed to work with seismic inertial loads (calculated as the product of the structure's mass and some suitably chosen earthquake input excitation) which are applied in all nodes of the structure. For a rigid base the effective foundation input motion $\{\ddot{u}_0^g(t)\}$ determines these inertial loads. As will be demonstrated in the following, the analysis is performed in two steps. In the first, called *kinematic interaction*, the mass of the structure is set equal to zero. Using rigid-body kinematics, the effective foundation input motion applied at the base will lead to the kinematic-interaction displacements throughout the structure. The internal forces of kinematic interaction for a rigid base vanish. The corresponding kinematic-interaction accelerations multiplied by the mass result in the inertial loads applied throughout the structure of the second step, the *inertial interaction*, for which a dynamic analysis including the structure's mass is performed. The total displacements and accelerations are equal to the sum of those of the kinematic- and inertial-interaction analyses. The internal forces and reactions follow from the result of the inertial-interaction analysis.

For a structure with a rigid base this procedure results in

$$\{\overset{t}{u}_s(t)\} = \{\overset{k}{u}_s(t)\} + \{\overset{i}{u}_s(t)\} \tag{6-26a}$$

$$\{\overset{t}{u}_0(t)\} = \{\overset{k}{u}_0(t)\} + \{\overset{i}{u}_0(t)\} \tag{6-26b}$$

where the superscripts k and i denote the components of kinematic and inertial interactions. As the motion of the ground enforcing the rigid-body kinematics along the base is the same as that of the kinematic-interaction part of the analysis (the presence of a massless structure does not change the response along the base),

$$\{\overset{k}{u}_0(t)\} = \{u_0^g(t)\} \tag{6-27a}$$

$$\{\overset{k}{u}_s(t)\} = [T_{s0}]\ \{u_0^g(t)\} \tag{6-27b}$$

apply, with the so-called quasi-static transformation matrix $[T_{s0}]$ following from rigid-body kinematics, analogously to the kinematic-constraint matrix $[A]$. Equations 6-26 and 6-27 are also valid for the accelerations and velocities. Substituting Eqs. 6-26 and 6-27 in Eq. 6-25 yields the equation of motion for inertial interaction

$$
\begin{bmatrix} [M_{ss}] & [M_{s0}] \\ & \\ [M_{0s}] & [M_{00}] \end{bmatrix} \begin{Bmatrix} \{\ddot{u}^i_s(t)\} \\ \\ \{\ddot{u}^i_0(t)\} \end{Bmatrix} + \begin{Bmatrix} \{F_s(t)\} \\ \\ \{F_0(t)\} \end{Bmatrix} +
$$

$$
\begin{Bmatrix} \{0\} \\ \\ [K_\infty]\{u^i_0(t)\} + [C]\{\dot{u}^i_0(t)\} + \int\limits_0^t [S^g_{r,00}(t-\tau)]\{u^i_0(\tau)\}d\tau \end{Bmatrix} = \tag{6-28}
$$

$$
- \begin{bmatrix} [M_{ss}] & [M_{s0}] \\ & \\ [M_{0s}] & [M_{00}] \end{bmatrix} \begin{bmatrix} [T_{s0}] \\ \\ [I] \end{bmatrix} \{\ddot{u}^g_0(t)\}
$$

The *load vector* on the right-hand side is equal to the *negative product of the mass of the structure and the kinematic-interaction accelerations*, the latter being determined by applying the effective foundation input acceleration $\{\ddot{u}^g_0(t)\}$ at the base and applying rigid-body kinematics. In this formulation, which leads to the displacements relative to those of kinematic interaction, the seismic acceleration $\{\ddot{u}^g_0(t)\}$ can be applied directly (and the seismic displacement $\{u^g_0(t)\}$ is not needed as in the total-displacement formulation in Eq. 6-25). Equation 6-28 is valid for a nonlinear structure, although the superposition principle (Eq. 6-26) is used in the derivation (in the kinematic-interaction analysis, the internal forces vanish!). The last term on the left-hand side can be processed as in any force-excited interaction problem, using a recursive evaluation of the convolution integral or a lumped-parameter model.

The kinematic and inertial motions are illustrated in Fig. 6-5. Summarizing from the knowledge of the free-field response $\{\ddot{u}^f_b(t)\}$, the effective foundation input motion $\{\ddot{u}^g_0(t)\}$ follows. (The corresponding time-domain analysis is discussed in Section 6.3.) The kinematic-interaction motion at the base is equal to the effective foundation input motion ($\{\ddot{u}^k_0(t)\} = \{\ddot{u}^g_0(t)\}$). Using rigid-body kinematics, the inertial loading to be applied throughout the structure in the inertial-interaction part of the analysis is determined. This physical interpretation of $\{\ddot{u}^g_0(t)\}$ is important. The structure is supported on the generalized spring; no support motion is applied in the inertial-interaction part. To determine the total displacements, those of the kinematic and inertial interactions have to be added. Figure 6-5 should be compared to the corresponding one illustrating the formulation in total motions using the same effective foundation input motion $\{u^g_0(t)\}$ (Fig. 6-4). Reference should also be made to the discussion of the effects of soil–structure interaction in Section 1.3 (Fig. 1-3b).

The examples presented in Fig. 2-39 and Fig. 3-30 are processed using the inertial-interaction analysis. For vertically propagating waves in the presence of a surface foundation, the effective foundation input motion is equal to that of the free field at the same location.

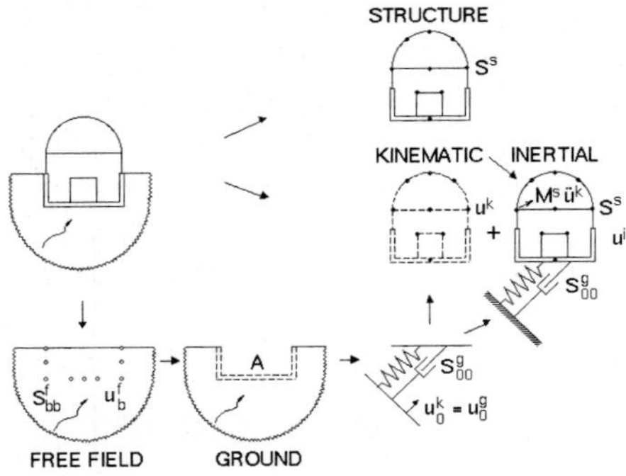

Fig. 6-5 Physical interpretation of kinematic and inertial interactions.

6.2 FREE-FIELD RESPONSE OF SITE

To be able to calculate the effective foundation input motion $\{u_0^g(t)\}$, the free-field motion $\{u_b^f(t)\}$ must be determined in the nodes b (Fig. 6-3) which subsequently will lie on the structure–soil interface (Fig. 6-1). The latter is shown in Fig. 6-6 for an embedded foundation. The site consists of layers of soil resting on flexible (bed) rock.

For seismic excitation the following approach is followed. When determining the free field's seismic environment, three interrelated aspects are considered. The first covers the selection of the *control motion*, which can, for example, consist of a historic earthquake record. The second governs the selection of the *control point*, where the motion is applied. The selected control point should be on the free field's ground surface (point A in Fig. 6-6) or at an assumed rock outcrop (point B) that is on the level of the rock, assuming that there is no soil on top. The third addresses the postulated *wave pattern*—that is, prescribing the types of waves, such as vertically incident body waves or surface waves, of which the motion consists.

In the following, only S-waves with a horizontal particle motion propagating vertically in the site with the shear-wave velocity c_s and P-waves with a vertical particle motion propagating vertically with the dilatational-wave velocity c_p are addressed. (For surface foundations, horizontally propagating surface waves with a prescribed apparent wave velocity can also be prescribed; but in this case, no site-response analysis is necessary. See Section 6.3.2.) This one-dimensional wave propagation can be modeled with a column of

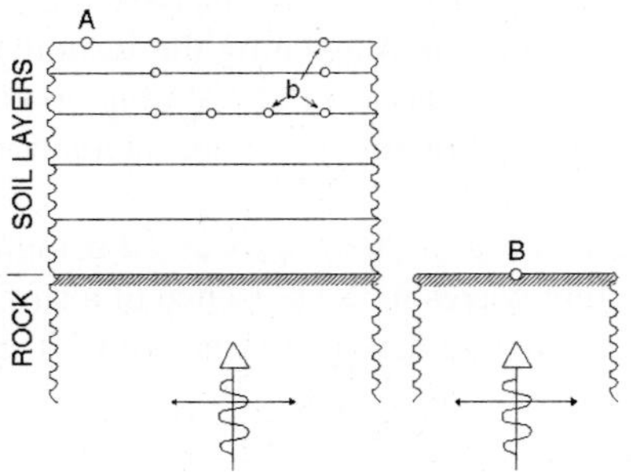

Fig. 6-6 Selection of control point for seismic input.

soil at the site (Fig. 6-7), consisting of n - 1 soil layers resting on flexible rock with n interfaces (nodes). In each layer j of depth d_j, the mass density ρ_j and the elastic modulus $\rho_j c_j^2$ (equal to shear modulus G_j for S-waves with $c_j = c_{sj}$, constrained modulus E_{cj} for P-waves with $c_j = c_{pj}$) are constant. The discrete model to calculate the free-field response directly in the time domain is also shown in Fig. 6-7 (for P-wave propagation). For each soil layer j, the spring coefficient is specified as $K_j = \rho_j c_j^2 / d_j$, and the masses from the j-th layer as $M_j = M_{j+1} = \rho d_j/2$. To consider material damping with the coefficient of friction $\tan \delta = 2\zeta_g$, a frictional element can be introduced parallel to the spring, as shown in Fig. 2-35 with the force (Eq. 2-256)

$$P_j(t) = K_j \left| \overline{u_j^f(t) - u_{j+1}^f(t)} \right| \tan \delta \ \mathrm{sgn}\left(\dot{u}_j^f(t) - \dot{u}_{j+1}^f(t)\right) \tag{6-29}$$

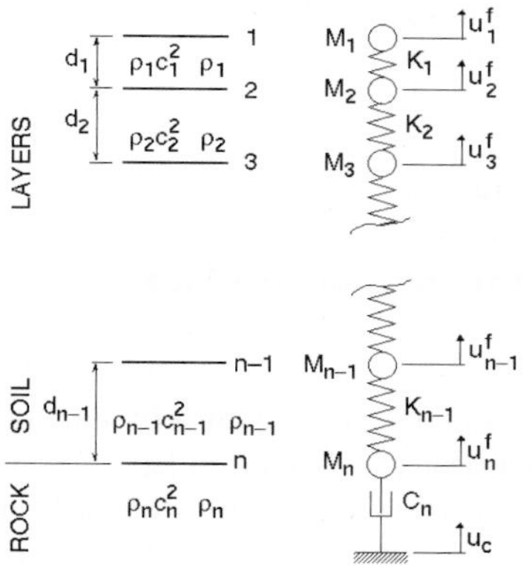

Fig. 6-7 Column of soil to represent free field and corresponding discrete model for vertically propagating P-waves.

The one-dimensional wave propagation in the rock halfspace leads to a dashpot with the coefficient $C_n = \rho_n c_n$ which is the exact representation of a semi-infinite prismatic column of unit area (see also Fig. 2-3). Assembling the corresponding static-stiffness matrix $[K]$, damping matrix $[C]$, and mass matrix $[M]$ leads to the equation of motion of the site

$$[M]\{\ddot{u}^f(t)\} + [C]\{\dot{u}^f(t)\} + [K]\{u^f(t)\} + \{P(t)\} = \{Q(t)\} \tag{6-30}$$

with the frictional forces from material damping $\{P(t)\}$. The only non-zero element in the damping matrix $[C]$ is the diagonal term corresponding to $\dot{u}_n^f$ with the value C_n. The right-hand side $\{Q(t)\}$ depends on the location of the control point.

If the control motion is specified at the free surface (point A in Fig. 6-6), the calculation proceeds from this point downwards with zero right-hand sides. The equation of motion at node j - 1 then leads to the motion at node j. The free-field motion at node j is thus independent of the properties of the site below this node.

For a prescribed control motion $u_c(t)$ (subscript c for control motion) at rock outcrop (point B), $\{Q(t)\}$ can be calculated analogously as the load vector in the basic equation of soil–structure interaction, Eq. 6-25. For the site, the soil layers and the rock represent the two substructures (Fig. 6-6), the first corresponding to the structure and the second to the soil in the soil–structure system. This determines the left-hand side of the equations of motion (Eq. 6-30). For one-dimensional wave propagation in the prismatic soil column

$[K_\infty]$ and $[S^g_{r,00}(t)]$ vanish in Eq. 6-25, which leads to the dashpot $C_n = \rho_n c_n$ as the only remaining term of the rock's contribution. The rock is regarded as the reference system of the soil in the basic equation of motion. The n-th element of the right-hand side will consist of the product of the dashpot C_n and the velocity $\dot{u}_c$ of the control motion, analogous to Eq. 6-25; all other elements are zero. The equation of motion at node n is thus formulated as

$$\frac{\rho_{n-1} d_{n-1}}{2} \ddot{u}^f_n(t) + \rho_n c_n \dot{u}^f_n(t) + \frac{\rho_{n-1} c^2_{n-1}}{d_{n-1}} \left(u^f_n(t) - u^f_{n-1}(t) \right) = \rho_n c_n \dot{u}_c(t) \qquad (6\text{-}31)$$

The free-field motion, which again follows from Eq. 6-30, is a function of the properties of the total system (site).

Figure 6-7 also applies to vertically propagating S-waves when the springs and dashpots act in the horizontal direction.

In addition to calculating the free-field displacements $\{u^f(t)\}$, the corresponding stresses and forces $\{P^f(t)\}$ can also be determined.

6.3 EFFECTIVE FOUNDATION INPUT MOTION

6.3.1 Vertical and Rocking Components of Rigid Surface Foundation Modeled with Subdisks

As discussed in Chapter 5, the vertical and rocking motions of a foundation on a homogeneous halfspace can be modeled with subdisks for which approximate Green's functions can be constructed. The corresponding dynamic-stiffness coefficients follow from Eq. 5-27.

An inclined body wave or a surface wave will lead to a free-field motion which propagates with an apparent velocity c_a in the horizontal x-direction along the surface. Its vertical component for harmonic excitation with frequency ω can be formulated as (Fig. 6-8)

$$u^f_z(x,t) = u^f_z(\omega) e^{i\omega\left(t - \frac{x}{c_a}\right)} \qquad (6\text{-}32)$$

Omitting the factor $e^{i\omega t}$, the amplitude describing the spatial variation is thus equal to

$$u^f_z(x,\omega) = u^f_z(\omega) e^{-i\frac{\omega}{c_a}x} \qquad (6\text{-}33)$$

The effective foundation input motion $\{u^g_0(\omega)\}$ consisting of a vertical displacement with amplitude $w^g_0(\omega)$ and a rocking motion with amplitude $\beta^g_0(\omega)$ is calculated based on Eq. 6-15. The nodes b coincide with the centers of the subdisks. Evaluating Eq. 6-33 in nodes

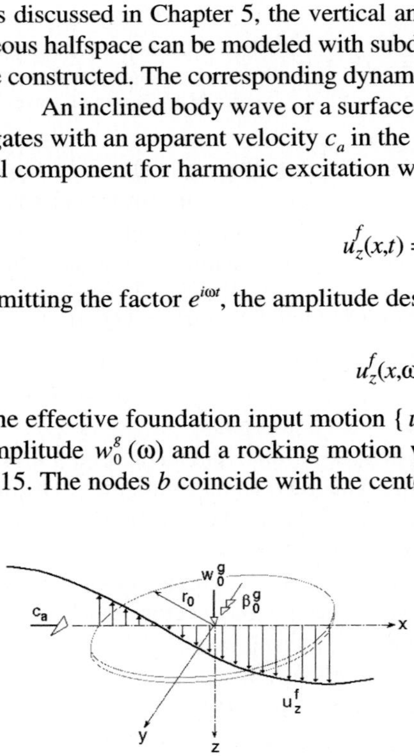

Fig. 6-8 Vertical component of horizontally propagating wave with corresponding effective foundation input motion (vertical and rocking).

b yields the amplitude vector of the free-field motion in the vertical direction $\{u_b^f(\omega)\}$. The matrix $[S_{bb}^f(\omega)]$ is equal to the dynamic-stiffness matrix with respect to the subdisks $[G(a_0)]^{-1}$ appearing in Eq. 5-27, whereby the dynamic-flexibility matrix $[G(a_0)]$ is defined in Eq. 5-23 ($a_0 = \omega r_0/c_s$). $[S_{00}^g(\omega)]$ corresponds to $[S(a_0)]$. This results in

$$\{u_0^g(\omega)\} = \left([A]^T [G(a_0)]^{-1} [A]\right)^{-1} [A]^T [G(a_0)]^{-1} \{u_b^f(\omega)\} \tag{6-34}$$

The effect of the horizontally propagating wave can be characterized by the frequency parameter $\omega r_0/c_a$, which can also be written as a function of the apparent wavelength λ_a in the horizontal direction $2\pi r_0/\lambda_a$.

The effective foundation input motion is calculated for a rigid disk of radius r_0 on the surface of a homogeneous halfspace with $\nu = 1/3$ (Fig. 6-9). The amplitudes of the vertical component $w_0^g(\omega)$ and of the rocking component $\beta_0^g(\omega)$ are complex; for the translational motion the real part dominates and for rocking the imaginary part. Because of the self-canceling (filtering) effect, $|w_0^g(\omega)|$ will be smaller than $|u_z^f(\omega)|$ for all $\omega r_0/c_a$. In general, the larger the apparent velocity c_a is, the closer $|w_0^g(\omega)|/|u_z^f(\omega)|$ is to unity ($c_a = \infty$ corresponds to vertically propagating waves). The rocking component $\beta_0^g(\omega)$, which is negligible for small and larger ratios $\omega r_0/c_a$, reaches a maximum at approximately $\omega r_0/c_a = 2$.

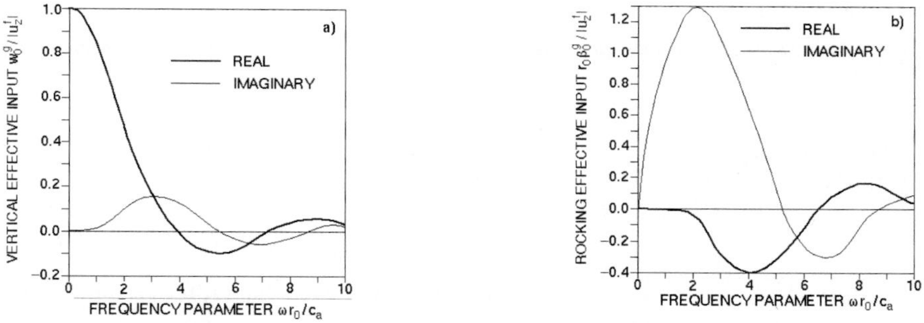

a. Vertical. **b.** Rocking.

Fig. 6-9 Effective foundation input motion of rigid disk for vertical component of horizontally propagating wave.

6.3.2 Rigid Surface Foundation Subjected to Horizontally Propagating Waves

As is apparent from Section 6.3.1 (Fig. 6-4), the free-field motion acts at the far end (the end that is not connected to the basemat) of a generalized spring modeling the soil, actually consisting of a spring–dashpot system. The response of the rigid basemat to this excitation is equal to the effective foundation input motion.

To evaluate approximately the effective foundation input motion of a surface foundation with a rigid basemat subjected to horizontally propagating waves, the dashpots can be omitted, and *continuously distributed springs* with constant values can be selected.

Horizontal component coinciding with direction of propagation. A simple geometric example is used to illustrate the procedure, which is straightforwardly generalized to any shape of the basemat. For a square rigid basemat of length $2a$ resting on distributed springs with a constant k (stiffness per area), the effective foundation input translation with amplitude $u_0^g(\omega)$ caused by the horizontal free-field displacement with amplitude $u_x^f(\omega)$ with a particule motion in the direction of propagation (x-axis) is determined (Fig. 6-10a). With the apparent velocity denoted as c_a, the amplitude of the spatial variation of the horizontal free-field displacement is described as

$$u_x^f(x,\omega) = u_x^f(\omega)\, e^{-i\frac{\omega}{c_a}x} \tag{6-35}$$

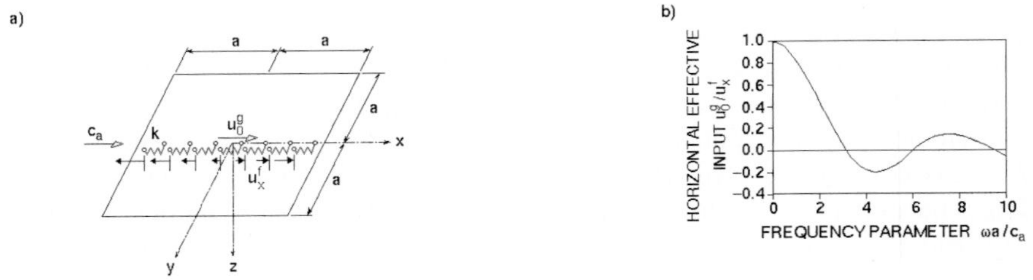

a. Rigid square basemat with distributed springs beneath.

b. Horizontal displacement.

Fig. 6-10 Effective foundation input motion for horizontal component of horizontally traveling wave coinciding with direction of propagation.

The amplitude of the horizontal component of the effective foundation input motion $u_0^g(\omega)$ follows from Eq. 6-15, which, as a scalar relation is equal to

$$u_0^g(\omega) = S_{00}^{g\ -1}\, A^T S_{bb}^f\, u_b^f(\omega) \tag{6-36}$$

The term $A^T S_{bb}^f\, u_b^f(\omega)$ is equal to the amplitude of the resultant force arising from the forces of the springs subjected to the prescribed support motion $u_x^f(x,\omega)$, and the horizontal stiffness coefficient $S_{00}^g = k4a^2$

$$u_0^g(\omega) = \frac{1}{k4a^2}\int_{-a}^{+a} k2a\, u_x^f(\omega) e^{-i\frac{\omega}{c_a}x}\, dx \tag{6-37}$$

This yields

$$\frac{u_0^g(\omega)}{u_x^f(\omega)} = \frac{1}{a}\int_0^a \cos\frac{\omega}{c_a}x\, dx = \frac{c_a}{\omega a}\sin\frac{\omega a}{c_a} \tag{6-38}$$

From the plot of $u_0^g(\omega)/u_x^f(\omega)$ versus $\omega a/c_a$ (Fig. 6-10b) the self-canceling effect of the free-field motion propagating under the structure is clearly visible. The details of how the soil is modeled will have only a small influence when calculating the effective foundation input motion. This is the case because in the equation for $u_0^g(\omega)$ (Eq. 6-36) the dynamic-stiffness coefficient and its inverse appear as factors.

Horizontal component perpendicular to direction of propagation. The effective foundation input motion consisting of the displacement with amplitude $v_0^g(\omega)$ and of torsional rotation (twisting) with amplitude $\gamma_0^g(\omega)$ caused by the horizontal free-field displacement with amplitude $u_y^f(\omega)$ with a particule motion perpendicular to the direction of propagation can be determined based on the same concept (Fig. 6-11a). The amplitude of translation $v_0^g(\omega)$ is calculated as above for $u_0^g(\omega)$ as

$$v_0^g(\omega) = \frac{1}{k4a^2}\int_{-a}^{+a} k2a\,u_y^f(\omega)\,e^{-i\frac{\omega}{c_a}x}\,dx \tag{6-39}$$

yielding

$$\frac{v_0^g(\omega)}{u_y^f(\omega)} = \frac{c_a}{\omega a}\sin\frac{\omega a}{c_a} \tag{6-40}$$

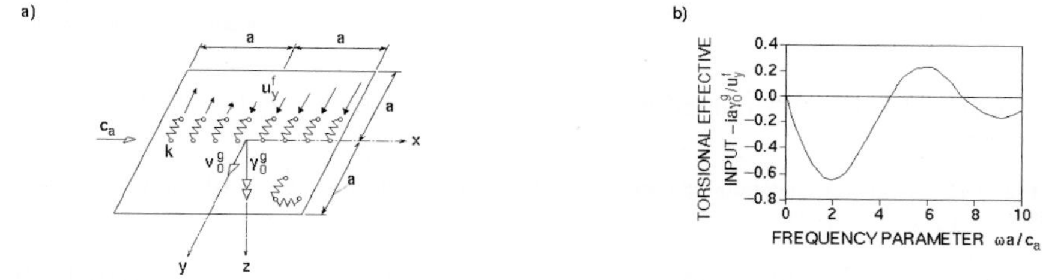

a. Rigid square basemat with distributed springs beneath.

b. Rotation (twisting).

Fig. 6-11 Effective foundation input motion for horizontal component of horizontally traveling wave perpendicular to direction of propagation.

The amplitude of rotation $\gamma_0^g(\omega)$ follows from Eq. 6-15 as the ratio of the amplitude of the resultant torsional moment occurring from the forces of the springs subjected to the prescribed support motion $u_y^f(x,\omega)$ to the torsional stiffness coefficient S_{00}^g (equal to the polar moment of inertia multiplied by k)

$$\gamma_0^g(\omega) = \frac{1}{k\frac{8}{3}a^4}\int_{-a}^{+a} x\,k2a\,u_y^f(\omega)\,e^{-i\frac{\omega}{c_a}x}\,dx \tag{6-41}$$

This yields

$$\frac{a\gamma_0^g(\omega)}{u_y^f(\omega)} = -i\frac{3}{2a^2}\int_0^a x\sin\frac{\omega}{c_a}x\,dx = -i\frac{3}{2}\frac{c_a}{\omega a}\left(\frac{c_a}{\omega a}\sin\frac{\omega a}{c_a} - \cos\frac{\omega a}{c_a}\right) \tag{6-42}$$

The factor ($-i$) means that $\gamma_0^g(\omega)$ lags behind $u_y^f(\omega)$ (and $v_0^g(\omega)$) by 90°. From the plot of $-ia\,\gamma_0^g(\omega)/u_y^f(\omega)$ versus $\omega a/c_a$ (Fig. 6-11b), the significant torsional motion with which the symmetric basemat is loaded is visible. The maximum occurs at $\omega a/c_a \sim 2$.

Vertical component. With the same spring model the vertical component of the free-field displacement with amplitude $u_z^f(\omega)$ of a horizontally propagating wave can be addressed (Fig. 6-12a). The effective foundation input motion consists of the vertical displacement with amplitude $w_0^g(\omega)$ and of the rocking motion with amplitude $\beta_0^g(\omega)$. And $w_0^g(\omega)$ follows as for $u_0^g(\omega)$ (Eq. 6-37) as

$$\frac{w_0^g(\omega)}{u_z^f(\omega)} = \frac{c_a}{\omega a} \sin \frac{\omega a}{c_a} \tag{6-43}$$

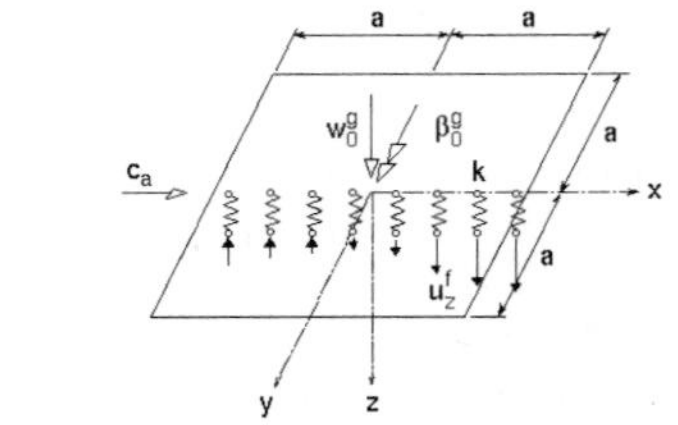

a. Rigid square basemat with distributed springs beneath.

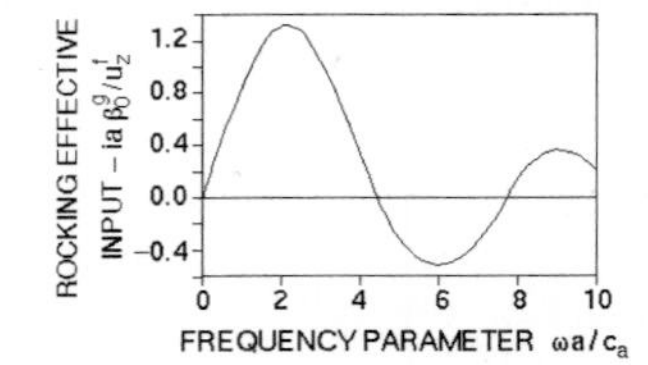

b. Rotation (rocking).

Fig. 6-12 Effective foundation input motion for vertical component of horizontally traveling wave.

The $\beta_0^g(\omega)$ is determined in the same way as $\gamma_0^g(\omega)$ (Eq. 6-41) is, whereby for the rocking component S_{00}^g is equal to the moment of inertia multiplied by k.

$$\beta_0^g(\omega) = -\frac{1}{k\frac{4}{3}a^4}\int_{-a}^{+a} x\, k 2a\, u_z^f(\omega) e^{-i\frac{\omega}{c_a}x}\, dx \tag{6-44}$$

This results in

$$\frac{a\beta_0^g(\omega)}{u_z^f(\omega)} = i3\frac{c_a}{\omega a}\left(\frac{c_a}{\omega a}\sin\frac{\omega a}{c_a} - \cos\frac{\omega a}{c_a}\right) \tag{6-45}$$

Apart from the sign, $\beta_0^g(\omega)$ (Fig. 6-12b) is twice as large as $\gamma_0^g(\omega)$ (Fig. 6-11b). Calculating an equivalent radius by equating the moments of inertia of the square and of the circle, the agreement with the more accurate result for the circular basemat with this radius presented in Fig. 6-9 is shown to be good.

Time-domain analysis. For all three cases presented in Figs. 6-10a, 6-11a, and 6-12a, a transient seismic record with a constant apparent velocity c_a can be processed directly in the time domain. It is sufficient to apply the time history at the far end of the springs and to calculate the corresponding response of the basemat. For instance, for the vertical component of a free-field motion propagating horizontally in the positive x–direction with the fixed apparent velocity c_a

$$u_z^f(x,t) = u^f\left(t - \frac{x}{c_a}\right) \tag{6-46}$$

the vertical effective foundation input displacement equals (see Eq. 6-37)

$$w_0^g(t) = \frac{1}{k4a^2} \int_{-a}^{+a} k2a\, u^f\left(t - \frac{x}{c_a}\right) dx \tag{6-47}$$

Introducing the new variable $\bar{x} = t - x/c_a$ yields

$$w_0^g(t) = \frac{c_a}{2a} \int_{t-a/c_a}^{t+a/c_a} u^f(\bar{x})\, d\bar{x} \tag{6-48}$$

The corresponding rocking effective foundation input rotation follows as

$$\beta_0^g(t) = -\frac{1}{k\frac{4}{3}a^4} \int_{-a}^{+a} xk2a\, u^f\left(t - \frac{x}{c_a}\right) dx \tag{6-49}$$

which leads to

$$\beta_0^g(t) = \frac{3}{2}\frac{c_a^2}{a^3}\left(\int_{t-a/c_a}^{t+a/c_a} \bar{x}\, u^f(\bar{x})\, d\bar{x} - t \int_{t-a/c_a}^{t+a/c_a} u^f(\bar{x})\, d\bar{x} \right) \tag{6-50}$$

6.3.3 Embedded Foundation Modeled with Stack of Disks

The effective foundation input motion of a rigid embedded cylindrical foundation modeled with a stack of disks can be determined for vertically propagating waves. Consistent with the formulation for the dynamic-stiffness matrix (Section 4.2) or the interaction force-displacement relationship (Section 4.3), the input motion can either be calculated for harmonic excitation or in the time domain.

Analysis for harmonic loading. As shown in Fig. 4-12, the soil region that will later be excavated is viewed as a stack of disks (embedded in the fullspace) with soil in between. The disks correspond to the nodes b where the free-field motion $\{u_b^f(\omega)\}$ is calculated (see also Fig. 6-3). To simplify the nomenclature, the subscript b is omitted in the following, as is the case in Section 4.2. The motion relative to this free-field motion causes the forces acting on the disks. In the presence of a free-field motion, the displacement-force relationship of the disks (Eq. 4-21) is formulated as

$$\{u^t(\omega)\} - \{u^f(\omega)\} = [G(\omega)]\,\{P(\omega)\} \tag{6-51}$$

where the superscript t denotes the total motion and $[G(\omega)]$ is the dynamic-flexibility matrix of the virgin continuous soil (soil system free-field, $[S^f(\omega)] = [G(\omega)]^{-1}$). Proceeding as in Section 4.2.2, where the rigid-body constraint expressed in Eq. 4-27 applies to the total motion, yields

$$\{P_0(\omega)\} = [S_{00}^g(\omega)]\,\{u_0^t(\omega)\} - [A]^T\,[S^f(\omega)]\,\{u^f(\omega)\} \tag{6-52}$$

The driving loads are equal to $\{P_0^g(\omega)\} = [A]^T[S^f(\omega)]\{u^f(\omega)\}$.

To calculate the effective foundation input motion, no forces are applied to the rigid foundation (Fig. 6-2). For $\{P_0(\omega)\} = 0$ and with $\{u_0^g(\omega)\} = \{u_0^t(\omega)\}$, Eq. 6-52 leads to

$$\{u_0^g(\omega)\} = [S_{00}^g(\omega)]^{-1}\,[A]^T\,[S^f(\omega)]\,\{u^f(\omega)\} \tag{6-53}$$

which is the same expression as specified in Eq. 6-15. Premultiplying Eq. 6-53 by $[S_{00}^g(\omega)]$ and then substituting in Eq. 6-52, Eq. 6-3 is recovered. The resultant interaction forces depend on the motion $\{u_0^t(\omega)\}$ - $\{u_0^g(\omega)\}$.

For vertically propagating P-waves, the effective foundation input motion consists of a vertical component which is calculated using the submatrices of $[S_{00}^g(\omega)]$, $[A]$, and $[S^f(\omega)]$ corresponding to the vertical motion. In the case of vertically propagating S-waves, the submatrix of $[S^f(\omega)]$ corresponding to the horizontal motion applies. In the $[A]$ and $[S_{00}^g(\omega)]$ matrices, both the horizontal and the rocking motions are considered, leading to a horizontal and a rocking component of the effective foundation input motion.

The extension to a single pile embedded in soil (Section 4.8) is straightforward.

Time-domain analysis. Following the procedure discussed in Section 4.3, the displacement-force relationship at time station n of the stack of disks (Fig. 4-20) in the presence of the free-field motion is formulated as (Eq. 4-42a).

$$\{u^t\}_n - \{u^f\}_n = \sum_{k=1}^{n} [G]_{n-k}\,\{P\}_k \tag{6-54}$$

The final equation relating the displacements of the embedded foundation to the resultant interaction forces (Eq. 4-49) is equal to

$$\{P_0\}_n = [A]^T\,[S^f]_0\,[A]\,\{u_0^t\}_n - [A]^T\,[S^f]_0\,\{\bar{u}\}_n - [M]\,\{\ddot{u}_0^t\}_n - [A]^T\,[S^f]_0\,\{u^f\}_n \tag{6-55}$$

The free-field motion thus leads to an additional load (last term on right-hand side). After determining the driving loads applied at the base, which are equal to $\{P_0^g\}_n = [A]^T[S^f]_0\{u^f\}_n$, the seismic analysis of a structure with an embedded foundation is straightforward. Setting $\{P_0\}_n = 0$ and with $\{u_0^g\}_n = \{u_0^t\}_n$, the effective foundation input motion $\{u_0^g\}_n$ follows from Eq. 6-55.

Example. The effective foundation input motion consisting of the horizontal component with amplitude $u_0^g(\omega)$ defined at the center of the basemat and of the rocking component with amplitude $\vartheta_0^g(\omega)$ for a cylindrical foundation of radius r_0 and embedment e in a halfspace with Poisson's ratio $\nu = 0.25$ is calculated for vertically propagating S-waves. The embedment ratios $e/r_0 = 0.5$, 1, and 2 are processed. This is the same example addressed in Section 4.2.3, where the discretization is discussed. The amplitude of the horizontal free-field motion can be described as

$$u^f(z,\omega) = u^f(\omega)\cos\frac{\omega}{c_s}z \tag{6-56}$$

with the depth z measured downwards from the free surface. The value at the surface equals $u^f(\omega)$. Evaluating Eq. 6-56 at the locations of the disks yields $\{u^f(\omega)\}$. The effective foundation input motion follows from Eq. 6-53. The real and imaginary parts of $u_0^g(\omega)$ and

$\vartheta_0^g(\omega)$ are plotted as a function of a_0 in Fig. 6-13. The results using embedded disks with the corresponding cone models are in excellent agreement with the rigorous solution of Luco and Wong [L4] denoted as exact in the figure.

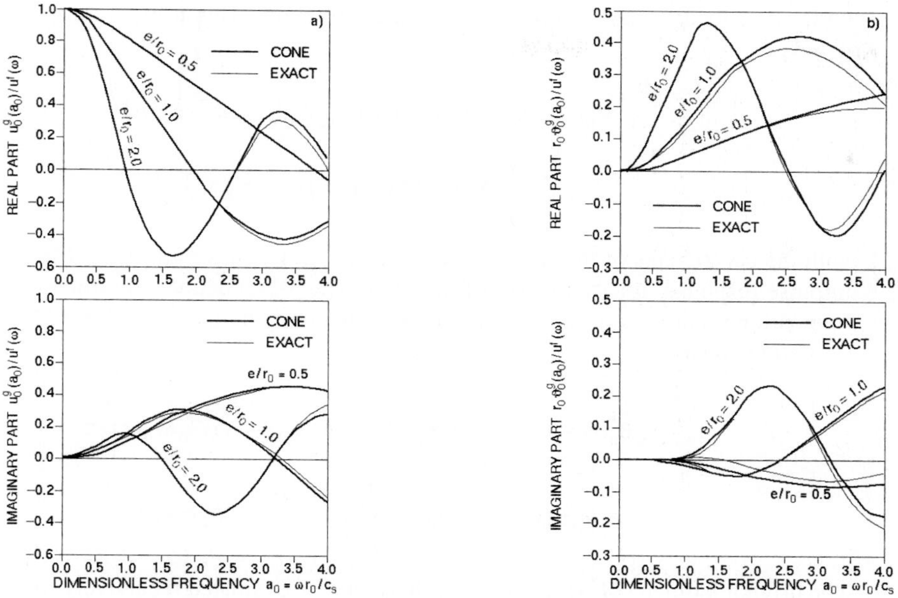

a. Horizontal displacement.

b. Rocking.

Fig. 6-13 Effective foundation input motion for harmonic loading of embedded cylindrical foundation.

6.4 SURFACE FOUNDATION MODELED WITH CONES AND UNFOLDED CONES

6.4.1 Homogeneous Soil Halfspace

After determining the effective foundation input motion, the driving loads applied to the basemat of the structure–soil system are calculated as follows. For the horizontal or vertical motion of an equivalent disk on a soil halfspace modeled with a translational cone, the interaction force-displacement relationship equals (Eq. 2-62)

$$P_0(t) = K\left(\overset{t}{u}_0(t) - \overset{g}{u}_0(t)\right) + C\left(\dot{\overset{t}{u}}_0(t) - \dot{\overset{g}{u}}_0(t)\right) \tag{6-57}$$

with the coefficients of the spring K and dashpot C specified in Eq. 2-63 and $u_0^g(t)$ denoting the translational component of the effective foundation input motion. Assuming, for simplicity, that the structure consists of a rigid block with mass m, the translational equation of motion of the structure–soil system for seismic analysis is written as

$$m\ddot{u}_0^t(t) + C\,\dot{u}_0^t(t) + K\,u_0^t(t) = K\,u_0^g(t) + C\dot{u}_0^g(t) \tag{6-58}$$

The right-hand side is the driving load $P_0^g(t)$.

For the rocking or torsional motion of an equivalent disk on a soil halfspace modeled with a rotational cone, the interaction moment-rotation relationship is formulated as (Eq. 2-145)

$$M_0(t) = K_\vartheta\left(\vartheta_0^t(t) - \vartheta_0^g(t)\right) + C_\vartheta\left(\dot{\vartheta}_0^t(t) - \dot{\vartheta}_0^g(t)\right) - \int_0^t h_1(t-\tau)\,C_\vartheta\left(\dot{\vartheta}_0^t(\tau) - \dot{\vartheta}_0^g(\tau)\right)d\tau \tag{6-59}$$

with the coefficients of the spring K_ϑ and the dashpot C_ϑ given in Eq. 2-146 and the unit-impulse response function $h_1(t)$ in Eq. 2-147. The $\vartheta_0^g(t)$ denotes the rotational effective foundation input motion. For a rigid block with a mass moment of inertia I_0, the equation of motion equals

$$I_0\ddot{\vartheta}_0^t(t) + C_\vartheta\dot{\vartheta}_0^t(t) + K_\vartheta\vartheta_0^t(t) - \int_0^t h_1(t-\tau)\,C_\vartheta\,\dot{\vartheta}_0^t(\tau)\,d\tau = K_\vartheta\vartheta_0^g(t) + C_\vartheta\dot{\vartheta}_0^g(t) - \int_0^t h_1(t-\tau)\,C_\vartheta\,\dot{\vartheta}_0^g(\tau)\,d\tau \tag{6-60}$$

The driving moment $M_0^g(t)$ on the right-hand can be evaluated recursively (Section 2.3.1).

As an example, the rocking motion of a simplified dynamic system consisting of a tall rigid structure on the surface of a soil halfspace is addressed. The same example is examined in Section 2.8.4. The rigid cylindrical structure with radius $r_0 = 20$ m and height $h = 60$ m exhibits a mass $m = 5.28 \cdot 10^7$ kg and a mass moment of inertia about the base diameter $I_0 = 6.86 \cdot 10^{10}$ kgm^2. The soil with mass density $\rho = 2 \cdot 10^3$ kg/m^3 and Poisson's ratio $\nu = 1/3$ has a dilatational wave-velocity $c_p = 400$ m/s.

The vertical component of the free-field motion on the free surface of the soil half-space $u^f(x,t)$ consists of the very simple excitation shown in Fig. 6-14. The symmetric piecewise linear acceleration $\ddot{u}^f(t)$ (Fig. 6-14a) leads after a time integration to the velocity $\dot{u}^f(t)$ consisting of two half-cycle pulses of opposite signs (Fig. 6-14b); and after performing another time integration, the corresponding displacement $u^f(t)$ made up of a single symmetric half-cycle pulse results (Fig. 6-14c). The total duration $T = 2s$ and $\ddot{u}^f_{max} = 0.15$ g are selected. This vertical component is assumed to propagate horizontally with an apparent velocity c_a equal to the Rayleigh-wave velocity = 0.9325 c_s (for $\nu = 1/3$) = 186.5 m/s. This free-field wave motion will lead to a rocking effective foundation input $\vartheta_0^g(t)$ and a vertical component which is, however, not addressed, as only the rocking motion of the dynamic system is of interest.

The aspect ratio of the rocking cone $z_0/r_0 = 2.356$ (Table 2-3B) yields $z_0 = 47.12$ m. The static-stiffness coefficient equals $K_\vartheta = 3\rho\, c_p^2 \pi r_0^4/(4z_0) = 2.56 \cdot 10^{12}$ Nm (Eq. 2-332) and the radiation dashpot $C_\vartheta = \rho c_p \pi r_0^4/4 = 1.01 \cdot 10^{11}$ Nms (Eq. 2-339b). As discussed in Section 2.8.4 the natural frequency of the dynamic system equals $\omega_r = 5.78/s$ (= 0.92 *Hz*) and the associated damping ratio equals 0.039.

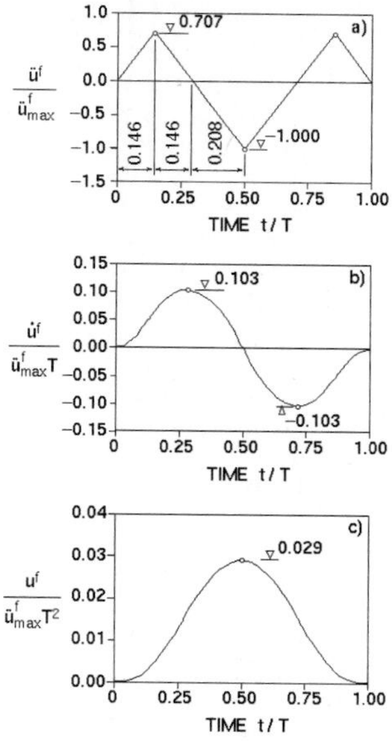

a. Piecewise linear acceleration.

b. Corresponding velocity consisting of two identical half-cycle pulses of opposite signs.

c. Corresponding displacement consisting of symmetric half-cycle pulse.

Fig. 6-14 Impulse excitation of free-field motion.

The frequency parameter on which the rocking effective foundation input depends equals for this natural frequency $\omega_r r_0/c_a = 0.62$. A reasonable ϑ^g_0 occurs (Fig. 6-9b, with $\vartheta^g_0 = -\beta^g_0$).

The rocking effective foundation input motion $\vartheta^g_0(t)$ is determined approximately in the time domain as described in Section 6.3.2. The half-length of the equivalent square foundation follows from equating the moments of inertia yielding $a = 0.5\sqrt[4]{3\pi}\, r_0 = 17.52$ m. From Eq. 6-50, $\vartheta^g_0(t)$ follows, with $\vartheta^g_0(t) = -\beta^g_0(t)$. The sign change becomes apparent by comparing the positive directions of $\vartheta_0(t)$ in Fig. 2-2b and of β^g_0 in Fig. 6-12. The rotation $\vartheta^g_0(t)$, velocity $\dot{\vartheta}^g_0(t)$, and acceleration $\ddot{\vartheta}^g_0(t)$ are plotted in Fig. 6-15. At time $t = 0$ the wave arrives at the center of the square foundation. The wave starts affecting the foundation when the wave encounters the edge at $t = -17.52/186.5 = -0.09$ s.

The equation of motion of the dynamic system specified in Eq. 6-60 is rearranged for use in an explicit algorithm formulated at time $n\Delta t$ as

$$\ddot{\vartheta}^t_{0n} = \frac{M^g_{0n} - K_\vartheta \vartheta^t_{0n} - C_\vartheta \tilde{\dot{\vartheta}}^t_{0n} + \int_0^{n\Delta t} h_1(n\Delta t - \tau)\, C_\vartheta\, \dot{\vartheta}^t_0(\tau)\, d\tau}{I_0} \tag{6-61}$$

with the driving moment

$$M^g_{0n} = K_\vartheta\, \vartheta^g_{0n} + C_\vartheta\, \dot{\vartheta}^g_{0n} - \int_0^{n\Delta t} h_1(n\Delta t - \tau)\, C_\vartheta\, \dot{\vartheta}^g_0(\tau)\, d\tau \tag{6-62}$$

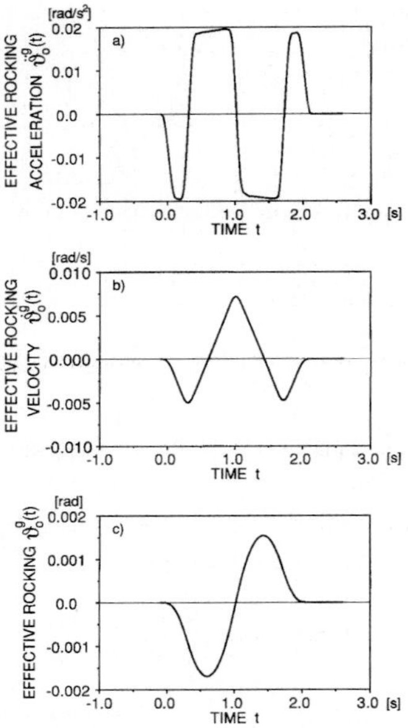

a. Acceleration.

b. Velocity.

c. Rotation.

Fig. 6-15 Rocking effective foundation input motion for vertical impulse excitation propagating horizontally.

The final rotation and the predicted velocity follow from the known values of the previous time $(n\text{-}1)\Delta t$

$$\vartheta^t_{0n} = \vartheta^t_{0n-1} + \Delta t\,\dot{\vartheta}^t_{0n-1} + \frac{\Delta t^2}{2}\,\ddot{\vartheta}^t_{0n-1} \tag{6-63a}$$

$$\tilde{\dot{\vartheta}}^t_{0n} = \dot{\vartheta}^t_{0n-1} + \frac{\Delta t}{2}\,\ddot{\vartheta}^t_{0\,n-1} \tag{6-63b}$$

After calculating $\ddot{\vartheta}^t_{0n}$ from Eq. 6-61, the velocity is corrected

$$\dot{\vartheta}^t_{0n} = \tilde{\dot{\vartheta}}^t_{0n} + \frac{\Delta t}{2}\,\ddot{\vartheta}^t_{0n} \tag{6-64}$$

The (negative value of the) convolution integral appearing in the numerator of Eq. 6-61 can be evaluated recursively (Eq. 2-150)

$$M^r_{0n} = a\,M^r_{0n-1} + b_0\left(-C_\vartheta\,\tilde{\dot{\vartheta}}^t_{0n}\right) + b_1\left(-C_\vartheta\,\dot{\vartheta}^t_{0n-1}\right) \tag{6-65}$$

with the definition

$$M^r_{0n} = -\int_0^{n\Delta t} h_1(n\Delta t - \tau)\,C_\vartheta\,\dot{\vartheta}^t_0(\tau)\,d\tau \tag{6-66}$$

The recursive coefficients a, b_0, and b_1 specified in Eq. 2-151 depend on $c_p\Delta t/z_0$.

The analysis is performed with $\Delta t = 0.02$ s. The relative rocking motion $\vartheta_0^t(t) - \vartheta_0^g(t)$ of the rigid block is plotted in Fig. 6-16a. As an intermediate result, the driving moment $M_0^g(t)$ is plotted in Fig. 6-16b.

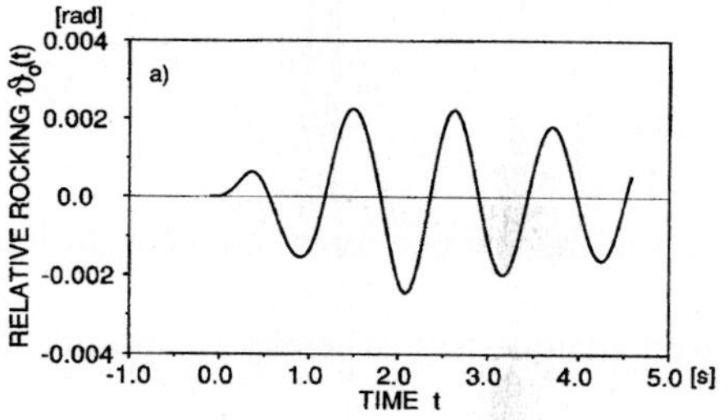

a. Relative rocking motion.

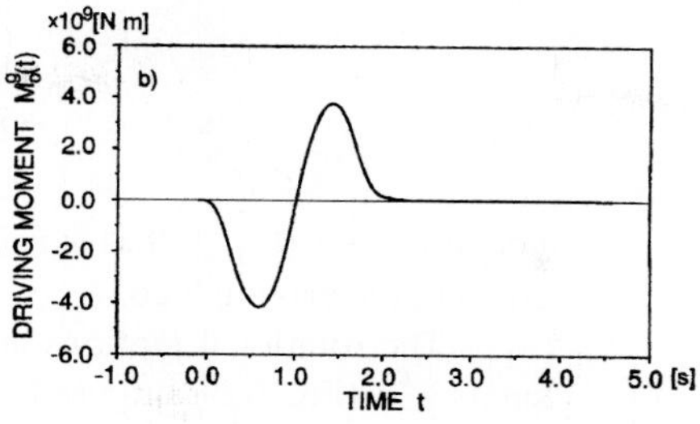

b. Driving moment at base.

Fig. 6-16 Response of rigid structure on soil halfspace.

The same example is calculated at the end of Section 6.5 with the discrete-element model of the soil.

6.4.2 Soil Layer on Rigid Rock

For the translational motions of an equivalent disk on a soil layer resting on rigid rock modeled with the unfolded layered cone (Section 3.1), the displacement of the halfspace modeled with an unlayered cone $\bar{u}_0(t)$ (generating function) follows as (Eqs. 3-24 and 3-26)

$$\bar{u}_0(t) = \sum_{j=0}^{k} e_j^K \left(u_0^t(t-jT) - u_0^g(t-jT) \right) \tag{6-67a}$$

$$\dot{\bar{u}}_0(t) = \sum_{j=0}^{k} e_j^K \left(\dot{u}_0^t(t-jT) - \dot{u}_0^g(t-jT) \right) \tag{6-67b}$$

with the stiffness echo constants e_j^K (Eq. 3-28) and the temporal parameter T (Eq. 3-9b). Substituting Eq. 6-67 in Eq. 2-62 the interaction force-displacement relationship results

$$P_0(t) = K\left(\sum_{j=0}^{k} e_j^K \left(u_0^t(t-jT) - u_0^g(t-jT) \right) \right) + C\left(\sum_{j=0}^{k} e_j^K \left(\dot{u}_0^t(t-jT) - \dot{u}_0^g(t-jT) \right) \right) \tag{6-68}$$

Assuming, for simplicity, that the structure consists of a rigid block with mass m, the equation of motion of the structure–soil system for seismic analysis is formulated as

$$m\ddot{u}_0^t(t) + C\dot{u}_0^t(t) + K u_0^t(t) =$$

$$- K\sum_{j=1}^{k} e_j^K u_0^t(t-jT) - C\sum_{j=1}^{k} e_j^K \dot{u}_0^t(t-jT)$$
$$+ K\sum_{j=0}^{k} e_j^K u_0^g(t-jT) + C\sum_{j=0}^{k} e_j^K \dot{u}_0^g(t-jT) \tag{6-69}$$

where $e_0^K = 1$ (Eq. 3-28a) is used. The sum of the last two terms on the right-hand side is equal to the driving load.

The rotational motions are processed analogously, whereby the pseudo-echo constants e_{rm}^K (Eq. 3-29) are used.

For the sake of illustration, the vertical seismic motion of a structure founded on the surface of a soil layer resting on rigid rock (Fig. 6-17a) is investigated. The structure with mass $m = 3.84 \cdot 10^7$ kg and mass $m_0 = m/4$ lumped at the foundation node has a fixed-base frequency of 6 Hz, resulting in the spring coefficient $k = (2\pi \cdot 6)^2$ m $= 5.46 \cdot 10^{10}$ N/m. The disk of radius $r_0 = 20$ m rests on the surface of a layer with mass density $\rho = 2.10^3$ kg/m^3, Poisson's ratio $\nu = 1/3$ and shear modulus $G = 8 \cdot 10^7$ N/m^2 ($c_p = 400$ m/s). Two radius-to-depth ratios r_0/d are investigated. For the shallow layer with $r_0/d = 1$, the cutoff frequency which, from a practical point of view, is equal to the vertical fundamental frequency of the layer, $c_p/(4d)$, will be larger than the fundamental frequency of the structure–soil system, resulting in no radiation damping occurring in this mode (see Section 3.8.1). For the deep layer with $r_0/d = 0.25$, the opposite applies. As the vertical free-field motion at the layer's free surface $u^g(t)$, the Rincon Hill, San Francisco, CA record of the Loma Prieta earthquake of 17 October 1989 is used. The acceleration, velocity, and displacement time histories are plotted in Fig. 6-17b.

The total dynamic system has two degrees of freedom: the vertical displacements of the structure's mass u^t and of the foundation node u_0^t. By studying the damped free vibrational behavior or by applying the procedure to determine the properties of an equivalent system discussed later in Section 7.3.1, the fundamental frequency $\tilde{\nu}$ and the corresponding damping ratio $\tilde{\zeta}$ of the total dynamic system modeling the structure and the soil can be estimated. For $r_0/d = 1$, $\tilde{\omega}/2\pi$ equals 2.6 Hz which is indeed smaller than the fundamental frequency of the layer $c_p/(4r_0) = 5$ Hz. The corresponding damping ratio $\tilde{\zeta}$ equals 0.03, which is very small for a vertical motion. For $r_0/d = 0.25$, $\tilde{\omega}/2\pi = 2.2$ Hz which is larger than the fundamental frequency of the layer 1.25 Hz. The soil layer's damping ratio at $\tilde{\omega}$ equals 0.60, resulting in an equivalent ratio of the total system $\tilde{\zeta} = 0.43$.

The explicit time-integration algorithm with the predictor-corrector scheme is applied. In the equation of motion of the foundation node with mass m_0, the contribution of the known spring force of the structure at the current time station $k(u_0^t - u^t)$must be added to the left-hand side of Eq. 6-69 before determining the acceleration $\ddot{u}_0^t$. The time step Δt equals 0.01 s. The structural distortion $u^t(t) - u_0^t(t)$ is plotted for the shallow layer ($r_0/d = 1$) and for the deep layer ($r_0/d = 0.25$) in Figs. 6-18a and 6-18b. The significant influence of radiation damping resulting in smaller peaks and a larger decay is clearly visible in Fig. 6-18b.

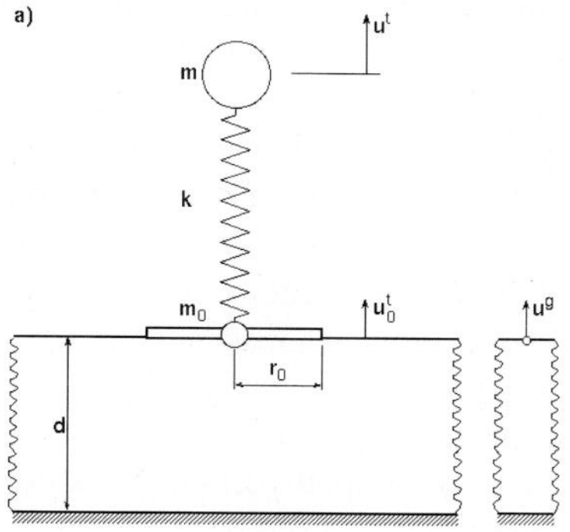

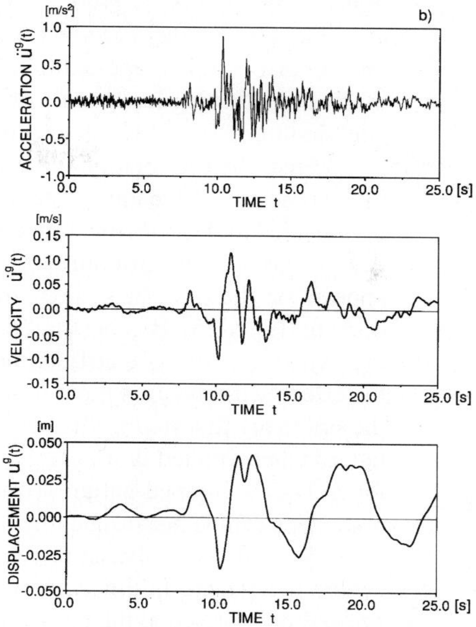

a. Dynamic system with two degrees of freedom.

b. Vertical seismic input motion at free surface.

Fig. 6-17 Vertical motion of structure on soil layer.

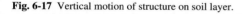

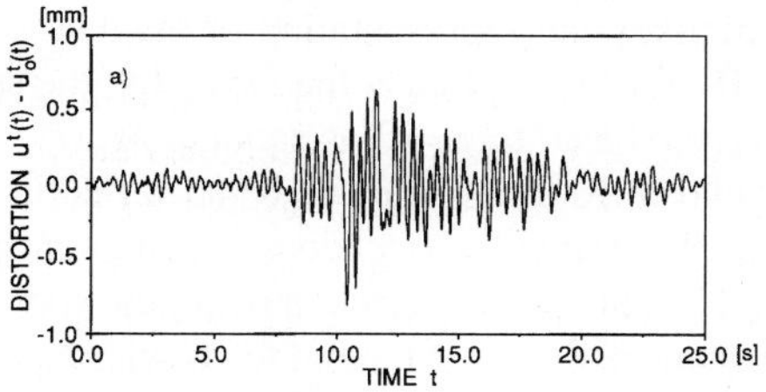

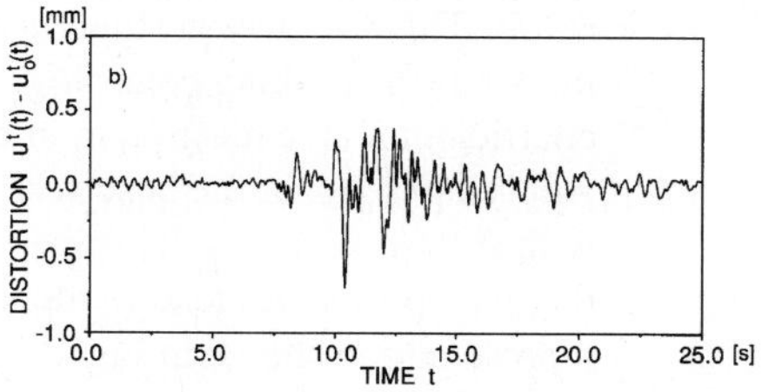

a. Fundamental frequency of dynamic system below fundamental frequency of soil layer (cutoff frequency).

b. Fundamental frequency of dynamic system above fundamental frequency of soil layer (cutoff frequency).

Fig. 6-18 Structural distortion.

6.5 FOUNDATION MODELED WITH LUMPED-PARAMETER MODEL

For practical applications the lumped-parameter model to represent quite complicated foundations is a powerful tool. It is inspired by the discrete-element model for a disk on a soil halfspace which corresponds to a cone. Seismic excitation can be processed directly in the time domain as easily as any load applied to the structure. However, a pitfall exists: The *effective seismic input motion cannot be applied to the far end of a lumped-parameter model of the soil that contains masses*! It is also unnecessary to calculate a special input motion that could be applied to this far end. When the lumped-parameter model contains masses, the driving loads must first be determined analyzing the soil model only and then applying these forces to the total dynamic system. This is explained in the following.

First, the formulation in total displacements is addressed. The driving loads applied to the base of the dynamic system are equal to the reaction forces of the soil when the base is subjected to the effective foundation input motion. This is apparent from the basic equation of motion for harmonic excitation (Eq. 6-4) where the product $[S_{00}^g(\omega)]\{u_0^g(\omega)\}$ appears on the right-hand side, and also from the corresponding relationship in the time domain (Eq. 6-25). By construction (Appendix B), the lumped-parameter model represents the dynamic-stiffness matrix of the soil defined at point 0 of the base (Fig. 6-2). Applying the effective foundation input motion $\{u_0^g(t)\}$ at point 0 of the lumped-parameter model of the soil in the first step will result in a reaction $\{P_0^g(t)\}$ that is equal to the driving load. The latter is then applied in the second step at point 0 to the total dynamic system with a fixed far end of the lumped-parameter model leading to the total displacements. The analyses in both steps can be performed directly in the time domain.

Second, as an alternative, the decomposition into kinematic and inertial interactions can be performed. In this case, the accelerations of the effective foundation input motion $\{\ddot{u}_0^g(t)\}$ are applied at the base at point 0 to the structure, and using rigid-body kinematics the inertial loads throughout the structure of the inertial-interaction analysis follow (Eq. 6-28 and Fig. 6-5).

Returning to the formulation in total displacements, it is important to note that the effective foundation input motion cannot be applied to the far end of the lumped-parameter model when the latter is constructed with masses. If this is done, incorrect driving loads result. This is easily verified using the monkey-tail representation of the discrete-element model of the rocking cone (Fig. 2-17b). If $\vartheta_0^g(t)$ is applied at the far end of the spring with coefficient K_ϑ the reaction moment for a fixed base ($\vartheta_0(t) = 0$) is equal to $M_0^g(t) = -K_\vartheta \vartheta_0^g(t)$. Applying $\vartheta_0^g(t)$ at the base with the far end fixed, the correct $M_0(t)$ specified in Eq. 2-161 with $\vartheta_0(t) = \vartheta_0^g(t)$ results. When the lumped-parameter model does not contain any masses, the two approaches lead to identical results. This can be checked using the spring-dashpot representation shown in Fig. 2-17a. Applying $\vartheta_0^g(t)$ at the far end leads with $\vartheta_0(t) = 0$ to the equations of motion (analogous to Eq. 2-153).

$$M_0^g = -K_\vartheta \vartheta_0^g - C_\vartheta \dot{\vartheta}_0^g + \frac{K_\vartheta}{3} \vartheta_1^g \tag{6-70a}$$

$$-\frac{K_\vartheta}{3} \vartheta_1^g - C_\vartheta \left(\dot{\vartheta}_1^g - \dot{\vartheta}_0^g \right) = 0 \tag{6-70b}$$

As $3C_\vartheta / K_\vartheta = z_0/c$ (Eq. 2-146), Eq. 6-70b yields

$$\frac{z_0}{c} \dot{\vartheta}_1^g + \vartheta_1^g = \frac{z_0}{c} \dot{\vartheta}_0^g \tag{6-71}$$

which is solved

$$\vartheta_1^g(t) = \frac{z_0}{c} \int_0^t h_1(t-\tau)\, \dot{\vartheta}_0^g(\tau) d\tau \tag{6-72}$$

Substituting Eq. 6-72 in Eq. 6-70a yields

$$M_0^g = -K_\vartheta\, \vartheta_0^g - C_\vartheta\, \dot{\vartheta}_0^g + \int_0^t h_1(t-\tau)\, C_\vartheta\, \dot{\vartheta}_0^g(\tau)\, d\tau \tag{6-73}$$

which is the same expression (with a different sign) as when $\vartheta_0^g(t)$ is applied at the other end of the model (Eq. 2-157 with $\vartheta_0(t) = \vartheta_0^g(t)$). As the discrete-element models of Fig. 2-17 are the building blocks, assembled in parallel, of any lumped-parameter model (Appendix B), the conclusion is also valid for the latter. The schematic representation of Fig. 6-4 where the effective foundation input motion is applied at the far end of the generalized spring is thus valid only when the soil is modeled with a lumped-parameter model consisting of springs and dashpots. When the lumped-parameter model is constructed with masses, the right-hand side illustrated in Fig. 6-4 is replaced by that of Fig. 6-19 showing the two steps discussed above. Of course, this procedure can also be used in the case of a lumped-parameter model without masses.

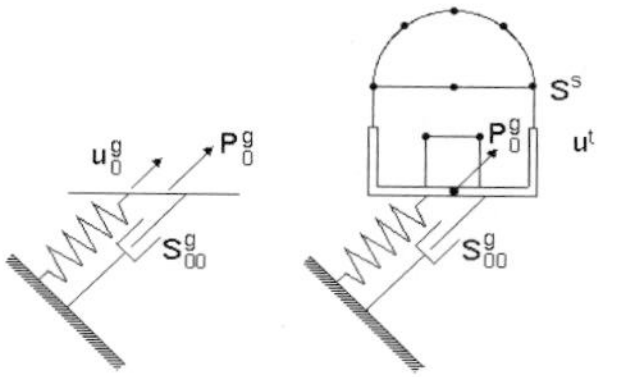

Fig. 6-19 Determination of driving loads and subsequent analysis of structure on generalized spring for soil–structure interaction analysis in total displacements.

For the rocking motion of a rigid structure on a soil halfspace represented by the spring-dashpot discrete-element model, the various formulations for seismic analysis are illustrated in Fig. 6-20. In the first approach (Fig. 6-20a), as no mass is present in this discrete-element model, the effective foundation rocking input $\vartheta_0^g(t)$ can be applied to its far end, resulting in the total rocking $\vartheta_0^t(t)$ of the structure–soil system in one analysis. In the first step of the second approach (Fig. 6-20b), the effective foundation rocking input $\vartheta_0^g(t)$ is applied to the base (where the structure will later be connected) of the discrete-element model, resulting in the reaction moment $M_0^g(t)$. In the second step, this driving moment $M_0^g(t)$ is applied to the base of the structure–soil system with a fixed far end of the discrete-element model yielding the total rocking $\vartheta_0^t(t)$. In the third approach (Fig. 6-20c), the total motion is decomposed into those of the kinematic and inertial interactions. The kinematic rocking is equal to the effective rocking input $\vartheta_0^g(t)$ and determines the inertial loads of the inertial-interaction analysis which is performed with a fixed far end of the discrete-element model (rigid support).

For the rocking motion of the rigid structure on a soil halfspace addressed at the end of Section 6.4.1, the driving moment $M_0^g(t)$ determined using the model shown in Fig. 6-20b is plotted in Fig. 6-16b.

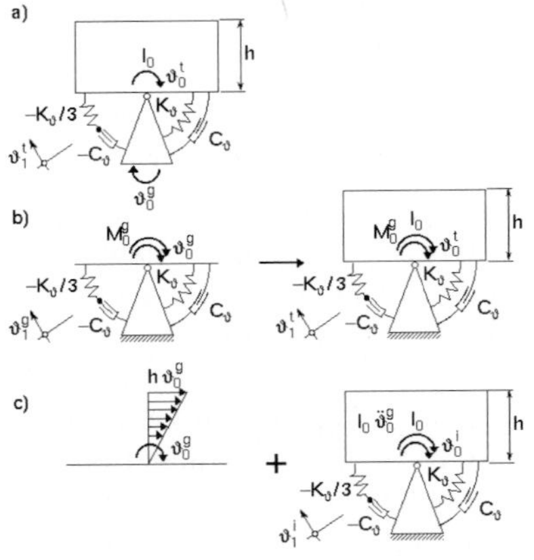

a. Total dynamic system with effective foundation input motion applied at far end of discrete-element model without mass resulting in total displacements.

b. Discrete-element model of soil with applied effective foundation input motion yielding reaction moment, which as driving moment acts on total dynamic system leading to total displacements.

c. Effective foundation input motion determining displacements of kinematic interaction and inertial loads in inertial interaction.

Fig. 6-20 Rigid structure on soil halfspace.

6.6 BASIC EQUATION OF MOTION OF DIRECT METHOD

As the direct method of dynamic soil–structure interaction analysis is not systematically developed in this book, only certain aspects of its equation of motion related to cones as a local transmitting boundary are briefly mentioned.

Assume that in a three-dimensional problem the artificial boundary surface introduced in Section 1.4 is doubly curved. In a specific boundary node b, a local transmitting boundary consisting of a cone with its axis perpendicular to the tangent plane can be constructed. An area ΔA with an equivalent radius Δr_0 of the artificial boundary surface can be assigned to each node b. The normals on the boundary of this area permit the apex of a cone to be estimated defining z_0. (If the artificial boundary is the surface of a sphere, then z_0 will be equal to the radius.) The apex thus follows from geometrical considerations and not from equating the static-stiffness coefficients. Introducing in each node b translational cones in the two tangential directions and in the normal direction with the same apex height as a local transmitting boundary leads for each of the three directions to the following nodal force-displacement relationship (analogous to Eq. 6-57 with Eq. 2-63).

$$P_b(t) = \frac{\rho c^2}{z_0}\pi\Delta r_0^2\left(u_b^t(t) - u_b^g(t)\right) + \rho c\pi\Delta r_0^2\left(\dot{u}_b^t(t) - \dot{u}_b^g(t)\right) \tag{6-74}$$

with $c = c_s$ for the tangential directions and $c = c_p$ for the normal direction. In principle, Eq. 6-20 without enforcing the rigid-body constraint also applies to the direct method. After assemblage of the property matrices which places the terms with the coefficients of $u_b^t(t)$ and $\dot{u}_b^t(t)$ on the left-hand side, the right-hand side, the driving loads, equals

$$P_b^g(t) = \frac{\rho c^2}{z_0}\pi\Delta r_0^2 u_b^g(t) + \rho c\pi\Delta r_0^2 \dot{u}_b^g(t) \tag{6-75}$$

The right-hand side corresponds to the term $S^g_{bb}(\omega)\ u^g_b(\omega)$ (for harmonic motion) which is equal to $S^f_{bb}(\omega)\ u^f_b(\omega)$ (Eq. 6-9). The latter can be expressed as (Eq. 6-18)

$$P^g_b(t) = \frac{\rho c^2}{z_0}\pi\Delta r_0^2\, u^f_b(t) + \rho c\pi\Delta r_0^2\, \dot{u}^f_b(t) + P^f_b(t) \tag{6-76}$$

Summarizing, the left-hand side of the basic equation of motion of the direct method consists of the assemblage of the property matrices of the structure with the adjacent bounded soil and of the unbounded soil (system ground), the latter being modeled with cones. The right-hand side for a formulation in total displacements involves the same properties of the cones (system ground) and the free-field response (displacements and stresses integrated to forces in the nodes b) on the artificial boundary only.

SUMMARY

1. The basic equation of motion of the substructure method for harmonic loading to analyze dynamic soil–structure interaction for seismic excitation is derived by merging the two substructures, the actual structure and the soil with excavation (system ground). The amplitudes of the soil's interaction forces are equal to

$$\{P_0(\omega)\} = [S^g_{00}(\omega)]\left(\{u^t_0(\omega)\} - \{u^g_0(\omega)\}\right)$$

This results in

$$\begin{bmatrix} [S_{ss}(\omega)] & [S_{s0}(\omega)] \\ [S_{0s}(\omega)] & [S^s_{00}(\omega)] + [S^g_{00}(\omega)] \end{bmatrix} \begin{Bmatrix} \{u^t_s(\omega)\} \\ \{u^t_0(\omega)\} \end{Bmatrix} = \begin{Bmatrix} \{0\} \\ [S^g_{00}(\omega)]\ \{u^g_0(\omega)\} \end{Bmatrix}$$

The amplitudes of the total displacements are associated with nodes within the structure $\{u^t_s(\omega)\}$ and on the (rigid) structure–soil interface $\{u^t_0(\omega)\}$. The corresponding coefficient matrix is formed by assembling the dynamic-stiffness matrices of the discretized structure $[S^s(\omega)]$ and of the unbounded soil with excavation with degrees of freedom along the structure–soil interface (base) $[S^g_{00}(\omega)]$. For a rigid base, the compatibility constraints are incorporated into the dynamic-stiffness matrices. The effective load amplitude vector (driving loads) $\{P^g_0(\omega)\}$ for this formulation in total displacements can be determined as the product of $[S^g_{00}(\omega)]$ and the vector of the effective foundation input motion $\{u^g_0(\omega)\}$. The nodes of the structure not in contact with the soil are unloaded. Alternatively, the load amplitude vector can be expressed as a function of the free-field response (amplitudes of displacements $\{u^f_b(\omega)\}$ and forces $\{P^f_b(\omega)\}$) only in those nodes b that subsequently will lie on the structure–soil interface. The earthquake free-field response in other points of the free field is thus not required.

The basic equation of motion can also be formulated in the time domain

$$\begin{bmatrix} [M_{ss}] & [M_{s0}] \\ [M_{0s}] & [M_{00}] \end{bmatrix} \begin{Bmatrix} \{\ddot{u}_s^t(t)\} \\ \{\ddot{u}_0^t(t)\} \end{Bmatrix} + \begin{Bmatrix} \{F_s(t)\} \\ \{F_0(t)\} \end{Bmatrix} +$$

$$\begin{Bmatrix} \{0\} \\ [K_\infty]\{u_0^t(t)\} + [C]\{\dot{u}_0^t(t)\} + \int_0^t [S_{r,00}^g(t-\tau)]\{u_0^t(\tau)\}d\tau \end{Bmatrix} =$$

$$\begin{Bmatrix} \{0\} \\ [K_\infty]\{u_0^g(t)\} + [C]\{\dot{u}_0^g(t)\} + \int_0^t [S_{r,00}^g(t-\tau)]\{u_0^g(\tau)\}d\tau \end{Bmatrix}$$

with the mass matrix $[M]$ and the (nonlinear) internal forces $\{F(t)\}$ of the structure, the singular part defined by $[K_\infty]$ and $[C]$, and the regular part $[S_{r,00}^g(t)]$ of the dynamic-stiffness matrix of the soil.

2. The total motion $\{u^t(t)\}$ can be split up into that arising from kinematic interaction $\{u^k(t)\}$ and that from inertial interaction $\{u^i(t)\}$. For a rigid base, $\{u_0^k(t)\} = \{u_0^g(t)\}$ and the kinematic motion throughout the structure follows as $\{u_s^k(t)\} = [T_{s0}]\{u_0^g(t)\}$ with $[T_{s0}]$ expressing rigid-body kinematics. The inertial-interaction analysis is governed by

$$\begin{bmatrix} [M_{ss}] & [M_{s0}] \\ [M_{0s}] & [M_{00}] \end{bmatrix} \begin{Bmatrix} \{\ddot{u}_s^i(t)\} \\ \{\ddot{u}_0^i(t)\} \end{Bmatrix} + \begin{Bmatrix} \{F_s(t)\} \\ \{F_0(t)\} \end{Bmatrix} +$$

$$\begin{Bmatrix} \{0\} \\ [K_\infty]\{u_0^i(t)\} + [C]\{\dot{u}_0^i(t)\} + \int_0^t [S_{r,00}^g(t-\tau)]\{u_0^i(\tau)\}d\tau \end{Bmatrix} =$$

$$-\begin{bmatrix} [M_{ss}] & [M_{s0}] \\ [M_{0s}] & [M_{00}] \end{bmatrix} \begin{bmatrix} [T_{s0}] \\ [I] \end{bmatrix} \{\ddot{u}_0^g(t)\}$$

The load vector on the right-hand side is equal to the negative product of the mass of the structure and the kinematic-interaction accelerations $\{\ddot{u}_0^k(t)\}$, $\{\ddot{u}_s^k(t)\}$.

3. Starting from the prescribed control motion acting in the selected control point and assuming the nature of the wave pattern, the free-field response is determined on the surface that subsequently will form the structure–soil interface. For vertically propagating S- and P-waves, a simple physical model exists—also when the control point is at a fictitious rock outcrop.
4. The effective foundation input motion for harmonic excitation is calculated as

$$\{u_0^g(\omega)\} = [S_{00}^g(\omega)]^{-1}\,[A]^T\,[S_{bb}^f(\omega)]\,\{u_b^f(\omega)\}$$

$[S_{bb}^f(\omega)]$ and $\{u_b^f(\omega)\}$ are the dynamic-stiffness matrix and motion of the free field in those nodes (disks) b which will later lie on the structure–soil interface; $[A]$ is the kinematic-constraint matrix.

In the time domain at time station n, the effective driving loads are equal to

$$\{P_0^g\}_n = [A]^T\,[S^f]_0\{u^f\}_n$$

For a vertically propagating S-wave, a rigid embedded foundation modeled with a stack of disks will also exhibit a rocking component.

$\{u_0^g(t)\}$ can be determined approximately for a surface foundation subjected to horizontally propagating waves based on continuously distributed springs.

5. To perform a seismic analysis for a surface foundation on soil modeled with cones, the required interaction force-displacement relationship involves the motion with respect to the effective foundation input motion.

translational cone for soil halfspace

$$P_0(t) = K\left(u_0^t(t) - u_0^g(t)\right) + C\left(\dot{u}_0^t(t) - \dot{u}_0^g(t)\right)$$

rotational cone for soil halfspace

$$M_0(t) = K_\vartheta\left(\vartheta_0^t(t) - \vartheta_0^g(t)\right) + C_\vartheta\left(\dot{\vartheta}_0^t(t) - \dot{\vartheta}_0^g(t)\right) - \int_0^t h_1(t-\tau)\,C_\vartheta\left(\dot{\vartheta}_0^t(\tau) - \dot{\vartheta}_0^g(\tau)\right)d\tau$$

translational unfolded layered cone for soil layer

$$P_0(t) = K\left(\sum_{j=0}^{k} e_j^K\left(u_0^t(t-jT) - u_0^g(t-jT)\right)\right) + C\left(\sum_{j=0}^{k} e_j^K\left(\dot{u}_0^t(t-jT) - \dot{u}_0^g(t-jT)\right)\right)$$

and analogously for the rocking unfolded layered cone.

6. When a lumped-parameter model for the soil is used, the first step is to apply the effective seismic input motion $\{u_0^g(t)\}$ at the base (where the structure will later be connected) of the soil model, leading to the reaction forces $\{P_0^g(t)\}$. The latter are then applied to the total dynamic model in the second step, yielding the dynamic response. When the lumped-parameter model does not contain masses, $\{u_0^g(t)\}$ can also be applied at the far end of the lumped-parameter model directly, and the total dynamic system analyzed in one step. As an alternative to these formulations in total displacements, the inertial loads as a function of the accelerations $[T_{s0}]\{\ddot{u}_0^g(t)\}$ and $\{\ddot{u}_0^g(t)\}$ acting in all nodes of the structure can be calculated, and an inertial-interaction analysis be performed.

7

Dynamic Soil–Structure Interaction

In Section 1.3 the main aspects of taking soil–structure interaction into consideration are discussed qualitatively. It is the goal of this chapter to investigate quantitatively how the response of a structure on flexible soil differs from that of the same structure on rigid rock. The simplest possible structure that can be modeled as a one-degree-of-freedom system is used. The examination is restricted to the inertial-interaction analysis, to the influence of the increased flexibility of the total dynamic system introduced through the presence of the soil (including the effect of the rocking motion) and of the larger effective damping, arising from radiation of energy, if applicable. The effect of the modification of the seismic input (free-field and kinematic-interaction responses) on the structural response is not addressed. The soil is modeled with cones, as described in Chapter 2 for the homogeneous halfspace and in Chapter 3 for the layer resting on rigid rock.

The equations of motion of the coupled structure–soil system for horizontal and rocking motions are formulated in Section 7.1, which allows the key aspects of the dynamic interaction of a structure with the soil to be demonstrated. The parameters of an equivalent one-degree-of-freedom system for the structure–soil model are derived in Section 7.2. A parametric study is then performed addressing the effective natural frequency and the effective damping ratio of this equivalent system. Emphasis is placed on the influence of the soil profile (halfspace versus layer resting on rigid rock). The vertical motion is also discussed. Finally, Section 7.3 investigates the parameters of an equivalent one-degree-of-freedom system for a redundant structure–soil model without rotation and for a rigid block on soil.

The effect of the interaction of the soil on the structure's response is best investigated for harmonic loading. The equations of motion are thus established in the frequency domain. After determining the effective frequency-independent natural frequency and damping ratio of the equivalent one-degree-of-freedom system, the analysis can, of course, be performed directly in the time domain. This procedure permits the effect of the various key parameters on the interaction response of the structure to be estimated. For an actual practical application, however, it is appropriate to determine the lumped-parameter model of the soil, which is then connected to the underside of the structure's dynamic model, which can be quite detailed. This procedure will be more accurate. The examples in the previous chapters demonstrate this application.

A more complete evaluation of the effects of dynamic soil–structure interaction, such as addressing the effect on higher structural modes and for horizontally propagating waves, is given in Wolf [W4].

7.1. COUPLED STRUCTURE–SOIL SYSTEM FOR HORIZONTAL AND ROCKING MOTIONS

7.1.1 Simplest Dynamic Model

The dynamic system shown in Fig. 7-1 is the simplest representation of a structure modeled as a one-degree-of-freedom system interacting with the flexible soil for the coupled horizontal and rocking motions excited by a horizontal earthquake. The analysis of this inertial interaction is performed for harmonic loading. The amplitude of the seismic free-field motion of the free surface from vertically propagating *S*-waves is denoted as $u^g(\omega)$ (*g* for ground).

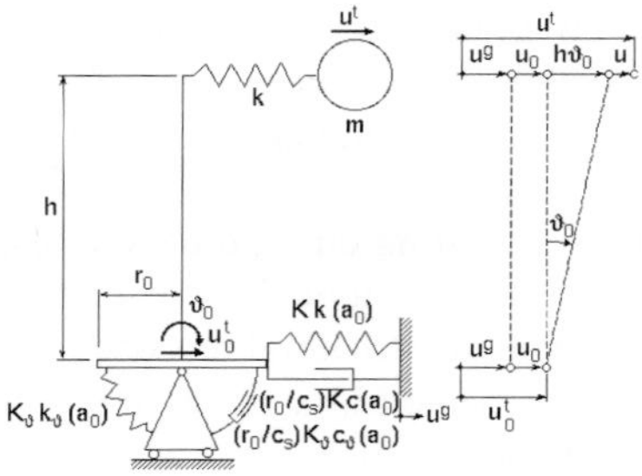

Fig. 7-1 Coupled dynamic model of structure and soil for horizontal and rocking motions.

The structure is modeled with a mass *m* and a static spring with coefficient *k* (representing the lateral stiffness) which is connected to a rigid bar of height *h*. The hysteretic damping with ratio ζ represents structural damping. This is the straightforward structural model of a single-story building frame or of a bridge whose girder can be regarded as rigid longitudinally compared to the (hinged) columns (Fig. 7-2). The values *m*, *k*, and *h* are easily determined. In addition, the idealized structure can also be interpreted as the model for a multistory structure which responds essentially in the fixed-base condition as if it were a single-degree-of-freedom system. The effective values *m* and *k* are associated with the fundamental mode of vibration of the structure fixed at its base, and *h* is the distance from the base to the centroid of the inertial loads. The fixed-base natural frequency ω_s of the structure equals

$$\omega_s = \sqrt{\frac{k}{m}} \tag{7-1}$$

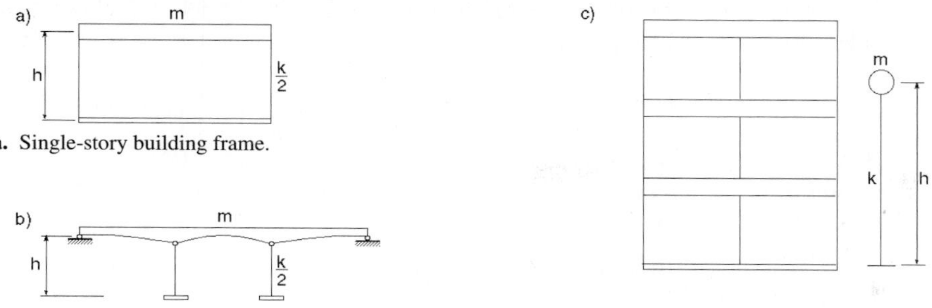

a. Single-story building frame.

b. Bridge with hinged columns.

c. Multistory structure responding in fixed-base condition.

Fig. 7-2 Structures that can be represented as a one-degree-of-freedom model.

In the force-displacement relationship of the structure, the spring coefficient k is multiplied by $1 + 2i\zeta$ (correspondence principle, see Section 2.4.1).

At the other end of the bar, at the basemat with equivalent radius r_0, the soil's dynamic stiffness is attached. For a surface foundation, the horizontal and rocking degrees of freedom are modeled independently. In each direction a spring with a frequency-dependent coefficient ($Kk(a_0)$ for the horizontal direction) and in parallel a dashpot with a frequency-dependent coefficient $[(r_0/c_s)\ Kc(a_0)]$ occur for harmonic loading, where K is the static-stiffness coefficient and c_s the shear-wave velocity. Using cones to model the halfspace, the spring and damping coefficients $k(a_0)$, $c(a_0)$ in the horizontal direction are specified in Eq. 2-73 and $k_\vartheta(a_0)$, $c_\vartheta(a_0)$ in rocking motion in Eq. 2-171 with $a_0 = \omega r_0/c_s$. Based on the unfolded layered cone models for the layer resting on rigid rock, $k(a_0)$, $c(a_0)$ follow from Eq. 3-39, and $k_\vartheta(a_0)$, $c_\vartheta(a_0)$ from Eq. 3-43, together with K and K_ϑ. After applying the correspondence principle, the spring and damping coefficients are also a function of the soil's hysteretic damping ratio ζ_g (Eq. 2-233), denoted by adding the subscript ζ_g. (Of course, in a time-domain analysis, the discrete-element and lumped-parameter models can be used, as shown for a halfspace at the bottom of Fig. 2-24.)

Although the system has three degrees of freedom—the total lateral displacements of the mass with amplitude $u^t(\omega)$, of the basemat with amplitude $u_0^t(\omega)$, and the rocking motion with amplitude $\vartheta_0(\omega)$—only one of them is "dynamic," as there is only one mass. This is the same number as that of the structure built in at its base (i.e., not considering soil–structure interaction). The model shown in Fig. 7-1 is thus well suited to identify the key parameters affecting soil–structure interaction and to study their effects. It is advantageous to split the total displacement amplitudes into their components

$$u^t(\omega) = u^g(\omega) + u_0(\omega) + h\vartheta_0(\omega) + u(\omega) \tag{7-2a}$$

$$u_0^t(\omega) = u^g(\omega) + u_0(\omega) \tag{7-2b}$$

where $u_0(\omega)$ is the amplitude of the base relative to the free-field motion $u^g(\omega)$, and $u(\omega)$ represents the amplitude of the relative displacement of the mass referred to a moving frame of reference attached to the rigid base, which is equal to the amplitude of the structural distortion (Fig. 7-1).

The interaction force-displacement relationship in the horizontal direction with the force amplitude $P_0(\omega)$ equals

$$P_0(\omega) = Kk_{\zeta_g}(a_0)\,u_0(\omega) + \frac{r_0}{c_s}Kc_{\zeta_g}(a_0)\,\dot{u}_0(\omega) = S_{\zeta_g}(a_0)\,u_0(\omega) \tag{7-3}$$

with the corresponding dynamic-stiffness coefficient $S_{\zeta g}(a_0)$. $P_0(\omega)$ depends on the relative motion $u_0(\omega)$. Analogously, the interaction moment-rotation relationship with the amplitude $M_0(\omega)$ is formulated as

$$M_0(\omega) = K_\vartheta k_{\vartheta\zeta_g}(a_0)\,\vartheta_0(\omega) + \frac{r_0}{c_s}K_\vartheta c_{\vartheta\zeta_g}(a_0)\,\dot{\vartheta}_0(\omega) = S_{\vartheta\zeta_g}(a_0)\vartheta_0(\omega) \tag{7-4}$$

7.1.2 Equations of Motion

Formulating the dynamic equilibrium of the node coinciding with the mass and the horizontal and rocking equilibrium equations of the total system leads using Eqs. 7-3 and 7-4 to

$$-m\omega^2\left[u(\omega) + u_0(\omega) + h\vartheta_0(\omega)\right] + k(1 + 2i\zeta)u(\omega) = m\omega^2 u^g(\omega) \tag{7-5a}$$

$$-m\omega^2\left[u(\omega) + u_0(\omega) + h\vartheta_0(\omega)\right] + S_{\zeta_g}(a_0)u_0(\omega) = m\omega^2 u^g(\omega) \tag{7-5b}$$

$$-mh\omega^2\left[u(\omega) + u_0(\omega) + h\vartheta_0(\omega)\right] + S_{\vartheta\zeta_g}(a_0)\vartheta_0(\omega) = mh\omega^2 u^g(\omega) \tag{7-5c}$$

Dividing Eqs. 7-5a and 7-5b by $m\omega^2$ and Eq. 7-5c by $mh\omega^2$ and substituting Eq. 7-1 results in the following symmetric equations of motion of the coupled system

$$\begin{bmatrix} \frac{\omega_s^2}{\omega^2}(1 + 2i\zeta) - 1 & -1 & -1 \\ -1 & \frac{S_{\zeta_g}(a_0)}{m\omega^2} - 1 & -1 \\ -1 & -1 & \frac{S_{\vartheta\zeta_g}(a_0)}{mh^2\omega^2} - 1 \end{bmatrix} \begin{Bmatrix} u(\omega) \\ u_0(\omega) \\ h\vartheta_0(\omega) \end{Bmatrix} = \begin{Bmatrix} 1 \\ 1 \\ 1 \end{Bmatrix} u^g(\omega) \tag{7-6}$$

Eliminating $u_0(\omega)$ and $h\vartheta_0(\omega)$ from Eq. 7-6 yields

$$u_0(\omega) = \frac{\omega_s^2\,(1 + 2i\zeta)}{\dfrac{S_{\zeta_g}(a_0)}{m}}\,u(\omega) \tag{7-7a}$$

$$h\vartheta_0(\omega) = \frac{\omega_s^2\,(1 + 2i\zeta)}{\dfrac{S_{\vartheta\zeta_g}(a_0)}{mh^2}}\,u(\omega) \tag{7-7b}$$

The $u(\omega)$ is expressed as

$$\left[1 + 2i\zeta - \frac{\omega^2}{\omega_s^2} - \frac{\omega^2(1+2i\zeta)}{\dfrac{S_{\zeta_g}(a_0)}{m}} - \frac{\omega^2(1+2i\zeta)}{\dfrac{S_{\vartheta\zeta_g}(a_0)}{mh^2}}\right]u(\omega) = \frac{\omega^2}{\omega_s^2}\,u^g(\omega) \tag{7-8}$$

7.1.3 Effect of Interaction with Soil on Response of Structure

Dimensionless parameters. For a specific excitation, the *response of the dynamic system will depend on the properties of the structure compared to those of the soil.* For the model illustrated in Fig. 7-1, the following dimensionless parameters are introduced:

1. The ratio of the stiffness of the structure to that of the soil is

$$\bar{s} = \frac{\omega_s h}{c_s} \tag{7-9a}$$

   As for certain tall buildings ω_s is approximately inversely proportional to h, $\omega_s h$ will be constant for this type of structure. For decreasing stiffness of the soil, $\bar{s}$ increases.
2. The slenderness ratio

$$\bar{h} = \frac{h}{r_0} \tag{7-9b}$$

3. The mass ratio

$$\bar{m} = \frac{m}{\rho r_0^3} \tag{7-9c}$$

   where ρ represents the mass density of the soil.
4. The Poisson's ratio ν of the soil.
5. The hysteretic-damping ratios of the structure ζ and of the soil ζ_g.
6. The ratio of depth to radius

$$\bar{d} = \frac{d}{r_0} \tag{7-9d}$$

   when the soil is not a homogeneous halfspace but is a layer resting on rigid rock.

Structural response. To introduce the frequency of the excitation ω, the ratio ω/ω_s is defined. The magnitudes of the structural distortion $|u(\omega)|$ (Eq. 7-8), of the displacement of the mass relative to the free field $|u_0(\omega) + h\vartheta_0(\omega) + u(\omega)|$, and of the total displacement of the base $|u_0(\omega) + u^g(\omega)|$ (Eq. 7-7), nondimensionalized, are plotted as a function of the excitation frequency ω/ω_s in Fig. 7-3. The following dimensionless parameters are used for the soil halfspace: $\bar{s} = 1$, $\bar{h} = 2/3$, $\bar{m} = 1.5$, $\nu = 1/3$, $\zeta = 0.025$, $\zeta_g = 0.05$. For comparison, the results are also shown when soil–structure interaction is neglected ($\bar{s} = 0$). The peak responses of the structural distortion and of the displacement of the mass are significantly smaller than those of the same structure on a rigid soil. The peaks that occur at a smaller frequency corresponding to a more flexible system are also broader, indicating that the damping ratio is larger when the soil is modeled. For a specific frequency, however, the response can be smaller or larger. For instance, for $\omega/\omega_s = 0.7$, taking soil–structure interaction into account increases the structural distortion and the displacement of the mass. Over a significant frequency range, the total displacement of the base is larger than the free-field motion. The response is also plotted for a soil layer on rigid rock with $\bar{d} = 1$. As for this case radiation damping is much smaller than for a homogeneous halfspace, the peak responses increase.

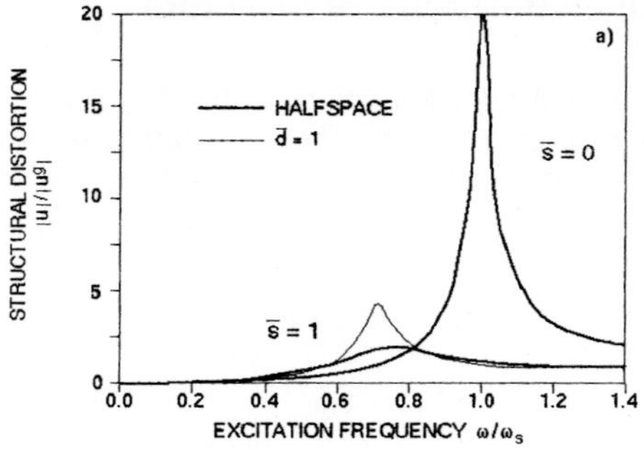

a. Structural distortion.

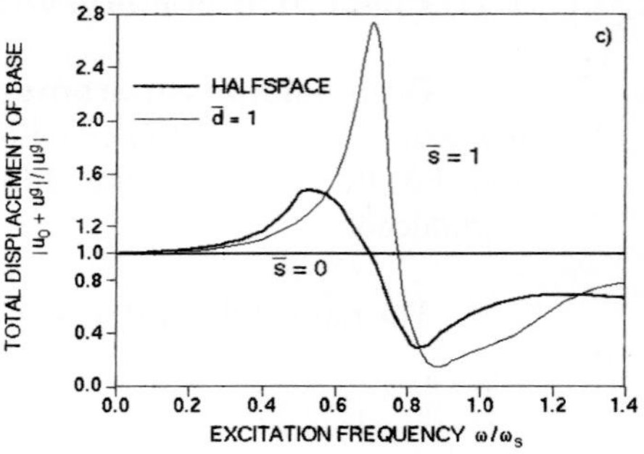

c. Total displacement of base.

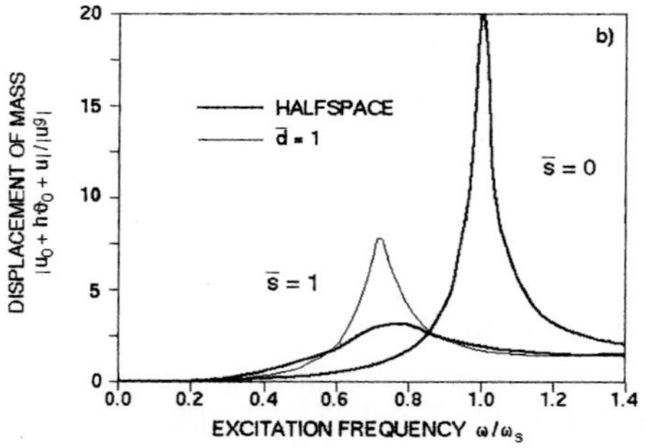

b. Displacement of mass relative to free field.

Fig. 7-3 Response of structure on soil halfspace and soil layer for harmonic loading based on coupled system.

7.2 EQUIVALENT ONE-DEGREE-OF-FREEDOM SYSTEM

7.2.1 Material Damping of Soil

To derive simple expressions that lead to physical insight, the effects of material damping and of radiation damping are separated in the dynamic-stiffness coefficients of the soil. This is achieved by considering the effect of material damping on the damping coefficients only, as described at the end of Section 2.4.1. Analogous to Eqs. 2-237 and 2-238, the dynamic-stiffness coefficients appearing in Eqs. 7-3 and 7-4 are approximated as

$$S_{\zeta_g}(a_0) = Kk(a_0)[1+2i\zeta_h(a_0)+2i\zeta_g] \tag{7-10}$$

$$S_{\vartheta\zeta_g}(a_0) = K_\vartheta k_\vartheta(a_0)[1+2i\zeta_r(a_0)+2i\zeta_g] \tag{7-11}$$

with the radiation-damping ratios of the undamped soil in the horizontal and rocking motions

$$\zeta_h(a_0) = \frac{a_0\, c(a_0)}{2k(a_0)} \tag{7-12}$$

$$\zeta_r(a_0) = \frac{a_0 c_\vartheta(a_0)}{2k_\vartheta(a_0)} \tag{7-13}$$

To evaluate the accuracy of this approximation, the dynamic-stiffness coefficient in the horizontal direction of a disk on the surface of a soil layer resting on rigid rock with $r_0 = d$ and Poisson's ratio $\nu = 1/3$ is calculated based on the unfolded layered cone model for a hysteretic damping ratio $\zeta_g = 0.05$ (Fig. 7-4). The exact solution corresponds to a direct application of the correspondence principle to Eq. 3-39; the first approximation is Eq. 2-236 where both the spring and damping coefficients are affected by material damping, and the second approximation is specified in Eq. 7-10 leading to the same spring coefficient as in the undamped case (Fig. 3-9a) and to the same damped damping coefficient as in the first approximation. In the low and intermediate frequency ranges of interest, the second approximation is sufficiently accurate. Notice that below the fundamental frequency of the layer in the horizontal direction corresponding to $a_0 = \pi/2$ the damping coefficient does not vanish for the damped case in contrast to the undamped case (Fig. 3-9a). A cutoff frequency does thus not exist for a soil layer with material damping.

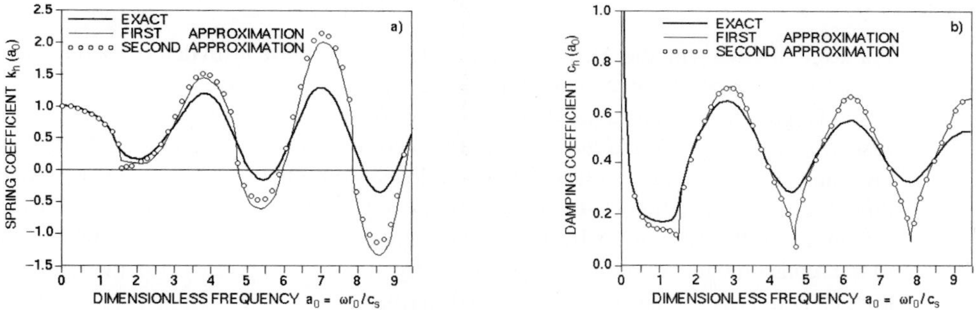

Fig. 7-4 Horizontal dynamic-stiffness coefficient for harmonic loading of disk on layer with material damping.

7.2.2 Equivalent Natural Frequency, Damping Ratio and Input Motion

Substituting Eqs. 7-10 and 7-11 for the dynamic-stiffness coefficients, the following ratios are expressed as

$$\frac{S_{\zeta_g}(a_0)}{m} = \omega_h^2(a_0)\,[1 + 2i\zeta_h(a_0) + 2i\zeta_g] \tag{7-14}$$

$$\frac{S_{\vartheta\zeta_g}(a_0)}{mh^2} = \omega_r^2(a_0)\,[1 + 2i\zeta_r(a_0) + 2i\zeta_g] \tag{7-15}$$

with

$$\omega_h(a_0) = \sqrt{\frac{Kk(a_0)}{m}} \tag{7-16}$$

$$\omega_r(a_0) = \sqrt{\frac{K_\vartheta k_\vartheta(a_0)}{mh^2}} \tag{7-17}$$

For an arbitrary ω, $\omega_h(a_0)$ and $\omega_r(a_0)$ should be regarded as abbreviations. The ω_h is equal to the natural frequency of the dynamic model of Fig. 7-1 corresponding to the horizontal direction, assuming that the structure is rigid ($k = \infty$) and that the foundation cannot rock ($K_\vartheta = \infty$). Analogously, ω_r equals the natural frequency corresponding to the rocking motion for a rigid structure ($k = \infty$) and with no horizontal motion of the foundation ($K = \infty$). For this interpretation, the a_0 appearing in the spring coefficient on the right-hand side must correspond to the natural frequency on the left-hand side, which is achieved by iteration (e.g., for Eq. 7-16 with $a_0 = \omega_h r_0/c_s$).

Substituting Eqs. 7-14 and 7-15 in Eq. 7-8 yields

$$\left[1 + 2i\zeta - \frac{\omega^2}{\omega_s^2} - \frac{\omega^2}{\omega_h^2(a_0)} \frac{1+2i\zeta}{1+2i\zeta_h(a_0) + 2i\zeta_g} - \frac{\omega^2}{\omega_r^2(a_0)} \frac{1+2i\zeta}{1+2i\zeta_r(a_0) + 2i\zeta_g}\right] u(\omega) = \frac{\omega^2}{\omega_s^2} u^g(\omega) \tag{7-18}$$

For a specified ω with the corresponding $u^g(\omega)$, the response $u(\omega)$ determined from solving this equation depends on ω_s, $\omega_h(a_0)$, $\omega_r(a_0)$, ζ, $\zeta_h(a_0)$, $\zeta_r(a_0)$, and ζ_g.

To gain further conceptual clarity, an *equivalent one-degree-of-freedom system representing the structure and the soil* is introduced, which is shown in Fig. 7-5. Its mass m is the same as the structure's mass in the coupled system of Fig. 7-1. The other properties of this replacement oscillator—the natural frequency $\tilde{\omega}$ determining the (static) spring coefficient $\tilde{k}$ and the ratio of the hysteretic damping $\tilde{\zeta}$—are selected such that when subjected to the replacement excitation $\tilde{u}^g(\omega)$ essentially the *same* response $u(\omega)$, the *structural distortion, results* as for the coupled system described by Eq. 7-18. A tilde (~) is used to denote the properties of this replacement oscillator. For harmonic motion, the equation of motion of the equivalent one-degree-of-freedom system is formulated as

$$[-m\omega^2 + \tilde{k}(1+2i\tilde{\zeta})]\, u(\omega) = m\omega^2 \tilde{u}^g(\omega) \tag{7-19}$$

Substituting

$$\tilde{\omega}^2 = \frac{\tilde{k}}{m} \tag{7-20}$$

in Eq. 7-19 results in

$$\left(1+2i\tilde{\zeta} - \frac{\omega^2}{\tilde{\omega}^2}\right) u(\omega) = \frac{\omega^2}{\tilde{\omega}^2} \tilde{u}^g(\omega) \tag{7-21}$$

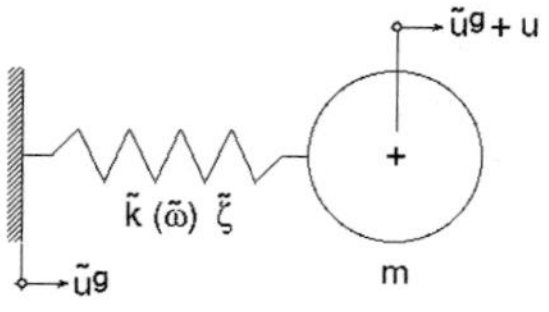

Fig. 7-5 Equivalent one-degree-of-freedom system (enforcing same structural distortion as in coupled dynamic model) with same mass, effective input motion, effective damping ratio and effective natural frequency determining effective spring.

The equivalent natural frequency $\tilde{\omega}$ is determined by noting that in an undamped system, the response is infinite at this natural frequency. This is verified for the replacement oscillator in Eq. 7-21, where for $\tilde{\zeta} = 0$ the coefficient on the left-hand side vanishes for $\omega = \tilde{\omega}$. Proceeding analogously with the coupled system; setting $\zeta = \zeta_h(a_0) = \zeta_r(a_0) = \zeta_g = 0$ in Eq. 7-18 and enforcing the coefficient of $u(\omega)$ to be zero results with $\omega = \tilde{\omega}$ in

$$\frac{1}{\tilde{\omega}^2} = \frac{1}{\omega_s^2} + \frac{1}{\omega_h^2(\tilde{a}_0)} + \frac{1}{\omega_r^2(\tilde{a}_0)} \tag{7-22}$$

with $\tilde{a}_0 = \tilde{\omega}\, r_0/c_s$. Equation 7-22 is thus satisfied for $\tilde{\omega}$ which must be considered when evaluating the abbreviations $\omega_h(\tilde{a}_0)$ and $\omega_r(\tilde{a}_0)$ (Eqs. 7-16 and 7-17). An iterative procedure with a few steps is, in general, necessary. The same Eq. 7-22 can be derived from Eq. 7-18 by defining the natural frequency $\tilde{\omega}$ as that frequency ω for which the real part of the coefficient on the left-hand side vanishes, whereby products of the damping ratios with respect to unity are neglected. It follows that the *equivalent fundamental frequency* $\tilde{\omega}$ of the soil–structure system *is always smaller than the fixed-base frequency of the structure* ω_s. Considering the soil thus makes the dynamic model more flexible.

To establish the equivalent damping ratio $\tilde{\zeta}$, Eq. 7-18 is formulated at resonance ($\omega = \tilde{\omega}$) and products of ζ, $\zeta_h(\tilde{a}_0)$, $\zeta_r(\tilde{a}_0)$, and ζ_g are neglected compared to unity. This operation transforms Eq. 7-18 to

$$\left[1+2i\zeta - \frac{\tilde{\omega}^2}{\omega_s^2} - \frac{\tilde{\omega}^2}{\omega_h^2(\tilde{a}_0)}\left(1 + 2i\zeta - 2i\zeta_h(\tilde{a}_0) - 2i\zeta_g\right) - \frac{\tilde{\omega}^2}{\omega_r^2(\tilde{a}_0)}\left(1 + 2i\zeta - 2i\zeta_r(\tilde{a}_0) - 2i\zeta_g\right)\right] u(\tilde{\omega}) = \frac{\tilde{\omega}^2}{\omega_s^2}\, u^g(\tilde{\omega}) \tag{7-23}$$

Substituting Eq. 7-22 into the left-hand side of Eq. 7-21, the corresponding relationship with $\omega = \tilde{\omega}$ for the equivalent system results.

$$\left[1+2i\tilde{\zeta} - \frac{\tilde{\omega}^2}{\omega_s^2} - \frac{\tilde{\omega}^2}{\omega_h^2(\tilde{a}_0)} - \frac{\tilde{\omega}^2}{\omega_r^2(\tilde{a}_0)}\right] u(\tilde{\omega}) = \tilde{u}^g(\tilde{\omega}) \tag{7-24}$$

The real parts of the left-hand sides of Eqs. 7-23 and 7-24 are the same. Equating the imaginary parts leads after using Eq. 7-22 to

$$\tilde{\zeta} = \frac{\tilde{\omega}^2}{\omega_s^2}\zeta + \left(1 - \frac{\tilde{\omega}^2}{\omega_s^2}\right)\zeta_g + \frac{\tilde{\omega}^2}{\omega_h^2(\tilde{a}_0)}\zeta_h(\tilde{a}_0) + \frac{\tilde{\omega}^2}{\omega_r^2(\tilde{a}_0)}\zeta_r(\tilde{a}_0) \tag{7-25}$$

The equivalent hysteretic damping ratio $\tilde{\zeta}$ determined at resonance is used over the whole range of frequency. The four different parts contributing to $\tilde{\zeta}$, the structural hysteretic material damping with ratio ζ, the hysteretic material damping of the soil with ratio ζ_g, the viscous radiation dampings of the undamped soil in the horizontal direction with ratio $\zeta_h(\tilde{a}_0)$ and in the rocking motion with ratio $\zeta_r(\tilde{a}_0)$ are clearly separated, are multiplied by weights, and are additive in this equation. When radiation damping is present, the *equivalent damping ratio* $\tilde{\zeta}$ of the *soil–structure system is larger than that of the structure alone*

ζ. Considering soil–structure interaction leads in this case to a dynamic system with larger damping. If no radiation damping occurs in the horizontal and rocking directions $[\zeta_h(\tilde{a}_0) = \zeta_r(\tilde{a}_0) = 0]$ and if the material-damping ratio of the structure is equal to that of the soil ($\zeta = \zeta_g$), $\tilde{\zeta} = \zeta$ results as is to be expected. As under normal circumstances ζ_g will not be smaller than ζ, the equivalent damping ratio $\tilde{\zeta}$ will, even in the absence of radiation damping, be somewhat larger than the damping ratio ζ of the structure, which would apply if soil–structure interaction were not taken into account.

Finally, comparing the right-hand sides of Eqs. 7-18 and 7-21, it follows that

$$\tilde{u}^g(\omega) = \frac{\tilde{\omega}^2}{\omega_s^2} u^g(\omega) \tag{7-26}$$

The *equivalent seismic input* $\tilde{u}^g(\omega)$ *will thus always be smaller than the seismic input motion of the coupled system* $u^g(\omega)$ determined from the earthquake.

Summarizing, the seismic response of the coupled system shown in Fig. 7-1 is, from a practical point of view, the same as that of the equivalent one-degree-of-freedom system presented in Fig. 7-5. This replacement oscillator with the same mass m is defined by the natural frequency $\tilde{\omega}$ (Eq. 7-22) and by the hysteretic-damping ratio $\tilde{\zeta}$ (Eq. 7-25) and is subjected to the ground-displacement amplitude $\tilde{u}_g(\omega)$ (Eq. 7-26). The displacement amplitude of the mass $u(\omega)$relative to $\tilde{u}^g(\omega)$ of this equivalent system, determined for each ω by solving Eq. 7-21 is equal to that of the structural distortion. The amplitude of the corresponding transverse-shear force in the structure equals $ku(\omega)$ and not $\tilde{k}\,u(\omega)$. The amplitudes of the horizontal displacement of the base $u_0(\omega)$ relative to the free-field motion $u^g(\omega)$ and of the rotation $\vartheta_0(\omega)$ follow from Eq. 7-7. Again, neglecting squares of damping terms compared to unity results in

$$u_0(\omega) = \frac{\omega_s^2}{\omega_h^2(a_0)} \left[1 + 2i\zeta - 2i\zeta_h(a_0) - 2i\zeta_g\right] u(\omega) \tag{7-27a}$$

$$h\vartheta_0(\omega) = \frac{\omega_s^2}{\omega_r^2(a_0)} \left[1 + 2i\zeta - 2i\zeta_r(a_0) - 2i\zeta_g\right] u(\omega) \tag{7-27b}$$

The motion of the mass point relative to that of the free field is described by

$$u_0(\omega) + h\vartheta_0(\omega) + u(\omega) =$$

$$u(\omega) + \omega_s^2\left[\left(\frac{1}{\omega_h^2(a_0)} + \frac{1}{\omega_r^2(a_0)}\right)(1 + 2i\zeta - 2i\zeta_g) - \frac{2i\zeta_h(a_0)}{\omega_h^2(a_0)} - \frac{2i\zeta_r(a_0)}{\omega_r^2(a_0)}\right] u(\omega) \tag{7-28}$$

It has to be calculated to design, for example, the gap between neighboring buildings or the movement of a bridge abutment.

The equivalent natural frequency $\tilde{\omega}$ and damping ratio $\tilde{\zeta}$ can be expressed for the soil halfspace and the soil layer resting on rigid rock as a function of the dimensionless parameters (Eq. 7-9) and the fixed-base natural frequency of the structure ω_s. For instance, for the disk on a halfspace with Poisson's ratio $\nu \le 1/3$ modeled with cones (Eqs. 2-73, 2-171, Table 2-3A), the terms present in Eqs. 7-22 and 7-25 are equal to

$$\omega_h^2 = \frac{8}{2-\nu}\frac{\bar{h}^2}{\bar{s}^2\,\bar{m}}\,\omega_s^2 \tag{7-29a}$$

$$\omega_r^2(\tilde{a}_0) = \frac{8}{3(1-\nu)}\frac{1}{\bar{s}^2\,\bar{m}}\left[1-\frac{1}{3}\frac{\left(\dfrac{\tilde{\omega}\bar{s}}{\omega_s\bar{h}}\right)^2}{\dfrac{512}{81\pi^2}\dfrac{1-2\nu}{(1-\nu)^3}+\left(\dfrac{\tilde{\omega}\,\bar{s}}{\omega_s\,\bar{h}}\right)^2}\right]\omega_s^2 \tag{7-29b}$$

$$\zeta_h(\tilde{a}_0) = \frac{\pi(2-\nu)}{16}\frac{\tilde{\omega}}{\omega_s}\frac{\bar{s}}{\bar{h}} \tag{7-29c}$$

$$\zeta_r(\tilde{a}_0) = \frac{9\sqrt{2\pi}}{64}\frac{(1-\nu)^{3/2}}{\sqrt{1-2\nu}}\,\frac{\left(\dfrac{\tilde{\omega}\bar{s}}{\omega_s\bar{h}}\right)^3}{\dfrac{512}{27\pi^2}\dfrac{1-2\nu}{(1-\nu)^3}+2\left(\dfrac{\tilde{\omega}\,\bar{s}}{\omega_s\,\bar{h}}\right)^2} \tag{7-29d}$$

7.2.3 Parametric Study

The ratio of the effective natural frequency of the equivalent one-degree-of-freedom system to that of the structure fixed at its base $\tilde{\omega}/\omega_s$ can be regarded as characterizing the effect of soil–structure interaction for a given site. It is to be expected that decreasing the stiffness of the soil ($\bar{s}$ increases) and augmenting the mass of the structure ($\bar{m}$ increases) increases the influence of soil–structure interaction, resulting in a decrease of $\tilde{\omega}/\omega_s$. This does actually occur as can be verified from Eq. 7-22 using Eqs. 7-29a and 7-29b. *Dynamic soil–structure interaction is thus important for stiff structures with a large mass supported on flexible soil.* In Fig. 7-6, $\tilde{\omega}/\omega_s$ and $\tilde{\zeta}$ are plotted as a function of $\bar{s}$, varying the slenderness ratio $\bar{h}$ for a soil halfspace modeled with cones. The mass ratio is selected as $\bar{m} = 1$, Poisson's ratio $\nu = 1/3$, and the hysteretic-damping ratios $\zeta = 0.025$ and $\zeta_g = 0.05$. For rock (c_s very large, $\bar{s} \to 0$), $\tilde{\zeta}$ is approximately equal to ζ for all $\bar{h}$. For squat structures ($\bar{h}$ small), whose mode shapes of the structure–soil system will predominantly consist of a translation, $\tilde{\zeta}$ is larger than for slender structures ($\bar{h}$ large), where the rocking motion is of paramount importance (see also discussion at end of Section 2.3.4). For $\bar{h} = 5$ hardly any radiation damping occurs, $\tilde{\zeta}$ being close to the material-damping ratio of the soil ζ_g for large $\bar{s}$. In Fig. 7-7 for a constant $\bar{h} = 2$, the mass ratio $\bar{m}$ is varied. The variables correspond to a typical reactor building of a nuclear-power plant. Extreme cases are also included to emphasize the effects.

Next, the influence of the depth d of the layer on the properties of the equivalent one-degree-of-freedom system is investigated. The disk on the surface of the soil layer resting on rigid rock is modeled with the unfolded layered cones. The corresponding dynamic-stiffness coefficients are specified in Section 3.3.5. For $\bar{h} = 2$, $\bar{m} = 1$, $\nu = 1/3$, $\zeta = 0.025$, and $\zeta_g = 0.05$, $\tilde{\omega}/\omega_s$ and $\tilde{\zeta}$ are plotted versus $\bar{s}$, varying the depth ratio $\bar{d}$ in Fig. 7-8. The ratios $\bar{d} = d/r_0 = 1$ and $= 4$ for the soil layer are investigated. The results for the halfspace ($\bar{d} = \infty$) are also shown for comparison. The ratio $\tilde{\omega}/\omega_s$ (Fig. 7-8a) depends only weakly on the site. For a negligible effect of soil–structure interaction (small $\bar{s}$), all curves for $\tilde{\zeta}$ (Fig. 7-8b)

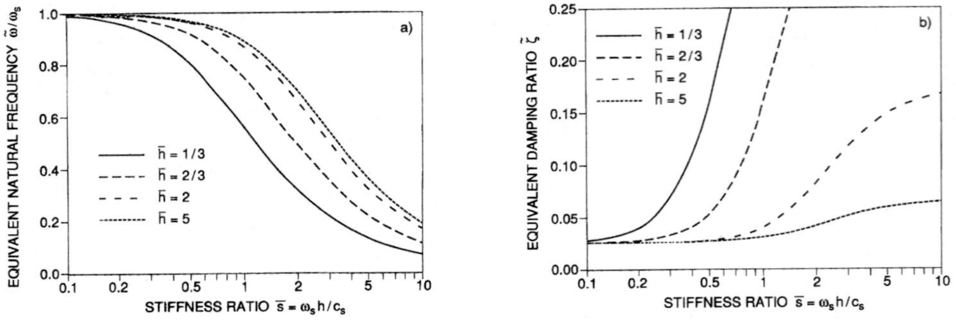

a. Natural frequency. **b.** Damping ratio.

Fig. 7-6 Properties of equivalent one-degree-of-freedom system varying slenderness ratio for soil halfspace.

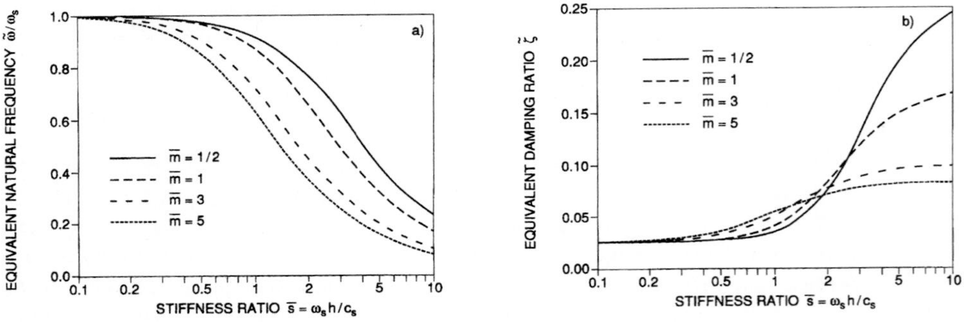

a. Natural frequency. **b.** Damping ratio.

Fig. 7-7 Properties of equivalent one-degree-of-freedom system varying mass ratio for soil halfspace.

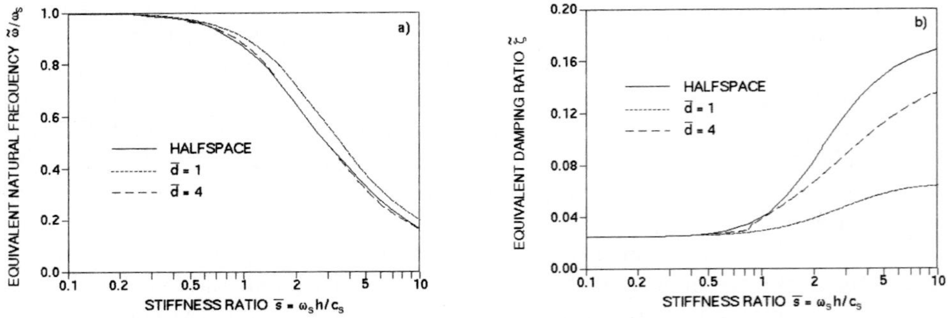

a. Natural frequency. **b.** Damping ratio.

Fig. 7-8 Properties of equivalent one-degree-of-freedom system varying depth of soil layer.

start at the structural damping ratio $\zeta = 0.025$. For the halfspace, $\tilde{\zeta}$ increases significantly for increasing $\bar{s}$, due to the large effect of radiation damping. For the other extreme site, $\bar{d} = 1$, *radiation damping is hardly activated*, as *the equivalent natural frequency* $\tilde{\omega}$ is essentially always *smaller than the natural frequency of the layer* in the horizontal direction (cut-off frequency in the undamped case). For significant soil–structure-interaction effect (large $\bar{s}$), the ratio $\tilde{\zeta}$ converges essentially to the material damping ratio of the soil $\zeta_g = 0.05$. For the intermediate site $\bar{d} = 4$, a transmission occurs.

Finally, the accuracy of the structure's response using the equivalent one-degree-of-freedom system (Fig. 7-5) compared to the results from the coupled dynamic model (Fig. 7-1) is investigated. The stiffness ratio equals $\bar{s} = 3$ for $\bar{d} = 4$ with the same other dimensionless parameters ($\bar{h} = 2$, $\bar{m} = 1$, $\nu = 1/3$, $\zeta = 0.025$, $\zeta_g = 0.05$). This results in $\tilde{\omega}/\omega_s = 0.490$ (Eq. 7-22) and $\tilde{\zeta} = 0.089$. (Eq. 7-25). The magnitudes of the structural distortion $|u(\omega)|$ (Eq. 7-21), of the displacement of the mass relative to the free field $|u_0(\omega) + h\vartheta_0(\omega) + u(\omega)|$ (Eq. 7-28) and of the total displacement of the base $|u_0(\omega) + u^g(\omega)|$ (Eq. 7-27a), non-dimensionalized, are shown as distinct points as a function of the excitation frequency ω/ω_s in Fig. 7-9. The corresponding results of the coupled model (Eqs. 7-8, 7-7) are presented as a continuous line. The equivalent one-degree-of-freedom system is very accurate.

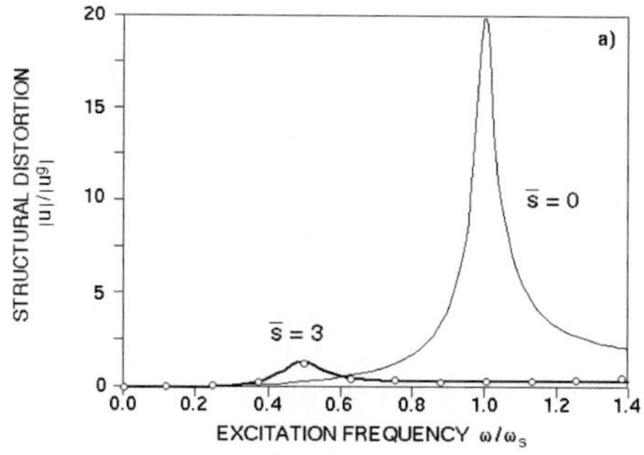

a. Structural distortion.

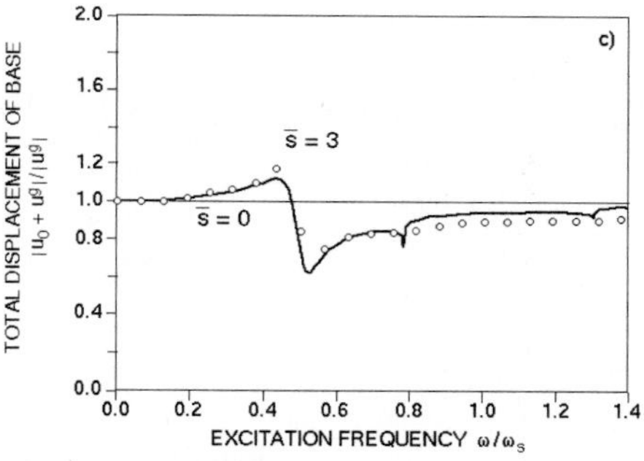

c. Total displacement of base.

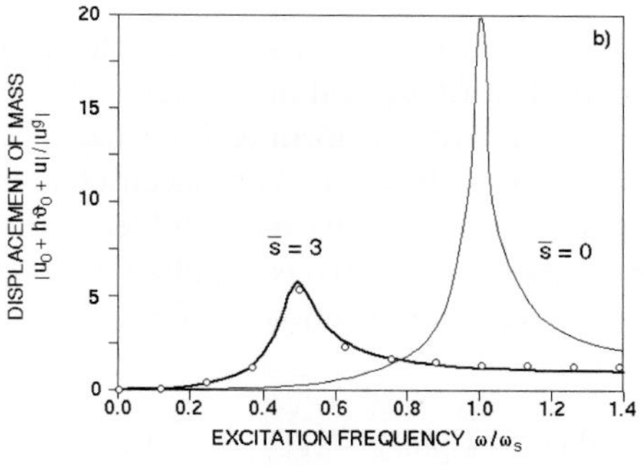

b. Displacement of mass relative to free field.

Fig. 7-9 Responses of structure on soil layer for harmonic loading based on coupled system and on equivalent one-degree-of-freedom system.

Notice that although for the equivalent oscillator the damping ratio $\tilde{\zeta}$ is generally larger than ζ (Eq. 7-25) and the effective seismic input $\tilde{u}^g(\omega)$ smaller than $u^g(\omega)$ (Eq. 7-26), the structural response for a specific ω can be larger or smaller than when soil–structure interaction is neglected ($\bar{s} = 0$). The curves in Fig. 7-9 for $\bar{s} = 3$ and $\bar{s} = 0$ intersect!

7.3 OTHER EQUIVALENT SYSTEMS

7.3.1 Dynamic Model for Vertical Motion

In Sections 7.1 and 7.2 the coupled horizontal and rocking motions of the dynamic structure–soil system for horizontal seismic excitation are investigated. To examine the vertical motion caused by a vertical earthquake with amplitude $u^g(\omega)$, the (even simpler) dynamic model shown in Fig. 7-10 can be used. The structure is modeled with a mass m and a static spring with coefficient k and with hysteretic structural damping ratio ζ. The fixed-base natural frequency ω_s in the vertical direction of the structure equals

$$\omega_s = \sqrt{\frac{k}{m}} \tag{7-30}$$

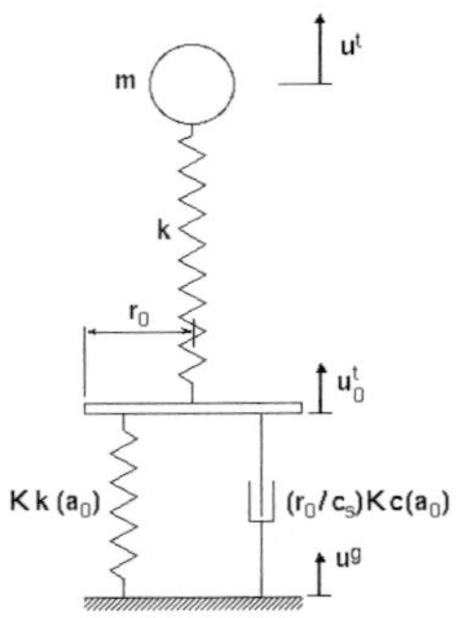

Fig. 7-10 Coupled dynamic model of structure and soil for vertical motion.

At the basemat with equivalent radius r_0 the soil's dynamic stiffness is attached, consisting of a spring with a frequency-dependent coefficient $Kk(a_0)$ and in parallel a dashpot with a frequency-dependent coefficient $(r_0/c_s)\ Kc(a_0)$ for harmonic loading. For a halfspace, modeled with a cone, the spring and damping coefficients $k(a_0)$, $c(a_0)$ are specified in Eq. 2-73; for a soil layer resting on rigid rock they are represented by the unfolded layered cone model in Eq. 3-39 (for the undamped case). Separating the effects of radiation damping and material damping with hysteretic ratio ζ_g as described in Section 7.2.1, the dynamic stiffness coefficient is approximated as

$$S_{\zeta_g}(a_0) = Kk(a_0)\ [1 + 2i\zeta_v(a_0) + 2i\zeta_g] \tag{7-31}$$

with the radiation-damping ratio of the undamped soil in the vertical direction

$$\zeta_v(a_0) = \frac{a_0 c(a_0)}{2k(a_0)} \tag{7-32}$$

Two degrees of freedom with the amplitudes of the total vertical displacement of the mass $u^t(\omega)$ and of the massless base $u_0^t(\omega)$ are introduced, whereby only one of them is dynamic. Decomposing the total displacement amplitudes into their components leads to

$$u^t(\omega) = u^g(\omega) + u_0(\omega) + u(\omega) \tag{7-33a}$$

$$u_0^t(\omega) = u^g(\omega) + u_0(\omega) \tag{7-33b}$$

with the structural distortion amplitude $u(\omega)$ and the displacement amplitude of the base $u_0(\omega)$ relative to the free-field motion $u^g(\omega)$.

Formulating dynamic equilibrium results in

$$-m\omega^2 \left[u(\omega) + u_0(\omega)\right] + k(1 + 2i\zeta)u(\omega) = m\omega^2 u^g(\omega) \tag{7-34a}$$

$$-m\omega^2 \left[u(\omega) + u_0(\omega)\right] + Kk(a_0)\left[1 + 2i\zeta_v(a_0) + 2i\zeta_g\right] = m\omega^2 u^g(\omega) \tag{7-34b}$$

or

$$\begin{bmatrix} \dfrac{\omega_s^2}{\omega^2}(1 + 2i\zeta) - 1 & -1 \\ -1 & \dfrac{\omega_v^2(a_0)}{\omega^2}\left[1 + 2i\zeta_v(a_0) + 2i\zeta_g\right] - 1 \end{bmatrix} \begin{Bmatrix} u(\omega) \\ u_0(\omega) \end{Bmatrix} = \begin{Bmatrix} 1 \\ 1 \end{Bmatrix} u^g(\omega) \tag{7-35}$$

with the abbreviation

$$\omega_v(a_0) = \sqrt{\frac{Kk(a_0)}{m}} \tag{7-36}$$

When the right-hand side of Eq. 7-36 is evaluated for the frequency ω_v appearing on the left-hand side, thus for $a_0 = \omega_v r_0/c_s$, then ω_v is equal to the natural frequency of the dynamic model of Fig. 7-10 for a rigid structure ($k = \infty$). Eliminating $u_0(\omega)$ from Eq. 7-35 yields

$$\left[1 + 2i\zeta - \frac{\omega^2}{\omega_s^2} - \frac{\omega^2}{\omega_v^2(a_0)}\frac{1+2i\zeta}{1 + 2i\zeta_v(a_0) + 2i\zeta_g}\right] u(\omega) = \frac{\omega^2}{\omega_s^2} u^g(\omega) \tag{7-37}$$

The equivalent one-degree-of-freedom system shown in Fig. 7-5 is also valid for the vertical motion. The corresponding equation of motion is specified in Eq. 7-21. The equivalent natural frequency $\tilde{\omega}$ is again derived by noting that the response $u(\tilde{\omega})$ is infinite in an undamped system. Setting $\zeta = \zeta_v(a_0) = \zeta_g = 0$ with $\omega = \tilde{\omega}$ in Eq. 7-37 and equating the coefficient of $u(\tilde{\omega})$ to zero leads to

$$\frac{1}{\tilde{\omega}^2} = \frac{1}{\omega_s^2} + \frac{1}{\omega_v^2(\tilde{a}_0)} \tag{7-38}$$

with $\tilde{a}_0 = \tilde{\omega} r_0/c_s$. The equivalent natural frequency $\tilde{\omega}$, which is smaller than ω_s, is determined by iteration. To establish the equivalent damping ratio $\tilde{\zeta}$, products of ζ, $\zeta_v(a_0)$, and ζ_g are neglected compared to unity. As the radiation damping ratio $\zeta_v(a_0)$ in the vertical direction is larger than in the horizontal $\zeta_h(a_0)$ and for rocking $\zeta_r(a_0)$, this assumption for the vertical motion is not as accurate as for the coupled horizontal and rocking motions. Formulated at resonance ($\omega = \tilde{\omega}$), Eq. 7-37 is written as

$$\left[1+2i\zeta - \frac{\tilde{\omega}^2}{\omega_s^2} - \frac{\tilde{\omega}^2}{\omega_v^2(\tilde{a}_0)}\left(1+2i\zeta-2i\zeta_v(\tilde{a}_0) - 2i\zeta_g\right)\right] u(\tilde{\omega}) = \frac{\tilde{\omega}^2}{\omega_s^2} u^g(\tilde{\omega}) \tag{7-39}$$

Substituting Eq. 7-38 into the left-hand side of Eq. 7-21, the corresponding relationship with $\omega = \tilde{\omega}$ for the equivalent system results

$$\left[1+2i\tilde{\zeta} - \frac{\tilde{\omega}^2}{\omega_s^2} - \frac{\tilde{\omega}^2}{\omega_v^2(\tilde{a}_0)}\right] u(\tilde{\omega}) = \tilde{u}^g(\tilde{\omega}) \tag{7-40}$$

Setting the imaginary parts on the left-hand sides of Eqs. 7-39 and 7-40 equal yields the equivalent damping ratio

$$\tilde{\zeta} = \frac{\tilde{\omega}^2}{\omega_s^2}\zeta + \left(1 - \frac{\tilde{\omega}^2}{\omega_s^2}\right)\zeta_g + \frac{\tilde{\omega}^2}{\omega_v^2(\tilde{a}_0)}\zeta_v(\tilde{a}_0) \tag{7-41}$$

Again, the contributions of the structural hysteretic damping with ratio ζ, of the hysteretic soil material damping with ratio ζ_g, and of the viscous radiation damping of the undamped soil with ratio $\zeta_v(\tilde{a}_0)$ are additive. Finally, it follows that the equivalent seismic input motion $\tilde{u}^g(\omega)$ equals

$$\tilde{u}^g(\omega) = \frac{\tilde{\omega}^2}{\omega_s^2} u^g(\omega) \tag{7-42}$$

The effective input motion for the equivalent one-degree-of-freedom system is again smaller than that applied to the coupled system.

It is also possible to express the response for a given excitation as a function of the following dimensionless parameters: $\bar{s} = \omega_s r_0/c_s$, $\bar{m} = m/(\rho r_0^3)$, ν, ζ, ζ_g, and $\bar{d} = d/r_0$ for a soil layer.

7.3.2 Redundant Coupled Structure–Soil System for Horizontal and Rocking Motions

The discretization of certain structures can lead to a simple dynamic model that is supported in a statically indeterminate way. The continuous bridge shown in Fig. 7-2b but with the columns built in (and not hinged) at the level of the girder for horizontal seismic excitation is such an example. Assuming the girder to be rigid compared to the stiffness of the columns results in a statically indeterminate model with one redundant shown in Fig. 7-11. The mass m with the total displacement amplitude $u^t(\omega)$ can move only horizontally without rotating. The base has two degrees of freedom, the horizontal translation with the amplitude $u_0^t(\omega)$ and the rocking with the amplitude $\vartheta_0^t(\omega)$. The column of length h (which for the sake of simplicity is assumed to be prismatic with the bending stiffness EI) has a static-stiffness coefficient k (equal to $12EI/h^3$) and hysteretic damping ratio ζ. The hysteretic-

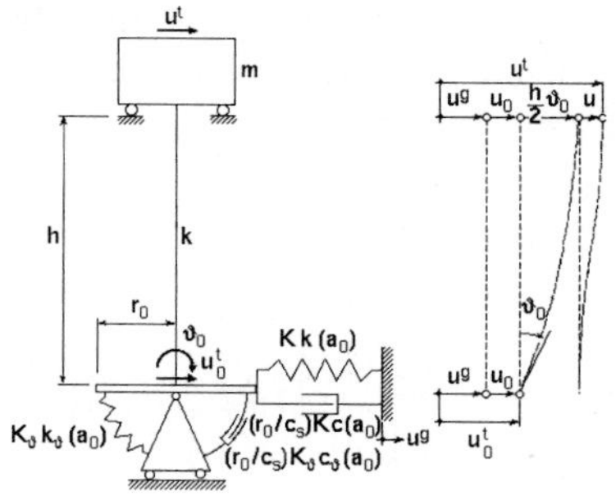

Fig. 7-11 Redundant coupled dynamic model of structure (with zero rotation of mass) and of soil for horizontal and rocking motions.

damping ratio of the soil, which again can be modeled with cones, is ζ_g. The amplitude of the horizontal seismic motion (free field) is denoted as $u^g(\omega)$.

The derivation of the equations of motion of the redundant coupled structure–soil system and of the parameters of the equivalent one-degree-of-freedom system follows closely the discussion in Sections 7.1 and 7.2, which is very detailed. Only the key relations are thus given in the following.

The fixed-base natural frequency of the structure is equal to

$$\omega_s = \sqrt{\frac{k}{m}} \tag{7-43}$$

The abbreviations

$$\omega_h(a_0) = \sqrt{\frac{Kk(a_0)}{m}} \tag{7-44}$$

$$\omega_r(a_0) = \sqrt{\frac{K_\vartheta k_\vartheta(a_0)}{mh^2}} \tag{7-45}$$

still apply. The radiation damping ratios are equal to

$$\zeta_h(a_0) = \frac{a_0 c(a_0)}{2k(a_0)} \tag{7-46}$$

$$\zeta_r(a_0) = \frac{a_0 c_\vartheta(a_0)}{2k_\vartheta(a_0)} \tag{7-47}$$

The displacement amplitudes are decomposed as (Fig. 7-11)

$$u^t(\omega) = u^g(\omega) + u_0(\omega) + \frac{h}{2}\vartheta_0(\omega) + u(\omega) \tag{7-48a}$$

$$u_0^t(\omega) = u^g(\omega) + u_0(\omega) \tag{7-48b}$$

The equations of motion for harmonic response are formulated as

$$
\begin{bmatrix}
\frac{\omega_s^2}{\omega^2}(1+2i\zeta)-1 & -1 & -1 \\
-1 & \frac{\omega_h^2(a_0)}{\omega^2}\left[1+2i\zeta_h(a_0)+2i\zeta_g\right]-1 & -1 \\
\frac{\omega_s^2}{2\omega^2}(1+2i\zeta)-1 & -1 & \frac{2\omega_r^2(a_0)}{\omega^2}\left[1+2i\zeta_r(a_0)+2i\zeta_g\right]+\frac{\omega_s^2}{6\omega^2}(1+2i\zeta)-1
\end{bmatrix}
$$

$$
\cdot \begin{Bmatrix} u(\omega) \\ u_0(\omega) \\ \frac{h}{2}\vartheta_0(\omega) \end{Bmatrix} = \begin{Bmatrix} 1 \\ 1 \\ 1 \end{Bmatrix} u^g(\omega) \tag{7-49}
$$

Neglecting products of the damping ratios compared to unity, the effective natural frequency $\tilde{\omega}$, damping ratio $\tilde{\zeta}$, and input motion $\tilde{u}^g(\omega)$ of the equivalent one-degree-of-freedom system are derived as

$$
\frac{1}{\tilde{\omega}^2} = \frac{1}{\omega_s^2} + \frac{1}{\omega_h^2(\tilde{a}_0)} + \frac{3}{\omega_s^2 + 12\,\omega_r^2(\tilde{a}_0)} \tag{7-50}
$$

$$
\tilde{\zeta} = \zeta - \frac{\tilde{\omega}^2}{\omega_h^2(\tilde{a}_0)}\left[\zeta - \zeta_g - \zeta_h(\tilde{a}_0)\right] - \frac{36\tilde{\omega}^2\,\omega_r^2(\tilde{a}_0)}{\left(\omega_s^2 + 12\omega_r^2(\tilde{a}_0)\right)^2}\left[\zeta - \zeta_g - \zeta_r(\tilde{a}_0)\right] \tag{7-51}
$$

$$
\tilde{u}^g(\omega) = \frac{\tilde{\omega}^2}{\omega_s^2}\, u^g(\omega) \tag{7-52}
$$

7.3.3 Coupled Rigid Block-Soil System

For a rigid structure (block) of finite dimensions on the soil's surface as shown in Fig. 2-24, coupling of the horizontal and rocking motions of the dynamic system occurs. In Section 2.3.4 the equations of motion are formulated for harmonic excitation (Eq. 2-206). An approximation for the fundamental frequency ω_1 (Eq. 2-212) can be determined using the uncoupled natural frequencies of the horizontal and rocking motions ω_h and ω_r (Eq. 2-210) in the same way as the equivalent natural frequency $\tilde{\omega}$ (Eq. 7-22, with $\omega_s = \infty$ for rigid structure) follows. The corresponding damping ratio ζ_1 (Eq. 2-219) is calculated as the sum of the contributions of the radiation damping ratios ζ_h (Eq. 2-211) and ζ_r (Eq. 2-193). This equation is the same as that for the equivalent damping ratio $\tilde{\zeta}$ (Eq. 7-25 with $\zeta = \zeta_g = 0$).

It is also possible to extend the formulation for the surface foundation (Fig. 2-24) to that of an embedded foundation (Fig. 4-31). In this case the stiffness and damping matrices $[K]$ and $[C]$ (Eq. 2-205b and c) exhibit off-diagonal terms.

SUMMARY

1. A simple coupled dynamic model consisting of a vertical rigid bar with the horizontal and rocking springs and dashpots with frequency-dependent coefficients $Kk(a_0)$, $(r_0/c_s)\,Kc(a_0)$, and $K_\vartheta k_\vartheta(a_0)$, $(r_0/c_s)\,K_\vartheta c_\vartheta(a_0)$ representing the soil attached at one end and at the other one, at a distance equal to the height h, a spring with coefficient k connected to a mass m, which models the structure, correctly captures the essential effects of soil–structure interaction for a horizontal seismic excitation. The dynamic-stiffness coefficients of the soil are calculated with cones for a halfspace and with unfolded layered cones for a layer resting on rigid rock.
2. The response of this coupled system for a prescribed horizontal seismic excitation $u^g(\omega)$ with a specified frequency is a function of the fixed-base natural frequency of the structure ω_s, of the natural frequencies for a rigid structure—assuming in addition that either the rocking or the horizontal spring is rigid—$\omega_h(a_0)$ and $\omega_r(a_0)$, the ratios for the viscous radiation damping in the horizontal and rocking directions $\zeta_h(a_0)$ and $\zeta_r(a_0)$ and of the ratios of the hysteretic damping ratios of the structure ζ and of the soil ζ_g.

$$\omega_s = \sqrt{\frac{k}{m}}$$

$$\omega_h(a_0) = \sqrt{\frac{Kk(a_0)}{m}}$$

$$\omega_r(a_0) = \sqrt{\frac{K_\vartheta k_\vartheta(a_0)}{mh^2}}$$

$$\zeta_h(a_0) = \frac{a_0 c(a_0)}{2k(a_0)}$$

$$\zeta_r(a_0) = \frac{a_0 c_\vartheta(a_0)}{2k_\vartheta(a_0)}$$

3. The coupled system can be replaced by an equivalent one-degree-of-freedom system enforcing the same structural distortion with the same mass

$$\frac{1}{\tilde{\omega}^2} = \frac{1}{\omega_s^2} + \frac{1}{\omega_h^2(\tilde{a}_0)} + \frac{1}{\omega_r^2(\tilde{a}_0)}$$

$$\tilde{\zeta} = \frac{\tilde{\omega}^2}{\omega_s^2}\zeta + \left(1 - \frac{\tilde{\omega}^2}{\omega_s^2}\right)\zeta_g + \frac{\tilde{\omega}^2}{\omega_h^2(\tilde{a}_0)}\zeta_h(\tilde{a}_0) + \frac{\tilde{\omega}^2}{\omega_r^2(\tilde{a}_0)}\zeta_r(\tilde{a}_0)$$

$$\tilde{u}^g(\omega) = \frac{\tilde{\omega}^2}{\omega_s^2}\,u^g(\omega)$$

with $\tilde{a}_0 = \tilde{\omega}\, r_0/c_s$. Its natural frequency $\tilde{\omega}$, which is a function of the three frequencies, is always smaller than the fixed-base frequency ω_s of the structure. Its damping ratio $\tilde{\zeta}$, which can be calculated by adding the contributions of the four weighted damping ratios, will, in general, be larger than the hysteretic damping ratio ζ of the structure. Its effective support motion $\tilde{u}^g(\omega)$ will be smaller than that of the coupled dynamic system $u^g(\omega)$.

4. The response is a function of the ratio of the stiffness of the structure to that of the soil $\omega_s h/c_s$, of the slenderness ratio h/r_0, of the ratio of the mass of the structure to that of the soil $m/(\rho r_0^3)$, of ν, and of ζ and ζ_g, as well as for a soil layer on rigid rock of the depth-to-radius ratio d/r_0.
5. Taking soil–structure interaction into account will reduce the peak structural distortion for harmonic excitation, while for a specific frequency of the excitation the result can be either smaller or larger than that of the fixed-base structure.

   The depth d of the soil layer strongly affects the structural response. For a shallow layer, the fundamental frequency $\tilde{\omega}$ of the equivalent one-degree-of-freedom system representing the structure and the soil is smaller than that of the site $\pi c_s/(2d)$. No radiation damping occurs ($\zeta_h(\tilde{a}_0) = \zeta_r(\tilde{a}_0) = 0$). The equivalent damping ratio $\tilde{\zeta}$ will thus be a weighted average of the material damping ratios of the structure ζ and of the soil ζ_g. For a layer with an intermediate depth, the fundamental frequency of the structure–soil system becomes, for a sufficiently soft soil, larger than that of the site, resulting in radiation damping which contributes significantly to the equivalent damping ratio. The deeper the layer is, the less soft the soil has to be to activate the radiation damping.
6. For the vertical motion of a structure on soil, the parameters of the equivalent one-degree-of-freedom system are

$$\frac{1}{\tilde{\omega}^2} = \frac{1}{\omega_s^2} + \frac{1}{\omega_v^2(\tilde{a}_0)}$$

$$\tilde{\zeta} = \frac{\tilde{\omega}^2}{\omega_s^2}\zeta + \left(1 - \frac{\tilde{\omega}^2}{\omega_s^2}\right)\zeta_g + \frac{\tilde{\omega}^2}{\omega_v^2(\tilde{a}_0)}\,\zeta_v(\tilde{a}_0)$$

$$\tilde{u}^g(\omega) = \frac{\tilde{\omega}^2}{\omega_s^2}\, u^g(\omega)$$

   with

$$\omega_v(a_0) = \sqrt{\frac{Kk(a_0)}{m}}$$

$$\zeta_v(a_0) = \frac{a_0 c(a_0)}{2k(a_0)}$$

7. For a redundant coupled structure–soil system where the mass of the structure can only displace horizontally but not rotate, the equivalent one-degree-of-freedom system for horizontal seismic excitation is defined by the parameters

$$\frac{1}{\tilde{\omega}^2} = \frac{1}{\omega_s^2} + \frac{1}{\omega_h^2(\tilde{a}_0)} + \frac{3}{\omega_s^2 + 12\omega_r^2(\tilde{a}_0)}$$

$$\tilde{\zeta} = \zeta - \frac{\tilde{\omega}^2}{\omega_h^2(\tilde{a}_0)}\left[\zeta - \zeta_g - \zeta_h(\tilde{a}_0)\right] - \frac{36\tilde{\omega}^2\omega_r^2(\tilde{a}_0)}{\left(\omega_s^2 + 12\omega_r^2(\tilde{a}_0)\right)^2}\left[\zeta - \zeta_g - \zeta_r(\tilde{a}_0)\right]$$

$$\tilde{u}^g(\omega) = \frac{\tilde{\omega}^2}{\omega_s^2}\, u^g(\omega)$$

A

Interaction Force-Displacement Relationship and Green's Function of Cone Model

In Appendix A the strength-of-materials theory of the truncated semi-infinite cone [M5, M11] is derived. This one-sided cone is used directly to model a disk on the surface of a homogeneous soil halfspace, or as a double-cone model represents a disk embedded in a fullspace. The trapped mass present for incompressible material for the vertical and rocking motions (see Section 2.1.4) can be shifted to the basemat and thus to the superstructure. This "trick" enables the analysis of the cone models in the same manner for all values of Poisson's ratio. Section A1 addresses the translational cone, whereby axial distortion (vertical motion) is examined; for the distortion in shear (horizontal motion) the derivation is analogous. Based on the equation of motion of rod (bar) theory, the interaction force-displacement relationship of the disk in the truncated cross section of the cone used in the stiffness formulation is developed, followed by the corresponding displacement-interaction force equation applied in the flexibility formulation. The Green's function is then discussed, the displacement along the axis of the cone caused by a unit impulse force acting on the disk. In Section A2 the same derivations are performed for the rotational cone. Again, rotational axial distortion corresponding to the rocking motion is addressed, but the results are also valid for rotational shear distortion as in torsional motion. All developments are first performed in the time domain and then, if used in the text, repeated for a harmonic load (frequency domain).

Section A3 examines the generalized wave pattern [M6] for the unfolded layered cone used to model a disk foundation on the surface of a soil layer.

A1 TRANSLATIONAL CONE

A1.1 Equation of Motion

The translational truncated semi-infinite cone with the apex height z_0 and the radius r_0 is shown for axial distortion in Fig. A-1, which is used to model the vertical degree of freedom of a disk of radius r_0 on the surface of a halfspace. The area A at depth z equals $A = (z^2/z_0^2)\, A_0$ with $A_0 = \pi r_0^2$, where z is measured from the apex. With c denoting the appropriate wave velocity of the compression-extension waves (dilatational waves) and ρ the mass density, ρc^2 is equal to the corresponding elastic modulus (constrained modulus). Also, u represents the axial displacement and N the axial force. Radial effects are disregarded. Formulating the equilibrium equation of an infinitesimal element (Fig. A-1) taking the inertial loads into account,

$$-N + N + N_{,z}\, dz - \rho A\, dz\, \ddot{u} = 0 \tag{A-1}$$

and substituting the force-displacement relationship,

$$N = \rho c^2 A\, u_{,z} \tag{A-2}$$

leads to the equation of motion in the time domain of the translational cone

$$u_{,zz} + \frac{2}{z} u_{,z} - \frac{\ddot{u}}{c^2} = 0 \tag{A-3}$$

which may be rewritten as a one-dimensional wave equation in zu

$$(zu)_{,zz} - \left(\frac{zu}{c^2}\right)^{\cdot\cdot} = 0 \tag{A-4}$$

The familiar solution of the wave equation in the time domain is

$$z\,u = z_0 f\left(t - \frac{z - z_0}{c}\right) + z_0\, g\left(t + \frac{z - z_0}{c}\right) \tag{A-5}$$

where f and g are arbitrary functions of the argument for outward-propagating (in the positive z-direction) and inward-propagating waves (in the negative z-direction), respectively.

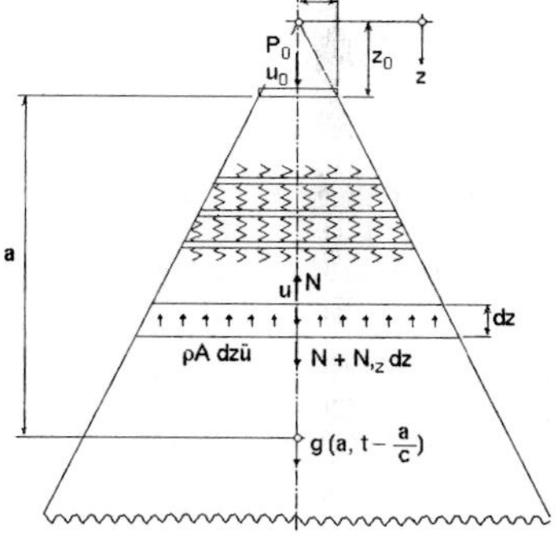

Fig. A-1 Translational truncated semi-infinite cone with nomenclature for vertical motion, axial-distortion mechanism and equilibrium of infinitesimal element.

In a radiation problem only outward-propagating waves are permissible ($g = 0$).

$$u(z, t) = \frac{z_0}{z} f\left(t - \frac{z - z_0}{c}\right) \qquad \text{(A-6)}$$

For harmonic loading of frequency ω resulting in the response $u(t) = u(\omega)e^{i\omega t}$, substituting $\ddot{u}(\omega) = -\omega^2 u(\omega)$ in Eq. A-4 leads to the wave equation in $zu(\omega)$ with the displacement amplitude $u(\omega)$

$$\left(zu(\omega)\right)_{,zz} + \frac{\omega^2}{c^2}\left(zu(\omega)\right) = 0 \qquad \text{(A-7)}$$

The solution omitting the inward-propagating wave varying as $e^{+i\omega(z-z_0)c}$ equals

$$u(\omega) = c_1 \frac{z_0}{z} e^{-i\frac{\omega}{c}(z-z_0)} \qquad \text{(A-8)}$$

with the integration constant c_1.

A1.2 Interaction Force-Displacement Relationship

For a prescribed displacement u_0 of the disk, the interaction force P_0 is calculated as follows. Enforcing the boundary condition $u(z = z_0) = u_0$ yields from Eq. A-6

$$f(t) = u_0 \qquad \text{(A-9)}$$

and thus the displacement decays in inverse proportion to the depth z

$$u(z,t) = \frac{z_0}{z} u_0\left(t - \frac{z - z_0}{c}\right) \qquad \text{(A-10)}$$

With

$$u_{,z} = -\frac{z_0}{z^2} u_0\left(t - \frac{z-z_0}{c}\right) - \frac{z_0}{zc} u_0'\left(t - \frac{z-z_0}{c}\right) \qquad \text{(A-11)}$$

where $u_0'(t - (z - z_0)/c)$ denotes differentiation of u_0 with respect to the argument $t - (z - z_0)/c$.

$$P_0 = -N(z = z_0) = -\rho c^2 A_0 u_{0,z} \qquad \text{(A-12)}$$

is rewritten as ($u_0'(z = z_0) = \dot{u}_0(z = z_0) = \dot{u}_0$)

$$P_0(t) = \frac{\rho c^2 A_0}{z_0} u_0(t) + \rho c A_o \dot{u}_0(t) \qquad \text{(A-13)}$$

or

$$P_0(t) = K u_0(t) + C \dot{u}_0(t) \qquad \text{(A-14)}$$

In this interaction force-displacement relationship K and C are the constant coefficients of the spring and dashpot:

$$K = \frac{\rho c^2 A_0}{z_0} \qquad \text{(A-15a)} \qquad\qquad C = \rho c A_0 \qquad \text{(A-15b)}$$

For harmonic loading,

$$P_0(\omega) = (K + i\,\omega\, C)\, u_0(\omega) \tag{A-16}$$

or

$$P_0(\omega) = S(\omega)\, u_0(\omega) \tag{A-17}$$

applies with the dynamic-stiffness coefficient $S(\omega)$

$$S(\omega) = K + i\,\omega\, C \tag{A-18}$$

Introducing the dimensionless frequency parameter b_0 defined with respect to the properties of the cone

$$b_0 = \frac{\omega\, z_0}{c} \tag{A-19}$$

$S(\omega)$ is formulated as

$$S(b_0) = K\left[k(b_0) + i\, b_0\, c(b_0)\right] \tag{A-20}$$

with the (dimensionless) spring and damping coefficients being independent of b_0 in this simple case

$$k(b_0) = 1 \qquad \text{(A-21a)} \qquad\qquad c(b_0) = 1 \qquad \text{(A-21b)}$$

A1.3 Displacement-Interaction Force Relationship

For a prescribed interaction force P_0 acting on the disk, the displacement u_0 is determined as follows. Dividing Eq. A-13 by K (Eq. A-15a) leads to the first-order differential equation in u_0

$$u_0(t) + \frac{z_0}{c}\,\dot{u}_0(t) = \frac{P_0}{K} \tag{A-22}$$

The general solution is the convolution integral

$$u_0(t) = \int_0^t h_1(t-\tau)\,\frac{P_0(\tau)}{K}\, d\tau \tag{A-23}$$

with the unit-impulse response function

$$h_1(t) = \frac{c}{z_0}\, e^{-\frac{c}{z_0}t} \qquad t \geq 0 \qquad \text{(A-24a)}$$

$$= 0 \qquad t < 0 \qquad \text{(A-24b)}$$

Equation A-23 derived purely on mathematical grounds can be interpreted physically as follows. The contribution of the infinitesimal pulse $(1/K)P_0(\tau)\, d\tau$ acting at time τ to the displacement $du_0(t)$ at time t $(t > \tau)$ depends on the unit-impulse response function evaluated for the time difference $t - \tau$, $h_1(t - \tau)$

$$du_0(t) = h_1(t-\tau)\,\frac{P_0(\tau)}{K}\, d\tau \tag{A-25}$$

The displacement-interaction force relationship (Eq. A-23) is the integral of Eq. A-25 over all pulses from $0 < \tau < t$.

For harmonic loading, the dynamic-flexibility coefficient $F(\omega)$ follows from Eqs. A-17 and A-18 through inversion:

$$u_0(\omega) = F(\omega)\, P_0(\omega) \tag{A-26}$$

with

$$F(\omega) = S^{-1}(\omega) = \frac{K - i\omega C}{K^2 + \omega^2 C^2} \tag{A-27}$$

or using Eqs A-16, A-15 and A-19

$$F(\omega) = \frac{1}{K} \frac{1}{1 + i\dfrac{\omega z_0}{c}} = \frac{1}{K} \frac{1}{1 + i b_0} \tag{A-28}$$

The unit-impulse response function $(1/K)\, h_1(t)$ and the dynamic-flexibility coefficient $F(\omega)$ form a Fourier-transform pair.

$$h_1(t) = \frac{K}{2\pi} \int_{-\infty}^{+\infty} F(\omega)\, e^{i\omega t}\, d\omega \tag{A-29}$$

A1.4 Green's Function

The Green's function in the time domain is defined as the displacement as a function of time $u(t)$ at a point, called the receiver, located at a distance a from the disk, the source (Fig. A-1), which is excited at time = 0 by a unit-impulse force P_0. The motion of the receiver at time t will have originated earlier at the source, at the retarded time $t - a/c$, with c = the appropriate wave velocity. Furthermore, the amplitude will have diminished after traveling the distance a. Thus the Green's function may be denoted as $u(t) = g\,(a, t - a/c)$. The first argument specifies the spatial distance from the source, the second argument the appropriate retarded time at the source.

Applying a unit-impulse force $P_0(\tau) = \delta(\tau)$ (with $\delta(\tau)$ denoting the Dirac-delta function) at time $\tau = 0$,

$$u_0(t) = \frac{1}{K} h_1(t) \tag{A-30}$$

follows from Eq. A-23, as $\int_0^t \delta(\tau)\, d\tau = 1$. Enforcing the boundary condition $u_0 = u\,(z = z_0)$ yields from Eq. A-6

$$f(t) = \frac{1}{K} h_1(t) \tag{A-31}$$

Replacing u by $g(a, t - a/c)$ and with $a = z - z_0$ leads to

$$g\left(a, t - \frac{a}{c}\right) = \frac{1}{K} \frac{1}{1 + \dfrac{a}{z_0}} h_1\left(t - \frac{a}{c}\right) \tag{A-32}$$

or after substituting Eq. A-24 the Green's function equals

$$g\left(a, t-\frac{a}{c}\right) = \frac{1}{K}\frac{1}{1+\dfrac{a}{z_0}}\frac{c}{z_0}e^{-\frac{c}{z_0}\left(t-\frac{a}{c}\right)} \qquad t-\frac{a}{c}\geq 0 \tag{A-33a}$$

$$= 0 \qquad t-\frac{a}{c} < 0 \tag{A-33b}$$

For harmonic loading, the displacement amplitude $u_0(\omega)$ caused by a unit interaction force amplitude $P_0(\omega) = 1$ follows from Eq. A-26 as

$$u_0(\omega) = F(\omega) \tag{A-34}$$

Enforcing the boundary condition $u_0(\omega) = u(z = z_0, \omega)$ in Eq. A-8 leads to

$$c_1 = u_0(\omega) \tag{A-35}$$

Replacing $u(\omega)$ by $g(a, \omega)$ results in

$$g(a, \omega) = F(\omega)\,\frac{z_0}{z}\,e^{-i\frac{\omega a}{c}} \tag{A-36}$$

or, after substituting Eq. A-28, the Green's function equals

$$g(a, \omega) = \frac{1}{K}\frac{1}{1+\dfrac{a}{z_0}}\frac{e^{-i\frac{\omega a}{c}}}{1+i\dfrac{\omega z_0}{c}} \tag{A-37}$$

The factor $e^{-i\omega a/c}$) arises from the time lag a/c. And $g(a, \omega)$ can be rewritten as

$$g(a, \omega) = \frac{1}{K}\frac{1}{1+\dfrac{a}{z_0}}H_1(\omega)\,e^{-i\frac{\omega a}{c}} \tag{A-38}$$

where

$$H_1(\omega) = \frac{1}{1+i\dfrac{\omega z_0}{c}} \tag{A-39}$$

The Fourier transform of $H_1(\omega)$ is $h_1(t)$. Introducing the dynamic-stiffness coefficient $S(\omega)$ specified in Eqs. A-20 and A-21,

$$g(a,\omega) = \frac{1}{S(\omega)}\frac{e^{-i\frac{\omega a}{c}}}{1+\dfrac{a}{z_0}} \tag{A-40}$$

follows.

As $g(a, t - a/c)$ and $g(a, \omega)$ form a Fourier-transform pair, the Green's function in the time domain can also be calculated as

$$g\left(a, t-\frac{a}{c}\right) = \frac{1}{2\pi}\int_{-\infty}^{+\infty} g(a, \omega)\,e^{i\omega t}\,d\omega \tag{A-41}$$

The Green's functions calculated above correspond to the one-sided cone. For the double-cone model used in the dynamic analysis of embedded and pile foundations, the right-hand sides are multiplied by 0.5, as K in Eqs. A-33 and A-37; A-38 is replaced by $2K$.

A2 ROTATIONAL CONE

A2.1 Equation of Motion

The rotational truncated semi-infinite cone is presented for rotational axial distortion in Fig. A-2. This model is used to represent the rocking degree of freedom of a disk on the surface of a halfspace. With $I_0 = (\pi/4)\, r_0^4$ the moment of inertia I at depth z equals $I = (z^4/z_0^4)\, I_0$. Also ϑ denotes the rotation and M the bending moment. Formulating the equilibrium equation of an infinitesimal element

$$-M + M + M,_z\, dz - \rho\, I\, dz\, \ddot{\vartheta} = 0 \tag{A-42}$$

and substituting the moment-rotation relationship

$$M = \rho\, c^2\, I\, \vartheta,_z \tag{A-43}$$

results in the equation of motion in the time domain of the rotational cone

$$\vartheta,_{zz} + \frac{4}{z}\, \vartheta,_z - \frac{\ddot{\vartheta}}{c^2} = 0 \tag{A-44}$$

Introducing a dimensionless potential function φ such that

$$\vartheta = \frac{z_0^2}{z}\, \varphi,_z \tag{A-45}$$

which when substituted in Eq. A-44 leads to

$$\varphi,_{zzz} + \frac{2}{z}\, \varphi,_{zz} - \frac{2}{z^2}\, \varphi,_z - \frac{1}{c^2}\, \ddot{\varphi},_z = 0 \tag{A-46}$$

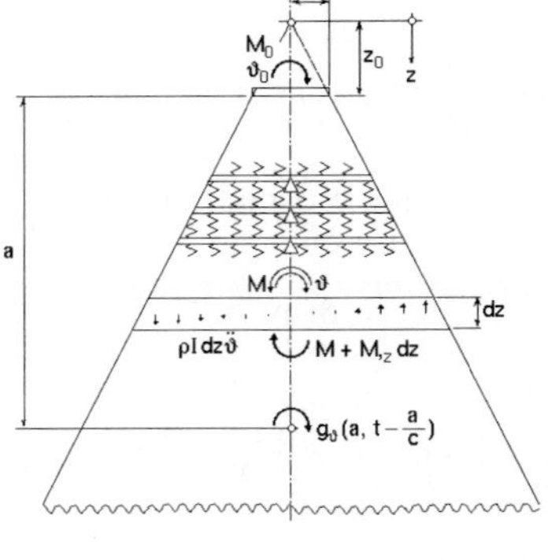

Fig. A-2 Rotational truncated semi-infinite cone with nomenclature for rocking motion, rotational axial-distortion mechanism and equilibrium of infinitesimal element.

or

$$\left[(z\,\varphi),_{zz} - \frac{(z\,\varphi)^{\cdot\cdot}}{c^2}\right],_z = \frac{1}{z}\left[(z\,\varphi),_{zz} - \frac{(z\,\varphi)^{\cdot\cdot}}{c^2}\right] \tag{A-47}$$

The equivalence of the previous two equations may be verified by expanding the derivatives of Eq. A-47. A solution is apparent by inspection

$$(z\,\varphi),_{zz} - \frac{(z\,\varphi)^{\cdot\cdot}}{c^2} = \text{const}\; z \tag{A-48}$$

The left-hand side is recognized as the wave equation in $z\,\varphi$. It turns out to be convenient to specify the constant on the right-hand side to be $3\Theta/z_0^2$. Then the solution of this modified wave equation takes the form

$$z\,\varphi = -z_0\left[\vartheta_1\left(t - \frac{z - z_0}{c}\right) - F\right] + \frac{\Theta\, z^3}{2\, z_0^2} + \Phi\, z \tag{A-49}$$

with additional constants of integration Φ and F. The minus sign preceding the term in brackets is chosen for convenience. As for the translational cone, $\vartheta_1(t - (z - z_0)/c)$ is any arbitrary function of the argument for outward-propagating waves. Inward-propagating waves are thus omitted. The correctness of the solution is verified by insertion of Eq. A-49 into Eq. A-48. Note that F is nothing more than a constant shift and may be incorporated into $\vartheta_1(t - (z - z_0)/c)$. With this simplification the potential function becomes

$$\varphi = -\frac{z_0}{z}\,\vartheta_1\left(t - \frac{z - z_0}{c}\right) + \frac{\Theta}{2}\frac{z^2}{z_0^2} + \Phi \tag{A-50}$$

The angle of rotation ϑ is retrieved via Eq. A-45

$$\vartheta = \frac{z_0^3}{z^3}\,\vartheta_1\left(t - \frac{z - z_0}{c}\right) + \frac{z_0}{c}\frac{z_0^2}{z^2}\,\vartheta_1'\left(t - \frac{z - z_0}{c}\right) + \Theta \tag{A-51}$$

where $\vartheta_1'(t - (z - z_0)/c)$ represents differentiation with respect to the argument $t - (z - z_0)/c$. Without loss of generality the constant shift Θ may be incorporated into ϑ (i.e., set equal to zero), resulting in

$$\vartheta(z,t) = \frac{z_0^3}{z^3}\,\vartheta_1\left(t - \frac{z - z_0}{c}\right) + \frac{z_0}{c}\frac{z_0^2}{z^2}\,\vartheta_1'\left(t - \frac{z - z_0}{c}\right) \tag{A-52}$$

This solution with outward-propagating waves for the rotational cone can be compared to that for the translational cone in Eq. A-6.

For harmonic motion the equation of motion corresponding to Eq. A-44 equals

$$\vartheta(\omega),_{zz} + \frac{4}{z}\,\vartheta(\omega),_z + \frac{\omega^2}{c^2}\,\vartheta(\omega) = 0 \tag{A-53}$$

The solution with outward-propagating waves is written as

$$\vartheta(\omega) = c_1 \left(\frac{z_0^3}{z^3} + i \frac{\omega}{c} \frac{z_0^3}{z^2} \right) e^{-i \frac{\omega}{c} (z - z_0)} \tag{A-54}$$

with the integration constant c_1.

A2.2 Interaction Moment-Rotation Relationship

For a specified rotation ϑ_0 of the disk, the variation of the angle of rotation ϑ with z and the interaction moment M_0 are calculated as follows. Enforcing the boundary condition $\vartheta\ (z = z_0) = \vartheta_0$ yields from Eq. A-52

$$\vartheta_1(t) + \frac{z_0}{c} \dot{\vartheta}_1(t) = \vartheta_0 \tag{A-55}$$

where at $z = z_0 \vartheta_1'(t) = \dot{\vartheta}_1(t)$. The general solution of this first-order differential equation is the convolution

$$\vartheta_1(t) = \int_0^t h_1(t - \tau)\, \vartheta_0(\tau)\, d\tau \tag{A-56}$$

with the unit-impulse response function $h_1(t)$ defined in Eq. A-24. Formulating Eq. A-55 with the argument $t - (z - z_0)/c$ instead of t and thus ϑ_1' instead of $\dot{\vartheta}_1$, which corresponds to a time shift, allows $\vartheta_1'(t - (z - z_0)/c)$ to be eliminated from Eq. A-52, leading to

$$\vartheta(z,t) = \frac{z_0^2}{z^2} \vartheta_0 \left(t - \frac{z - z_0}{c} \right) + \left(\frac{z_0^3}{z^3} - \frac{z_0^2}{z^2} \right) \vartheta_1 \left(t - \frac{z - z_0}{c} \right) \tag{A-57}$$

At the surface ($z = z_0$) the contribution of the second term vanishes and $\vartheta(z = z_0) = \vartheta_0$. In the underground region, however, part of the rotation is due to the convolution ϑ_1; and the amplitude decays in inverse proportion to a combination of the square and the cube of depth z.

Two limiting cases are of particular interest. For static loading (ϑ_0 = constant), $\vartheta_1 = \vartheta_0$ (Eq. A-55), which yields from Eq. A-57 the static limit

$$\vartheta(z) = \frac{z_0^3}{z^3} \vartheta_0 \tag{A-58}$$

On the other hand, for high-frequency harmonic motion the convolution ϑ_1 tends to zero (as is verified from Eq. A-55) and Eq. A-57 reverts to the high-frequency limit

$$\vartheta\ (z,t) = \frac{z_0^2}{z^2} \vartheta_0 \left(t - \frac{z - z_0}{c} \right) \tag{A-59}$$

Turning to M_0, the derivative $\vartheta,_z$ is calculated via the chain rule from Eq. A-52

$$\vartheta,_z = -\frac{3}{z_0} \frac{z_0^4}{z^4} \vartheta_1 \left(t - \frac{z - z_0}{c} \right) - \frac{3}{c} \frac{z_0^3}{z^3} \vartheta_1' \left(t - \frac{z - z_0}{c} \right) - \frac{z_0}{c^2} \frac{z_0^2}{z^2} \vartheta_1'' \left(t - \frac{z - z_0}{c} \right) \tag{A-60}$$

At $z = z_0$, with $\vartheta_{0,z} = \vartheta_{,z}(z = z_0)$

$$\vartheta_{0,z} = -\frac{3}{z_0}\vartheta_1(t) - \frac{3}{c}\dot{\vartheta}_1(t) - \frac{z_0}{c^2}\ddot{\vartheta}_1(t) \tag{A-61}$$

applies. Using Eq. A-55, the first two terms on the right-hand side are equal to $-3/z_0\, \vartheta_0$. To eliminate $\ddot{\vartheta}_1$, Eq. A-55 is differentiated with respect to time

$$\dot{\vartheta}_1(t) + \frac{z_0}{c}\ddot{\vartheta}_1(t) = \dot{\vartheta}_0 \tag{A-62}$$

This first-order differential equation in $\dot{\vartheta}_1$ can be solved as

$$\dot{\vartheta}_1(t) = \int_0^t h_1(t-\tau)\, \dot{\vartheta}_0(\tau)\, d\tau \tag{A-63}$$

Substituting for $\ddot{\vartheta}_1$ from Eqs. A-62 in Eq. A-61 and then eliminating $\dot{\vartheta}_1$ using Eq. A-63 leads to

$$\vartheta_{0,z} = -\frac{3}{z_0}\vartheta_0 - \frac{1}{c}\dot{\vartheta}_0 + \frac{1}{c}\int_0^t h_1(t-\tau)\, \dot{\vartheta}_0(\tau)\, d\tau \tag{A-64}$$

Finally, with

$$M_0 = -M(z = z_0) = -\rho c^2 I_0\, \vartheta_{0,z} \tag{A-65}$$

the interaction moment-rotation relationship is formulated as

$$M_0(t) = \frac{3\rho c^2 I_0}{z_0}\vartheta_0(t) + \rho c I_0\, \dot{\vartheta}_0(t) - \rho c I_0 \int_0^t h_1(t-\tau)\, \dot{\vartheta}_0(\tau)\, d\tau \tag{A-66}$$

or

$$M_0(t) = K_\vartheta\, \vartheta_0(t) + C_\vartheta\, \dot{\vartheta}_0(t) - \int_0^t h_1(t-\tau)\, C_\vartheta\, \dot{\vartheta}_0(\tau)\, d\tau \tag{A-67}$$

K_ϑ and C_ϑ are the constant coefficients of the rotational spring and dashpot

$$K_\vartheta = \frac{3\rho c^2 I_0}{z_0} \tag{A-68a}$$

$$C_\vartheta = \rho c I_0 \tag{A-68b}$$

The convolution which is preceded by a negative sign with h_1 specified in Eq. A-24 involves the rotational velocity $\dot{\vartheta}_0$, which leads to the rotational static-stiffness coefficient K_ϑ appearing in the interaction moment-rotation relationship. If the convolution is performed with the rotation ϑ_0,

$$M_0(t) = \frac{2}{3}\, K_\vartheta\, \vartheta_0(t) + C_\vartheta\, \dot{\vartheta}_0(t) + \int_0^t h_1(t-\tau)\, \frac{K_\vartheta}{3}\vartheta_0(\tau)\, d\tau \tag{A-69}$$

is derived. This form is not attractive because of the factor 2/3 with which the static-stiffness coefficient is multiplied to form the spring coefficient.

For harmonic loading, enforcing the boundary condition $\vartheta_0(\omega) = \vartheta(z = z_0, \omega)$ yields the integration constant c_1 in Eq. A-54. The rotational amplitude with depth z follows as

$$\vartheta(z, \omega) = \vartheta_0(\omega) \frac{\dfrac{z_0^3}{z^3} + i\dfrac{\omega}{c}\dfrac{z_0^3}{z^2}}{1 + i\dfrac{\omega}{c} z_0} e^{-i\frac{\omega}{c}(z-z_0)} \tag{A-70}$$

The equation

$$M_0(\omega) = -\rho\, c^2 I_0\, \vartheta_0(\omega)_{,z} \tag{A-71}$$

leads with b_0 defined in Eq. A-19 to

$$M_0(\omega) = S_\vartheta(b_0)\, \vartheta_0(\omega) \tag{A-72}$$

where the dynamic-stiffness coefficient

$$S_\vartheta(b_0) = K_\vartheta\, \left[k_\vartheta(b_0) + i\, b_0\, c_\vartheta(a_0)\right] \tag{A-73}$$

with

$$k_\vartheta(b_0) = 1 - \frac{1}{3}\frac{b_0^2}{1 + b_0^2} \tag{A-74a}$$

$$c_\vartheta(b_0) = \frac{1}{3}\frac{b_0^2}{1 + b_0^2} \tag{A-74b}$$

A2.3 Rotation-Interaction Moment Relationship

For a prescribed interaction moment M_0 acting on the disk, the rotation ϑ_0 is calculated as follows. Enforcing the boundary condition (Eq. A-65) and substituting Eq. A-61 leads to

$$\frac{z_0^2}{3\, c^2}\ddot{\vartheta}_1(t) + \frac{z_0}{c}\dot{\vartheta}_1(t) + \vartheta_1(t) = \frac{M_0}{K_\vartheta} \tag{A-75}$$

To derive the unit-impulse response function $h_2(t)$ for the rotation ϑ_0, the loading M_0/K_ϑ is equal to the Dirac-delta function $\delta(t)$.

$$\frac{z_0^2}{3\, c^2}\ddot{\vartheta}_1(t) + \frac{z_0}{c}\dot{\vartheta}_1(t) + \vartheta_1(t) = \delta(t) \tag{A-76}$$

The solution of this second-order differential equation can be derived by noting that the latter has the same form as the dynamic equation of motion of a one-degree-of-freedom system with viscous damping.

$$m\, \ddot{h}(t) + c\, \dot{h}(t) + k\, h(t) = \delta(t) \tag{A-77}$$

The unit-impulse response function $h(t)$ is equal to the damped vibration for an initial velocity $\delta(t)dt/m = 1/m$

$$h(t) = \frac{1}{\omega_d\, m} e^{-\omega_0 \zeta t} \sin \omega_d t \tag{A-78}$$

with the natural frequency ω_0, the damping ratio ζ, and the damped frequency ω_d

$$\omega_0 = \sqrt{\frac{k}{m}} \tag{A-79a}$$

$$\zeta = \frac{c}{2\sqrt{km}} \tag{A-79b}$$

$$\omega_d = \omega_0 \sqrt{1-\zeta^2} \tag{A-79c}$$

From Eq. A-78, $\vartheta_1(t)$ follows. Comparing Eq. A-77 to Eq. A-76,

$$\vartheta_1(t) = 2\sqrt{3}\frac{c}{z_0} e^{-\frac{3}{2}\frac{c}{z_0}t} \sin\frac{\sqrt{3}}{2}\frac{c}{z_0}t \tag{A-80}$$

Substituting in Eq. A-52 for $z = z_0$, the unit-impulse response function $h_2(t)$ which is equal to $\vartheta_0(t)$ is written as

$$h_2(t) = \frac{c}{z_0} e^{-\frac{3}{2}\frac{c}{z_0}t}\left(3\cos\frac{\sqrt{3}}{2}\frac{c}{z_0}t - \sqrt{3}\sin\frac{\sqrt{3}}{2}\frac{c}{z_0}t\right) \quad t \geq 0 \tag{A-81a}$$

$$= 0 \quad t < 0 \tag{A-81b}$$

The rotation-interaction moment relationship is specified as the convolution integral (Eq. A-75)

$$\vartheta_0(t) = \int_0^t h_2(t-\tau)\frac{M_0(\tau)}{K_\vartheta} d\tau \tag{A-82}$$

For harmonic loading, the dynamic-flexibility coefficient $F_\vartheta(\omega)$ follows from Eq. A-72 through inversion

$$\vartheta_0(\omega) = F_\vartheta(\omega) M_0(\omega) \tag{A-83}$$

where (Eqs. A-73, A-74)

$$F_\vartheta(\omega) = S_\vartheta^{-1}(\omega) = \frac{1}{K_\vartheta}\frac{3(1+i b_0)}{3+3i b_0 + (i b_0)^2} \tag{A-84}$$

In Eq. A-19 b_0 is defined. As $(1/K_\vartheta)\, h_2(t)$ and $F_\vartheta(\omega)$ form a Fourier-transform pair, $h_2(t)$ can also be determined as

$$h_2(t) = \frac{K_\vartheta}{2\pi}\int_{-\infty}^{+\infty} F_\vartheta(\omega)\, e^{i\omega t} d\omega \tag{A-85}$$

A2.4 Green's Function

Exciting the disk, the source, at time $t = 0$ by a unit-impulse moment M_0, the rotation as a function of time $\vartheta(t)$ at the receiver, located at a distance a from the source (Fig. A-2), is equal to the Green's function $g_\vartheta\,(a, t - a/c)$. As for the translational cone (Section A1.4), the Green's function is a function of a which governs the attenuation of the wave and of the retarded time $t - a/c$.

Applying a unit-impulse moment $M_0(\tau) = \delta(\tau)$, the rotation of the disk ϑ_0 determined from Eq. A-82 equals

$$\vartheta_0(t) = \frac{1}{K_\vartheta} h_2(t) \qquad \text{(A-86)}$$

The equation of $\vartheta(z, t)$ specified in Eq. A-52 leads after enforcing the boundary condition $\vartheta_0 = \vartheta(z = z_0)$ to the first-order differential equation in $\vartheta_1(t)$ (Eq. A-55) with the solution given in Eq. A-56. Substituting Eq. A-86 into this equation yields

$$\vartheta_1(t) = \frac{1}{K_\vartheta} \int_0^t h_1(t-\tau)\, h_2(\tau)\, d\tau \qquad \text{(A-87)}$$

Another unit-impulse response function, $h_3(t)$, is defined as

$$h_3(t) = \int_0^t h_1(t-\tau)\, h_2(\tau)\, d\tau \qquad \text{(A-88)}$$

with, after substituting $h_1(t)$ from Eq. A-24 and $h_2(t)$ from Eq. A-81,

$$h_3(t) = 2\sqrt{3}\frac{c}{z_0} e^{-\frac{3}{2}\frac{c}{z_0}t} \sin\frac{\sqrt{3}}{2}\frac{c}{z_0}t \qquad t \geq 0 \qquad \text{(A-89a)}$$

$$= 0 \qquad t < 0 \qquad \text{(A-89b)}$$

Substituting Eq. A-87 with Eq. A-88 and Eq. A-86 in Eq. A-57 and replacing ϑ with g_ϑ $(a, t - a/c)$ and with $a = z - z_0$ leads to the Green's function

$$g_\vartheta\left(a, t-\frac{a}{c}\right) = \frac{1}{K_\vartheta}\frac{1}{\left(1+\frac{a}{z_0}\right)^2}\left[h_2\left(t-\frac{a}{c}\right) + \left(-1+\frac{1}{1+\frac{a}{z_0}}\right)h_3\left(t-\frac{a}{c}\right)\right] \qquad \text{(A-90)}$$

Both $h_2(t - a/c)$ and $h_3(t - a/c)$ are specified in Eqs. A-81 and A-89 with t replaced by $t - a/c$.

For harmonic loading, the rotation amplitude $\vartheta_0(\omega)$ caused by a unit interaction moment amplitude $M_0(\omega) = 1$ follows from Eq. A-83 as

$$\vartheta_0(\omega) = F_\vartheta(\omega) \qquad \text{(A-91)}$$

Enforcing this boundary condition $\vartheta_0(\omega) = \vartheta(z = z_0, \omega)$ in Eq. A-54 yields

$$c_1 = \frac{1}{1 + i\frac{\omega z_0}{c}} \vartheta_0(\omega) \qquad \text{(A-92)}$$

Replacing $\vartheta(\omega)$ by $g_\vartheta(a,\omega)$ results in

$$g_\vartheta(a, \omega) = F_\vartheta(\omega) \frac{1}{1 + i\frac{\omega z_0}{c}} \left(\frac{z_0^3}{z^3} + i\frac{\omega}{c}\frac{z_0^3}{z^2}\right) e^{-i\frac{\omega a}{c}} \qquad \text{(A-93)}$$

or after substituting Eq. A-84 the Green's function equals

$$g_\vartheta(a, \omega) = \frac{3}{K_\vartheta}\left(\frac{1}{\left(1+\frac{a}{z_0}\right)^3} + i\frac{\omega z_0}{c}\frac{1}{\left(1+\frac{a}{z_0}\right)^2}\right)\frac{1}{3+3\,i\frac{\omega z_0}{c}+\left(i\frac{\omega z_0}{c}\right)^2}\,e^{-i\frac{\omega a}{c}} \quad \text{(A-94)}$$

Here $e^{-i\omega a/c}$ represents the time-lag factor. Thus, $g_\vartheta(a,\omega)$ can be rewritten as

$$g_\vartheta(a,\omega) = \frac{1}{K_\vartheta}\frac{1}{\left(1+\frac{a}{z_0}\right)^2}\left[H_2(\omega) + \left(-1+\frac{1}{1+\frac{a}{z_0}}\right)H_3(\omega)\right]e^{-i\frac{\omega a}{c}} \quad \text{(A-95)}$$

where

$$H_2(\omega) = \frac{3\left(1+i\frac{\omega z_0}{c}\right)}{3+3\,i\frac{\omega z_0}{c}+\left(i\frac{\omega z_0}{c}\right)^2} \quad \text{(A-96)}$$

$$H_3(\omega) = \frac{3}{3+3\,i\frac{\omega z_0}{c}+\left(i\frac{\omega z_0}{c}\right)^2} \quad \text{(A-97)}$$

The Fourier transforms of $H_2(\omega)$ and $H_3(\omega)$ are $h_2(t)$ and $h_3(t)$, respectively. Introducing the dynamic-stiffness coefficient $S_\vartheta(\omega)$ specified in Eqs. A-73 and A-74 results in

$$g_\vartheta(a,\omega) = \frac{e^{-i\frac{\omega a}{c}}}{S_\vartheta(\omega)}\left[\frac{1}{\left(1+\frac{a}{z_0}\right)^2} + \left\{\frac{1}{\left(1+\frac{a}{z_0}\right)^3} - \frac{1}{\left(1+\frac{a}{z_0}\right)^2}\right\}\frac{1}{1+i\frac{\omega z_0}{c}}\right] \quad \text{(A-98)}$$

Again, as $g(a, t - a/c)$ and $g_\vartheta(a,\omega)$ form a Fourier-transform pair, the Green's function in the time domain can also be determined as

$$g_\vartheta\left(a, t-\frac{a}{c}\right) = \frac{1}{2\pi}\int_{-\infty}^{+\infty} g_\vartheta(a,\omega)\, e^{i\omega t}\, d\omega \quad \text{(A-99)}$$

The Green's functions specified above correspond to the one-sided cone. For the double-cone model used in the dynamic analysis of embedded and pile foundations, the right-hand sides are multiplied by 0.5, as K_ϑ in Eqs. A-90 and A-94, A-95, A-98 is replaced by $2\,K_\vartheta$.

A3 GENERALIZED WAVE PATTERN OF UNFOLDED LAYERED CONE

As discussed in Section 3.1, a generalized wave pattern $w(z, t)$ may be postulated for an unfolded layered cone, in which the amplitude of motion decays in inverse proportion to the Nth power of the distance from the apex:

$$w(z,t) = \overbrace{\frac{z_0^N}{(z_0+z)^N}\bar{w}_0\left(t-\frac{z}{c}\right)}^{\text{incident wave}}$$

$$+\sum_{j=1}^{k}(-1)^j\left[\overbrace{\frac{z_0^N\bar{w}_0\left(t-\frac{2jd}{c}+\frac{z}{c}\right)}{(z_0+2jd-z)^N}}^{\text{upwave from bedrock}}+\overbrace{\frac{z_0^N\bar{w}_0\left(t-\frac{2jd}{c}-\frac{z}{c}\right)}{(z_0+2jd+z)^N}}^{\text{downwave from surface}}\right] \tag{A-100}$$

The barred time history $\bar{w}_0(t)$ generates the motion. For the unfolded layered cone in rotation, there are three such $\bar{w}_0$ functions. The first corresponds to $\bar{\vartheta}_0$ with $N = 2$; the second corresponds to $\bar{\vartheta}_1$ with $N = 3$; and the third corresponds to $-\bar{\vartheta}_1$, with $N = 2$. Since they superpose independently, attention may be restricted to any one of these three separate generating functions. To verify Eq. A-100, it suffices to consider the first few terms of the sum:

$$w = \frac{z_0^N}{(z_0+z)^N}\bar{w}_0\left(t-\frac{z}{c}\right) - \frac{z_0^N\bar{w}_0\left(t-\frac{2d}{c}+\frac{z}{c}\right)}{(z_0+2d-z)^N} - \frac{z_0^N\bar{w}_0\left(t-\frac{2d}{c}-\frac{z}{c}\right)}{(z_0+2d+z)^N}$$
$$+\frac{z_0^N\bar{w}_0\left(t-\frac{4d}{c}+\frac{z}{c}\right)}{(z_0+4d-z)^N} + \frac{z_0^N\bar{w}_0\left(t-\frac{4d}{c}-\frac{z}{c}\right)}{(z_0+4d+z)^N} - \dots \tag{A-101}$$

The displacement w_d at bedrock ($z = d$) is

$$w_d = \frac{z_0^N}{(z_0+d)^N}\bar{w}_0\left(t-\frac{d}{c}\right) - \frac{z_0^N}{(z_0+d)^N}\bar{w}_0\left(t-\frac{d}{c}\right)$$
$$-\frac{z_0^N}{(z_0+3d)^N}\bar{w}_0\left(t-\frac{3d}{c}\right) + \frac{z_0^N}{(z_0+3d)^N}\bar{w}_0\left(t-\frac{3d}{c}\right) + \dots \tag{A-102}$$

Since the terms cancel pair-for-pair, the displacement w_d at bedrock is zero; that is, the layer is, as it should be, built-in at its base.

The strain is equal to the spatial derivative of w. Application of the chain rule to Eq. A-101 yields

$$w_{,z} = -\frac{z_0^N}{(z_0+z)^N}\left[\frac{N}{z_0+z}\bar{w}_0\left(t-\frac{z}{c}\right)+\frac{1}{c}\bar{w}_0'\left(t-\frac{z}{c}\right)\right]$$
$$-\frac{z_0^N}{(z_0+2d-z)^N}\left[\frac{N}{z_0+2d-z}\bar{w}_0\left(t-\frac{2d}{c}+\frac{z}{c}\right)+\frac{1}{c}\bar{w}_0'\left(t-\frac{2d}{c}+\frac{z}{c}\right)\right]$$
$$+\frac{z_0^N}{(z_0+2d+z)^N}\left[\frac{N}{z_0+2d+z}\bar{w}_0\left(t-\frac{2d}{c}-\frac{z}{c}\right)+\frac{1}{c}\bar{w}_0'\left(t-\frac{2d}{c}-\frac{z}{c}\right)\right]-\dots \tag{A-103}$$

In this expression $\bar{w}_0'$ denotes differentiation with respect to the complete argument of the generating function, that is $\bar{w}_0' = \partial\bar{w}_0(\,)/\partial(\,)$ with for example $(\,) = t - 2d/c + z/c$. The strain at the surface ($z = 0$) is

$$
\begin{aligned}
w_{0,z} = & -\frac{N}{z_0}\,\bar{w}_0(t) - \frac{1}{c}\,\dot{\bar{w}}_0(t) \\
& - \frac{z_0{}^N}{(z_0+2d)^N}\left[\frac{N}{z_0+2d}\,\bar{w}_0\left(t-\frac{2d}{c}\right) + \frac{1}{c}\,\dot{\bar{w}}_0\left(t-\frac{2d}{c}\right)\right] \\
& + \frac{z_0{}^N}{(z_0+2d)^N}\left[\frac{N}{z_0+2d}\,\bar{w}_0\left(t-\frac{2d}{c}\right) + \frac{1}{c}\,\dot{\bar{w}}_0\left(t-\frac{2d}{c}\right)\right] - \ldots
\end{aligned}
\tag{A-104}
$$

in which the $\bar{w}_0'$ derivatives revert to the velocity $\dot{\bar{w}}_0$. Due to the free-surface boundary condition implicit in Eq. A-104, the strain contributions resulting from the reflected waves cancel pair-for-pair. The strain at the surface (first two terms in Eq. A-104) is identical to that of an unlayered cone subjected to the generating motion $\bar{w}_0$ and $\dot{\bar{w}}_0$.

The relationship between the generating motion $\bar{w}_0$ of the associated unlayered cone and the true surface motion w_0 of the unfolded layered cone is found by evaluating Eq. A-100 at the surface ($z = 0$):

$$
w_0(t) = \bar{w}_0(t) + 2\sum_{j=1}^{k}(-1)^j\,\frac{\bar{w}_0(t-jT)}{(1+j\kappa)^N}
\tag{A-105}
$$

with echo interval $T = 2d/c$ and geometric parameter $\kappa = 2d/z_0$. Introducing echo constants e_j^F allows Eq. A-105 to be expressed concisely as

$$
w_0(t) = \sum_{j=0}^{k} e_j^F\,\bar{w}_0(t-jT)
\tag{A-106}
$$

with

$$
e_0^F = 1
\tag{A-107a}
$$

and for $j \geq 1$

$$
e_j^F = \frac{2(-1)^j}{(1+j\kappa)^N}
\tag{A-107b}
$$

For the rotational cone the various contributions associated with $N = 2$ and $N = 3$ are simply superposed to determine $\vartheta_0(t)$.

B

Consistent Lumped-Parameter Model

A lumped-parameter model to represent the unbounded (linear) soil in a soil-structure-interaction analysis consists of one or several springs, dashpots, and masses with frequency-independent real coefficients. In such a discrete physical model besides the foundation node additional nodes with degrees of freedom can be introduced. The finite-element discretization of the bounded structure and the lumped-parameter model of the unbounded soil will both lead to stiffness, damping, and mass matrices of finite dimensions, which can be assembled straightforwardly to describe the behavior of the total dynamic system.

The discrete-element representations of the translational and rotational cones are the building blocks of the lumped-parameter model addressed in this text. The basic lumped-parameter model is constructed by placing three discrete-element models of the rotational cone in parallel. The unknown coefficients of the springs, dashpots, and the mass are determined using curve-fitting applied to the dynamic stiffness.

In Appendix B a generalization is described leading to a systematic procedure to construct consistent lumped-parameter models. This method is treated in Wolf [W9], where more aspects are addressed. For the sake of simplicity, only the scalar case is examined in the following. For the matrix case, Wolf [W10] should also be consulted.

The procedure, which leads to physical insight, exhibits many advantages: The only approximation (replacing the rigorous dynamic stiffness by a ratio of two polynomials) can be evaluated visibly at the beginning of the method, independent of any choice of the lumped-parameter model. No nonlinear system of equations has to be solved, in contrast to other procedures. No unfamiliar discrete-time manipulations such as the z-transformation are used. A whole family of lumped-parameter models guaranteeing real coefficients for the springs, dashpots, and masses can be constructed by selecting different degrees of the polynomials. The stiffness, damping, and mass matrices corresponding to the lumped-parameter model are automatically symmetrical. The lumped-parameter model is exact for the static case and for the asymptotic value of infinite frequency. As the lumped-parameter model can be interpreted as a so-called realization in linear system theory, a strong mathematical foundation addressing stability exists.

Section B1 addresses the mathematical preliminaries consisting of approximating the dynamic stiffness in the frequency domain using the least squares method as a ratio of two polynomials, which is then expressed as a partial-fraction expansion. The various discrete-element models representing the terms in this partial-fraction expansion are derived in Section B2. As an example the semi-infinite rod on an elastic foundation, the simplest dispersive system with a cutoff frequency, is represented by this consistent lumped-parameter model in Section B3.

B1 PARTIAL-FRACTION EXPANSION OF DYNAMIC STIFFNESS FOR HARMONIC LOADING

B1.1 Dynamic-Stiffness Coefficient

The dynamic-stiffness coefficient of the unbounded soil for harmonic loading $S(a_0)$, relating the displacement amplitude $u_0(a_0)$ to the interaction force amplitude $P_0(a_0)$, is written as

$$S(a_0) = K\left[k(a_0) + ia_0\, c(a_0)\right] \tag{B-1}$$

with the static-stiffness coefficient K and the dimensionless frequency $a_0 = \omega r_0/c_s$ (r_0 = characteristic length of the basemat (foundation), c_s = shear-wave velocity).

$S(a_0)$ is decomposed into the singular part $S_s(a_0)$, which is equal to its asymptotic value for $a_0 \rightarrow \infty$, and the remaining regular part $S_r(a_0)$

$$S(a_0) = S_s(a_0) + S_r(a_0) \tag{B-2}$$

where

$$\frac{S_s(a_0)}{K} = k + i\,a_0\, c \tag{B-3}$$

with the constants k and c, which are equal to the nondimensional coefficients of the spring and dashpot. For a plane basemat-soil interface, k vanishes. $Kia_0c = i\omega Kr_0/c_s c$ is equal to $i\omega A_0\rho c_s$ for the translational degree of freedom tangential to the basemat-soil interface ("horizontal"), from which c is determined (A_0 = area of foundation, ρ = mass density). For the translational degree of freedom perpendicular to the basemat-soil interface ("vertical") the dilatational-wave velocity c_p replaces c_s in the last expression. The incompressible case is disregarded. For the rotational degrees of freedom, the moment of inertia I_0 appears instead of A_0.

B1.2 Polynomial-Fraction Approximation

The regular part $S_r(a_0)$ is approximated as a ratio of two polynomials P and Q in ia_0

$$\frac{S_r(a_0)}{K} \simeq \frac{S_r(ia_0)}{K} = \frac{P(ia_0)}{KQ(ia_0)} = \frac{1 - k + p_1\, i a_0 + p_2\,(i a_0)^2 + \ldots + p_{M-1}\,(ia_0)^{M-1}}{1 + q_1 i a_0 + q_2(ia_0)^2 + \ldots + q_M\,(ia_0)^M} \tag{B-4}$$

The total approximated dynamic-stiffness coefficient is exact for the static limit $a_0 \rightarrow 0$ and the high-frequency limit $a_0 \rightarrow \infty$ (doubly-asymptotic approximation). The degree of the polynomial in the numerator is selected as one less that in the denominator M.

The $2M$ - 1 unknown real coefficients p_i, q_i are determined using a curve-fitting technique with a complex S_r based on the least squares method. Denoting the exact (complex) value of the dynamic-stiffness coefficient discretized in the j-th data point a_{0j} as $S_r(a_{0j})$, the square of the following Euclidian norm $\in$ is made a minimum.

$$\in^2 = \left\| w(a_{0j})\left(S_r(a_{0j})\, Q(ia_{0j}) - P(ia_{0j})\right)\right\|^2 \tag{B-5}$$

where

$$\begin{aligned}\in^2 &= \sum_{j=1}^{J} w(a_{0j})|S_r(a_{0j})\, Q(ia_{0j}) - P(ia_{0j})|^2 \\ &= \sum_{j=1}^{J} w(a_{0j})\left(S_r(a_{0j})\, Q(ia_{0j}) - P(ia_{0j})\right) \left(S_r^*(a_{0j}) Q^*(ia_{0j}) - P^*(ia_{0j})\right)\end{aligned} \tag{B-6}$$

The weights are denoted as $w(a_{0j})$. J is the number of data points (frequencies) used in the curve-fitting process. An asterisk as superscript denotes the complex conjugate value. $\in^2$ is a real quadratic form in p_i, q_i. For a minimum of $\in^2$,

$$\frac{\partial \in^2}{\partial p_i} = 0 \qquad i = 1, 2, \ldots M{-}1 \tag{B-7a}$$

$$\frac{\partial \in^2}{\partial q_i} = 0 \qquad i = 1, 2, \ldots M \tag{B-7b}$$

applies, which leads to a system of $2M$ - 1 linear symmetric equations with real coefficients for the $2M$ - 1 real unknowns p_i, q_i.

No other approximation is performed in the total procedure to construct lumped-parameter models.

B1.3 Partial-Fraction Expansion

The ratio of the polynomials in Eq. B-4 can be expressed as a partial-fraction expansion of the form

$$\frac{S_r(ia_0)}{K} = \sum_{\ell=1}^{M} \frac{A_\ell}{ia_0 - s_\ell} \tag{B-8}$$

where s_ℓ are the roots of $Q(ia_0)$; the poles of $S_r(ia_0)$, and the A_ℓ are the residues at the poles

$$A_\ell = (ia_0 - s_\ell)\left.\frac{S_r(ia_0)}{K}\right|_{ia_0 = s_\ell} \tag{B-9}$$

If some of the s_ℓ are complex, they will appear in complex conjugate pairs $s_{\ell,\ell+1} = s_{1\ell} \pm is_{2\ell}$, whereby the corresponding A_ℓ are also complex conjugate pairs $A_{\ell,\ell+1} = A_{1\ell} \pm iA_{2\ell}$. By adding two corresponding first-order terms, a second-order term with real coefficients results. For L conjugate pairs, Eq. B-8 can be written as

$$\frac{S_r(ia_0)}{K} = \sum_{\ell=1}^{L} \frac{\beta_{1\ell}\, ia_0 + \beta_{0\ell}}{(ia_0)^2 + \alpha_{1\ell}\, ia_0 + \alpha_{0\ell}} + \sum_{\ell=1}^{M-2L} \frac{A_\ell}{ia_0 - s_\ell} \tag{B-10}$$

with

$$\alpha_{0\ell} = s_{1\ell}^2 + s_{2\ell}^2 \quad \text{(B-11a)} \qquad \alpha_{1\ell} = -2s_{1\ell} \quad \text{(B-11b)}$$

$$\beta_{0\ell} = -2\left(A_{1\ell}\, s_{1\ell} + A_{2\ell}\, s_{2\ell}\right) \quad \text{(B-11c)} \qquad \beta_{1\ell} = 2A_{1\ell} \quad \text{(B-11d)}$$

All coefficients are real.

For the sake of simplicity, only first-order poles are assumed. Multiple-order poles can, however, also be processed. The two terms corresponding to second-order poles s_ℓ, $A_\ell/(ia_0 - s_\ell)$, and $A_{\ell+1}/(ia_0 - s_\ell)^2$ can be added to form a second-order term. It is of course also possible to add two first-order terms with real poles to form a second-order term.

Summarizing, the total (approximated) dynamic-stiffness coefficient can be written in the so-called parallel form as

$$\frac{S(ia_0)}{K} = k + ia_0\, c + \sum_{\ell=1}^{L} \frac{\beta_{1\ell}\, ia_0 + \beta_{0\ell}}{(ia_0)^2 + \alpha_{1\ell}\, ia_0 + \alpha_{0\ell}} + \sum_{\ell=1}^{M-2L} \frac{A_\ell}{ia_0 - s_\ell} \tag{B-12}$$

B1.4 Stability Criterion and Rate of Energy Transmission

When the real parts of all roots s_ℓ of the polynomial $Q(ia_0)$ are negative, stability of the lumped-parameter model representing the unbounded soil (for a prescribed displacement u_0) is guaranteed. This is easily verified by addressing the Fourier transformation of Eq. B-8 which is equal to

$$\frac{S_r(\bar{t})}{K} = \frac{c_s}{r_0} \sum_{\ell=1}^{M} A_\ell e^{s_\ell \bar{t}} \tag{B-13}$$

with the dimensionless time $\bar{t} = tc_s/r_0$. The exponential function will decay for a real part of s_ℓ which is negative.

This criterion is, however, not sufficient for the stability of the total dynamic system including the structure supported by the soil, as encountered in a typical soil-structure-interaction analysis. To achieve this, the rate of energy transmission (power) of the unbounded domain's model (the soil) through the (foundation) basemat-soil interface must be nonnegative—no energy can be created within this model. This rate of energy transmission $N(\omega)$ as the product of the real parts of force and velocity amplitudes, averaged over a period $2\pi/\omega$, equals

$$N(\omega) = \frac{\omega}{2\pi} \int_0^{2\pi/\omega} Re\left(P_0(\omega)e^{i\omega t}\right) Re\left(\dot{u}_0(\omega)e^{i\omega t}\right) dt \tag{B-14}$$

Substituting the interaction force amplitude $P_0(\omega)$ expressed as the product of the dynamic-stiffness coefficient $S(\omega)$ (Eq. B-1) and the displacement amplitude $u_0(\omega)$ results after straightforward algebra in

$$N(\omega) = \frac{\omega}{2} u_0^*(\omega)\, Im\big(S(\omega)\big)\, u_0(\omega) \tag{B-15}$$

For $N(\omega) \geq 0$, $Im(S(\omega)) = a_0 K c(a_0)$—the damping coefficient $c(a_0)$ of the total dynamic stiffness of the soil must be nonnegative for all frequencies.

To achieve negative s_ℓ and $c(a_0) \geq 0$, experience shows that the weights $w(a_{0j})$ in the least squares method of curve-fitting (Eq. B-5) in the lower-frequency range (especially below the cutoff frequency, if present) have to be chosen much larger than for the high frequencies, typically for selected points by a factor as high as 10^3 - 10^5.

B2 DISCRETE-ELEMENT MODELS FOR PARTIAL-FRACTION EXPANSION

B2.1 Overview

Each term of the partial-fraction expansion of the total dynamic-stiffness coefficient (Eq. B-12) is represented independently from the others by a discrete-element model. For the first- and second-order terms the discrete-element models are not unique. The discrete-element models with some internal degrees of freedom will consist of frequency-independent springs and dashpots only, which is attractive to the engineer for a dynamic model of the supporting soil of a structure. The corresponding equations of motion are described by first-order differential equations. It is customary in structural dynamics to work with masses, in addition to springs and dashpots, and thus with second-order differential equations. Such discrete-element models for the second-order term are also addressed, whereby the number of internal degrees of freedom is halved compared to that of the model using only springs and dashpots. Only a selection of possible discrete-element models is presented in the following.

The discrete-element models of all terms of the partial-fraction expansion are arranged in parallel, with the coupling occurring only at the foundation node enforcing the compatibility of the displacements and the summation of the interaction forces. This results in the lumped-parameter model. The discrete-element models thus form the building blocks of the dynamic-stiffness coefficient's lumped-parameter model.

B2.2 Constant and Linear Terms

The constant and linear terms $k + ia_0c$ (multiplied by K) with the two known coefficients k, c of the singular part of the dynamic-stiffness coefficient (Eq. B-12) lead to a spring with a coefficient κK and in parallel to a dashpot with a coefficient $\gamma(r_0/c_s)K$ referred to time with a dimension (Fig. B-1). K denotes the static-stiffness coefficient, and r_0 and c_s the characteristic length and the shear-wave velocity. The κ and γ represent unknown dimensionless coefficients which are determined by matching the force-displacement relationship specified by the singular term of Eq. B-12 with that determined by the discrete model shown in

Fig. B-1 for harmonic loading. Applying a displacement with amplitude $u_0(a_0)$ at the foundation node 0 leads to the interaction force amplitude $P_0(a_0)$

$$\frac{P_0(a_0)}{K} = \left(\kappa + i a_0\, \gamma\right) u_0(a_0) \tag{B-16}$$

which yields

$$\kappa = k \quad \text{(B-17a)} \qquad\qquad \gamma = c \quad \text{(B-17b)}$$

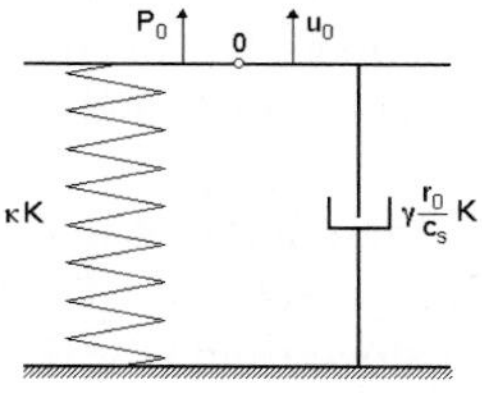

Fig. B-1 Discrete-element model for constant and linear terms (singular term).

B2.3 First-Order Term

The discrete-element model corresponding to a typical first-order term with two known coefficients s, A, $A/(ia_0 - s)$ (multiplied by K), of the partial-fraction expansion in Eq. B-12 is shown in Fig. B-2a. Besides a spring with a coefficient $-\kappa K$, an element in parallel is present which consists of a spring with a coefficient κK and a dashpot with a coefficient $\gamma(r_0/c_s)K$ in series. One internal degree of freedom u_1 in node 1 is thus introduced. The two coefficients κ and γ are determined as follows, applying the loading case $u_0(a_0)$. Formulating equilibrium in node 1 for harmonic loading results in

$$\kappa K\left[u_1(a_0) - u_0(a_0)\right] + i a_0\, \gamma\, K u_1(a_0) = 0 \tag{B-18}$$

The interaction force-displacement relationship in the foundation node 0 is specified as

$$P_0(a_0) = \kappa K\left[u_0(a_0) - u_1(a_0)\right] - \kappa K u_0(a_0) \tag{B-19}$$

Determining $u_1(a_0)$ from Eq. B-18 and substituting in Eq. B-19 leads to

$$\frac{P_0(a_0)}{K} = \frac{-\dfrac{\kappa^2}{\gamma}}{i a_0 + \dfrac{\kappa}{\gamma}}\, u_0(a_0) \tag{B-20}$$

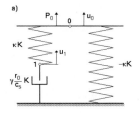

a. Spring-dashpot system.

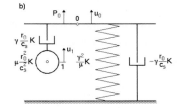

b. Dashpot-mass system (monkey-tail model).

Fig. B-2 Discrete-element model for first-order term.

Equating the (dimensionless) dynamic-stiffness coefficient in Eq. B-20 to the first-order term results in

$$\kappa = \frac{A}{s} \qquad \text{(B-21a)} \qquad\qquad \gamma = -\frac{A}{s^2} \qquad \text{(B-21b)}$$

As A and s are real, κ and γ will also be real, although not necessarily positive. As s is negative for a stable system and A turns out to be positive in many cases (as for the discrete-element model of the rotational cone), the spring and the dashpot connected to the internal degree of freedom will both have negative coefficients. It is of course no problem to handle them mathematically, although springs and dashpots with negative coefficients do not exist in reality.

To avoid this conceptual difficulty, the mass-dashpot interconnection as a sort of "monkey tail" in parallel to a spring and dashpot with the coefficients indicated in Fig. B-2b can be used as an alternative. Proceeding analogously in the derivation

$$\gamma = \frac{A}{s^2} \qquad \text{(B-22a)} \qquad\qquad \mu = -\frac{A}{s^3} \qquad \text{(B-22b)}$$

follows. For a negative s and a positive A, γ and μ are positive. This discrete-element system is not easy to comprehend physically; intuitively one might fear that the mass would eventually fall off, which of course does not happen.

B2.4 Second-Order Term

The discrete-element model corresponding to a typical second-order term with four known coefficients α_0, α_1, β_0, β_1, $(\beta_1 ia_0 + \beta_0)/((ia_0)^2 + \alpha_1 ia_0 + \alpha_0)$ (multiplied by K), of the partial-fraction expansion in Eq. B-12 is shown in Fig. B-3a. An element which consists of a spring with a coefficient $\kappa_2 K$ and in parallel of a dashpot with a coefficient $\gamma_2(r_0/c_s)K$ is present in series with the discrete-element model of the first-order term with a spring coefficient $\kappa_1 K$ and a dashpot coefficient $\gamma_1(r_0/c_s)K$. Two nodes with internal degrees of freedom u_1 and u_2 are introduced. Four coefficients κ_1, κ_2, γ_1, γ_2 have to be determined. For the harmonic loading case of a prescribed $u_0(a_0)$ equilibrium is formulated in nodes 1 and 2. Solving for $u_1(a_0)$ and $u_2(a_0)$ and substituting in the force-displacement relationship in node 0 results in

$$\frac{P_0(a_0)}{K} = \frac{-\kappa_1^2\dfrac{\gamma_1+\gamma_2}{\gamma_1\gamma_2}\, ia_0 - \dfrac{\kappa_1^2\kappa_2}{\gamma_1\gamma_2}}{(ia_0)^2 + \left(\kappa_1\dfrac{\gamma_1+\gamma_2}{\gamma_1\gamma_2} + \dfrac{\kappa_2}{\gamma_2}\right) ia_0 + \dfrac{\kappa_1\kappa_2}{\gamma_1\gamma_2}}\, u_0(a_0) \qquad \text{(B-23)}$$

Equating the constant and linear terms in the numerator and denominator to those of the second-order term leads to

$$\kappa_1 = -\frac{\beta_0}{\alpha_0} \qquad \text{(B-24a)} \qquad\qquad \kappa_2 = \frac{\beta_0}{\alpha_0^2}\,\frac{(-\alpha_0\beta_1+\alpha_1\beta_0)^2}{\alpha_0\beta_1^2 - \alpha_1\beta_0\beta_1 + \beta_0^2} \qquad \text{(B-24b)}$$

$$\gamma_1 = \frac{\alpha_0\beta_1-\alpha_1\beta_0}{\alpha_0^2} \qquad \text{(B-24c)} \qquad\qquad \gamma_2 = \frac{\beta_0^2}{\alpha_0^2}\,\frac{-\alpha_0\beta_1+\alpha_1\beta_0}{\alpha_0\beta_1^2 - \alpha_1\beta_0\beta_1 + \beta_0^2} \qquad \text{(B-24d)}$$

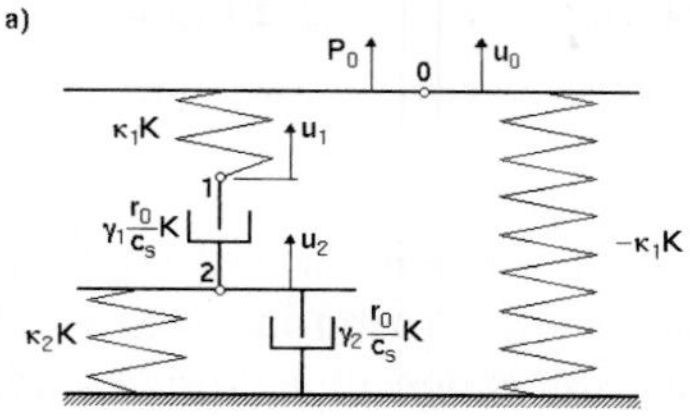

a. Spring-dashpot system with two internal unknowns.

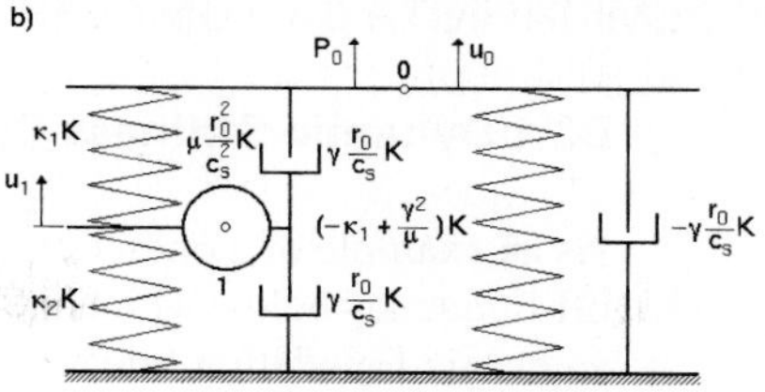

b. Spring-dashpot-mass system with one internal unknown.

Fig. B-3 Discrete-element model for second-order term.

All coefficients κ_1, ... γ_2 are real.

As an alternative, the discrete-element model of Fig. B-3b, corresponding to a typical second-order term of Eq. B-12, consists of springs, dampers, and a mass with only one internal degree of freedom u_1. The node 1 with a mass with a coefficient $\mu(r_0^2/c_s^2)K$ is connected to the foundation node 0 through a spring with a coefficient $\kappa_1 K$ and in parallel a dashpot with a coefficient $\gamma(r_0/c_s)K$. Node 1 is supported by a spring with a coefficient $\kappa_2 K$ and by the same dashpot with coefficient $\gamma(r_0/c_s)K$. Proceeding as before, the following interaction force-displacement relationship is derived for the discrete-element model.

$$\frac{P_0(a_0)}{K} = \frac{2\left(-\frac{\kappa_1\gamma}{\mu} + \frac{\gamma^3}{\mu^2}\right) i a_0 - \frac{\kappa_1^2}{\mu} + \frac{(\kappa_1+\kappa_2)\gamma^2}{\mu^2}}{(ia_0)^2 + 2\frac{\gamma}{\mu} i a_0 + \frac{\kappa_1+\kappa_2}{\mu}} u_0(a_0) \tag{B-25}$$

Equating the coefficients in Eq. B-25 to those of the second-order term leads to a quadratic equation for μ

$$a\mu^2 + b\mu + c = 0 \quad \text{(B-26a)} \qquad \text{with} \quad a = \alpha_1^4 - 4\alpha_0\alpha_1^2 \quad \text{(B-26b)}$$

$$b = -8\alpha_1\beta_1 + 16\beta_0 \quad \text{(B-26c)} \qquad c = 16\frac{\beta_1^2}{\alpha_1^2} \quad \text{(B-26d)}$$

Two solutions will result. For $b^2 - 4ac \geq 0$, μ will be real:

$$\alpha_0\beta_1^2 - \alpha_1\beta_0\beta_1 + \beta_0^2 \geq 0 \tag{B-27}$$

The other coefficients are calculated as

$$\kappa_1 = \frac{\mu\alpha_1^2}{4} - \frac{\beta_1}{\alpha_1} \quad \text{(B-28a)} \qquad \kappa_2 = \mu\alpha_0 - \kappa_1 \quad \text{(B-28b)}$$

$$\gamma = \frac{\mu}{2}\alpha_1 \quad \text{(B-28c)}$$

For a real μ, the other coefficients κ_1, κ_2, γ will also be real.

Finally, it is worth remembering that also two terms with real poles of the partial-fraction expansion (and not only those with complex conjugate poles) can be added to generate a second-order term. The potential for halving the number of internal degrees of freedom is thus present.

B3 SEMI-INFINITE BAR ON ELASTIC FOUNDATION

B3.1 Dynamic-Stiffness Coefficient

As an example of a one-degree-of-freedom system with dispersive waves exhibiting a cut-off frequency below which the radiation damping vanishes, the semi-infinite rod resting on an elastic foundation (area A, modulus of elasticity E, mass density ρ, spring stiffness per unit length k_g, Fig. B-4a) is addressed. This is a very stringent test for a lumped-parameter model, for which the system of dashpots does not dissipate energy in a finite frequency range (from zero to the cutoff frequency).

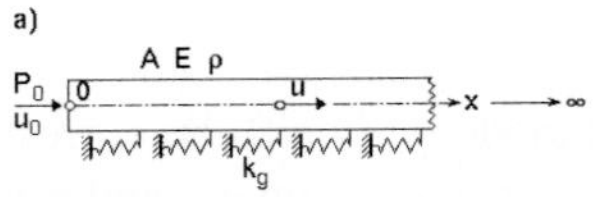

a. Geometry with nomenclature.

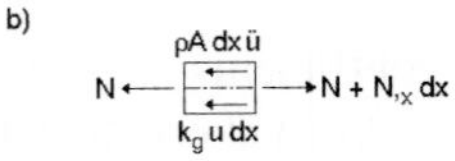

b. Equilibrium of infinitesimal element.

Fig. B-4 Semi-infinite rod resting on elastic foundation.

The dynamic-stiffness coefficient $S(a_0)$, relating $u_0(a_0)$ to $P_0(a_0)$, is derived as follows. N represents the normal force and u the axial displacement. Formulating equilibrium (Fig. B-4b)

$$-N + N + N_{,x}\, dx - k_g\, u\, dx - A\, \rho\, \ddot{u}\, dx = 0 \tag{B-29}$$

and substituting the force-displacement relationship

$$N = E\, A\, u_{,x} \tag{B-30}$$

leads to the equation of motion

$$u_{,xx} - \frac{1}{r_0^2} u - \frac{\ddot{u}}{c_l^2} = 0 \tag{B-31}$$

with the characteristic length

$$r_0 = \sqrt{\frac{E\,A}{k_g}} \tag{B-32}$$

and the rod velocity

$$c_l = \sqrt{\frac{E}{\rho}} \tag{B-33}$$

For harmonic loading with the dimensionless frequency

$$a_0 = \frac{\omega\, r_0}{c_l} \tag{B-34}$$

Eq. B-31 is formulated as

$$r_0^2 u(a_0)_{,xx} - u(a_0) + a_0^2\, u(a_0) = 0 \tag{B-35}$$

The solution is given by

$$u(a_0) = c_1\, e^{-i\frac{\sqrt{a_0^2-1}}{r_0}x} + c_2\, e^{+i\frac{\sqrt{a_0^2-1}}{r_0}x} \tag{B-36}$$

When determining the dynamic-stiffness coefficient at $x = 0$ in the semi-infinite system, only outwardly propagating waves exist. Setting $c_2 = 0$ and enforcing $u(x = 0, a_0) = u_0(a_0)$ leads to

$$u(a_0) = u_0(a_0)\, e^{-i\frac{\sqrt{a_0^2-1}}{r_0}x} \tag{B-37}$$

Substituting Eq. B-37 in

$$P_0(a_0) = -EA_0\, u_0(a_0)_{,x} \tag{B-38}$$

leads to the interaction force-displacement relationship

$$P_0(a_0) = i\sqrt{EAk_g}\,\sqrt{a_0^2-1}\, u_0(a_0) \tag{B-39}$$

With the static-stiffness coefficient

$$K = \sqrt{EAk_g} \tag{B-40}$$

the dynamic-stiffness coefficient is formulated as

$$S(a_0) = K\sqrt{1-a_0^2} \tag{B-41}$$

Below the cutoff frequency $a_0 = 1$, the imaginary part of $S(a_0)$ representing radiation damping vanishes.

As specified in Eq. B-2, $S(a_0)$ is decomposed into the singular part

$$\frac{S_s(a_0)}{K} = ia_0 \tag{B-42}$$

and the remaining regular part

$$\frac{S_r(a_0)}{K} = \sqrt{1-a_0^2} - ia_0 \tag{B-43}$$

B3.2 Lumped-Parameter Model with Two Internal Degrees of Freedom

First, $S_r(a_0)$ is approximated as a ratio of a second- to third-degree polynomial in ia_0 ($M = 3$) in Eq. B-4. The curve-fitting is performed for $0 < a_0 < 3.5$. The approximate total dynamic-stiffness coefficient $S_s(a_0) + S_r(ia_0)$ is compared with the exact value of Eq. B-41 in Fig. B-5. As in Eq. B-1, $k(a_0)$ and $c(a_0)$ are defined. Considering that M is only equal to 3, the agreement is surprisingly good, even outside the range used in the curve-fitting procedure. For all a_0, $c(a_0)$ is positive.

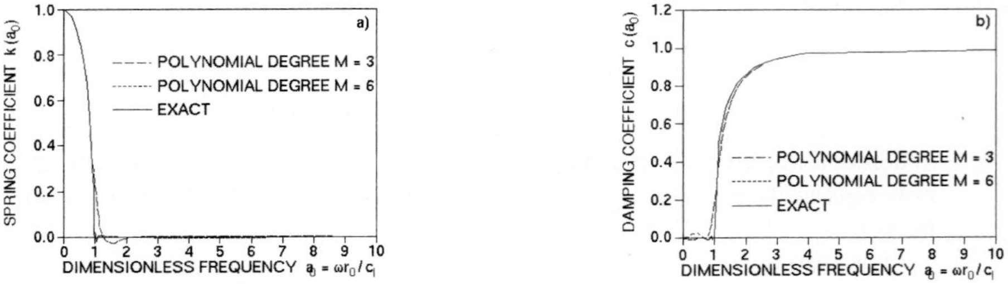

Fig. B-5 Dynamic-stiffness coefficient for harmonic loading based on lumped-parameter model (degree of polynomial in denominator =3 and =6).

After the partial-fraction expansion (Eq. B-8),

$$\frac{S_r(i a_0)}{K} = \frac{-0.022 + 0.225i}{ia_0 - (-0.393 - 0.929i)} + \frac{-0.022 - 0.225i}{ia_0 - (-0.393 + 0.929i)} + \frac{0.564}{ia_0 - (-0.930)} \tag{B-44}$$

holds. Note that the real parts of all roots are negative. Combining the first two terms with complex conjugate pairs leads to (Eq. B-10)

$$\frac{S_r(i a_0)}{K} = \frac{-0.045ia_0 + 0.428}{(ia_0)^2 + 0.786i a_0 + 1.017} + \frac{0.564}{ia_0 - (-0.930)} \tag{B-45}$$

The last term in Eq. B-45 corresponds to the first-order discrete-element model shown in Fig. B-2a. The corresponding coefficients follow from Eq. B-21 as

$$\kappa = -0.606 \quad \text{(B-46a)} \qquad \gamma = -0.651 \quad \text{(B-46b)}$$

The coefficients of the spring and dashpot adjacent to node 1 are indeed negative. The first term on the right-hand side of Eq. B-45 leads to the second-order discrete-element model of Fig. B-3b. The coefficients specified in Eqs. B-26 and B-28 are equal to (with the positive sign in front of the square root in the equation for μ)

$$\mu = -0.008 \quad \text{(B-47a)} \qquad \kappa_1 = 0.056 \quad \text{(B-47b)}$$

$$\kappa_2 = -0.064 \quad \text{(B-47c)} \qquad \gamma = -0.003 \quad \text{(B-47d)}$$

The singular term $S_s(a_0)$ results in (Eq. B-17)

$$\kappa = 0 \quad \text{(B-48a)} \qquad \gamma = 1 \quad \text{(B-48b)}$$

for the discrete-element model of Fig. B-1.

By arranging the discrete-element models of the singular term, of the one first-order term and the one second-order term in parallel, the lumped-parameter model with two internal degrees of freedom shown in Fig. B-6a is constructed. The springs and dashpots connecting the foundation node directly to the rigid support are combined. The eight coefficients are equal to (specifying 6 significant digits)

$$K_1 = -0.605785K \quad \text{(B-49a)} \qquad K_2 = 0.0559794K \quad \text{(B-49b)}$$

$$K_3 = -0.0639044K \quad \text{(B-49c)} \qquad K_4 = 0.548601K \quad \text{(B-49d)}$$

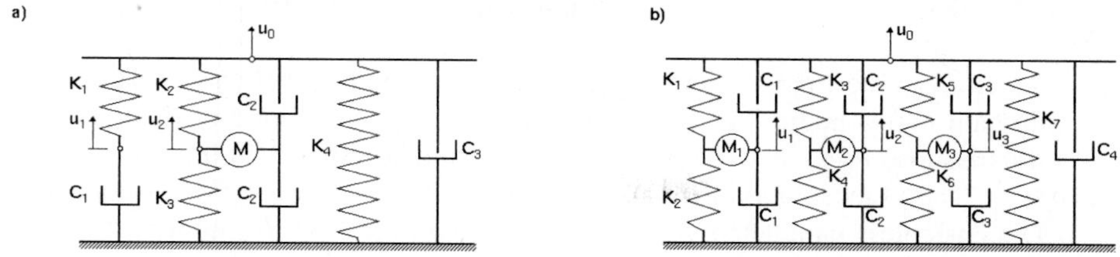

a. Degree of polynomial in denominator = 3.

b. Degree of polynomial in denominator = 6.

Fig. B-6 Lumped-parameter model.

$$C_1 = -0.651197\frac{r_0}{c_l}K \qquad \text{(B-49e)}$$

$$C_2 = -0.00306417\frac{r_0}{c_l}K \qquad \text{(B-49f)}$$

$$C_3 = 1.003064\frac{r_0}{c_l}K \qquad \text{(B-49g)}$$

$$M = -0.00779299\frac{r_0^2}{c_l^2}K \qquad \text{(B-49h)}$$

B3.3 Lumped-Parameter Model with Three Internal Degrees of Freedom

Second, $S_r(a_0)$ is approximated as a ratio of a fifth- to sixth-degree polynomial in ia_0 (M = 6) in Eq. B-4. The curve-fitting is again performed for $0 < a_0 < 3.5$. As expected, the agreement of the approximate total dynamic-stiffness coefficient $S_s(a_0) + S_r(ia_0)$ with the exact value of Eq. B-41 is excellent (Fig. B-5).

The roots of the polynomial in the denominator appear as three complex conjugate pairs. The partial-fraction expansion of $S_r(ia_0)$ thus consists of the sum of three second-order terms (Eq. B-10 with L = 3) The coefficients of each second-order term determine those of the corresponding discrete-element model of Fig. B-3b. Assembling these three discrete-element models and that of the singular part (Fig. B-1 with Eq. B-48) in parallel leads to the lumped-parameter model of Fig. B-6b with three internal degrees of freedom. The 14 coefficients of the springs, dashpots, and masses are equal to

$$K_1 = +0.188961K \qquad \text{(B-50a)}$$

$$K_2 = -2.60797K \qquad \text{(B-50b)}$$

$$K_3 = +0.137339K \qquad \text{(B-50c)}$$

$$K_4 = -0.266934K \qquad \text{(B-50d)}$$

$$K_5 = -0.520779K \qquad \text{(B-50e)}$$

$$K_6 = +0.258331K \qquad \text{(B-50f)}$$

$$K_7 = +0.000782341K \qquad \text{(B-50g)}$$

$$C_1 = -0.107464\frac{r_0}{c_l}K \qquad \text{(B-50h)}$$

$$C_2 = -0.0327325\frac{r_0}{c_l}K \qquad \text{(B-50i)}$$

$$C_3 = -0.214319\frac{r_0}{c_l}K \qquad \text{(B-50j)}$$

$$C_4 = +1.35452\frac{r_0}{c_l}K \qquad \text{(B-50k)}$$

$$M_1 = -2.35702\frac{r_0^2}{c_l^2}K \qquad \text{(B-50l)}$$

$$M_2 = -0.123794\frac{r_0^2}{c_l^2}K \qquad \text{(B-50m)}$$

$$M_3 = -0.254981\frac{r_0^2}{c_l^2}K \qquad \text{(B-50n)}$$

B3.4 Dynamic-Flexibility Coefficient in Time Domain

As an example involving all frequencies, the dynamic-stiffness coefficient in the time domain $F(t)$ is addressed: the displacement $u_0(t)$ as a function of time caused by a unit-impulse load P_0 equal to the Dirac-delta function $\delta(t)$ applied at time zero.

The exact solution $F(t)$ can be determined as the inverse Fourier transform of $F(\omega)$. Dimensionless parameters for frequency a_0 and time $\bar{t} = tc_s/r_0$ are used in this derivation. $F(a_0)$ is the inverse of $S(a_0)$. Using Eq. B-41

$$F(a_0) = \frac{1}{K}\frac{1}{\sqrt{1-a_0^2}} \tag{B-51}$$

results. $F(\omega)$ and $F(t)$ form a Fourier-transform pair

$$F(t) = \frac{1}{2\pi}\int_{-\infty}^{+\infty} F(\omega)e^{i\omega t}\,d\omega \tag{B-52a}$$

or in nondimensional parameters

$$F(\bar{t}) = \frac{c_l}{2\pi\, r_0}\int_{-\infty}^{+\infty} F(a_0)e^{ia_0\bar{t}}\,da_0 \tag{B-52b}$$

Substituting Eq. B-51 in Eq. B-52b and using Fourier integral tables yields

$$F(\bar{t}) = \frac{c_l}{r_0 K}J_0(\bar{t}) \tag{B-53}$$

$J_0(\bar{t})$ is the Bessel function of the zero-th order. This exact solution is plotted in nondimensional form in Fig. B-7.

The calculations using the lumped-parameter models for $M = 3$ and 6 shown in Figs. B-6a and B-6b are performed based on an explicit algorithm with a predictor-corrector scheme. The time step $\Delta t = 0.1\, r_0/c_\ell$ is selected. The load P_0 is assumed to act only over the first time step. While the numerical results shown in Fig. B-7 for $M = 3$ indicate that the corresponding lumped-parameter model exhibits a tendency to be too strongly damped, the accuracy of the lumped-parameter model with $M = 6$ is excellent: Its numerical solution coincides within drawing accuracy with the exact values. This lumped-parameter model thus passes this challenging test with an impulse loading exhibiting all frequencies acting on a system with dispersive waves and a cutoff frequency extremely well.

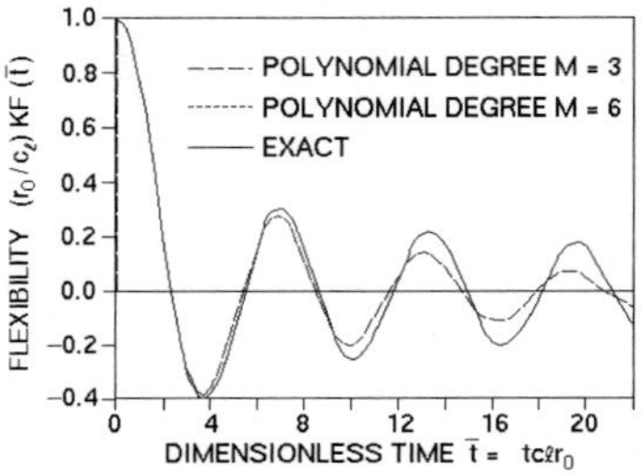

Fig. B-7 Dynamic-flexibility coefficient in time domain.

C

Recursive Evaluation of Convolution Integral

C1 OVERVIEW

A typical convolution integral relating the input time history $x(\tau)$ $0 \le \tau \le t$ to the output $y(t)$ at time t is written as

$$y(t) = \int_0^t h(t-\tau)\, x(\tau)\, d\tau \tag{C-1}$$

where $h(t)$ is the unit-impulse response function. In a flexibility or Green's function formulation (Duhamel's integral), the $x(\tau)$ corresponds to the force and $y(t)$ to the displacement. In a stiffness formulation involving a convolution of the velocity, the $x(\tau)$ is this velocity and $y(t)$ represents (the regular part of) the interaction force.

The evaluation of the convolution integral, which has to be performed for each time station, from its definition (Eq. C-1) is inefficient. The total number of operations will be proportional to the square of the total number of time stations. In addition, the input time history from $t = 0$ onwards has to be stored during the computation. The calculation may be expediated by the recursive procedure. In this recursive evaluation of the convolution integral in the time domain the output y_n at time $t = n\Delta t$ is computed from the input x_n and the I and J past most recent values of the output and the input

$$y_n = \sum_{i=1}^{I} a_i\, y_{n-i} + \sum_{i=0}^{J} b_i\, x_{n-i} \tag{C-2}$$

with the recursion coefficients a_i and b_i to be determined. Thus, this procedure does not correspond to a truncation of the convolution integral that only retains recent past input values. The recursive evaluation is actually highly accurate. Assuming that the input varies piecewise linearly between consecutive time stations, it is exact for the computation of the convolutions occurring with the translational and rotational cones. It will be shown that the recursive evaluation is also exact if the (regular part of the) dynamic-stiffness coefficient for harmonic loading can be formulated as a ratio of two polynomials in $i\omega$ as presented in Eq. B-4. This is the case for a discrete system consisting of a finite number of springs, dashpots, and masses. It can also be achieved for any dynamic-stiffness coefficient of an unbounded continuous system by using a curve-fitting technique based on the least squares method (Section B1.2). In particular, the consistent lumped-parameter models presented for a foundation on the surface of or embedded in a soil layer resting on rigid rock could be replaced equivalently for the computation of the interaction forces by a corresponding recursive evaluation.

Many procedures exist to determine the recursive coefficients a_i, b_i in Eq. C-2 [W6]. This can be performed either in the time domain [W7] or in the frequency domain [W8, M3], whereby unfamiliar discrete-time manipulations such as the z-transformation are used in the derivation. In the following, only a simple procedure using elementary mathematics working in the frequency domain is discussed. Its starting point is the same as that for the construction of the consistent lumped-parameter model described in Appendix B. This recursive evaluation can thus also be interpreted as a so-called realization in linear system theory.

The input-output relationship in the time domain (Eq. C-1) is formulated in the frequency domain as

$$y(\omega) = H(\omega)\, x(\omega) \tag{C-3}$$

The amplitudes of the input and output are $x(\omega)$ and $y(\omega)$. The transfer function $H(\omega)$ and the unit-impulse response function $h(t)$ form a Fourier transform pair

$$h(t) = \frac{1}{2\pi} \int_{-\infty}^{+\infty} H(\omega) e^{i\omega t}\, d\omega \tag{C-4}$$

$H(\omega)$ is established directly for harmonic loading. In the general case, $H(\omega)$ (nondimensionalized appropriately) is approximated as a ratio $H(ia_0)$ of two polynomials in ia_0 ($a_0 = \omega r_0/c_s$, with r_0 = characteristic length of foundation (basemat), c_s = shear-wave velocity)

$$H(a_0) \simeq H(ia_0) = \frac{p_0 + p_1\, ia_0 + p_2(ia_0)^2 + \ldots + p_{M-1}(ia_0)^{M-1}}{1 + q_1 ia_0 + q_2(ia_0)^2 + \ldots + q_M(ia_0)^M} \tag{C-5}$$

The degree of the polynomial in the numerator is selected as one less than in the denominator M. The real coefficients p_i, q_i are determined using a curve-fitting technique based on the least squares method, as described in Section B1.2. A linear system of symmetric equations has to be solved. The transfer function $H(ia_0)$ either represents the regular part of the dynamic-stiffness coefficient $S_r(ia_0)$ (Eq. B-4) or the dynamic-flexibility coefficient $F(ia_0)$, which is equal to the inverse of the total dynamic-stiffness coefficient $S(ia_0)$. (In the case of the regular part of the dynamic-stiffness coefficient enforcing the total static-stiffness coefficient determines p_0, see Eq. B-4.) Using Eqs. B-2, B-3, and B-4, it is easily verified that the degree of the polynomial in the numerator of $F(ia_0)$ is one less than in the denominator $M + 1$. The form is thus the same as specified in Eq. C-5, with the degree of the polynomials increased by one.

The ratio of the polynomials in Eq. C-5 is expressed as a partial-fraction expansion of the form

$$H(ia_0) = \sum_{\ell=1}^{M} \frac{A_\ell}{ia_0 - s_\ell} \tag{C-6}$$

The roots of the denominator are s_ℓ; the poles of $H(ia_0)$, and the A_ℓ are the residues at the poles

$$A_\ell = (ia_0 - s_\ell)H(ia_0)\Big|_{ia_0 = s_\ell} \tag{C-7}$$

If some of the s_ℓ are complex, they will appear in complex conjugate pairs $s_{\ell,\,\ell+1} = s_{1\ell} \pm is_{2\ell}$, and the A_ℓ will also be complex conjugate pairs $A_{\ell,\,\ell+1} = A_{1\ell} \pm \mathrm{i}A_{2\ell}$. By adding such two first-order terms, a second-order term with real coefficients results. For L conjugate pairs, Eq. C-6 is rewritten as

$$H(ia_0) = \sum_{\ell=1}^{L} \frac{\beta_{1\ell} ia_0 + \beta_{0\ell}}{(ia_0)^2 + \alpha_{1\ell} ia_0 + \alpha_{0\ell}} + \sum_{\ell=1}^{M-2L} \frac{A_\ell}{ia_0 - s_\ell} \tag{C-8}$$

with

$$\alpha_{0\ell} = s_{1\ell}^2 + s_{2\ell}^2 \quad \text{(C-9a)} \qquad \alpha_{1\ell} = -2s_{1\ell} \quad \text{(C-9b)}$$

$$\beta_{0\ell} = -2(A_{1\ell}s_{1\ell} + A_{2\ell}s_{2\ell}) \quad \text{(C-9c)} \qquad \beta_{1\ell} = 2A_{1\ell} \quad \text{(C-9d)}$$

All coefficients in Eq. C-8 are real.

In this so-called parallel form each term of the partial-fraction expansion is processed independently, leading to a recursive equation. It will be demonstrated in Section C2 that a typical first-order term in the frequency domain leads to the recursive equation in the time domain

$$y_n^\ell = a^\ell\, y_{n-1}^\ell + b_0^\ell x_n + b_1^\ell\, x_{n-1}, \tag{C-10}$$

and in Section C3 that a second-order term results in

$$y_n^\ell = a_1^\ell\, y_{n-1}^\ell + a_2^\ell\, y_{n-2}^\ell + b_0^\ell\, x_n + b_1^\ell\, x_{n-1} + b_2^\ell\, x_{n-2} \tag{C-11}$$

Note that the input x does not depend on ℓ. The final result is then equal to the sum of the contributions of all terms—the sum of the $M - 2Ly_n^\ell$ terms in Eq. C-10 is added to the sum of the Ly_n^ℓ terms in Eq. C-11, which leads to the convolution integral y_n.

If the input varies piecewise linearly between the time stations, the recursive equations specified in Eqs. C-10 and C-11 are exact.

The number of operations is proportional to the total number of time stations. The time history of the input from 0 to $n - 3$ is not stored.

The recursive algorithm is self-starting from quiescent initial conditions at time $t = 0$. The quiet past implies that in the evaluation of Eq. C-11 for $n = 1$, the samples y_0^ℓ, y_{-1}^ℓ, x_0, and x_{-1} are all zero, so that $y_1^\ell = b_0^\ell x_1$.

The stability criterion and the condition for a nonnegative rate of energy transmission discussed for the lumped-parameter model in Section B1.4 remain valid for the recursive evaluation.

C2 FIRST-ORDER TERM

Dropping the index ℓ for the sake of conciseness, the transfer function in the frequency domain of a typical first-order term in Eq. C-8 with two known (real) coefficients s, A is written as

$$H(ia_0) = \frac{A}{ia_0 - s} \tag{C-12}$$

with the corresponding input-output relationship

$$y(\omega) = \frac{\frac{c_s}{r_0} A}{i\omega - \frac{c_s}{r_0} s} x(\omega) \tag{C-13}$$

In the time domain, the ordinary differential equation of first order in time

$$\dot{y}(t) - \frac{c_s}{r_0} s\, y(t) = \frac{c_s}{r_0} A x(t) \tag{C-14}$$

results. For $x(t) = \delta(t)$, the Dirac-delta function, the unit-impulse response function in the time domain equals

$$h(t) = \frac{c_s}{r_0} A\, e^{\frac{c_s}{r_0} s t} \tag{C-15}$$

Equation C-15 can also be derived by substituting Eq. C-12 in Eq. C-4.

The solution of Eq. C-14 at time station $n\Delta t$, y_n depends on its value at time station $(n - 1)\Delta t$, y_{n-1}, and on the variation of the input $x(\tau)$ from $(n - 1)\Delta t$ to $n\Delta t$. For a piecewise linear variation of $x(\tau)$ over the n-th time step, with $(n - 1)\Delta t \leq \tau \leq n\Delta t$ and $x_n = x(n\Delta t)$

$$x(\tau) = \left(n - \frac{\tau}{\Delta t}\right) x_{n-1} + \left(1 - n + \frac{\tau}{\Delta t}\right) x_n \tag{C-16}$$

Thus, y_n must be able to be expressed as

$$y_n = a\, y_{n-1} + b_0\, x_n + b_1\, x_{n-1} \tag{C-17}$$

The recursion coefficients a, b_0, b_1 are constants to be determined, which depend on Δt.

To determine a, no input is assumed to act—the free vibration characterized by $h(t)$ is examined. Using Eq. C-15

$$h_n = h(n\Delta t) = \frac{c_s}{r_0} A\, e^{\frac{c_s}{r_0} s n \Delta t} = e^{\frac{c_s}{r_0} s \Delta t} \frac{c_s}{r_0} A\, e^{\frac{c_s}{r_0} s (n-1) \Delta t} = e^{\frac{c_s}{r_0} s \Delta t} h_{n-1} \tag{C-18}$$

follows, which results, when comparing with Eq. C-17, in

$$a = e^{\frac{c_s}{r_0} s \Delta t} \qquad \text{(C-19)}$$

Thus a depends only on s (and not on A).

To determine b_0, b_1, the convolution integral (Eq. C-1) is formulated with vanishing input over all time steps up to and including the $(n-1)$-th

$$y_n = \int_{(n-1)\Delta t}^{n\Delta t} h(n\Delta t - \tau)\, x(\tau)\, d\tau \qquad \text{(C-20)}$$

Substituting Eqs. C-15 and C-16 in Eq. C-20 leads after comparison with Eq. C-17 with $y_{n-1} = 0$ for this loading case to

$$b_0 = \frac{A}{s}\left(\frac{e^{\frac{c_s}{r_0} s\Delta t} - 1}{\frac{c_s}{r_0} s\Delta t} - 1\right) \qquad \text{(C-21a)}$$

$$b_1 = \frac{A}{s}\left(\frac{-e^{\frac{c_s}{r_0} s\Delta t} + 1}{\frac{c_s}{r_0} s\Delta t} + e^{\frac{c_s}{r_0} s\Delta t}\right) \qquad \text{(C-21b)}$$

As a special case, the convolution integral involving the unit-impulse response function $h_1(t)$ (Eq. A-24) is addressed

$$h_1(t) = \frac{c}{z_0} e^{-\frac{c}{z_0} t} \qquad t \geq 0 \qquad \text{(C-22a)}$$

$$= 0 \qquad t < 0 \qquad \text{(C-22b)}$$

Note that $h_1(t)$ appears in the displacement-interaction force relationship (Eq. A-23) and in the Green's function (Eq. A-32) of the translational cone and in the interaction moment-rotation relationship (Eq. A-67) of the rotational cone. The corresponding transfer function (Eq. A-39) which is indeed of first order, is equal to

$$H_1(\omega) = \frac{1}{i\frac{\omega z_0}{c} + 1} \qquad \text{(C-23)}$$

with the dimensionless frequency parameter defined with respect to the properties of the cone. With $r_0 = z_0$, $c_s = c$, the root s equals -1 and the residue A is equal to 1. The corresponding recursive coefficients follow from Eqs. C-19 and C-20 as

$$a = e^{-\frac{c\Delta t}{z_0}} \qquad \text{(C-24a)}$$

$$b_0 = 1 + \frac{e^{-\frac{c\,\Delta t}{z_0}} - 1}{\frac{c\Delta t}{z_0}} \tag{C-24b}$$

$$b_1 = \frac{-e^{-\frac{c\,\Delta t}{z_0}} + 1}{\frac{c\Delta t}{z_0}} - e^{-\frac{c\,\Delta t}{z_0}} \tag{C-24c}$$

C3 SECOND-ORDER TERM

The transfer function in the frequency domain of a typical second-order term in Eq. C-8 equals

$$H(ia_0) = \frac{A_1 + iA_2}{ia_0 - (s_1 + is_2)} + \frac{A_1 - iA_2}{ia_0 - (s_1 - is_2)} = \frac{\beta_1\, ia_0 + \beta_0}{(ia_0)^2 + \alpha_1 ia_0 + \alpha_0} \tag{C-25}$$

with four real coefficients α_0, α_1, β_0, β_1 which are expressed as a function of four other real coefficients s_1, s_2, A_1, A_2 in Eq. C-9.

Substituting Eq. C-25 in Eq. C-4 leads to the unit-impulse response function

$$h(t) = 2\frac{c_s}{r_0} e^{\frac{c_s}{r_0} s_1 t} \left(A_1 \cos \frac{c_s}{r_0} s_2 t - A_2 \sin \frac{c_s}{r_0} s_2 t\right) \tag{C-26}$$

Proceeding with each of the complex conjugate first-order terms in Eq. C-25 as in Section C2, it can easily be shown that the recursive equation will be of the form

$$y_n = a_1 y_{n-1} + a_2 y_{n-2} + b_0 x_n + b_1 x_{n-1} + b_2 x_{n-2} \tag{C-27}$$

For a piecewise linear input, this equation will again lead to the exact value of the convolution integral.

To determine a_1, a_2, the free vibration given by $h(t)$ is examined. Formulating Eq. C-27 for vanishing input at $n\Delta t$ and at $(n-1)\,\Delta t$ yields

$$h_n = a_1\, h_{n-1} + a_2 h_{n-2} \tag{C-28a}$$

$$h_{n-1} = a_1\, h_{n-2} + a_2\, h_{n-3} \tag{C-28b}$$

with $h_n = h(n\Delta t)$ specified in Eq. C-26. Solving Eq. C-28 for a_1, a_2 leads to

$$a_1 = 2\, e^{\frac{c_s}{r_0} s_1 \Delta t} \cos \frac{c_s}{r_0} s_2 \Delta t \tag{C-29a}$$

$$a_2 = -e^{2\frac{c_s}{r_0} s_1 \Delta t} \tag{C-29b}$$

Neither a_1, a_2 depend on A_1, A_2.

The following concise derivation leads to b_0, b_1, b_2. The unit-impulse function $h(t)$ is the response (output) to the Dirac-delta function $\delta(t)$ (input). The first time integral of $\delta(t)$ is equal to the Heavyside function, the second integral represents the ramp function $x(t)$:

$$x(t) \quad = t \qquad t \geq 0 \tag{C-30a}$$

$$= 0 \qquad t < 0 \tag{C-30b}$$

It follows that the second time integral of $h(t)$, denoted as $r(t)$, with vanishing initial conditions $r(0) = \dot{r}(0) = 0$ is the response (output) to the ramp function (input). Using Eq. C-26

$$r(t) = f_0 \frac{r_0}{c_s} + f_1\, t + f_2 \frac{r_0}{c_s} e^{\frac{c_s}{r_0} s_1 t} \sin \frac{c_s}{r_0} s_2 t + f_3 \frac{r_0}{c_s} e^{\frac{c_s}{r_0} s_1 t} \cos \frac{c_s}{r_0} s_2 t \qquad t \geq 0 \tag{C-31a}$$

$$r(t) = 0 \qquad t < 0 \tag{C-31b}$$

with

$$f_0 = -2 \frac{\left(s_1^2 - s_2^2\right) A_1 + 2 s_1 s_2 A_2}{\left(s_1^2 + s_2^2\right)^2} \tag{C-32a}$$

$$f_1 = -2 \frac{s_1 A_1 + s_2 A_2}{s_1^2 + s_2^2} \tag{C-32b}$$

$$f_2 = 2 \frac{2 s_1 s_2 A_1 - \left(s_1^2 - s_2^2\right) A_2}{\left(s_1^2 + s_2^2\right)^2} \tag{C-32c}$$

$$f_3 = 2 \frac{\left(s_1^2 - s_2^2\right) A_1 + 2 s_1 s_2 A_2}{\left(s_1^2 + s_2^2\right)^2} \tag{C-32d}$$

Samples of the ramp function input are $x_{-1} = x_0 = 0$, $x_1 = \Delta t$, $x_2 = 2\Delta t$, $x_3 = 3\Delta t$. The corresponding samples of the ramp response, computed from Eq. C-31 are $y_{-1} = y_0 = 0$, $y_1 = r(\Delta t)$, $y_2 = r(2\Delta t)$, $y_3 = r(3\Delta t)$.

Evaluating the recursion equation (Eq. C-27) leads to the following three equations: for $n = 1$

$$r(\Delta t) = b_0 \Delta t \tag{C-33a}$$

for $n = 2$

$$r(2\Delta t) = a_1 r(\Delta t) + 2 b_0 \Delta t + b_1 \Delta t \tag{C-33b}$$

and for $n = 3$

$$r(3\Delta t) = a_1 r(2\Delta t) + a_2 r(\Delta t) + 3 b_0 \Delta t + 2 b_1 \Delta t + b_2 \Delta t \tag{C-33c}$$

Solving for b_0, b_1, b_2 results in

$$b_0 = \frac{r(\Delta t)}{\Delta t} \tag{C-34a}$$

$$b_1 = \frac{r(2\Delta t) - (2 + a_1)\, r(\Delta t)}{\Delta t} \tag{C-34b}$$

$$b_2 = \frac{r(3\Delta t) - (2 + a_1)r(2\Delta t) + (1 + 2a_1 - a_2)\, r(\Delta t)}{\Delta t} \qquad \text{(C-34c)}$$

As a special case, the convolution integrals involving the unit-impulse functions $h_2(t)$ (Eq. A-81) and $h_3(t)$ (Eq. A-89) are examined.

$$h_2(t) = \frac{c}{z_0} e^{-\frac{3}{2}\frac{c}{z_0}t}\left(3 \cos \frac{\sqrt{3}}{2}\frac{c}{z_0}t - \sqrt{3} \sin \frac{\sqrt{3}}{2}\frac{c}{z_0}t\right) \qquad t \geq 0 \qquad \text{(C-35a)}$$

$$= 0 \qquad t < 0 \qquad \text{(C-35b)}$$

$$h_3(t) = 2\sqrt{3}\frac{c}{z_0} e^{-\frac{3}{2}\frac{c}{z_0}t} \sin \frac{\sqrt{3}}{2}\frac{c}{z_0}t \qquad t \geq 0 \qquad \text{(C-36a)}$$

$$= 0 \qquad t < 0 \qquad \text{(C-36b)}$$

The $h_2(t)$ appears in the rotation-interaction moment relationship (Eq. A-82) and in addition together with $h_3(t)$ in the Green's function (Eq. A-90) of the rotational cone. The corresponding transfer functions (Eqs. A-96, A-97), which are indeed of second order, are equal to

$$H_2(\omega) = \frac{3\left(i\frac{\omega z_0}{c} + 1\right)}{\left(i\frac{\omega z_0}{c}\right)^2 + 3\frac{i\omega z_0}{c} + 3} \qquad \text{(C-37)}$$

$$H_3(\omega) = \frac{3}{\left(i\frac{\omega z_0}{c}\right)^2 + 3\frac{i\omega z_0}{c} + 3} \qquad \text{(C-38)}$$

From Eq. C-25, the roots s equal $-1.5 \pm i\sqrt{3}/2$. For $H_2(\omega)$, the residues A are equal to $1.5 \pm i\sqrt{3}/2$; for $H_3(\omega)$, A are equal to $\mp i\sqrt{3}$. From Eq. C-29

$$a_1 = 2\, e^{-\frac{3}{2}\frac{c\Delta t}{z_0}} \cos \frac{\sqrt{3}}{2}\frac{c\Delta t}{z_0} \qquad \text{(C-39a)}$$

$$a_2 = -\, e^{-3\frac{c\Delta t}{z_0}} \qquad \text{(C-39b)}$$

follow. The ramp functions are equal to

$$r_2(t) = t - \frac{2\sqrt{3}}{3}\frac{z_0}{c} e^{-\frac{3}{2}\frac{ct}{z_0}} \sin \frac{\sqrt{3}}{2}\frac{ct}{z_0} \qquad t \geq 0 \qquad \text{(C-40a)}$$

$$= 0 \qquad t < 0 \qquad \text{(C-40b)}$$

$$r_3(t) = -\frac{z_0}{c} + t + \frac{\sqrt{3}}{3}\frac{z_0}{c} e^{-\frac{3}{2}\frac{ct}{z_0}}\left(\sin \frac{\sqrt{3}}{2}\frac{ct}{z_0} + \sqrt{3} \cos \frac{\sqrt{3}}{2}\frac{ct}{z_0}\right) \qquad t \geq 0 \qquad \text{(C-41a)}$$

$$= 0 \qquad t < 0 \qquad \text{(C-41b)}$$

The recursive coefficients b_0, b_1, and b_2 are then calculated from Eq. C-34.

As another special case the transfer function

$$H_4(\omega) = \frac{\gamma}{\left(i\frac{\omega z_0}{c}\right)^2 + \gamma\frac{i\omega z_0}{c} + \gamma} \tag{C-42}$$

appearing in the flexibility formulation of a one-degree-of-freedom system (Section 2.2.2) is examined. The γ denotes a positive real value. For $\gamma = 3$ $H_3(\omega)$ (Eq. C-38) follows. From Eq. C-25 it follows that for $4/\gamma - 1 \geq 0$ the roots s equal $-\gamma/2 \pm i(\gamma/2)\sqrt{4/\gamma - 1}$ and the residues $A = \mp i/\sqrt{4/\gamma - 1}$. Equation C-26 leads to

$$h_4(t) = \frac{2c}{z_0}\frac{1}{\sqrt{\frac{4}{\gamma}-1}}e^{-\frac{c}{z_0}\frac{\gamma}{2}t}\sin\left(\frac{c}{z_0}\frac{\gamma}{2}\sqrt{\frac{4}{\gamma}-1}\,t\right) \qquad t \geq 0 \tag{C-43a}$$

$$= 0 \qquad t < 0 \tag{C-43b}$$

From Eq. C-29

$$a_1 = 2e^{-\frac{c}{z_0}\frac{\gamma}{2}\Delta t}\cos\left(\frac{c}{z_0}\frac{\gamma}{2}\sqrt{\frac{4}{\gamma}-1}\,\Delta t\right) \tag{C-44a}$$

$$a_2 = -e^{-\frac{c}{z_0}\gamma\Delta t} \tag{C-44b}$$

follow. The ramp function (Eq. C-31) equals

$$r_4(t) = -\frac{z_0}{c} + t + \frac{1-\frac{2}{\gamma}}{\sqrt{\frac{4}{\gamma}-1}}\frac{z_0}{c}e^{-\frac{c}{z_0}\frac{\gamma}{2}t}\sin\left(\frac{c}{z_0}\frac{\gamma}{2}\sqrt{\frac{4}{\gamma}-1}\,t\right) \tag{C-45a}$$

$$+\frac{z_0}{c}e^{-\frac{c}{z_0}\frac{\gamma}{2}t}\cos\left(\frac{c}{z_0}\frac{\gamma}{2}\sqrt{\frac{4}{\gamma}-1}\,t\right) \qquad t \geq 0 \tag{C-45b}$$

$$= 0 \qquad t < 0$$

The remaining recursive coefficients b_0, b_1, and b_2 are then again determined from Eq. C-34. For the case $4/\gamma - 1 < 0$ with real roots s, the following equations apply

$$h_4(t) = \frac{2c}{z_0}\frac{1}{\sqrt{1-\frac{4}{\gamma}}}e^{-\frac{c}{z_0}\frac{\gamma}{2}t}\sinh\left(\frac{c}{z_0}\frac{\gamma}{2}\sqrt{1-\frac{4}{\gamma}}\,t\right) \qquad t \geq 0 \tag{C-46a}$$

$$= 0 \qquad t < 0 \tag{C-46b}$$

$$a_1 = 2e^{-\frac{c}{z_0}\frac{\gamma}{2}\Delta t}\cosh\left(\frac{c}{z_0}\frac{\gamma}{2}\sqrt{1-\frac{4}{\gamma}}\,\Delta t\right) \tag{C-47a}$$

$$a_2 = -e^{-\frac{c}{z_0}\gamma\Delta t} \tag{C-47b}$$

$$r_4(t) = -\frac{z_0}{c} + t + \frac{1-\frac{2}{\gamma}}{\sqrt{1-\frac{4}{\gamma}}} \frac{z_0}{c} e^{-\frac{c}{z_0}\frac{\gamma}{2}t} \sinh \frac{c}{z_0}\frac{\gamma}{2}\sqrt{1-\frac{4}{\gamma}}\, t \tag{C-48a}$$

$$+ \frac{z_0}{c} e^{-\frac{c}{z_0}\frac{\gamma}{2}t} \cosh \frac{c}{z_0}\frac{\gamma}{2}\sqrt{1-\frac{4}{\gamma}}\, t \qquad t \geq 0$$

$$= 0 \qquad t < 0 \tag{C-48b}$$

For the first-order term corresponding to the unit-impulse response function $h_1(t)$ (Eq. C-22) with $a_1 = a$ specified in Eq. C-24a and a_2=0, Eqs. C-34a, C-344b can also be used to determine b_0, b_1 ($b_2 = 0$), whereby

$$r_1(t) = t - \frac{z_0}{c}\left(1 - e^{-\frac{ct}{z_0}}\right) \qquad t \geq 0 \tag{C-49a}$$

$$= 0 \qquad t < 0 \tag{C-49b}$$

is used.

D

Dynamic Stiffness of Foundation on or Embedded in Layered Soil Halfspace

In Appendix D the concept using cones to determine the dynamic behavior both of a homogeneous halfspace and of a layer on a halfspace, discussed extensively in the text, is generalized to the case of a layered halfspace. This permits the dynamic-stiffness coefficient for each degree of freedom to be determined for a foundation on the surface of, or embedded in, a layered soil halfspace. The analysis is performed in physical coordinates, and no transformation to the wave-number domain as in the rigorous method is required. The concept specified in Section D1 consists of first determining the so-called backbone cone of the site, which determines the radii of the disks at the upper and lower interfaces of each layer. The dynamic-stiffness matrices of all layers with the corresponding disks are then calculated and subsequently assembled together with that of the corresponding disk on the underlying homogeneous halfspace to form the dynamic-stiffness matrix of the site. Eliminating the displacement amplitudes at all interfaces, except that of the original disk, yields its own dynamic-stiffness coefficient. In Section D2 the dynamic-stiffness matrix of a single isolated layer (with the corresponding disks at the upper and lower interfaces) is derived. Section D3 demonstrates that assembling the dynamic-stiffness matrix of such a layer with the same material properties as the underlying halfspace, and the dynamic-stiffness coefficient of the corresponding disk on the latter, does indeed result (after elimination of the displacement amplitude at the interface) in the dynamic-stiffness coefficient of the disk on the same homogeneous halfspace. Section D4 addresses a disk on the surface of a layer resting on a flexible halfspace and a disk on an inhomogeneous halfspace with the shear modulus increasing with depth. Good agreement with rigorous solutions results.

Together with the methods described in Chapter 6 the seismic analysis of a surface or embedded foundation in a horizontally layered site can be calculated concisely.

The procedure described in the following can serve as the basis for further research, which should allow certain shortcomings to be eliminated in the future. This appendix is based on Wolf and Meek [W17].

D1 BACKBONE CONE AND LAYERED HALFSPACE

The horizontally stratified site with the material properties varying with depth shown in Fig. D-1 is discretized with $n - 1$ layers of constant properties resting on a homogeneous halfspace with index n. The nodes coinciding with the interfaces of the layers are numbered from 1 at the free surface to n at the top of the underlying homogeneous halfspace.

The procedure to calculate the dynamic-stiffness coefficient of a disk on the free surface is addressed (Fig. D-1a). The vertical degree of freedom is used for illustration. The method is straightforwardly applicable to the other degrees of freedom (horizontal, rocking, and torsional), which are independent from one another. The radius of the surface disk is denoted as r_0. The *backbone cone* of the site determining the radii of the disks at the interfaces is constructed as follows. The radius r_1 of the upper disk of the first layer which is located at the first interface is identical to r_0. For Poisson's ratio ν of the first layer the aspect ratio z_{01}/r_1 follows from Table 2-3A. The radius of the lower disk follows from geometry as $r_2 = r_1(1 + d_1/z_0)$ with the depth of the first layer d_1. This radius r_2 is equal to that of the upper disk of the second layer located at the second interface. Proceeding analogously for the layers located further down, the radius r_{j+1} of the lower disk of layer j which is located at interface $j + 1$ follows as

$$r_{j+1} = \frac{z_{0j} + d_j}{z_{0j}} r_j \tag{D-1}$$

The last radius, r_n, is that of the underlying homogeneous halfspace. In general, the aspect ratios of the cone frustums depending on ν will change with depth, resulting in a backbone cone with piecewise linear segments. For constant ν the backbone cone exhibits no slope discontinuities.

The displacements in the nodes on the axis of the backbone cone with amplitudes $u_j(\omega)$ are introduced. For the layer j with an upper disk with radius r_j and a lower disk with r_{j+1}, $u_j(\omega)$ and $u_{j+1}(\omega)$ are related to the corresponding forces with amplitudes $P_j(\omega)$ and $P_{j+1}(\omega)$ as

$$\begin{Bmatrix} P_j(\omega) \\ P_{j+1}(\omega) \end{Bmatrix} = [S_j(\omega)] \begin{Bmatrix} u_j(\omega) \\ u_{j+1}(\omega) \end{Bmatrix} \tag{D-2}$$

The dynamic-stiffness matrix of the layer $[S_j(\omega)]$ derived in Section D-2 depends besides on the motion on the radii r_j, r_{j+1}, the thickness d_j and the material properties of the layer. For the underlying homogeneous halfspace

$$P_n(\omega) = S_n(\omega)\, u_n(\omega) \tag{D-3}$$

applies with the dynamic-stiffness coefficient $S_n(\omega)$ of the cone (Eqs. 2-69, 2-70 for translation and Eqs. 2-168, 2-169 for rotation).

Assembling the dynamic-stiffness matrices of the layers and the underlying homogeneous halfspace yields the discretized dynamic-equilibrium equations

$$[S(\omega)]\ \{u(\omega)\} = \{Q(\omega)\} \tag{D-4}$$

The dynamic-stiffness matrix associated with the backbone cone is denoted as $[S(\omega)]$; $\{u(\omega)\}$ is the vector of the displacement amplitudes with elements $u_1(\omega)$ to $u_n(\omega)$ and $\{Q(\omega)\}$ the vector of the external load amplitudes. As in the assembling process $[S_j(\omega)]$ of two adjacent layers are partly overlapped, $[S(\omega)]$ is tridiagonal. It is important to realize that the load amplitude vector $\{Q(\omega)\}$ contains only a single non-zero element, the load at the source disk with radius r_0, located in this case at the surface.

To determine the dynamic-stiffness coefficient $S(\omega)$ of a surface disk on a layered halfspace, a load with unit amplitude $P_0(\omega) = 1$ is applied. With $\{Q(\omega)\} = [1, 0, 0, ..., 0]^T$, $\{u(\omega)\}$ follows from Eq. D-4. The reciprocal of the first element $u_0(\omega) = u_1(\omega)$ in $\{u(\omega)\}$ is equal to the dynamic-stiffness coefficient $S(\omega)$

$$P_0(\omega) = S(\omega)\ u_0(\omega) \tag{D-5}$$

The procedure to calculate the dynamic-stiffness coefficient $S(\omega)$ of a disk embedded in a layered halfspace (Fig. D-1b) is analogous. Starting from the source disk with radius r_0, the backbone cone is constructed downwards and upwards as before. It is advantageous to divide the layer that includes the embedded disk into two sublayers as shown in the figure. The reciprocal of the displacement amplitude of the embedded disk $u_0(\omega)$ caused by a load with unit amplitude $P_0(\omega) = 1$ applied at the same disk, determined by solving the discretized dynamic-equilibrium equations of the site, is equal to $S(\omega)$. The displacements with amplitudes $\{u(\omega)\}$ throughout the site for the load with amplitude $P_0(\omega)$ also follow. By subdividing the layers accordingly, the displacement amplitudes can be calculated in any point on the axis of the backbone cone.

As described in Section 4.2, to calculate the dynamic-stiffness matrix of an embedded cylindrical foundation, the soil region that will later be excavated is viewed as a sandwich consisting of embedded rigid disks. For each loaded embedded disk using the procedure described in the preceding paragraph, the displacement amplitudes at the location of all embedded disks can be calculated. This leads to a column of the

Fig. D-1 Backbone cone consisting of frustums with varying aspect ratios in horizontally layered halfspace.

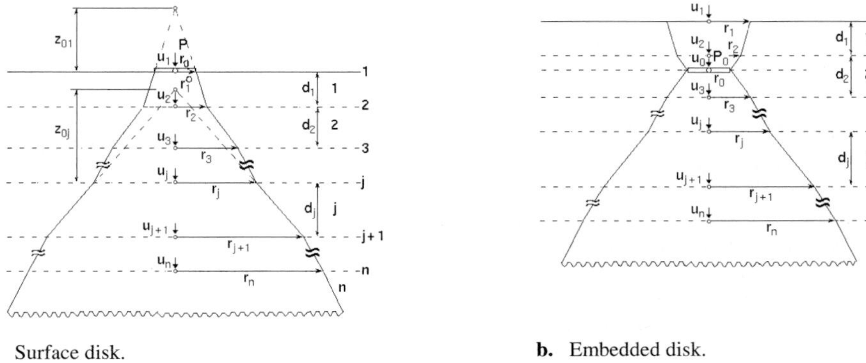

a. Surface disk. **b.** Embedded disk.

dynamic-flexibility matrix $[G(\omega)]$ defined in Eq. 4-21. The rest of the analysis is the same as described for the embedded foundation in Section 4.2 and for the single pile in Section 4.8.2.

D2 DYNAMIC-STIFFNESS MATRIX OF LAYER

D2.1 Translational Motion

The horizontal layer of depth d with constant material properties shown in Fig. D-2a represents the basic element for analyzing a layered site. Again, the vertical degree of freedom is used for illustration, involving the translational cone. The backbone cone's frustum results in a radius r_1 of the upper disk with displacement and force amplitudes $u_1(\omega)$, $P_1(\omega)$ at interface 1 and in a radius $r_2 = r_1(1 + d/z_0)$ (Eq. D-1) of the lower disk with $u_2(\omega)$, $P_2(\omega)$ at interface 2. As indicated by Figs. D-2b and D-2c the upper and lower disks are present in either/or fashion—at the source location of the loads. The remaining parts of the interfaces are regarded as free surfaces. The corresponding motivation, which is related to the necessity of modeling the free surface at the site, is illustrated in Section D3. The dynamic-stiffness matrix $[S(\omega)]$ relates $u_1(\omega)$, $u_2(\omega)$ to $P_1(\omega)$, $P_2(\omega)$ as

$$\begin{Bmatrix} P_1(\omega) \\ P_2(\omega) \end{Bmatrix} = [S(\omega)] \begin{Bmatrix} u_1(\omega) \\ u_2(\omega) \end{Bmatrix} \tag{D-6}$$

a)

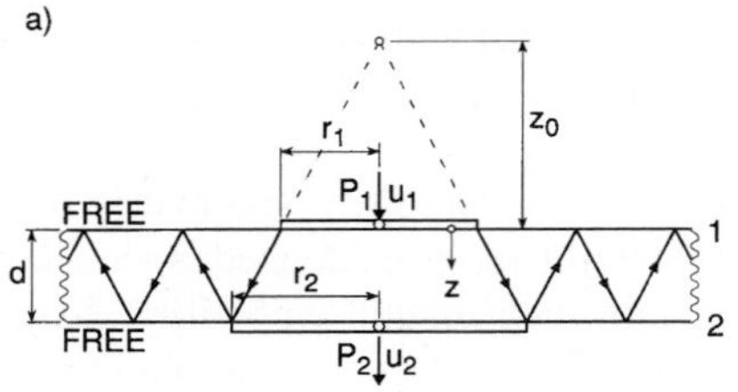

a. Dynamic-stiffness matrix of layer with corresponding disks.

c)

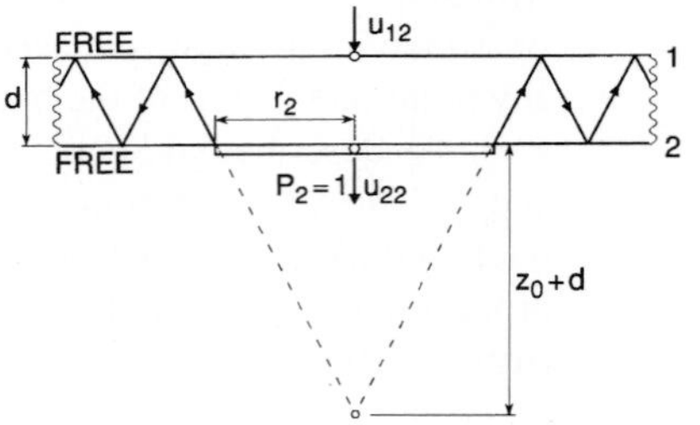

c. Dynamic-flexibility coefficients for load on lower disk.

b)

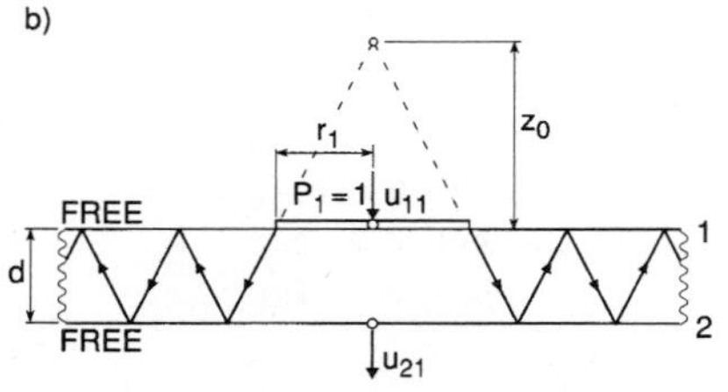

b. Dynamic-flexibility coefficients for load on upper disk.

Fig. D-2 Layer with free surfaces and two disks with wave pattern in refolded cones.

To determine $[S(\omega)]$, the dynamic-flexibility coefficients for a loaded upper disk (Fig. D-2b) and lower disk (Fig. D-2c) are calculated. The force with unit amplitude $P_1(\omega) = 1$ acting on the upper disk yields the displacements on the axis of the frustum with amplitudes $u_{11}(\omega)$ at interface 1 and $u_{21}(\omega)$ at interface 2. The resulting displacement pattern in the layer is similar to that in the unfolded cone discussed in connection with Figs. 3-2, 3-4 and 3-5, with the exception that the boundary at depth d, the interface 2, is free and not fixed. The only modification in the wave pattern specified in Eq. 3-7b consists of replacing the factors $(-1)^j$ by $(+1)^j$; or stated differently, the reflection coefficient $-\alpha$ is equal to +1 in Eq. 3-77 for this layer "on air." Alternatively, the same wave pattern can be derived from the superposition of the Green's functions corresponding to the double-cone models of anti-symmetrically loaded mirror-image disks embedded in the fullspace. In this case the derivation in Section 4.1.4 with the appropriate sign changes still applies, as does Fig. 4-8, but with the loads on all disks acting downwards as shown in Fig. D-3. By symmetry the lower surface at depth d is stress free. The upper disk is denoted as disk 1. For harmonic loading, the Green's function $g(a,\omega)$ of the translational double cone with distance a is specified in Eq. 4-4. For $P_1(\omega) = 1$ (Fig. D-2b), the displacement amplitudes at $z = 0$ and $z = d$ are equal to (Fig. D-3)

$$u_{11}(\omega) = 2g(o,\omega) + 4g(2d,\omega) + 4g(4d,\omega) + 4g(6d,\omega) + \ldots \quad \text{(D-7a)}$$

$$u_{21}(\omega) = 4g(d,\omega) + 4g(3d,\omega) + 4g(5d,\omega) + 4g(7d,\omega) + \ldots \quad \text{(D-7b)}$$

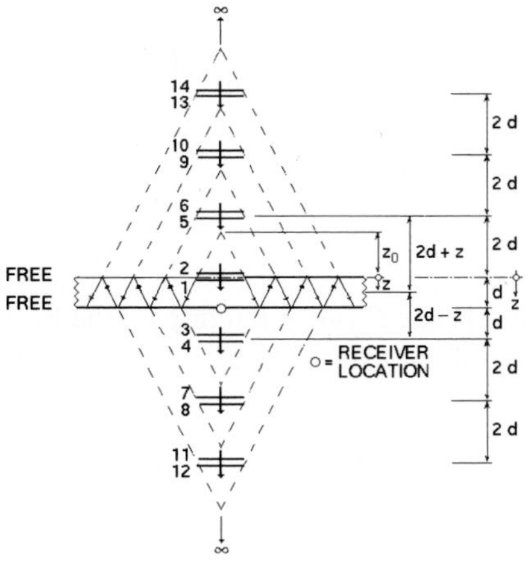

Fig. D-3 Arrangement of mirror-image disks embedded in full space with loads acting in same direction to model disk on surface of layer "on air."

As two disks (numbers 1 and 2) at an infinitesimal separation are present in the fullspace at interface 1 (Fig. 4-8), the factor 2 appears on the right-hand side of Eq. D-7a. Farther away, two such pairs of disks occur as mirror images at a specific distance, resulting in the factor 4 in the remaining terms.

Note that in the "stack model" (Fig. D-3) no rigid disks are present at the receiver location at the base of the layer. (See also Fig. D-2.) The displacement at the base level is maximum at the node along the axis and decreases with radial distance. The maximal nodal displacement is *not* the work-equivalent average displacement across the base. This implies that the dynamic-flexibility and stiffness matrices of the layer will *not* be symmetric.

Substituting Eq. 4-4 in Eq. D-7 leads to

$$u_{11}(\omega) = \frac{T_{11}(\omega)}{S_1(\omega)} \tag{D-8a}$$

$$u_{21}(\omega) = \frac{T_{21}(\omega)}{S_1(\omega)} \tag{D-8b}$$

with the transfer functions

$$T_{11}(\omega) = 1 + 2\sum_{j=1}^{\infty} \frac{e^{-ij\omega T}}{1 + j\dfrac{2d}{z_0}} \tag{D-9a}$$

$$T_{21}(\omega) = 2\sum_{j=1}^{\infty} \frac{e^{-i(j-0.5)\omega T}}{1 + (j-0.5)\dfrac{2d}{z_0}} \tag{D-9b}$$

and the travel time $T = 2d/c$ specified in Eqs. 3-9b (c = appropriate wave velocity, $c = c_p$ for the vertical motion when $\nu \le 1/3$). The dynamic-stiffness coefficient $S_1(\omega)$ of the single cone modeling a disk of radius r_1 on a halfspace with the properties of the layer equals (Eqs. 2-69, 2-70)

$$S_1(\omega) = K_1\left(1 + i\frac{\omega z_0}{c}\right) \tag{D-10}$$

The aspect ratio z_0/r_1 follows from Table 2-3A and the static-stiffness coefficient equals (Eq. 2-63a)

$$K_1 = \frac{\rho c^2 \pi r_1^2}{z_0} \tag{D-11}$$

with the density ρ.

Analogously, the force with unit amplitude $P_2(\omega) = 1$ acting on the lower disk (Fig. D-2c) yields the following displacements on the axis with the amplitudes at interfaces 1 and 2

$$u_{12}(\omega) = \frac{T_{12}(\omega)}{S_2(\omega)} \tag{D-12a}$$

$$u_{22}(\omega) = \frac{T_{22}(\omega)}{S_2(\omega)} \tag{D-12b}$$

where

$$T_{12}(\omega) = 2\sum_{j=1}^{\infty} \frac{e^{-i(j-0.5)\omega T}}{1 + (j-0.5)\dfrac{2d}{z_0 + d}} \tag{D-13a}$$

$$T_{22}(\omega) = 1 + 2\sum_{j=1}^{\infty} \frac{e^{-ij\omega T}}{1 + j\dfrac{2d}{z_0 + d}} \tag{D-13b}$$

and

$$S_2(\omega) = K_2\left[1 + i\frac{\omega(z_0 + d)}{c}\right] \tag{D-14}$$

$$K_2 = \frac{\rho c^2 \pi r_2^2}{z_0 + d} \tag{D-15}$$

Notice that the distance from the center of the lower disk to the apex equals $z_0 + d$ (Fig. D-2c).

The displacement-force relationship for harmonic loading is formulated as (Eqs. D-8, D-12)

$$\begin{Bmatrix} u_1(\omega) \\ u_2(\omega) \end{Bmatrix} = \begin{bmatrix} \dfrac{T_{11}(\omega)}{S_1(\omega)} & \dfrac{T_{12}(\omega)}{S_2(\omega)} \\ \dfrac{T_{21}(\omega)}{S_1(\omega)} & \dfrac{T_{22}(\omega)}{S_2(\omega)} \end{bmatrix} \begin{Bmatrix} P_1(\omega) \\ P_2(\omega) \end{Bmatrix} \tag{D-16}$$

The coefficient matrix on the right-hand side represents the dynamic-flexibility matrix, which, as explained previously, is not symmetric. For $\omega \to 0$ (static case), the coefficients diverge.

Inverting Eq. D-16 yields

$$\begin{Bmatrix} P_1(\omega) \\ P_2(\omega) \end{Bmatrix} = \frac{S_1(\omega)S_2(\omega)}{T_{11}(\omega)T_{22}(\omega) - T_{12}(\omega)T_{21}(\omega)} \begin{bmatrix} \dfrac{T_{22}(\omega)}{S_2(\omega)} & -\dfrac{T_{21}(\omega)}{S_1(\omega)} \\ -\dfrac{T_{12}(\omega)}{S_2(\omega)} & \dfrac{T_{11}(\omega)}{S_1(\omega)} \end{bmatrix} \begin{Bmatrix} u_1(\omega) \\ u_2(\omega) \end{Bmatrix} \tag{D-17}$$

The coefficient matrix is equal to the translational dynamic-stiffness matrix of the layer (Eq. D-6)

$$[S(\omega)] = \frac{S_1(\omega)}{T_{11}(\omega)T_{22}(\omega) - T_{12}(\omega)T_{21}(\omega)} \begin{bmatrix} T_{22}(\omega) & -\dfrac{S_2(\omega)}{S_1(\omega)}T_{21}(\omega) \\ -T_{12}(\omega) & \dfrac{S_2(\omega)}{S_1(\omega)}T_{11}(\omega) \end{bmatrix} \tag{D-18}$$

$[S(\omega)]$ is complex, indicating that radiation of energy in the horizontal direction of the layer does take place. $[S(\omega)]$ is also non-symmetric. For $\omega \to 0$ (static case), the real part converges. $[S(\omega)]$ is a function of $r_1(r_2)$, d, c, ν, ρ.

$S_1(\omega)$ and $S_2(\omega)$ are the dynamic-stiffness coefficients of disks with radii r_1 and r_2 on a homogenous halfspace with the properties of the layer. Instead of using the cone model to calculate these values (Eqs. D-10, D-14), the rigorous elastodynamic solutions could be applied. This alternative is justified in Section D3.

To model the underlying homogeneous halfspace (Fig. D-1), the dynamic-stiffness coefficient of the translational cone (Eq. D-10) with the corresponding radius r_n is used.

For the horizontal degree of freedom, the same equations apply with $c = c_s$ and z_0/r_1 again following from Table 2-3A.

D2.2 Rotational Motion

The procedure for the rotational degree of freedom (rocking $c = c_p$ for $\nu \leq 1/3$, torsional $c = c_s$, z_0/r_1 from Table 2-3A) is analogous. For easy reference, the equations are summarized in the following. The rotational dynamic-stiffness matrix $[S_\vartheta(\omega)]$ relates the amplitude of the rotations $\vartheta_1(\omega)$, $\vartheta_2(\omega)$ to those of the moments $M_1(\omega)$, $M_2(\omega)$.

$$\begin{Bmatrix} M_1(\omega) \\ M_2(\omega) \end{Bmatrix} = [S_\vartheta(\omega)] \begin{Bmatrix} \vartheta_1(\omega) \\ \vartheta_2(\omega) \end{Bmatrix} \tag{D-19}$$

From the Green's function for harmonic loading (Eq. 4-6 in the form of Eq. A-98) the transfer functions are specified as

$$T_{11}(\omega) = 1 + 2\sum_{j=1}^{\infty} e^{-ij\omega T} \left[\frac{1}{\left(1 + j\frac{2d}{z_0}\right)^2} + \left\{ \frac{1}{\left(1 + j\frac{2d}{z_0}\right)^3} - \frac{1}{\left(1 + j\frac{2d}{z_0}\right)^2} \right\} \frac{1}{1 + i\frac{\omega z_0}{c}} \right] \tag{D-20a}$$

$$T_{21}(\omega) = 2\sum_{j=1}^{\infty} e^{-i(j-0.5)\omega T} \cdot \left[\frac{1}{\left(1 + (j-0.5)\frac{2d}{z_0}\right)^2} + \left\{ \frac{1}{\left(1 + (j-0.5)\frac{2d}{z_0}\right)^3} - \frac{1}{\left(1 + (j-0.5)\frac{2d}{z_0}\right)^2} \right\} \frac{1}{1 + i\frac{\omega z_0}{c}} \right] \tag{D-20b}$$

$$T_{12}(\omega) = 2\sum_{j=1}^{\infty} e^{-i(j-0.5)\omega T} \cdot \left[\frac{1}{\left(1 + (j-0.5)\frac{2d}{z_0 + d}\right)^2} + \left\{ \frac{1}{\left(1 + (j-0.5)\frac{2d}{z_0 + d}\right)^3} - \frac{1}{\left(1 + (j-0.5)\frac{2d}{z_0 + d}\right)^2} \right\} \frac{1}{1 + i\frac{\omega(z_0 + d)}{c}} \right] \tag{D-20c}$$

$$T_{22}(\omega) = 1 + 2\sum_{j=1}^{\infty} e^{-ij\omega T} \cdot \left[\frac{1}{\left(1 + j\frac{2d}{z_0 + d}\right)^2} + \left\{ \frac{1}{\left(1 + j\frac{2d}{z_0 + d}\right)^3} - \frac{1}{\left(1 + j\frac{2d}{z_0 + d}\right)^2} \right\} \frac{1}{1 + i\frac{\omega(z_0 + d)}{c}} \right] \tag{D-20d}$$

and the dynamic-stiffness coefficients of a single cone modeling a disk on a halfspace (Eqs. 2-168, 2-169)

$$S_{\vartheta_1}(\omega) = K_{\vartheta_1}\left[1 - \frac{1}{3}\frac{\left(\frac{\omega z_0}{c}\right)^2}{1+\left(\frac{\omega z_0}{c}\right)^2} + i\frac{\omega z_0}{3c}\frac{\left(\frac{\omega z_0}{c}\right)^2}{1+\left(\frac{\omega z_0}{c}\right)^2}\right] \tag{D-21a}$$

$$S_{\vartheta_2}(\omega) = K_{\vartheta_2}\left[1 - \frac{1}{3}\frac{\left[\frac{\omega(z_0+d)}{c}\right]^2}{1+\left[\frac{\omega(z_0+d)}{c}\right]^2} + i\frac{\omega(z_0+d)}{3c}\frac{\left[\frac{\omega(z_0+d)}{c}\right]^2}{1+\left[\frac{\omega(z_0+d)}{c}\right]^2}\right] \tag{D-21b}$$

with the static-stiffness coefficients (Eq. 2-146a)

$$K_{\vartheta_1} = \frac{3\rho c^2 I_0}{z_0} \tag{D-22a}$$

$$K_{\vartheta_2} = \frac{3\rho c^2 I_0}{z_0 + d} \tag{D-22b}$$

I_0 equals $\pi r_0^4/4$ for rocking and $\pi r_0^4/2$ for the torsional degree of freedom. The rotational dynamic-stiffness matrix of the layer is formulated as

$$[S_\vartheta(\omega)] = \frac{S_{\vartheta_1}(\omega)}{T_{11}(\omega)T_{22}(\omega) - T_{12}(\omega)T_{21}(\omega)}\begin{bmatrix} T_{22}(\omega) & -\frac{S_{\vartheta_2}(\omega)}{S_{\vartheta_1}(\omega)}T_{21}(\omega) \\ -T_{12}(\omega) & \frac{S_{\vartheta_2}(\omega)}{S_{\vartheta_1}(\omega)}T_{11}(\omega) \end{bmatrix} \tag{D-23}$$

To model the underlying homogeneous halfspace, the dynamic-stiffness coefficient of the rotational cone (Eq. D-21a) with the corresponding radius r_n is applied.

D2.3 Example

As an example, the vertical and rocking dynamic-stiffness coefficients of the upper disk (interface 1) for $r_1 = d$ and $\nu = 1/3$ are calculated. From Table 2-3A, z_0/d follows. And c equals c_p. From Eq. D-18 and using Eq. D-10, the vertical dynamic-stiffness coefficient, nondimensionalized with the static-stiffness coefficient K_1, equals

$$\frac{S_{11}(\omega)}{K_1} = \frac{\left(1 + i\frac{\omega z_0}{c_p}\right)T_{22}(\omega)}{T_{11}(\omega)T_{22}(\omega) - T_{12}(\omega)T_{21}(\omega)} = k_{v11}(a_0) + ia_0c_{v11}(a_0) \tag{D-24}$$

with $T_{11}(\omega)$, $T_{21}(\omega)$, $T_{12}(\omega)$, $T_{22}(\omega)$ specified in Eqs. D-9 and D-13. The spring and damping coefficients $k_{v11}(a_0)$, $c_{v11}(a_0)$ are plotted as a solid dark line (FREE–FREE) as a function of $a_0 = \omega r_0/c_s$ in Fig. D-4. The dynamic-stiffness coefficient $S_{11}(\omega)$ represents the force amplitude at the top of the layer when the base *disk* is held fixed, the remainder of the base layer being *stress free*. For comparison, the solution with the *entire interface* 2 fixed based on the unfolded layered cone (Fig. 3-2, Eq. 3-39, Fig. 3-9b) is also shown as a solid light line (FREE–FIXED). The rigorous solution of Kausel [K1] based on the theory of elastodynamics denoted as exact is also presented. For rocking, the corresponding plots are shown in

Fig. D-5 with the solution for interface 2 fixed taken from Fig. 3-10b (Eq. 3-43). The two solutions based on cones are quite close. Note that for $a_0 \rightarrow 0$ the spring coefficient converges while the damping coefficient for the layer "on air" becomes negative. It should be remembered that $c_{11}(a_0)$ is multiplied by $a_0 < 1$ in this range, leading to a small imaginary part which, in practice, should be set equal to zero. The other dynamic-stiffness coefficients not shown exhibit a similar tendency, with the exception that $c_{22}(a_0)$ remains positive for $a_0 \rightarrow 0$. After assembling the dynamic-stiffness matrices, these undesirable effects will tend to cancel, as is demonstrated in Section D3.

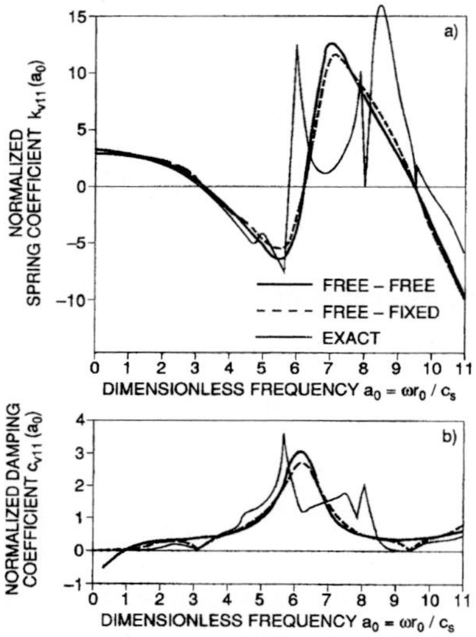

Fig. D-4 Vertical dynamic-stiffness coefficient of upper disk on layer.

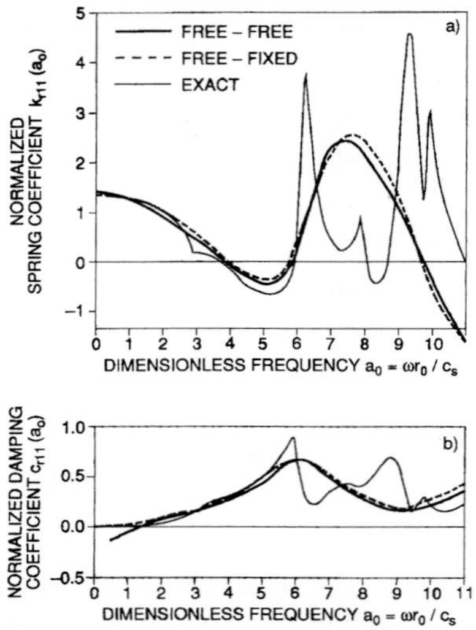

Fig. D-5 Rocking dynamic-stiffness coefficient of upper disk on layer.

D3 CLONING PROPERTY

To determine the dynamic-stiffness coefficient of a disk on a homogeneous halfspace, the upper region of the soil could be divided artificially into a number of layers. The layers and the underlying halfspace have the same material properties as the original halfspace. After constructing the backbone cone, each layer can be modeled as a cone frustum and the underlying halfspace as a cone. The corresponding dynamic-stiffness matrices are then assembled, yielding that of the subdivided halfspace. All degrees of freedom with the exception of that of the disk on the free surface are subsequently eliminated, resulting in the dynamic-stiffness coefficient of the disk on the *subdivided* halfspace. This coefficient should ideally be identical to the dynamic-stiffness coefficient of the disk on a *homogeneous* halfspace modeled directly with a cone.

This procedure is illustrated using one layer for the vertical degree of freedom in Fig. D-6. The dynamic-stiffness coefficient $S_2(\omega)$ of the disk with radius r_2 on the underlying halfspace is specified in Eq. D-14 and that of the layer of depth d with the upper and lower disks with radii r_1, r_2, $[S(\omega)]$, in Eq. D-18. After assembling the dynamic-stiffness coefficients, the discretized dynamic-equilibrium equations of the subdivided halfspace equal with the displacement amplitudes $u_1(\omega)$, $u_2(\omega)$

$$\frac{S_1(\omega)}{T_{11}(\omega)T_{22}(\omega)-T_{12}(\omega)T_{21}(\omega)}\begin{bmatrix} T_{22}(\omega) & -\dfrac{S_2(\omega)}{S_1(\omega)}T_{21}(\omega) \\ -T_{12}(\omega) & \dfrac{S_2(\omega)}{S_1(\omega)}T_{11}(\omega) \end{bmatrix}\begin{Bmatrix} u_1(\omega) \\ u_2(\omega) \end{Bmatrix} + \begin{Bmatrix} 0 \\ S_2(\omega)u_2(\omega) \end{Bmatrix} = \begin{Bmatrix} P_0(\omega) \\ 0 \end{Bmatrix} \tag{D-25}$$

$P_0(\omega)$ is the amplitude of the force applied to the surface disk. Eliminating $u_2(\omega)$ from Eq. D-25 yields

$$\frac{S_1(\omega)(1+T_{22}(\omega))}{T_{11}(\omega)+T_{11}(\omega)T_{22}(\omega)-T_{12}(\omega)T_{21}(\omega)}u_1(\omega)=P_0(\omega) \tag{D-26}$$

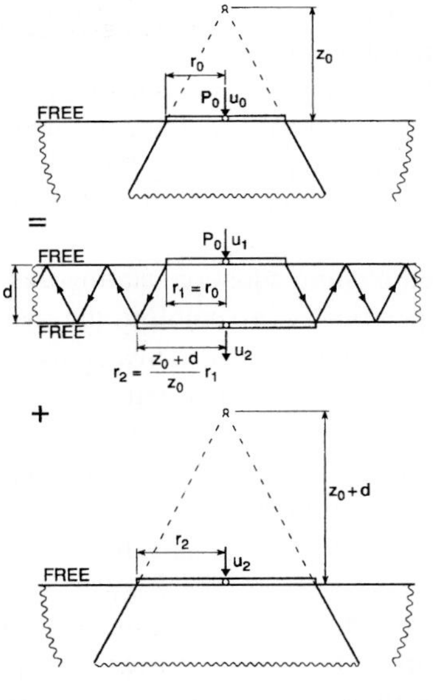

Fig. D-6 Cloning of cones.

Notice that the dynamic-stiffness coefficient $S_2(\omega)$ cancels out. This equation must correspond to the force-displacement relationship of the disk with radius r_0 (= r_1) on the halfspace

$$S_1(\omega)u_0(\omega) = P_0(\omega) \tag{D-27}$$

With $u_0(\omega) = u_1(\omega)$, matching the left-hand sides of Eqs. D-26 and D-27 leads to the requirement

$$T_{11}(\omega)-T_{22}(\omega)+T_{11}(\omega)T_{22}(\omega)-T_{12}(\omega)T_{21}(\omega)=1 \tag{D-28}$$

in which now the dynamic-stiffness coefficient $S_1(\omega)$ has canceled out. As the transfer functions (Eqs. D-9, D-13) contain infinite sums, an analytical proof of the validity of this identity, if at all possible, is cumbersome. However, by summing 100,000 terms Eq. D-28 has been verified in the range $0 < a_0 < 11$ and for $0.1 < d/r_0 < 100$. The fact that Eq. D-28 also holds for $a_0 \to 0$, indicates that the canceling effect of the damping coefficients during assemblage discussed at the end of Section D2.3 does actually occur. Numerical tests show that Eq. D-28 is also satisfied for the rotational degree of freedom with the transfer functions specified in Eq. D-20.

Because the dynamic-stiffness coefficients $S_1(\omega)$ and $S_2(\omega)$ do not appear in Eq. D-28, they need not be determined from cone models. In particular, the rigorous elastodynamical solutions for $S(\omega)$ of the disk on the elastic halfspace may be used to construct the layer's dynamic-stiffness matrices, Eq. D-18 and D-23 (retaining the transfer functions $T_{11}(\omega)$, $T_{21}(\omega)$, $T_{12}(\omega)$, and $T_{22}(\omega)$ derived from the approximate Green's functions). Due to the identity Eq. D-28, such layers with identical material properties will recombine perfectly to yield the rigorous dynamic-stiffness coefficient of the disk on the homogeneous halfspace.

For stratified soil with varying properties the construction of the layer's dynamic-stiffness matrices using the rigorous solutions for $S(\omega)$ instead of those for cones will lead to increased accuracy. Evidence is provided at the end of Section 3.6.4, Fig. 3-24. This promising approach deserves further investigation.

The crucial identity Eq. D-28 only holds when the transfer functions $T_{11}(\omega)$ and so on are defined for free–free boundary conditions as shown in Fig. D-6, thus matching the stress conditions at the bottom surface of the layer and the top surface of the underlying cone. Not surprisingly, the wave pattern corresponding to a totally fixed interface 2 (Figs. 3-9b, 3-10b) violates Eq. D-28.

Equation D-28 represents the so-called *cloning* equation, stating that the properties of the cone reproduce themselves. More specifically, assembling the dynamic-stiffness coefficient $S_2(a_{02})$ of the cone representing the disk of radius r_2 on an unbounded halfspace (with its dimensionless frequency $a_{02} = \omega r_2/c_s$) and the dynamic-stiffness matrix of the cone frustum (with disks of radii r_2, r_1 modeling the layer) yields the dynamic-stiffness coefficient $S_1(a_{01})$ of a cone representing the disk of radius r_1 on the same unbounded halfspace with its dimensionless frequency $a_{01} = \omega r_1/c_s$. The functions $S_1(a_{01})/K_1$ and $S_2(a_{02})/K_2$ are, however, not identical. The r_1 and r_2 are not the same and as a result a_{01} and a_{02} have, for a specific ω, different numerical values.

The cloning property obviously holds for slices of a homogeneous tapered bar (isolated cone) without horizontal wave spreading. The important implication of Eq. D-28 is that cloning remains valid when the opportunity for horizontal wave spreading is included (see Fig. D-6). Such horizontal wave spreading occurs whenever the impedance match of subsequent layers is imperfect—when the layers have different material properties, as in the stratified site.

D4 EXAMPLES

First, a disk of radius r_0 on a soil layer of depth d resting on a flexible rock halfspace is addressed (Fig. D-7). The same example is treated via the reflection coefficients in Section

3.6.4. The system with a layer that is stiffer and denser than the halfspace is specified by the following ratios (index L for layer, R for rock): $d/r_0 = 1$, $G_L/G_R = 5$ (shear modulus G), $\rho_L/\rho_R = 1.25$ (mass density ρ) and $\nu_L = \nu_R = 1/3$. The site is modeled with one cone frustum for the layer resting on a cone representing the halfspace. The dynamic-stiffness coefficient is nondimensionalized as

$$S(a_0) = K[k(a_0) + ia_0 c(a_0)] \tag{D-29}$$

where K is the static-stiffness coefficient of the disk on a homogeneous halfspace with the properties of the layer and $a_0 = \omega r_0 / c_s^L$. For vertical and rocking motions, the dynamic-stiffness coefficients are plotted in Figs. D-8 and D-9. The results (dashed line) are practically identical to those computed using the frequency-dependent reflection coefficient (dotted line, Figs. 3-16, 3-17). The rigorous solution of Waas [W2] is also shown.

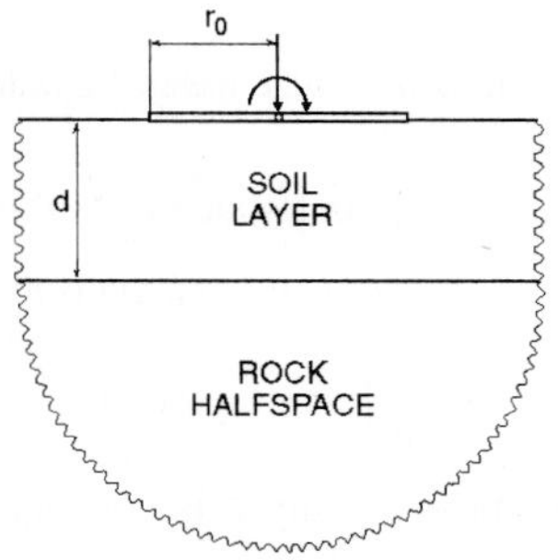

Fig. D-7 Disk on surface of soil layer resting on flexible rock halfspace.

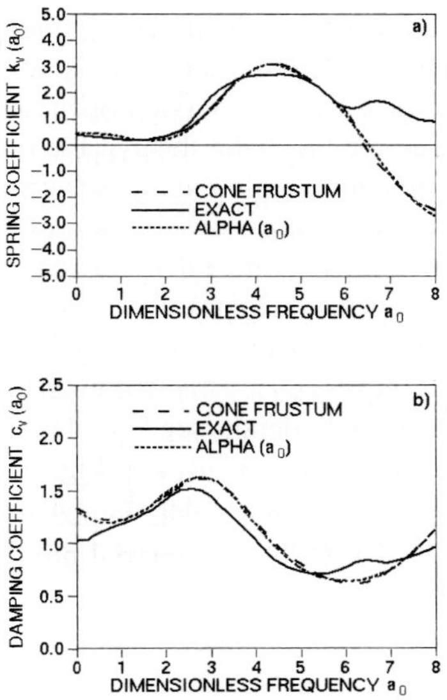

Fig. D-8 Vertical dynamic-stiffness coefficient for disk on soil layer stiffer than rock halfspace.

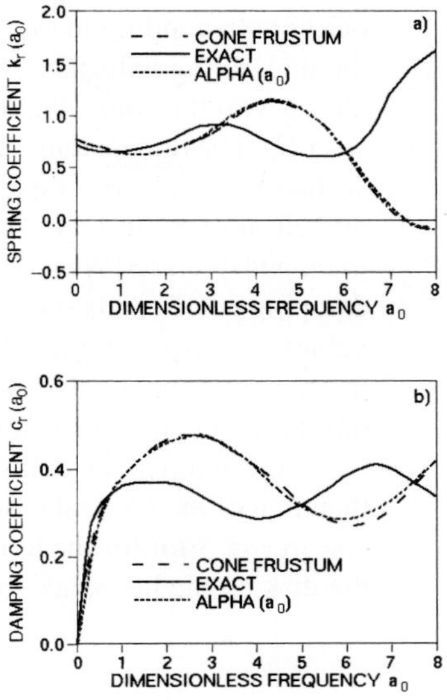

Fig. D-9 Rocking dynamic-stiffness coefficient for disk on soil layer stiffer than rock halfspace.

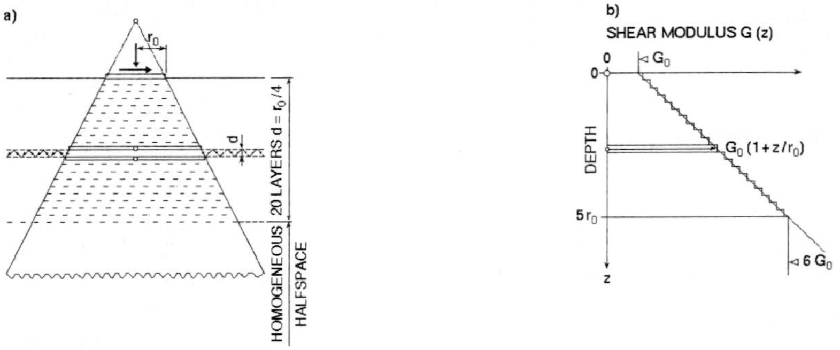

a. Discretization with 20 cone frustums with increasing radius and cone for underlying homogeneous halfspace.

b. Linearly increasing shear modulus with depth.

Fig. D-10 Disk on surface of inhomogeneous halfspace with shear modulus increasing with depth.

Second, a disk of radius r_0 on an inhomogeneous halfspace is examined (Fig. D-10). The shear modulus $G(z)$ increases with depth z as

$$G(z) = G_0\left(1 + \frac{z}{r_0}\right) \tag{D-30}$$

with the shear modulus at the surface G_0. The mass density ρ is constant. Poisson's ratio equals $\nu = 1/3$. The selected dynamic system consists of 20 layers with increasing G and thickness $d = r_0/4$ extending to a depth $5r_0$, where a homogeneous halfspace with the G (= $6G_0$) corresponding to this depth is placed. Each layer is modeled with a cone frustum and the underlying halfspace with a cone. The dynamic-stiffness coefficient is nondimensionalized with the static-stiffness coefficient of the disk with r_0 on a homogeneous halfspace with G_0. The dimensionless frequency a_0 is also defined with the shear-wave velocity at the surface $c_s = \sqrt{G_0/\rho}$. The dynamic-stiffness coefficients are plotted for the vertical and horizontal motions in Figs. D-11 and D-12. The rigorous numerical solution of Waas, Hartman, and Werkle [W1], denoted as exact, is also shown. Notice that whereas the rigorous damping coefficients start at $a_0 = 0$ at zero, the cone-frustum approach again yields negative values of $c_v(a_0)$ and $c_h(a_0)$ for $a_0 < 0.3$. Because negative damping is physically unrealistic, the negative $c(a_0)$-values should be set to zero *a posteriori*. This irritating feature is not observed in situations where the stiffness decreases with depth (see Figs. D-8 and D-9).

It is interesting to observe that the accuracy of the horizontal case (Fig. D-12) is better than that of the vertical case (Fig. D-11). The probable reason is that the dynamic-stiffness coefficient $S(\omega)$ for the horizontal cone is a better representation of the rigorous result for the disk on the halfspace than in the vertical case.

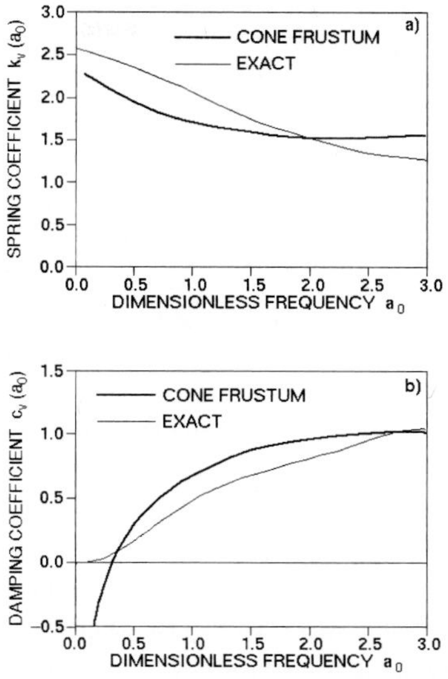

Fig. D-11 Vertical dynamic-stiffness coefficient for disk on inhomogeneous halfspace.

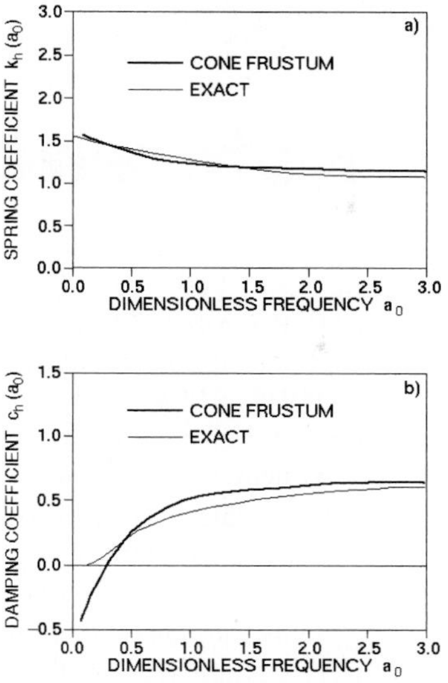

Fig. D-12 Horizontal dynamic-stiffness coefficient for disk on inhomogeneous halfspace.

References

[A1] R.J. Apsel and J.E. Luco, Impedance Functions for Foundations Embedded in a Layered Medium: An Integral Equation Approach, *Earthquake Engineering and Structural Dynamics*, 15 (1987): 213–231.

[D1] R. Dobry and G. Gazetas, Simple Method for Dynamic Stiffness and Damping of Floating Pile Groups, *Géotechnique*, 38 (1988): 557–574.

[E1] G. Ehlers, The Effect of Soil Flexibility on Vibrating Systems, *Beton und Eisen*, 41 (1942): 197–203 [in German].

[E2] J.M. Emperador and J. Dominguez, Dynamic Response of Axisymmetric Embedded Foundations, *Earthquake Engineering and Structural Dynamics*, 18 (1989): 1105–1117.

[G1] G. Gazetas and R. Dobry, Simple Radiation Damping Model for Piles and Footings, *Journal of Engineering Mechanics*, ASCE, 110 (1984), 937–956.

[G2] G. Gazetas, Simple Physical Methods for Foundation Impedances, in *Dynamic Behaviour of Foundations and Buried Structures* (Developments in Soil Mechanics and Foundation Engineering, Vol. 3), edited by P.K. Banerjee and R. Butterfield, Chapter 2, pp. 45–93 (London: Elsevier Applied Science, 1987).

[G3] G. Gazetas, Foundation Vibrations, in *Foundation Engineering Handbook*, 2nd Edition, edited by H.-Y. Fang, Chapter 15, pp. 553–593 (New York: Van Nostrand Reinhold, 1991).

[G4] G. Gazetas and N. Makris, Dynamic Pile-Soil-Pile Interaction, Part I: Analysis of Axial Vibration, *Earthquake Engineering and Structural Dynamics*, 20 (1991): 115–132.

[G5] G. Gazetas, Formulas and Charts for Impedances of Surface and Embedded Foundations, *Journal of Geotechnical Engineering*, ASCE, 117 (1991): 1363–1381.

[J1] L.B.W. Jolley, *Summation of Series*. New York: Dover, 1961.

[K1] E. Kausel, personal communication, 1990.

[K2] A.M. Kaynia and E. Kausel, Dynamic Behavior of Pile Groups, *Proceedings of 2nd International Conference on Numerical Methods of Offshore Piling*, Austin, TX, 1982: 509–532.

[L1] J.E. Luco and A.H. Hadjian, Two-Dimensional Approximations to the Three-Dimensional Soil-Structure Interaction Problem, *Nuclear Engineering and Design*, 31 (1974): 195–203.

[L2] J.E. Luco, Impedance Functions for a Rigid Foundation on a Layered Medium, *Nuclear Engineering and Design*, 31 (1974): 204–217.

[L3] J.E. Luco and A. Mita, Response of a Circular Foundation on a Uniform Halfspace to Elastic Waves, *Earthquake Engineering and Structural Dynamics*, 15 (1987): 105–118.

[L4] J.E. Luco and H.L. Wong, Seismic Response of Foundations Embedded in a Layered Half-space, *Earthquake Engineering and Structural Dynamics*, 15 (1987): 233–247.

[M1] N. Makris and G. Gazetas, Dynamic Pile-Soil-Pile Interaction, Part II: Lateral and Seismic Response, *Earthquake Engineering and Structural Dynamics*, 21 (1992): 145–162.

[M2] J.W. Meek and A.S. Veletsos, Simple Models for Foundations in Lateral and Rocking Motions, *Proceedings of the 5th World Conference on Earthquake Engineering*, Rome, 1974, Vol. 2: 2610–2613.

[M3] J.W. Meek, Recursive Analysis of Dynamical Phenomena in Civil Engineering, *Bautechnik*, 67 (1990: 205–210 [in German].

[M4] J.W. Meek and J.P. Wolf, Insights on Cutoff Frequency for Foundation on Soil Layer, *Earthquake Engineering and Structural Dynamics*, 20 (1991): 651–665.

[M5] J.W. Meek and J.P. Wolf, Cone Models for Homogeneous Soil, *Journal of Geotechnical Engineering*, ASCE, 118 (1992): 667–685.

[M6] J.W. Meek and J.P. Wolf, Cone Models for Soil Layer on Rigid Rock, *Journal of Geotechnical Engineering*, ASCE, 118 (1992): 686–703.

[M7] J.W. Meek and J.P. Wolf, Cone Models for Nearly Incompressible Soil, *Earthquake Engineering and Structural Dynamics*, 22 (1993): 649–663.

[M8] J.W. Meek and J.P. Wolf, Why Cone Models Can Represent the Elastic Half-space, *Earthquake Engineering and Structural Dynamics*, 22 (1993): 759–771.

[M9] J.W. Meek and J.P. Wolf, Approximate Green's Function for Surface Foundations, *Journal of Geotechnical Engineering*, ASCE 119 (1993): 1499–1514.

[M10] J.W. Meek and J.P. Wolf, Material Damping for Lumped-Parameter Models of Foundations, *Earthquake Engineering and Structural Dynamics*, 23 (1994), in press.

[M11] J.W. Meek and J.P. Wolf, Cone Models for an Embedded Foundation, *Journal of Geotechnical Engineering*, ASCE, 120 (1994): 60–80.

[M12] G.F. Miller and H. Pursey, The Field and Radiation Impedance of Mechanical Radiators on the Free Surface of a Semi-Infinite Isotropie Solid, *Proceedings Royal Society*, A 223 (1954): 521–541.

[M13] G.F. Miller and H. Pursey, On the Partition of Energy between Elastic Waves in a Semi-Infinite Solid, *Proceedings Royal Society*, A 233 (1955): 55–69.

[N1] National Earthquake Hazards Reduction Program (NEHRP), *Recommended Provisions for the Development of Seismic Regulations for New Buildings.* Building Seismic Safety Council, Washington, DC, 1986.

[P1] A. Pais and E. Kausel, Approximate Formulas for Dynamic Stiffnesses of Rigid Foundations, *Soil Dynamics and Earthquake Engineering*, 7 (1988): 213–227.

[P2] R.Y.S. Pak and A.T. Gobert, Forced Vertical Vibration of Rigid Discs with Arbitrary Embedment, *Journal of Engineering Mechanics*, ASCE, 117 (1991): 2527–2548.

[P3] H.G. Poulos and E.H. Davis, *Pile Foundation Analysis and Design.* New York: Wiley, 1980.

[R1] F.E. Richart, R.D. Woods and J.R. Hall, *Vibrations of Soils and Foundations.* Englewood Cliffs, NJ: Prentice-Hall, 1970.

[R2] R. Rücker, Dynamic Behaviour of Rigid Foundations of Arbitrary Shape on a Halfspace, *Earthquake Engineering and Structural Dynamics*, 10 (1982): 675–690.

[T1] J.L. Tassoulas, *Elements for the Numerical Analysis of Wave Motion in Layered Media* (Research Report R81-2). Department of Civil Engineering, Massachusetts Institute of Technology, Cambridge, MA: 1981.

[V1] A.S. Veletsos and Y.T. Wei, Lateral and Rocking Vibration of Footings, *Journal of the Soil Mechanics and Foundation Division*, ASCE, 97 (1971): 1227–1248.

[V2] A.S. Veletsos and B. Verbic, Vibration of Viscoelastic Foundations, *Earthquake Engineering and Structural Dynamics*, 2 (1973): 87–102.

[V3] A.S. Veletsos and B. Verbic, Basic Response Functions for Elastic Foundations, *Journal of the Engineering Mechanics Division*, ASCE, 100 (1974): 189–202.

[V4] A.S. Veletsos and V.D. Nair, Response of Torsionally Excited Foundations, *Journal of the Geotechnical Engineering Division*, ASCE, 100 (1974): 476–482.

[W1] G. Waas, H.G. Hartmann and H. Werkle, Damping and Stiffnesses of Foundations on Inhomogeneous Media, *Proceedings of 9th World Conference on Earthquake Engineering*, Vol. 3, Tokyo-Kyoto, 1988: 343–348.

[W2] G. Waas, personal communication, 1992.

[W3] R.V. Whitman, Soil-Platform Interaction, *Proceeding of Conference on Behaviour of Offshore Structures*, Vol. 1, Norwegian Geotechnical Institute, Oslo, 1976: 817–829.

[W4] J.P. Wolf, *Dynamic Soil-Structure Interaction.* Englewood Cliffs, NJ: Prentice-Hall, 1985.

[W5] J.P. Wolf, and D.R. Somaini, Approximate Dynamic Model of Embedded Foundation in Time Domain, *Earthquake Engineering and Structural Dynamics*, 14 (1986): 683–703.

[W6] J.P. Wolf, *Soil-Structure-Interaction Analysis in Time Domain.* Englewood Cliffs, NJ: Prentice-Hall, 1988.

[W7] J.P. Wolf and M. Motosaka, Recursive Evaluation of Interaction Forces of Unbounded Soil in the Time Domain, *Earthquake Engineering and Structural Dynamics*, 18 (1989): 345–363.

[W8] J.P. Wolf and M. Motosaka, Recursive Evaluation of Interaction Forces of Unbounded Soil in the Time Domain from Dynamic-Stiffness Coefficients in the Frequency Domain, *Earthquake Engineering and Structural Dynamics*, 18 (1989): 365–376.

[W9] J.P. Wolf, Consistent Lumped-Parameter Models for Unbounded Soil: Physical Representation, *Earthquake Engineering and Structural Dynamics*, 20 (1991): 11–32.

[W10] J.P. Wolf, Consistent Lumped-Parameter Models for Unbounded Soil: Frequency-Independent Stiffness, Damping and Mass Matrices, *Earthquake Engineering and Structural Dynamics*, 20 (1991): 33–41.

[W11] J.P. Wolf and A. Paronesso, Lumped-Parameter Model for Foundation on Layer, *Proceedings of 2nd International Conference on Recent Advances in Geotechnical Earthquake Engineering and Soil Dynamics*, Vol. I, St. Louis, MO, 1991, 895–905.

[W12] J.P. Wolf, J.W. Meek and Ch. Song, Cone Models for a Pile Foundation, *Piles under Dynamic Loads*, Edited by S. Prakash. Geotechnical Special Publication No. 34, ASCE (1992): 94–113.

[W13] J.P. Wolf and A. Paronesso, Lumped-Parameter Model for a Rigid Cylindrical Foundation Embedded in a Soil Layer on Rigid Rock, *Earthquake Engineering and Structural Dynamics*, 21 (1992): 1021–1038.

[W14] J.P. Wolf and J.W. Meek, Cone Models for a Soil Layer on Flexible Rock Half-space, *Earthquake Engineering and Structural Dynamics*, 22 (1993): 185–193.

[W15] J.P. Wolf and J.W. Meek, Insight on 2D-versus 3D-Modelling of Surface Foundations via Strength-of-Materials Solutions for Soil Dynamics, *Earthquake Engineering and Structural Dynamics*, 23 (1994): 91–112.

[W16] J.P. Wolf and J.W. Meek, Rotational Cone Models for a Soil Layer on Flexible Rock Half-space, *Earthquake Engineering and Structural Dynamics*, 23 (1994), in press.

[W17] J.P. Wolf and J.W. Meek, Dynamic Stiffness of Foundation on or Embedded in Layered Soil Halfspace, *Earthquake Engineering and Structural Dynamics Using Cone Frustrums*, 23 (1994), in press.

[W18] H.L. Wong and J.E. Luco, Tables of Influence Functions for Square Foundations on Layered Media, *Soil Dynamics and Earthquake Engineering*, 4 (1985): 64–81.

[W19] H.L. Wong and J.E. Luco, Dynamic Interaction between Rigid Foundations in a Layered Half-space, *Soil Dynamics and Earthquake Engineering*, 5 (1986): 149–158.

Index